Edited by Vasile I. Parvulescu,
Monica Magureanu, and Petr Lukes

Plasma Chemistry and Catalysis in Gases and Liquids

Related Titles

Rauscher, H., Perucca, M., Buyle, G. (eds.)

Plasma Technology for Hyperfunctional Surfaces

Food, Biomedical, and Textile Applications

2010
Hardcover
ISBN: 978-3-527-32654-9

Kawai, Y., Ikegami, H., Sato, N., Matsuda, A., Uchino, K., Kuzuya, M., Mizuno, A. (eds.)

Industrial Plasma Technology

Applications from Environmental to Energy Technologies

2010
Hardcover
ISBN: 978-3-527-32544-3

Heimann, R. B.

Plasma Spray Coating

Principles and Applications

2008
Hardcover
ISBN: 978-3-527-32050-9

Hippler, R., Kersten, H., Schmidt, M., Schoenbach, K. H. (eds.)

Low Temperature Plasmas

Fundamentals, Technologies and Techniques

2008
Hardcover
ISBN: 978-3-527-40673-9

d'Agostino, R., Favia, P., Kawai, Y., Ikegami, H., Sato, N., Arefi-Khonsari, F. (eds.)

Advanced Plasma Technology

2008
Hardcover
ISBN: 978-3-527-40591-6

Edited by Vasile I. Parvulescu, Monica Magureanu, and Petr Lukes

Plasma Chemistry and Catalysis in Gases and Liquids

WILEY-VCH

WILEY-VCH Verlag GmbH & Co. KGaA

The Editors

Prof. Dr. Vasile I. Parvulescu
University of Bucharest
Faculty of Chemistry
Regina Elisabetha Bld. 4-12
030016 Bucharest
Romania

Dr. Monica Magureanu
Nat. Inst. for Lasers,
Plasma and Radiation Physics
Atomistilor Str. 409
077125 Bucharest-Magurele
Romania

Dr. Petr Lukes
Institute of Plasma Physics AS CR, v.v.i.
Dept. of Pulse Plasma Systems
Za Slovankou 3
182 00 Prague
Czech Republic

Library of Congress Card No.: applied for

British Library Cataloguing-in-Publication Data
A catalogue record for this book is available from the British Library.

Bibliographic information published by the Deutsche Nationalbibliothek
The Deutsche Nationalbibliothek lists this publication in the Deutsche Nationalbibliografie; detailed bibliographic data are available on the Internet at <http://dnb.d-nb.de>.

Cover Design Formgeber, Eppelheim
Typesetting Laserwords Private Limited, Chennai, India
Printing and Binding Markono Print Media Pte Ltd, Singapore

Print ISBN: 978-3-527-33006-5
ePDF ISBN: 978-3-527-64955-6
ePub ISBN: 978-3-527-64954-9
mobi ISBN: 978-3-527-64953-2
oBook ISBN: 978-3-527-64952-5

Contents

Preface

Plasma-chemical and plasma-catalytic processes associated with low-temperature plasma generated by electrical discharges in gas, liquid, and gas–liquid environments have recently generated considerable interest. Nonthermal plasmas offer a unique way to initiate chemical reactions in the gas phase as well as in liquids, which have potential for practical utilization in different environmental, biological, or medical applications, and also in energy topics or molecular synthesis. Since plasma-chemical processes are rather nonselective, combination with catalysis can provide improved selectivity, by steering the reactions in the desired direction. Catalyst activation by plasma is different from that in case of conventional heating, and therefore the knowledge of plasma-catalyst interaction represents a key issue both from the fundamental point of view, for the understanding of reaction mechanisms involved in the plasma-catalytic process, and obviously, from the point of view of applications. Promising results have been obtained in environmental applications, where it was found that nonequilibrium plasma generated in electrical discharges at atmospheric pressure and room temperature can be successful in destroying a wide range of air pollutants. Serious attention is also directed to plasma-catalytic applications for hydrogen production, which plays a key role in fuel cell technology, as well as for the conversion of natural gas into syngas or into higher hydrocarbons, which can be used as fuel for transportation and raw material in chemical industry. In this direction, the control of catalyst properties by preparation or treatment techniques, as well as their modifications during plasma-catalytic reactions, catalyst stability, and regeneration processes, are important issues. Another vital issue for environmental research is water pollution. During the past 20 years, promising results have been obtained for the degradation of water pollutants and inactivation of various microorganisms using nonequilibrium plasma generated by electrical discharges in liquids and gas–liquid environments. These discharges have been shown to initiate various chemical and physical processes that have potential for practical utilization in different environmental, biological, or medical applications. For example, electrical discharges were successfully applied to degrade and inactivate a number of organic compounds and microorganisms in water. There are also first successful biomedical applications of discharge plasma in liquids.

This book provides an overview of the basic principles of plasma-chemical and plasma-catalytic processes generated by electrical discharges in gas, liquid, and gas–liquid environments, which is addressed by experts in the fields of plasma physics, plasma chemistry, and plasma catalysis. The book is divided into four major sections containing altogether nine chapters that cover the state of the art of this topic in both fundamental and applied aspects.

The first section contains two introductory chapters (Chapters 1 and 2). The first chapter provides an introduction to the fundamental aspects of nonthermal plasma generated by various types of electrical discharges operating in gas at atmospheric pressure and its properties. Chapter 2 focuses on the analysis of the intrinsic characteristics of the catalysts used in plasma-catalytic processes. The control of catalyst properties by preparation and treatment techniques and factors controlling the catalyst stability and regeneration processes represent other issues analyzed in this chapter. All these aspects are important criteria for the selection of appropriate catalysts for the desired applications.

The Chapters 3–5 give an extensive overview of the plasma-catalytic processes associated with low-temperature electrical discharge plasma in gases and their application for air pollution abatement. Chapter 3 is devoted to nitrogen oxides remediation (deNO_x) by plasma-assisted catalysis. Chapters 4 and 5 are dedicated to the decomposition of volatile organic compounds (VOCs) in air using plasma-catalytic systems. Results obtained in different plasma-catalytic systems are discussed, and the interactions between plasma and catalysts as well as the mechanisms responsible for NO_x and VOC remediation are addressed.

The Chapters 6–8 present the state-of-art fundamental and applied knowledge on plasma-chemical processes associated with nonequilibrium plasma generated by electrical discharges in liquids and gas–liquid environments. In these chapters, for the first time, a comprehensive overview of the elementary chemical and physical phenomena in low-temperature plasma in liquid and gas–liquid environments is provided, including fundamental mechanisms of plasma generation by electrical discharges in water and gas–liquid environments, chemistry and reaction kinetics of primary and secondary species generated by plasma in water and gas–liquid interfaces, mechanisms of interaction of plasma with chemical and biological content in water, plasma-catalytic processes in water and gas-liquid environments, and environmental and biomedical applications of plasma in water and gas–liquid environments.

Chapter 9 focuses on applications of nonthermal plasma and plasma-catalytic processes in energy conversion. An overview of the current state of hydrogen and syngas production, applications, and technical requirements is presented. Detailed discussions are provided with respect to steam reforming, partial oxidation, and carbon dioxide dry reforming, including coupling to higher hydrocarbons and plasma pyrolysis, as well as combined processes, highlighting the key issues to determine practical and economic viability.

This book is equally addressed to scientists and engineers with research interests in the fields of plasma, chemistry, catalysis, pollution abatement, synthesis of new materials, or energy conversion techniques. It may also be a very good support for students and Ph.D. students performing research in one of these fields.

Monica Magureanu
Petr Lukes
Vasile I. Parvulescu

List of Contributors

François Baudin
Université Pierre et Marie Curie
(Paris 6)
Laboratoire Réactivité de Surface
UMR CNRS 7609
4 Place Jussieu
75252 Paris Cedex 05
France

Jean-Louis Brisset
University of Rouen
Department of Chemistry
Faculty of Sciences
F-76821
Mont Saint-Aignan Cdx
France

Peter Bruggeman
Eindhoven University of
Technology
Department of Physics
Den Dolech 2
5612 AZ Eindhoven
The Netherlands

Patrick Da Costa
Institut Jean Le Rond d'Alembert
Université Pierre et Marie Curie
(Paris 6) CNRS UMR 7190
2 Place de la Gare de Ceinture
78210 Saint Cyr l'Ecole
France

Gérald Djéga-Mariadassou
Université Pierre et Marie Curie
(Paris 6)
Laboratoire Réactivité de Surface
UMR CNRS 7609
4 Place Jussieu
75252 Paris Cedex 05
France

Ute Ebert
Centrum Wiskunde &
Informatica (CWI)
Science Park 123
1098 XG Amsterdam
The Netherlands

Moritz Heintze
Centrotherm Photovoltaics AG
Johannes-Schmid-Strasse 8
89143 Blaubeuren
Germany

Ahmed Khacef
Laboratoire GREMI
CNRS-Université d'Orléans UMR 7344
14 rue d'Issoudun BP 6744
45067 Orléans Cedex 02
France

Christophe Leys
Ghent University
Department of Applied Physics
Faculty of Engineering and Architecture
Research Unit Plasma Technology
Jozef Plateaustraat 22
9000 Ghent
Belgium

Bruce R. Locke
Florida State University
Department of Chemical and Biomedical Engineering
FAMU-FSU College of Engineering
2525 Pottsdamer Street
Tallahassee
FL 32310
USA

Petr Lukes
Institute of Plasma Physics AS CR, v.v.i.
Department of Pulse Plasma Systems
Za Slovankou 3
182 00 Prague
Czech Republic

Monica Magureanu
National Institute for Lasers Plasma and Radiation Physics
Department of Plasma Physics and Nuclear Fusion
Atomistilor Street 409
077125 Bucharest-Magurele
Romania

Rino Morent
Ghent University
Department of Applied Physics
Faculty of Engineering and Architecture
Research Unit Plasma Technology
Jozef Plateaustraat 22
9000 Ghent
Belgium

Sander Nijdam
Eindhoven University of Technology
Department of Physics
Den Dolech 2
5612 AZ Eindhoven
The Netherlands

Vasile I. Parvulescu
University of Bucharest
Department of Organic Chemistry
Biochemistry and Catalysis
Boulevard Regina Elisabeta 4-12
Bucharest 030016
Romania

Eddie van Veldhuizen
Eindhoven University of Technology
Department of Physics
Den Dolech 2
5612 AZ Eindhoven
The Netherlands

1
An Introduction to Nonequilibrium Plasmas at Atmospheric Pressure

Sander Nijdam, Eddie van Veldhuizen, Peter Bruggeman, and Ute Ebert

1.1
Introduction

1.1.1
Nonthermal Plasmas and Electron Energy Distributions

Plasmas are increasingly used for chemical processing of gases such as air, combustion exhaust, or biofuel; for treatment of water and surfaces; as well as for sterilization, plasma deposition, plasma medicine, plasma synthesis and conversion, cleaning, and so on. These plasmas are never in thermal equilibrium – actually, we know of no exemption – and this fact has two main reasons.

1) It is easier to apply electromagnetic fields than to uniformly heat and confine a plasma. However, electromagnetic fields naturally transport charged species whose concentrations and energies therefore naturally vary in space, particularly, close to the walls of the container. Generically, the species in such a plasma are not in thermal equilibrium.
2) It is energy efficient to not feed energy equally into all degrees of freedom within a gas or plasma, such as into the thermal displacement, rotation, and vibration of neutral molecules, but only into those degrees of freedom that can efficiently create the desired final reaction products for the particular application. Therefore it is frequently preferable to accelerate only electrons to high velocities and let them excite and ionize molecules by impact while keeping the gas cold. If the electron energy distribution is appropriate, some reactions can be triggered very specifically.

In this manner, the nonthermal nature of the plasmas that are created electromagnetically is made into an asset. By varying gas composition, electrode and wall configuration, and circuit characteristics more energy can be channeled into specific excitations and reactions. Recent examples include the optimization of the

Plasma Chemistry and Catalysis in Gases and Liquids, First Edition.
Edited by Vasile I. Parvulescu, Monica Magureanu, and Petr Lukes.

pulsed power source for ozone generation in streamer corona reactors [1], or dual frequency RF-generated plasmas [2].

To elaborate the physical understanding further, Mark Kushner has proposed a workshop at the Gaseous Electronics Conference (GEC) 2011 on how the electron energy distribution within a discharge can be tailored for a specific application. A joint approach to this question by theory and experiment now seems within reach because of the large progress of theory in recent years.

1.1.2 Barrier and Corona Streamer Discharges – Discharges at Atmospheric Pressure

The past has mainly seen an experimental approach by trial and error, also guided by some physical understanding. Within the limited space available here, we will review some setups and their physical mode of operation. A common theme is the avoidance of plasma thermalization in the form of arcs and sparks. Variations over two basic approaches are used very commonly and will make the main theme of this review: the corona discharge and the barrier discharge. In a barrier discharge, large currents are suppressed by dielectric barriers on the electrodes. Basically, the discharge evolves only up to the moment when so much charge is deposited on the insulator surfaces that the field over the gas is screened. In a corona discharge, the discharge expands from a needle or wire electrode into outer space where the electric field decreases and finally does not support a discharge anymore. The discharge then has to feed its current into the high-ohmic region of the nonionized gas, which limits the current as well. These two basic principles have seen many variations in the past years and decades. For example, in corona discharges, short and highly ramped voltage pulses create much more efficient streamers that do not cease due to the spatial decrease of the electric field away from the curved electrode but due to the final duration of the voltage pulse.

Both discharge types can (but need not) operate at atmospheric pressure. This poses an advantage as well as a challenge. The advantage lies in the fact that no expensive and complex vacuum systems are required. This makes the design of any reactor a lot simpler, not only when the operating gas is air but also when other gases (such as argon or helium) are used. The challenge consists of the observation that characteristic length scales within the discharge can be much smaller than the discharge vessel and that the discharge can therefore form complex structures, rather than a more or less uniform plasma. These structures have to be understood and used appropriately. For instance, the initial evolution of streamer discharges follows similarity laws [3]: when the gas density is changed, the same voltage will create essentially the same type of streamer, but on different length and timescales. Therefore, streamer fingers and trees grow in a similar manner at 10 μbar as at 1000 mbar, but 10 μbar corresponds to an atmospheric altitude of 83 km where the so-called sprite streamers have a diameter of at least ~10 m, while at 1000 mbar, the minimal streamer diameter is ~100 μm and conveniently fits into typical experiments.

1.1.3 Other Nonthermal Discharge Types

There is a large variety of nonthermal plasmas. They can be classified into different discharge types, although definitions used by different authors vary significantly. The plasmas or discharges can be classified according to their time dependence (transient or stationary), importance of space charge effects or of heating of the neutral gas species, and presence of a surface close to the discharge. The most important nonthermal plasmas along with their energization method and typical applications are listed in Table 1.1.

This table is intended to give a general idea, but it is far from complete. A further complication is that definitions are used in different ways. For example, in Ref. 8, Braun *et al.* use what they call a microdischarge for ozone generation, whereas the microdischarges as intended in Table 1.1 are much smaller. The microwave discharge made by Hrycak *et al.* [28] qualifies much more for the term *plasmajet* than for microdischarge. More information on the different types of microdischarges is given in [29]; some examples of the use of microdischarges are given in Section 1.4.4.

In many transient discharges, the different discharge types can occur after each other. For example, a discharge can start as an avalanche and then become a streamer, which can develop into a glow and finally into an arc discharge. When applying a DC field between two metal electrodes, a discharge at high pressure will become a thermal arc if the power supply can deliver the current. Nonthermal discharges are, by definition, almost always transient.

Table 1.1 Overview of nonthermal discharge types and their most common applications.

Type of discharge	Gap (mm)	Plasma	Energization	Typical application	References
Corona	10–300	Filaments	Pulsed/DC	Gas cleaning/dust precipitation	[4, 5]
Corona with barrier	10–30	Filaments	Pulsed	Gas and water cleaning	[6, 7]
Plates/cylinders with barrier	1–5	Filaments	AC	Ozone generation/ large surface treatment/ excimer lamps	[8–12]
Barrier with packed bed	3–10	Filaments	AC	Chemicals conversion	[13–15]
Plates with barrier	1–5	Diffuse	AC	Surface treatment/deposition	[16, 17]
Surface discharge	1–5	Filaments	AC	Surface treatment/deposition	[18, 19]
Surface barrier	1–5	Filaments	Pulsed	Aerodynamic control	[20–22]
Plasma jets	0.5–10	Diffuse	AC/RF	Local surface	[18, 23–25]
Microdischarge	0.1–1	Diffuse	AC/RF	Chemicals conversion/ light generation	[26, 27]

An essential feature of a cold nonthermal discharge is its short duration. Therefore, the largely varying timescales of the processes inside the discharge must be considered. The excitation timescales, which often range from picoseconds to a few microseconds, are clearly not the timescale necessary for preventing thermalization as thermalization occurs in millisecond-order timescales. The critical timescale is basically the characteristic time of the glow-to-spark transition. This transition time can highly depend on conditions such as voltage amplitude and gas composition but is often in the order of a (few) hundred nanoseconds [30]. Dielectric barrier discharges (DBDs) are a well-known example of how (dielectric) barriers can reduce current density and n_e to keep the gas temperature of the discharge low.

Like streamer and avalanche discharges, Townsend and glow discharges are cold discharges. They usually occur as a stationary discharge but have to be preceded by another discharge such as a streamer or avalanche discharge to ignite. In Townsend and glow discharges, electrons are emitted from the electrode and are then multiplied in the gap. In the case of a Townsend discharge, the electron multiplication takes place in the whole gap, while in a glow discharge, space charge concentrates the multiplication in the cathode sheath region. Electrons are freed from the cathode by the temperature of the cathode itself or by secondary emission either due to the impact of energetic positive ions or due to photons or heavy neutrals.

Several cold atmospheric pressure discharges operate in helium. This is not a coincidence as He has a thermal heat conductivity that is about 10 times larger than that of most other gases, which renders heat removal from the discharge to be more efficient. Other methods for efficient heat removal include strongly forced convection cooling in flow stabilized discharges and creation of discharge with a large area-to-volume ratio (microplasmas, see also further) to make the heat losses to the walls more efficient.

1.1.3.1 Transition to Sparks, Arcs, or Leaders

Avalanches, Townsend, streamer, and glow discharges are examples of cold discharges. This means that the heavy particle temperature is not much above room temperature and definitely far below the electron temperature ($T_e \gg T_i \approx T_n$ where e,i, and n stand for electron, ion, and neutral, respectively). At even higher currents, at higher pressures, or with longer pulse durations, these discharges can transform into spark, arc, or leader discharges. These are hot discharges, the heavy particle temperature is close to the electron temperature and can reach thousands of Kelvin ($T_e \gtrsim T_i \approx T_n$). In applications, heating of the gas is often unwanted, and therefore, cold discharges are preferred in many plasma treatment applications.

1.1.4 Microscopic Discharge Mechanisms

1.1.4.1 Bulk Ionization Mechanisms

The main ionization mechanism in electric discharges is impact ionization; in attaching gases such as air, impact ionization is counteracted by electron attachment.

Other mechanisms that create free electrons such as photoionization or electron detachment from negative ions are discussed in Section 1.2.4.1. Impact ionization occurs when electrons are accelerated in a high local electric field. At a certain kinetic energy, they can ionize background gas atoms or molecules and create more electrons. In air, this occurs by the following reactions:

$$O_2 + e \longrightarrow O_2^+ + 2e \tag{1.1}$$

$$N_2 + e \longrightarrow N_2^+ + 2e \tag{1.2}$$

In the so-called local field approximation (i.e., when the reaction rate is approximated as depending on the local electron density and local electric field only) [31, 32], the number of electrons generated per unit length per electron is called the *Townsend impact ionization coefficient* $\alpha_i(|E|) = \sigma_i(|E|) \cdot n_0$. Here $\boldsymbol{E}$ is the electric field, σ_i the cross section for electron impact ionization, and n_0 is the background gas density. An old and much used approximation is

$$\alpha_i\,(|E|) = \alpha_0 \exp(-\boldsymbol{E}_0/|E|) \tag{1.3}$$

This notation illustrates that the Townsend coefficient is characterized by two parameters: $\boldsymbol{E}_0$ characterizes the electric field where impact ionization is important; this electric field is proportional to the gas density n_0. α_0 characterizes the inverse of the ionization length at these fields. More precisely, $1/\alpha_i(|E|)$ is the mean length that an electron drifts in the field $\boldsymbol{E}$ before it creates an electron–ion pair by impact. Therefore, in geometries smaller than this length, no gas discharge can occur. Both the electron mean free path, between any collision, and the ionization length scale with inverse gas density.

The electron loss rate due to electron attachment on attaching gas components has a similar functional dependence as the impact ionization rate, but different parameters. One needs to distinguish between dissociative attachment

$$e + O_2 \longrightarrow O + O^- \quad \text{(in air)} \tag{1.4}$$

and three-body attachment

$$e + O_2 + M \longrightarrow O_2^- + M \quad \text{(in air)} \tag{1.5}$$

where M is an arbitrary third-body collider, for example, N_2 or O_2. As a third body is required here to conserve energy and momentum, the importance of three-body attachment relative to dissociative attachment increases with density. Dissociative attachment scales with gas density in the same manner as the impact ionization reaction, while three-body attachment is favored at higher gas density. On the other hand, dissociative attachment becomes more important at higher electric fields, even at standard temperature and pressure. For detailed discussions of the derivation of these reaction coefficients, we refer to [33–36].

The *breakdown field* is defined as the field where impact ionization and electron attachment precisely balance; at higher electric fields, an ionization reaction sets in. The spatial and temporal evolution of the discharge depends on the distribution of electrons and electric fields; this is discussed in more detail below.

1.1.4.2 Surface Ionization Mechanisms

Next to the bulk gas, the presence of a dielectric or metallic surface can also affect the discharge significantly. It will modify the electric field configuration, and it is able to provide electrons. Dielectrics can also store surface charges [37] and prevent charge carrier flow through the surface.

Electrons can be freed from a surface by high fields or by secondary emission on impact of ions [38], fast neutrals, or (UV) photons [39]. Photons can be generated in the bulk of the discharge and then free an electron from the surface. Electron emission can be enhanced by the local electric field at the surface or by higher surface temperatures. The freed electrons can form the start of an avalanche, which enables the discharge to initiate or propagate (over the surface). See Section 1.4.3 for a more elaborate discussion on this topic.

1.1.5 Chemical Activity

The main advantage of nonthermal plasmas is their high chemical efficiency. As little or no heat is produced, nearly all input energy is converted to energetic electrons. This is in contrast to thermal plasmas in which the heating itself leads to higher thermal losses and thereby can be a waste of energy, which reduces the chemical efficiency of these hot plasmas [40] and can damage walls and other nearby surfaces (such as the substrate in a surface processing application). Furthermore, higher gas temperatures will change the reaction kinetics which, amongst others, may lead to breakdown of ozone and increased formation of NO_x. Of course, the different reaction kinetics of higher gas temperatures can also be beneficial for some chemical reactions such as destruction of hydrocarbons.

The fast electrons produced in a nonthermal plasma can have energies of the order 10 eV or even higher and can therefore trigger many different chemical processes. Besides fast electrons, energetic photons can also play a role in the reactions in a nonthermal plasma. One important example of such a reaction is photoionization in air, which is discussed in detail in Section 1.2.4.1. However, the primary source of all reactions is electron impact on the bulk gas molecules, which leads to many reactive species that can than further react with more stable species. Examples of the reactive species are OH, O, and N radicals; excited N_2 molecules; and atomic and molecular ions (e.g., O^+, O_2^+).

One of the main paths of chemical activity in nonthermal plasmas in air is ozone production. This is generally believed to be a two-step process as described by Chang *et al.* [41] and Ono and Oda [42].

1) First, free oxygen radicals are produced by inelastic electron impact.

$$O_2 + e \longrightarrow O^+ + O + 2e \tag{1.6}$$

$$O_2 + e \longrightarrow O + O + e \tag{1.7}$$

$$O_2 + e \longrightarrow O^- + O \tag{1.8}$$

2) Then, ozone is created by reactions of these free radicals.

$$O + O_2 + M \longrightarrow O_3 + M \quad M = O_2 \text{ or } N_2 \tag{1.9}$$

Ozone can be produced with a wide range of electrode and discharge topologies, many of which are treated below; the most popular are dielectric barrier discharges. An early example is the ozone generator of Siemens made in 1857. The most important application of this device was ozone production for disinfection of water. Even now, this device is used, with only minor modifications [43]. But corona discharges can create O radicals (and thereby ozone) with very high energy efficiency as well [1], as will be discussed in more detail further below. In commercial ionizers, pure oxygen is often used as the starting gas because the nitrogen that is present in air can lead to the formation of NO_x (a general term used for NO and NO_2 and sometimes other nitrogen–oxygen compounds) with the following reactions [44]:

$$N + O_2 \longrightarrow NO + O \tag{1.10}$$

$$O + N_2 \longrightarrow NO + N \tag{1.11}$$

where the O radicals come from Eqs. (1.6–1.8) and the N radicals are produced by [45]

$$N_2 + e \longrightarrow N + N + e \tag{1.12}$$

The produced NO can further react with NO_2 as described in [45, 46]

$$O + NO + M \longrightarrow NO_2 + M \tag{1.13}$$

$$2NO + O_2 \longrightarrow 2NO_2 \tag{1.14}$$

$$2NO + O_2 \longrightarrow NO_2 + N + O_2 \tag{1.15}$$

However, nonthermal plasmas can also remove NO from gas streams. The main path for the removal of NO from air at low NO concentrations is (Eq. (1.12)) followed by [47]

$$N + NO \longrightarrow N_2 + O \tag{1.16}$$

A second type of radical that is important in nonthermal plasmas is OH. This is produced in moist gases (e.g., moist air) by the following reaction [48]:

$$H_2O + e \longrightarrow H + OH + e \tag{1.17}$$

Note that apart from electron-induced dissociation, dissociative electron recombination of water containing ions can also efficiently produce OH.

$$H_3O^+ + e \longrightarrow OH + H_2 \tag{1.18}$$

The rate of this reaction for nonthermal discharges with T_e in the range 1–2 eV is sometimes even faster than electron dissociation [49]. Several secondary reactions are also believed to play an important role in the production of OH

$$H_2O + O(^1D) \longrightarrow 2OH \tag{1.19}$$

$$H_2O + N_2(A) \longrightarrow OH + H + N_2(X) \tag{1.20}$$

where $O(^1D)$ is an excited state of atomic oxygen, $N_2(A)$ is a metastable nitrogen molecule and $N_2(X)$ is a nitrogen molecule in the ground state. It is clear that Eq. (1.17) occurs only in the ionizing phase, while Eqs. (1.18–1.20) also occur in the recombining phase when the electron temperature is equal to the gas temperature.

Which reactions dominate depends on the electron energy (which is dependent on topology, voltage shape, and amplitude, etc.) and the composition of the gas. In general, thermal discharges mostly produce NO_x, while nonthermal discharges produce ozone instead and can remove NO_x when concentrations are high. At low NO_x concentrations also, nonthermal discharges can lead to the net production of NO_x. A comparison of NO_x production by sparks and corona discharges was performed by Rehbein and Cooray [50]. They found that sparks produce about 2 orders of magnitude more NO_x per Joule than corona discharges. Overviews of different reactive species and the conditions in which they are important are given by Eliasson and Kogelschatz [51] and Kim [43].

Besides NO_x removal, which was discussed above, a host of other species can be removed from gas streams by nonthermal plasmas. Examples are volatile organic compounds (VOCs), chlorofluorocarbons (CFCs), SO_2, odors, and living cells (in disinfection or sterilization).

Most charges in a nonthermal discharge in air are initially produced by the direct impact ionization of nitrogen

$$N_2 + e \longrightarrow N_2^+ + e + e \tag{1.21}$$

with a threshold ionization energy of 15.58 eV or of oxygen (Eq. (1.1)) with a threshold ionization energy of 12.07 eV. According to Aleksandrov and Bazelyan [52], N_2^+ and O_2^+ will quickly change to other species according to the following scheme (for dry air under standard conditions):

$$N_2^+ \longrightarrow N_4^+ \longrightarrow O_2^+ \longrightarrow O_4^+ \tag{1.22}$$

After some tens of nanoseconds, the positive ions are dominated by O_4^+. Electrons are quickly attached to molecular oxygen by reactions given in Eqs. (1.4) and (1.5).

1.1.6
Diagnostics

In all nonthermal plasmas, fast electrons excite species. Many of the excited species can fall back to lower excited levels or the ground level and thereby emit a photon. These photon emissions are by far the most important property of cold discharges that are studied experimentally. They are used for imaging and for optical emission spectroscopy. Spectra of cold discharges in air are dominated by the emissions of the second positive systems of N_2 (SPSs, upper states $B^3\Pi_g$ and $C^3\Pi_u$). The SPS is often used to obtain the rotational temperature, which is mostly a good indication of the gas temperature [53].

For strongly pulsed and high field discharges and also in discharges in, for example, He with air impurities, the first negative system of N_2^+ (FNS, upper state $B^2\Sigma_u^+$) readily occurs. Relative intensity comparisons of the SPS and this FNS have been performed by many authors and are used to determine the electric field in nitrogen-containing discharges. This method is employed, for example, by Kozlov *et al.* [54] for laboratory scale discharges and by Liu *et al.* [55] for sprites.

There are many other rotational bands of different molecules that can be used to obtain rotational temperatures, which are mostly a good indication of the gas temperature. Especially popular is the UV emission band of OH(A–X) around 309 nm [53]. However, it has recently been found that the rotational population distribution is not always in equilibrium with the gas temperature and sometimes leads to overestimates [56].

Electron densities above 10^{20} m^{-3} can be determined by measuring the Stark broadening of the hydrogen Balmer lines. Especially the Balmer β line is very popular. It is important to note that it is necessary to carefully take into account all broadening mechanisms including van der Waals broadening, which can become quite important for low-temperature atmospheric pressure plasmas. A detailed description can be found in [53].

Besides (passive) optical emission spectroscopy, there are many other techniques to study nonthermal plasmas. Apart from standard voltage and current waveform measurements, several electrical probes exist, especially developed for low pressure plasmas, although it is often difficult and very complicated to apply them on atmospheric pressure plasmas. The active laser spectroscopy techniques have developed into a wide field. The techniques most commonly applied to atmospheric pressure plasmas include laser-induced fluorescence (LIF) and two-photon-absorption laser-induced fluorescence (TALIF), which are good ways to obtain information on the chemical composition of radicals. With proper calibration, even absolute densities can be obtained [57, 58]. Other well-known laser-based techniques are based on scattering of photons. Thomson scattering can give direct information on the electron density and temperature [59, 60]. Rayleigh and Raman scattering provide information on gas density and temperatures. The conceptually simplest active technique is absorption spectroscopy (often also performed with lasers). This technique is used to determine absolute densities of certain species, often in the ground state (e.g., OH). Radical density fluxes can also be obtained by appearance potential mass spectrometry [61]. Mass spectrometry also gives the possibility to measure the ion flux of one of the electrodes directly and determine the ion composition of the plasma [62].

1.2 Coronas and Streamers

1.2.1 Occurrence and Applications

Streamers are the earliest stage of electric breakdown of large nonionized regions. They precede sparks and create the path for lightning leaders; they also occur as enormous sprite discharges, far above thunderclouds. Streamers and the subsequent electric breakdown are a threat to most high-voltage technology.

However, streamers are also used in a variety of applications and are appreciated for their energy-efficient plasma processing. The following is an (incomplete) application list:

- Gas and water cleaning: The chemical active species that are produced by streamers can break up unwanted molecules in industrially polluted gas and water streams. Contaminants that can be removed include organic compounds (including odors), NO_x, SO_2, and tar [3, 6, 63, 64].
- Ozone generation: By simply applying a streamer discharge in air, first O^* radicals and then ozone is created. The low temperature in a streamer discharge limits the destruction of the produced ozone. The ozone can be used for different purposes such as disinfection of medical equipment, sanitizing of swimming pools, manufacturing of chemical compounds, and more [4].
- Particle charging: A negative DC corona discharge can charge dust particles in a gas flow. These charged dust particles can now be extracted from the gas by electrostatic attraction. Such a system is called an electrostatic precipitator (ESP) and is used in the utility, iron/steel, paper manufacturing, and cement and ore-processing industries. Similar charging methods are used in copying machines and laser printers [4, 65].

A corona discharge is (an often DC-driven) discharge in which many streamers are initiated from one electrode and, depending on the conditions, may or may not reach another electrode. The name corona comes from the crownlike appearance of the many streamer channels around the primary (driven) electrode.

Traditionally, DC corona discharges are classified in several different forms depending on the field polarity and electrode configuration [41]. In case of a positive point-plane discharge, one can recognize the burst pulse corona, streamer corona, glow corona, and spark for an increase in applied voltage. In a negative point-plane corona, this is replaced by a Trichel pulse corona, a pulseless corona, and again, a spark.

Since the 1980s, corona discharges are separated into two different categories: continuous and pulsed. Continuous corona discharges occur at DC or low-frequency AC voltages. If the circuit providing the voltage can support high currents, these will transform into a stationary glow or spark discharge. Therefore, continuous corona discharges can only occur if the current is limited. One example is a continuous corona discharge around high-voltage power lines, where the large gap to the ground limits the current. A recent example of work on DC-excited corona discharges is by Eichwald *et al.* [66].

The current of a continuously excited corona is often spiked because the discharge is not really continuous but is self-repetitive in nature. In such a self-repetitive corona, the discharge stops itself due to the buildup of space charge near the electrode tip. Only after this space charge has disappeared by diffusion and drift will a new discharge occur [67].

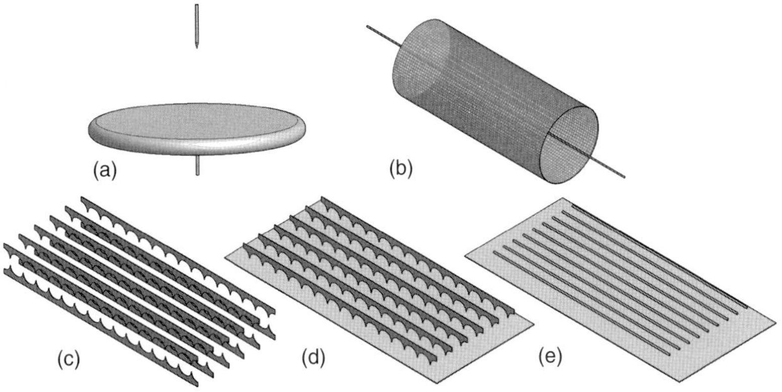

Figure 1.1 Schematic depictions of popular electrode geometries in corona reactors: (a) point-plane, (b) wire-cylinder, (c) double sawblade, (d) sawblade-plane, and (e) wire-plane. The high voltage is applied to the following parts: (a) top needle, (b) central wire, and (c–e) top sawblade/wires. The other parts are grounded.

A pulsed corona is produced by applying a short (usually submicrosecond) voltage pulse to an electrode. Its practical advantages are that the short duration of the pulse ensures that no transition to spark takes place, therefore it can be used at voltages and currents higher than that at continuous corona can be used.

Shang and Wu [68] have shown that a positive-polarity-pulsed corona removes more NO than a negative polarity discharge. van Heesch *et al.* [1] show that negative coronas have a higher efficiency in the production of O^* radicals (about a factor of 2 higher).

In laboratory studies of corona discharges, the most popular geometry is a point-plane geometry (Figure 1.1a), where a needle is placed above a grounded plane. The high voltage (pulse) is applied to the needle electrode. However, for industrial applications, this geometry is not sufficient, as it does not fill the whole gas volume with the discharge. The most popular geometries in industrial applications are the wire-cylinder, wire-plate, and the saw-blade geometries [41, 69]. See Figure 1.1b–e for schematic images of these geometries.

The wire-cylinder geometry is probably used the most. It ensures a quite homogeneous distribution of the discharge and is easy to implement in a gas-flow system. Often, multiple wire-cylinder reactors are mounted in parallel with regard to the gas flow to enable high gas throughput.

1.2.2
Main Properties of Streamers

Streamers are rapidly extending ionized fingers that can appear in gases, liquids, and solids. They are generated by high electric fields but can penetrate into areas where the background electric field is below the ionization threshold due to the

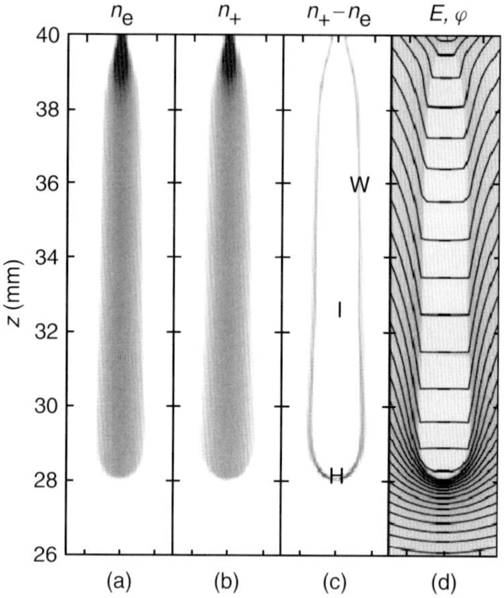

Figure 1.2 Structure of positive streamers shown by zooming into the relevant region of a simulation by Ratushnaya *et al.* The panels show (a) electron density n_e, (b) ion density n_+, (c) space charge density $(n_+ - n_e)$, as well as (d) electric field strength E and equipotential lines φ. The letters in (c) indicate the streamer regions: H – streamer head, I – interior, and W – wall of the streamer channel. (Source: Image from Ref. [71].)

strong field enhancement at their tip. The mechanism of field enhancement is illustrated in Figure 1.2, which shows the simulation of a positive streamer in air at standard temperature and pressure; for details we refer the readers to [70, 71]. The plots show electron and ion density, space charge, and field distribution. The plots can be understood as follows. Panels (a) and (b) show that the interior of the streamer channel consists of a conducting plasma with roughly the same electron and ion densities. The electric field (panel d) in this ionized area is largely screened by the thin space charge layer shown in panel (c).[1)] In front of the ionized finger, the space charge layer is strongly curved, and therefore, it significantly enhances the electric field in the nonionized area ahead of it. This self-organization mechanism due to space charge effects makes the streamer a well-defined nonlinear structure; gas heating is negligible in most cases.

As described in a previous streamer review for geophysicists [3], the electrons in the high-field zone at the streamer head are very far from equilibrium. The electron energy distribution can develop a long tail at high energies, and it is now known that electrons at the tip of negative streamers can even run away

1) We remark that in the older literature and in many books the space charge is smeared out over the complete streamer head and only simulations in the past 25 years have shown that it is concentrated in a thin layer. This is important for the streamer electrodynamics.

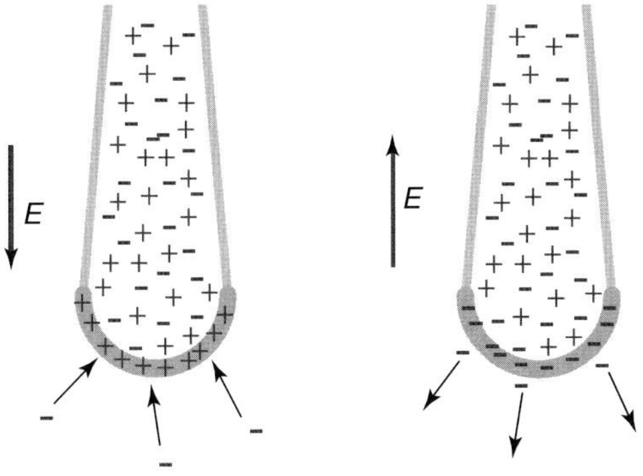

Figure 1.3 Illustration of downward propagating positive (a) and negative (b) streamers. The plus symbols indicate positive ions, while the minus symbols indicate negative ions or free electrons.

[31, 72–75], if the field enhancement is above 180 kV cm^{-1} in STP air, corresponding to 720 Td. This is the current explanation for the hard X-rays emitted during the early streamer-leader phase of MV-driven pulses [76]. How to optimize the electron energy distribution for a particular plasma processing purpose is a current research question.

The fact that streamer velocities and diameters can vary substantially between different electrode geometries and electric circuits is by now well established [5, 77, 78]. Simulations show that the maximum of the enhanced electric field also varies substantially, as reviewed recently in [79].

The maximal field determines the ionization rate inside the streamer [31, 70, 80] and, therefore, the excitation rates for gas processing purposes. The search for optimal processing conditions determined by both the electron energy distribution and the ionization rate is currently underway, both theoretically and through the development of optimized electric circuits. Here it should be mentioned that very short voltage rise times create much thicker [5, 77, 78] and more efficient streamers [1].

An important distinction is between positive and negative streamers, where the polarity refers to the net charge at their tips (Figure 1.3). They are also known as *cathode- or anode-directed streamers*. A negative streamer moves in the electron drift direction, and as the streamer velocity is frequently comparable to the local electron drift velocity,[2)] its motion can be explained by purely local mechanisms. On the contrary, a positive streamer moves, in most cases, even faster [78]. The reason for this counterintuitive behavior lies in the fact that the relative immobility of the ions in the space charge layer around the positive streamer keeps the streamer finger thin and focused; therefore the electric field at the tip can be much higher [81]. The mechanism allowing positive streamers to propagate is explained below.

2) The older Russian literature frequently states that the streamer velocity would be much larger than the electron drift velocity, but there the local field enhancement and local drift velocities are not characterized very carefully.

Concerning theory and simulations, there are currently three models: (i) Monte Carlo and (ii) hybrid models that follow the single-electron dynamics within a streamer, but are still constrained to rather short streamers, fluid, or density models, which now also start to treat the interaction of streamers, but cannot resolve the electron energy distribution, and (iii) moving boundary models where the thin space charge layer around the streamer is treated as a moving boundary. Currently, reviews of all three model classes have been published or are under review; we refer the reader for details to [71, 82, 83].

1.2.3
Streamer Initiation or Homogeneous Breakdown

When a discharge starts to develop, there are only few free charge carriers present, and therefore the electric field is not modified by space charge effects yet. The discharge is then said to be in the avalanche phase where free charge carriers multiply in regions where the electric field is above the breakdown value.

The discharge can then evolve either in a more homogeneous or a more streamerlike manner. If the initial ionization seed is very localized (e.g., because it evolves out of a single electron or because a macroscopic seed is ejected form a pointed needle electrode), or if the electric field is above breakdown only in a small part of space (again, e.g., close to a needle electrode), a localized structure such as a streamer that carries a field enhancement forward at its tip can emerge. On the other hand, if there is a higher level of preionization and if the electric field is at most places above the breakdown value, a more homogeneous discharge will emerge [84].

If a single electron or a very localized seed is placed in a homogeneous field above the breakdown value, Raether and Meek estimated in the late 1930s, that space charge effects set in and a streamer initiates when the total number of free electrons reaches 10^8-10^9 in air at standard temperature and pressure [85, 86]. However, this estimate is independent of the electric field. Taking into account that an electron avalanche grows with a slower rate in a weaker field, but that their diffusive broadening is essentially the same, a correction to the so-called Raether–Meek criterion was developed by Montijn and Ebert [87].

However, in most streamer experiments and applications, streamers are generated from a tip- or wirelike structure and not in a homogeneous field. At such a (sharp) tip or wire, the electric field will be greatly enhanced, which makes it easier to initiate a streamer. After initiation, the streamer can propagate into the rest of the gap where the background field may be too low for streamer initiation, but high enough for streamer propagation (discussed in the next section). Such a geometry with field enhancement greatly reduces the required voltages for streamer initiation, which makes experiments and applications smaller, cheaper, and easier to operate.

The lowest voltage at which a streamer can initiate from a pointed electrode is called the *inception voltage*; it depends on electrode shape and material as well as on gas composition and density and (up to now) has no direct interpretation in terms of microscopic discharge properties yet.

1.2.4 Streamer Propagation

After initiation, a streamer will propagate under the influence of an external electric field augmented by its self-generated field, as already discussed in Section 1.2.2. To sustain the extension of the plasma channel by impact ionization in the high-field zone, enough free electrons need to be present there. In negative streamers, the electrons drift from the ionized region in the direction of streamer propagation and reach the high-field zone. However, in positive streamers, the electrons cannot come from the streamer itself. Therefore, for positive streamer propagation, "fresh" electrons are needed in front of the streamer head. The possible sources of these free electrons are discussed below.

As was discussed in Section 1.1.5, the positive charges indicated in Figure 1.3 will mainly consist of positive molecular ions and the negative charges indicated in the streamer tails in air in Figure 1.3 will be negative molecular oxygen ions, limiting the total conductivity. Therefore, streamers in pure nitrogen can become longer than those in air under similar conditions as less electron attachment occurs if current flow from behind is required. The negative charges in the streamer head, as well as the moving charges in front of the streamer heads, will be mostly free electrons.

Owing to the electric screening layer around the curved streamer head, the electric field ahead of it is usually much higher than the external or background field.

1.2.4.1 Electron Sources for Positive Streamers

Positive streamers need a constant source of free electrons in front of them in order to propagate. Because of the electronegativity of molecular oxygen, free electrons in air quickly attach to oxygen by Eqs. (1.4) and (1.5) if the electric field is below $\sim 30\ \mathrm{kV\,cm^{-1}}$. If this is the case, a high field is needed to detach the electrons so that they can be accelerated. The exact level of the detachment field depends on the vibrational excitation of the molecule. According to Pancheshnyi [88] and Wormeester *et al.* [89], a good value of the instant detachment field under standard conditions in air is $38\ \mathrm{kV\,cm^{-1}}$.

Photoionization In most streamer models, air is the medium and the major source of electrons in front of the streamer head is taken as photoionization. In air, photoionization occurs when a UV photon in the 98–102.5 nm range, emitted by an excited nitrogen molecule, ionizes an oxygen molecule, thereby producing a free electron.

$$N_2^* \longrightarrow N_2 + \gamma_{98-102.5\ \mathrm{nm}} \tag{1.23}$$

$$O_2 + \gamma_{98-102.5\ \mathrm{nm}} \longrightarrow O_2^+ + e \tag{1.24}$$

As the emitted photon can ionize an oxygen molecule some distance away from its origin, this is a nonlocal effect, therefore, excited nitrogen molecules in the streamer head can create free electrons in front of the streamer head (as well as in other places around the streamer head). The average distance that a UV photon can travel depends on the density of the absorbing species, oxygen in this case. In atmospheric pressure air under standard conditions, this distance will be about 1.3 mm [90].

Background Ionization Besides photoionization, there is another source that can provide free electrons in front of a positive streamer head: background ionization. Background ionization is ionization that is already present in the gas before the streamer starts, or at least, it is not produced by the streamer. It can have different sources. In ambient air, radioactive compounds (e.g., radon) from building materials and cosmic rays are the most important sources of background ionization. They lead to a natural background ionization level of 10^9–10^{10} m^{-3} at the ground level (Pancheshnyi [88]).

Another source of background ionization can be leftover ionization from previous discharges. This is especially important in repetitive discharge types such as DC corona discharges or repetitive pulsed discharges. Already at a slow repetition rate of about 1 Hz, leftover charges can lead to background ionization densities of the order of 10^{11} m^{-3}. Background ionization can also be created by external UV radiation sources, X-ray sources, addition of radioactive compounds to the gas or surfaces, electron or ion beam injection, and more.

Independent of the source of background ionization, in air, the created electrons will always quickly be bound by oxygen. This means that they will have to be detached by the high field of the streamer before they can be accelerated and form avalanches.

1.2.5 Initiation Cloud, Primary, Secondary, and Late Streamers

Recent imaging with high spatial and temporal resolution has shown how a streamer tree starts from a needle electrode, which in most cases is positively charged [91–93]. The discharge starts with a small ball of light around the needle tip that was called the *initiation cloud*. This ball expands and forms a shell; this shell can be interpreted as a radially expanding ionization front, and in the case of a negative needle tip in air, its maximal radius fits the theoretical estimates well [93]. For positive voltages, it has been verified that the size l of the initiation cloud scales with gas density n_0 according to the similarity laws ($l \propto 1/n_0$) but it also depends on gas composition and, of course, on the applied voltage. For example, in air, the initiation cloud is much larger (up to a factor 10 or more) than in pure nitrogen [91]. In fact, what on time integrated images of the discharge seems like a light emitting cloud is in fact often a smaller cloud that transforms into a thin expanding shell.

Eventually, the expanding shell breaks up into multiple streamer channels, except when the gap is so small that the initiation cloud extends into roughly half the gap distance; in that case, it usually destabilizes into one channel only. These first streamers emerging from the initiation cloud are called *primary streamers*. Example

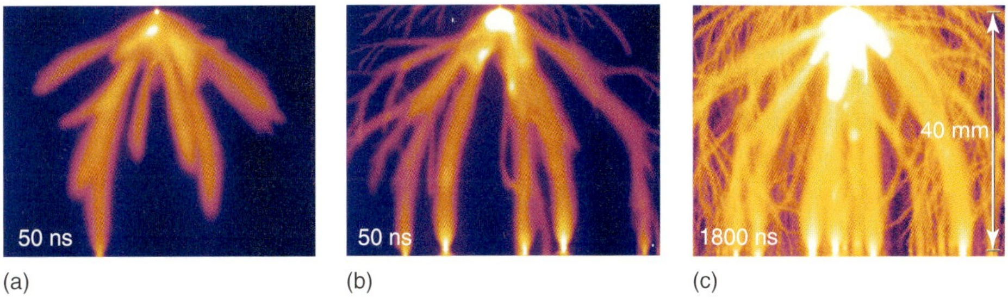

Figure 1.4 Streamer discharges in a 40 mm gap in atmospheric air with a 54 kV pulse, 30 ns risetime, and half-width of about 70 ns. The images are acquired with short (a,b) and long (c) exposure times. The exact image start delay is varied between (a) and (b) (exact values unknown). (Source: Images by Tanja Briels, originally published in Figure 6 of Ref. [77].)

of such streamers are shown in Figure 1.4a,b. For long gaps, low voltages, or short pulse durations, the primary streamers often do not reach the other side and extinguish somewhere between the electrodes.

Briels *et al.* [77, 94] characterize different streamer types with very different diameters and velocities, although they realize and later show [78] that there is no phase transition between these types. For voltages between 5 and 95 kV, the streamer diameters vary by more than an order of magnitude and the velocities by almost 2 orders of magnitude. The relation between velocity and diameter is discussed in [79, 81]. The streamers with minimal diameter (the so-called minimal streamers) are never seen to branch. This minimal diameter depends on density, roughly in agreement with the similarity laws [3], but it does not depend on the background field or other pulse parameters. This concept was proposed by Ebert *et al.* [95]. The thick streamers grow only if the voltage rises sufficiently fast. Only then there is sufficient voltage initially on the pointed electrode to develop a very wide ionization cloud that can eject fat streamers.

After the primary streamer, more light-emitting discharge phenomena can occur. If the same streamer channels reilluminate rather immediately, one speaks of a secondary streamer, while if a streamer follows a different track at some later time, one speaks of a late streamer.

Secondary streamers have been described, for example, by Marode [96], Sigmond [97], Ono and Oda [98], or Winands *et al.* [5]. Sigmond remarks that moving secondary streamer fronts in centimeter-scale gaps in atmospheric air does not perturb the smoothly decaying streamer current and that they are only reported in air. Ono and Oda [98] have compared primary and secondary streamers; they were created in air in a needles-to-plane geometry with gaps of 13 mm length and voltages of 13–37 kV (compare with the 37–77 mm, 25–45 kV wire plane discharge of Winands *et al.*). They observe that emission from the FNS of N_2^+ (391.4 nm) is only observed in primary streamers and not in secondary streamers. This is attributed to the fact that electron energies required for propagation of primary streamers are higher than those for secondary streamers as primary streamers have to create ionization, while secondary streamers propagate along the ionized channel created by the primary streamers. Furthermore, they find that secondary streamers only occur at higher voltages (15 kV in air and 20 kV in pure nitrogen). van Heesch *et al.* [1] found that the O^* radical yield from primary steamers is up to two times higher than that from secondary streamers. They explain this by higher local electric fields and electron energies in the primary streamers.

The literature presents different suggestions for the physical mechanism of secondary streamers. Marode [96] suggests that secondary streamers correspond to a moving equivalent of the positive column of a glow discharge. Sigmond [97] suggests that the ionized column created after the primary streamer has crossed the gap decays into one region with high and another region with low electric field due to an attachment instability. The electrodynamic consistency of these calculations is under examination at present. A different mechanism is suggested by recent simulations of Liu [99] and Luque and Ebert [80]. They find that inside a streamer that requires a growing charge in its tip – because it accelerates and expands or because it propagates into a region with higher gas density – a secondary ionization wave can set in, and that the electric field inside this wave reaches approximately the breakdown field. This process can set in before the primary streamer has reached an electrode. We note that in the experiments of Winands *et al.* [5] where long secondary streamers were observed, the primary streamers were accelerating and expanding as well, just like in the simulations of Liu.

A third streamer category, besides primary or secondary streamers, is the so-called late streamers. They occur only for long enough pulses and are, in fact, the primary streamers that either start later than the dominant streamers or are so slow that they seem to have started later. Late streamers propagate along completely different paths than the other (primary) streamers before them. They are often very thin, which is related to their slow propagation velocity (see, e.g., Briels *et al.* [78]). In most cases, they do not appear from the sharp electrode tip itself but instead from the (less sharp) edges of the electrode or electrode holder because the tip is already screened by a glow region and therefore no longer enhances the electric field sufficiently. Examples of these late streamers are visible in Figure 1.4b,c. In Figure 1.4b, the late streamers have just started and are visible on the top of the image. In Figure 1.4c, a much longer camera exposure is used. Therefore the primary streamers are now overexposed as their secondary and glow phase is also included in this exposure. However, many (thin) late streamers are clearly visible crisscrossing all corners of the image.

1.2.6
Streamer Branching and Interaction

Most streamer discharges contain more than one streamer channel. Therefore, interactions between streamers are important when studying streamer behavior. One important aspect is streamer branching where one streamer channel splits into two (or more) channels. Other interactions are attraction and repulsion of streamer channels. Furthermore, neighboring channels influence each others field configuration. If attraction occurs, this may lead to streamer merging or (re-)connection. Discussion and measurements regarding streamer merging and (re-)connection are given by Nijdam *et al.* [100, 101].

Branching is observed in most streamer discharges, except when the gap is so short that the streamer has reached the other side before it has branched. Furthermore, streamers of minimal diameter (so-called minimal streamers, see below) also do not branch but eventually extinguish. This is the main argument why streamer discharges are never real fractals.

The mechanism of streamer branching has been under investigation for quite a long time now. It is certainly due to a Laplacian instability of the thin space charge layer visible in Figures 1.2 and 1.3; this instability bears strong mathematical similarities with viscous fingering [102]. For a recent review of the analytical, numerical, and experimental results, we refer to [71]. The Laplacian instability can actually set in without any stochastic effects [102, 103]. However, the branching instability can be accelerated by electron density fluctuations in the lowly ionized region ahead of the streamer [104]; these fluctuations are due to the discrete quantum nature of the electrons. Indeed, these fully three-dimensional recent simulations for positive streamers in air (with the standard photo-ionization model) show a ratio of streamer branching length to streamer diameter similar to that obtained in experiments [91, 100].

The acceleration of branching through electron density fluctuations is consistent with older concepts, which can be traced back to Raether [86] and Loeb and Meek in 1940 [105]. However, in these older sketches, the fact that the streamer has to develop a thin space charge layer before it can destabilize was missed. The older concept that can be found in many books emphasizes the spatially well-separated avalanches ahead of the streamer as direct precursors of different branches. Such avalanches have now indeed been seen in very pure gases [89, 106]. However, the photoionization density in air is much too large to create individual avalanches [107].

van Veldhuizen and Rutgers [108] have experimentally investigated streamer branching in argon and ambient air for different discharge geometries and pulse characteristics. They find that streamers in a point-wire discharge branch about 10 times more often (in the middle of the gap) than in a discharge between a plane with a protrusion and another plane.

A very different branching mechanism is branching at macroscopic inhomogeneities such as bubbles (for streamers in liquids). This mechanism was recently described in detail by Babaeva and Kushner [109].

A proper understanding of streamer branching, on the one hand, and streamer thickness and efficiency, on the other hand, is required to understand which volume fraction of gas is being processed in a streamer corona reactor. The streamer interaction mechanisms discussed above are an important ingredient for building models of a complete streamer discharge. However, a complete model based on measurements or theoretical understanding of the microscopic processes is not yet available. There are a number of models for streamer trees that start from phenomenological assumptions of streamer channel properties as a whole.

All currently available models neglect the large variation of streamer diameters and velocities in pulsed corona reactors. The first phenomenological model for a complete discharge tree was proposed by Niemeyer *et al.* [110]; it approximates sliding surface discharges and creates fractal structures. This model includes streamer branching in a purely phenomenological manner and assumes that all streamers are equal and that the interior is completely screened from the electric field. Since then, a number of authors have developed this model further, in chemical physics, geophysics [111], and electrical engineering [112]. At present, the challenge lies in extending such models to all recently identified microscopic ingredients such as branching statistics, streamer diameters and velocities, and interior electric fields coupled to the external circuit.

1.3
Glow Discharges at Higher Pressures

1.3.1
Introduction

The classic low-pressure glow discharge has been studied extensively for several decades. The discharge is typically produced in a low-pressure (order of 1 mbar) noble gas between two electrodes that are separated from 1 cm up to 1 m. The light emission pattern of a low-pressure glow discharge is described in all standard books [113] and includes a cathode glow, cathode dark space, negative glow, Faraday dark space, the positive column, the anode dark space, and the anode glow.

The sheath region of a glow discharge has a high electric field because of charge separation between fast electrons and slow positive ions (creating the so-called cathode fall). The fast electrons emitted by the cathode and accelerated by the high field multiply by impact ionization on the sheath edge. In many glow discharges, most space between the electrodes is occupied by the positive column, a region with a relatively low, constant electric field. See also Šijacic and Ebert [114] for a detailed description and numerical model of the Townsend to glow discharge transition. In their one-dimensional model (equivalent to a plate–plate discharge), they found that depending on $p \cdot d$ (pressure times distance) and the secondary emission coefficient of the cathode γ, the transition can occur according to the subcritical behavior described in books (with a negative current–voltage characteristic (CVC) from Townsend to glow) or for smaller values of $p \cdot d$, it can also behave supercritical or have some intermediate "mixed" behavior.

In spite of the fact that it is easy to produce glow discharges at low pressure (applying typically a few hundred volts DC), with increasing pressure, the glow discharge has the tendency to become unstable and constrict: a glow-to-spark transition occurs. Thus, at atmospheric pressure, it is necessary to use special geometries, electrodes, or excitation methods to obtain diffuse glow discharges. Spark/arc formation is a restriction for the generation conditions of nonthermal (cold) atmospheric pressure plasmas in general.

High-pressure glow discharges have been studied for several years because they are scalable to large areas while remaining relatively uniform. This is especially interesting for surface interactions under controlled conditions without the necessity of vacuum equipment. Studies of atmospheric pressure glow discharges (APGDs) go back to von Engel *et al.* [115]. High-pressure glow discharges and also the instabilities that occur have been studied in the context of the construction of lasers [113]. More recently, these discharges are produced to obtain homogeneous treatment of materials and large-volume homogeneous discharges [116, 117].

A possibility to prevent the direct transition from a Townsend to a filamentary discharge is increasing the preionization in the gas [84]. Basically, the electrical field is reduced by the interaction of the avalanches, which does not allow the Meek criterion to be reached. The avalanche-to-streamer transition and the start of filamentation of the discharge is more suppressed.

1.3.2
Properties

It must be noted that several authors use the label glow discharge, in general, for a discharge that looks homogeneous to the naked eye. A more strict use of the term *glow discharge* is often appropriate, especially because discharges of a filamentary nature, such as certain DBD discharges, can look very diffuse when time averaged, while the properties and chemistry can be quite different from diffuse discharges. In spite of several differences between the low-pressure glow discharge and APGD, there are several similarities that motivate the use of same label glow discharge at atmospheric pressure also.

The similarities with low-pressure glow discharges include the following:

- The reduced current density (J/n_0^2) is independent of density (or pressure) and applied voltage.
- The characteristic light emission pattern of the glow discharges.
- There is constant electrical field in the positive column.
- The discharge voltage is independent of the current when corrected for the temperature rise, constriction of the positive column, and current dependence of the cathode–anode voltage drop.
- The electron temperature is much higher than the gas temperature.
- The glow discharge operates at the Stoletov point; that is, the thickness of the cathode fall region is adjusted so that the conditions to operate in the minimum of the Paschen curve are reached.
- The burning voltage and cathode voltage drop is significantly larger than in the case of arc discharges.

The main differences with low-pressure discharges are the following:

- The dimensions of the characteristic light emission pattern of the glow discharges scale (inversely) with pressure and are considerably smaller (typically tens or hundreds of meters at atmospheric pressure instead of centimeters at millibar pressure).

- Owing to the high pressure, gas heating can be considerable up to a few thousand kelvin while most low-pressure glows are close to room temperature.
- Owing to gas heating, scaling laws always need to be written as a function of density and not pressure, as is mostly done in the old literature (for low-pressure discharges).
- At low pressure, the electron losses are dominated by diffusion, while in the high-pressure case, due to the high collisionality bulk processes (such as dissociative electron-ion recombination) become important.
- The sheath is highly collisional at atmospheric pressure, which means that the ion energies impacting the electrode are considerably smaller than at low pressure.

The electron density of diffuse APGDs is estimated to be in the range $10^{17}-10^{19}$ m^{-3}. This is too low for accurate line-broadening measurements. Only few measurements exist that are based on microwave absorption [118] and millimeter wave interferometry [119]. They give values of $4-7 \times 10^{17}$ and 8×10^{18} m^{-3}, respectively. Other values are often derived from modeling or estimates from current densities and are not very accurate. Gas temperatures range from room temperature up to 3000 K [56, 120, 121]. The electron energy distribution is highly non-Boltzmann. High-energy electrons are produced in the cathode region, penetrate in the bulk, and sustain the discharge. The electron energy in the bulk is of course much lower (often the values of an effective electron temperature of 1–5 eV circulate in the literature), but often, a high-energy component originating from the cathode region in small electrode gaps is present [122].

1.3.3
Studies

Standard glow discharges have to be stabilized with a negative feedback, for example, by including a resistor in series. The series resistor can prevent current runaway as the resistor causes the voltage across the discharge gap to decrease with increasing current for constant applied voltage. Similar behavior can be obtained by using a capacitor or inductance in series with the discharge gap. The latter has been shown by Aldea *et al.* [123], who used it to stabilize large area APGDs for material treatment applications.

The lumped resistor approach can work, but using a resistive electrode or a dielectric barrier between the electrodes causes a distributed resistor or capacitor, which can even enhance the diffusivity of the discharge. Atmospheric pressure glow discharges stabilized by resistive electrodes are studied by Laroussi *et al.* [124]. Also, water electrodes (which are, of course, resistive in nature) are used to generate glow discharges, as has been studied by Andre *et al.* [125], Lu and Laroussi [126], and Bruggeman *et al.* [127]. Bruggeman *et al.* have shown that in the case of a liquid electrode, there is a significant polarity effect. In the case of a water cathode, the discharge is filamentary close to the cathode because of instabilities of the liquid surface caused by the strong electrical field in the cathode layer. When the discharge is generated between a liquid anode and a metal cathode, a diffuse glow

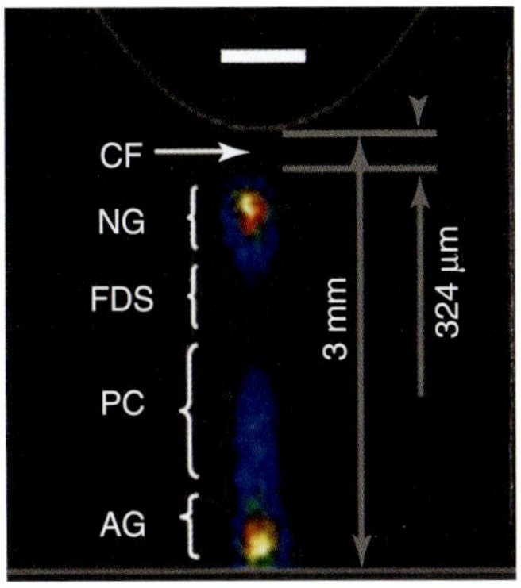

Figure 1.5 Example of an atmospheric pressure air glow discharge in a metal pin (top) water electrode (bottom) geometry. The typical structures of the low-pressure glow discharge are clearly visible, although on a sub-millimeter scale. (CF) cathode fall; (NG) negative glow; (FDS) Faraday dark space; (PC) positive column; (AG) anode glow. (Source: Taken with permission from Ref. [20].)

is observed, which has the same characteristic emission pattern as the low-pressure discharge, but on a (sub) millimeter length scale [120] (Figure 1.5). Diffuse DBD discharges have been investigated by many authors. Nonetheless, in this case, a diffuse glow discharge is not always found. The discharge often looks diffuse but consists of filamentary microdischarges, as will be discussed in more detail in the section 1.4 on DBD discharges. For higher frequencies (hundreds of kilohertz) and in gases such as He and N_2, diffuse glow discharges can be obtained [128, 129]. Note that sometimes the addition of a trace gas turns a filamentary discharge into a diffuse discharge, which indicates a clear dependence of filamentation on the chemistry of a discharge. Massines *et al.* also investigated low current discharges without the development of space charge in DBD configurations in the context of material treatment. This Townsend mode is a low-intensity diffuse plasma, but only for higher current densities and after the development of space charge, glow discharge structure with significant emission in the cathode region (negative glow) is observed. An example of a diffuse and a filamentary discharge in a parallel plate DBD geometry is shown in Figure 1.6.

DC glow and microglow discharges between two metal electrodes were investigated by Staack *et al.* [121]. The microglow discharges remain stable because of the high surface–volume ratio and thus efficient heat removal. For discharges on a micrometer scale, the positive column is not present. This increases the stability of the discharge, as well as a positive column has the tendency to contract, for example, due to significant heating and a heating ionization instability. That is, for a fixed $\boldsymbol{E}$ field $\boldsymbol{E}/n_0$ increases with increasing temperature, which means that the ionization rate, and consequently, the electrical conductivity and the heating, also increases. This again leads to an increase of $\boldsymbol{E}/n_0$ and consequently runaway behavior. Large-scale glow discharges have the tendency to contract radially.

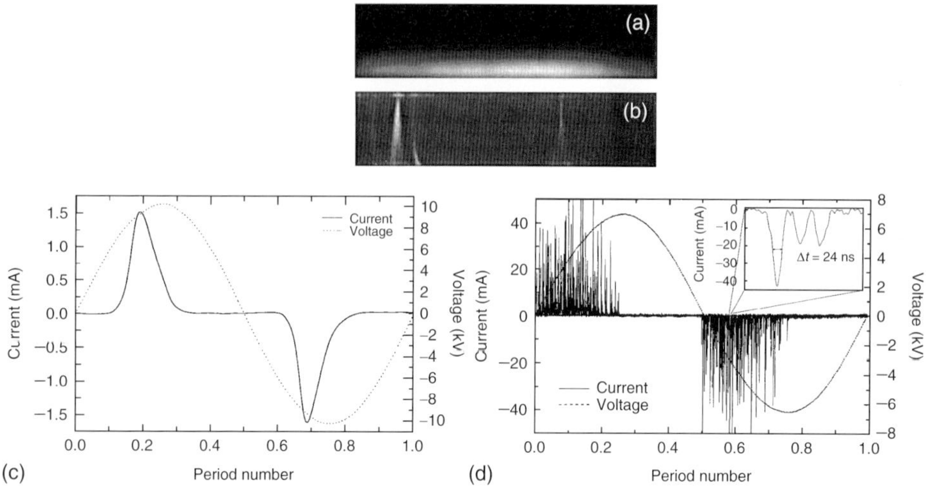

Figure 1.6 The glow mode (a) and the filamentary (micro-discharge) mode (b) in a parallel plate dielectric barrier. The corresponding current (c) and voltage (d) waveforms are shown as well. It is clearly visible that in the filamentary mode, the current pulses of microdischarges can be observed, while in the glow mode, there is a broad current peak for every voltage cycle. (Source: Taken with permission from Ref. [130].)

The CVC of microglow discharges can also be positive in contrast with most other glow discharges. This is in agreement with [114] where the CVC of one-dimensional glow discharges is calculated in the simplest possible model for different gap lengths, and it is found that in short gaps, there is no falling CVC, that is, no negative differential conductivity (NDC), in agreement with early measurements. Raizer *et al.* [131] gives a small correction to [114]. The calculation is done assuming a constant γ of secondary emission.

Šijacic *et al.* [132–134] treat a system where a planar discharge (between Townsend and glow) is sandwiched with a planar high-ohmic semiconductor between planar electrodes to which a DC voltage is supplied. It is believed that a negative differential conductivity would be necessary for spontaneous oscillations [135], but that is not true – a falling CVC of the discharge in the gap as a whole should not be confused with a local NDC. In [132, 133], we analyze the oscillations in [134] and also spatiotemporal patterns.

The Loughborough group has studied RF APGDs in He in a parallel plate metal electrode geometry [136]. It is interesting that this discharge, just like at low pressure, operates in two modes, the alpha and the gamma mode. In the alpha mode, the discharge is sustained by bulk ionization, while in the gamma mode, the discharge is sustained by the electrons generated at the cathode [137]. The discharge often has the tendency to contract radially in the gamma mode.

Several microplasmas can also be categorized as glow discharges. Many different configurations exist, and the reader is referred for a thorough review to [27]. The hollow cathode discharges, for example, can operate in a glow mode at low currents and in an abnormal glow mode when the entire cathode electrode area is covered. In the abnormal glow discharge, the voltage increases with current as the increase in electrode area cannot compensate anymore for the current increases at constant discharge voltage. For microdischarges, the electron density can be of the order of 10^{21} m^{-3} or even higher. Gas temperatures are often 1000 K or larger. Microdischarges are very efficient for producing excimer light.

Atmospheric pressure jets, which are now very popular for biomedical-oriented research, are also often considered to be microplasmas. These plasmas operate at a lower gas temperature. They are mostly constructed in a DBD configuration (see further). In addition, the air flow, and often the He carrier gas, causes an even larger reduction in temperature compared to other typical microplasmas.

The groups of Akishev and Leys investigated the stabilization of high-current negative-glow corona discharges. It is found that with a gas flow of the order of 10 $m\,s^{-1}$ the transition from the glow to spark can be postponed to larger currents [138, 139]. In this extended operation range, the discharge deviates from the CVC of a negative corona ($I = kV(V - V_0)$) due to the growing importance of space charge. With flow stabilization, it is possible to produce large-volume plasmas in centimeter gaps.

From the point of view of diagnostics, interesting work has been done on the chemical activity of nanosecond-pulsed discharges (which can also be of the glow type) by Stancu *et al.* [58]. Optical studies of these discharges have been done, for example, by Bruggeman *et al.* [120], Laux *et al.* [53], Machala *et al.* [140], and Schulz-von der Gathen *et al.* [141]. More recently, the ion flux to the electrodes of an RF-excited APGD in $He{-}H_2O$ has also been studied [62]. However, a lot of these results remain disperse and several open questions remain, especially for discharges operating in gas mixtures containing molecular gases, which is typically the case for applications. These open questions have been partly attacked by modeling such as is intensively made at the groups from, for example, Loughborough University [122] and Queens University, Belfast [142, 143]. Recently, attempts to study an extensive chemistry in these discharges with a zero-dimensional model are published by Liu *et al.* for $He{-}O_2$ and $He{-}H_2O$ diffuse-glow RF discharges [144, 145].

1.3.4 Instabilities

Glow discharge instabilities can occur at the anode [146], at the cathode [147–149], and in the gap [150, 151]. In order to get a glow-to-spark transition, a certain amount of energy needs to be dissipated in the discharge during the glow phase [152]. In cases when the constriction starts at the cathode, this is believed to be in the cathode fall region. It is clear from the different locations where the instability starts that the instabilities are strongly influenced by the exact discharge geometry and electrode properties. Suleebka *et al.* [153] studied constriction of a high-pressure glow discharge in hydrogen and concluded that the electrode history and the amplitude of the overvoltage significantly alter the initial location of the constriction. In the first series of experiments, they found that the constriction of the glow always started at the cathode. When the electrodes become conditioned (after formation of an oxide layer), the constriction is initiated at the anode, in the midgap, and again at the cathode with increasing overvoltages. Recently, the glow-to-spark-transition is investigated in a metal pin–water electrode geometry, which indicated that in the case of low conductivity of the water electrode broadened sparks are observed [151]. This is a nice example of the stabilization of resistive electrodes on the constriction of diffuse glow discharges even after a contraction occurred in the bulk of the discharge.

Recently, Pai *et al.* [154] investigated nanosecond-pulsed atmospheric pressure discharges that can occur in a corona, glow, and spark mode. The glow-to-spark transition has been described by the thermal instability mechanism. It must be mentioned that no general criterion for the contraction of APGDs exists.

1.4 Dielectric Barrier and Surface Discharges

1.4.1 Basic Geometries

Electric discharges in air under ambient conditions have a strong tendency to develop instabilities. These instabilities can develop in space, that is, streamers or filaments, and λ also in temperature, that is, sparks or arcs. Both effects are undesired if one wants to use a volume discharge for chemical conversions. Sparks can be suppressed relatively easily by limiting the current. One way to limit the current is to use a series resistance. This can be done by adding a resistor in the power supply lead or by using a semiconductor for electrode material. A disadvantage of the resistive method is that it leads to loss of electrical energy that is converted into heat. A second way to limit the current is the use of dielectric barriers. The breakdown field strength of a dielectric can be up to hundred times higher than that of air. Therefore, when a streamer reaches the dielectric layer on top of the electrode it will extinguish. Figure 1.7a–c shows the basic shape with two flat electrodes and one or two dielectric barriers. Coaxial shapes, as shown in Figure 1.7d, are very common in ozonizers and other cases of gas treatment.

In the early days, the common dielectric material was glass, which could be the reason why cylinders were often used. Large systems are made with hundreds of long cylinders in parallel [51]. Ozonizers with a capacity of 100 kg O_3 per h have been made this way [4]. Nowadays plastics and ceramics are more common. Sufficient breakdown strength of the dielectric layer is mandatory. But a thicker layer requires a higher voltage, so a compromise must be made here. The dielectric material must be extremely free of voids to avoid damage in the long term. As in high-voltage cable, partial discharges are initiated in voids, which lead to degradation of the dielectric [155]. It can take up to many years before damage occurs. Therefore this effect is mostly unnoticed in laboratory experiments but crucial for robustness of commercial applications.

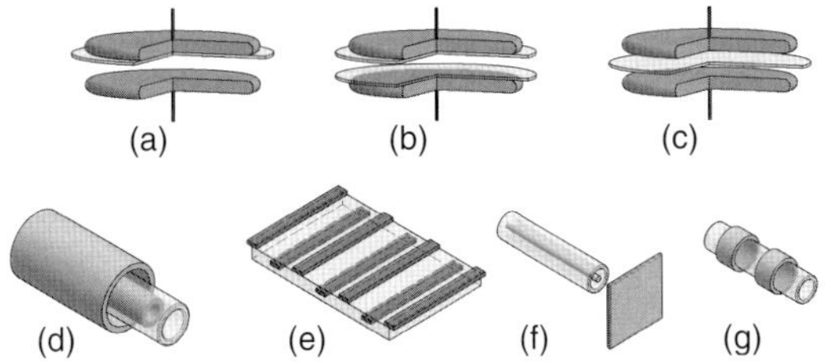

Figure 1.7 Dielectric barrier discharge configurations: (a–c) plane barriers, mostly used in research, (d) coaxial design, often used in ozonizers, (e) (partly) buried electrodes for surface discharge, and (f,g) plasma jets as used in, for example, biomedical applications.

Barrier discharges work best with an electrode separation of a few millimeters. The discharge becomes too inhomogeneous with larger gaps. A good example of this effect is shown in [156]. The spatial nonuniformities of the barrier discharge are a subject of study even now. Recent work on this pattern formation can be found in [157]. In the 1990s, research was begun on how to avoid the inhomogeneities. A good explanation on how to obtain the Townsend, glow, or filamentary discharge was given by Massines and Gouda [116] (Figure 1.6). The transition of the glow discharge into filaments can be avoided with electronic stabilization, as demonstrated by Aldea *et al.* [123]. By now, APGDs have been achieved in barrier discharges in several gases [16, 17].

Alternative geometries have gained a lot of interest in the recent years. The first are dielectric sheaths with electrodes on the top or inside; an example is given in Figure 1.7e where the electrodes are alternatively inside and on the top of the dielectric. All electrodes inside go to one connector of the power supply and all electrodes on the top to the other. The discharge develops form the electrodes on the top over the dielectric surface. A different layout is shown in Figure 1.8. This is a photo of a stack of plastic plates covered on both sides with meshes that are connected to a transformer. The discharge develops at the mesh wires. Very large systems are easily made with this method at relatively low cost. Geometries with electrodes buried in the dielectric were pioneered already in the 1980s by Masuda *et al.* [158], and many alternatives exist nowadays [16, 18]. A nice example of this type of discharge is the plasma display panel (PDP) as used in televisions [159]. Panasonic claims to have made the world's largest plasma TV with a diagonal of 152 in., that is, almost 4 m.

An even newer concept is the cold atmospheric pressure (CAP) plasma jet, two possible shapes are shown in Figure 1.7f,g. This type of plasma source is made for local treatment and is especially popular in biomedical applications [160]. An important aspect is that the gas stays close to room temperature. The electric power consumption ranges from far below 1 W to about 10 W [161], in this case with 13.56 MHz excitation frequency and electrode configuration of Figure 1.7f. Figure 1.9 shows a photo of an RF plasma jet made at the Technische Universiteit, Eindhoven. It has a 1 mm pin electrode inside a 2-mm glass tube with a helium

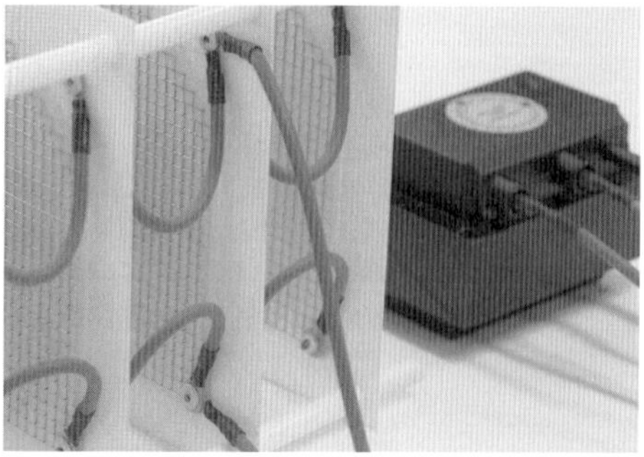

Figure 1.8 PlasmaNorm® demonstration unit, that is, a dielectric barrier discharge made from plastic sheets with a metal mesh on both sides. The power supply is a 50 Hz high-voltage transformer. (Source: Courtesy of Circlair Benelux BV.)

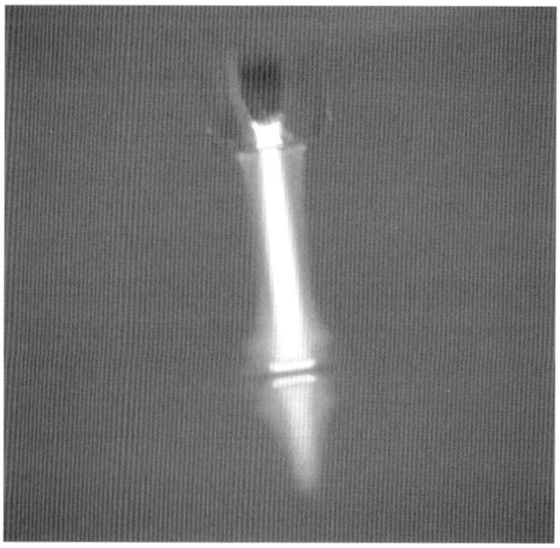

Figure 1.9 RF helium plasma jet hitting a glass substrate. (Source: Photo: Sven Hofman, TU/e.)

flow of $1\,\mathrm{l\,min^{-1}}$. The power consumption by the plasma is 1 W. Lower frequencies are also used, for example, in the kilohertz range [162], mostly with ring electrodes as in Figure 1.7g. Even microwave frequencies of 2.54 GHz [28] are applied in coaxial configurations where the outer cylinder is metal. The microwave power can be up to tens of watts, which will lead to considerable gas heating.

Recent reviews on plasmas for biomedical applications are made by Ehlbeck *et al.* [163] and Lee *et al.* [164]. The plasma jet geometry is also used in combination with nanosecond pulses; in this way, the so-called plasma bullets are created. This method was probably first discussed by Laroussi and Lu [24]; a detailed study is reported by Jarrige *et al.* [165].

1.4.2
Main Properties

The conventional barrier discharge as shown in Figure 1.7a–c is a mixture of a volume and a surface discharge. The first free electron that initiates an avalanche is most likely in the volume. When the applied field is high enough, this avalanche develops into a streamer that travels toward both electrodes since this field is homogeneous. A gap distance of the order of 1 mm is already sufficiently large at a pressure of 1 bar, which is the standard in practically all applications. Most barrier discharges are excited with sine waves, so the streamer velocities will be near the minimum value of 10^5 m s^{-1} [78]. This implies that a streamer reaches the electrodes within 10 ns. After that it can continue across the surface of the dielectric as long as the field strength is high enough. However, the streamer heads deposit charge on the surface where they land and this charge counteracts the applied field. So the discharge extinguishes naturally, and there is no risk for breakdown into a spark as long as the dielectric material is unimpaired. This is the main advantage of the barrier discharge.

The charges can stay on the surface for many seconds, that is, much longer than the repetition time of the applied voltage pulses, which is in the millisecond range. Therefore, a second pulse of the same polarity will still notice counteraction of the surface charge on the dielectric and the discharge will not ignite again. For this reason, barrier discharges are operated with alternating voltage sine waves so that the applied field adds up to the space charge field, which makes it even easier for a new discharge to develop. The frequencies used mostly range from 50 Hz to 100 kHz, but frequencies up to many megahertz are encountered especially in miniature versions, the so-called microdischarges [26]. This name is used, however, for many different geometries and does not always refer to discharges on the micrometer scale.

The question arises how barrier discharges differ from other cold atmospheric plasmas and do they have typical characteristics that make them suitable for certain applications. Table 1.1 gives a coarse overview on these items.

An important question is what makes a certain discharge suitable for a certain application? From Table 1.1, one can conclude that it is basically the shape that is determining: large gaps to treat gases and small gaps and jets for surface treatment. But now the question arises whether or not there are more fundamental differences between these two types of discharges. In the early years of 2D streamer calculations, there was a general idea that corona discharges had a higher electron energy than barrier discharges [8, 34, 166–169]. Numbers mentioned for average electron energies in the streamer heads were 5–15 eV for coronas and 2–6 eV for barriers. The main difference between both discharges is the shape of the applied electric field. In corona's, it is very high near the sharp electrode and below inception in the bulk of the gas; in barrier discharges, the field is more or less homogeneous and above inception in the whole gap.

More recent simulations and experiments (such as those discussed in Section 1.2.2 also) support these general conclusions. From experiments it was found that the T_e and n_e of filamentary DBD discharges, which normally operate close to room temperature, are 1–2 eV and 10^{20}–10^{22} m^{-3}, respectively [170–172]. This is not exactly the same as the older calculations but still supports its general conclusions that in a DBD discharge the electron temperature is lower but the electron density is higher than in a corona discharge. However, in combination with the wide range of circumstances of all discharges being involved, general conclusions cannot be drawn on how to select a certain discharge for a specific application. For the time being, this will remain an empirical process.

1.4.3
Surface Discharges and Packed Beds

As said before, a barrier discharge can be a combination of a volume discharge and a surface discharge or it is only a surface discharge. Specific geometries are made to provoke only surface discharges, such as shown in Figure 1.7e. Similar to barrier discharges, the surface discharge can appear in the shape of filaments or as a homogeneous glow. The surface discharge is also known from high-voltage equipment where it is definitely undesired because it can lead to serious damage [173]. It is generally known that flashover occurs more easily over a surface than through a volume with gas under ambient conditions. The condition of the surface plays a major role here: scratches, dirt, or moisture facilitate the breakdown process. However, only a few fundamental processes that can cause this difference are known: electric charges that are somehow deposited can change the applied field strength [37] and electrons can be released from the surface either by ions [38] or by UV photons [39]. Measurements that reveal differences between surface and volume discharge are very limited. An example is the point-plane geometry studied by Sobota *et al.*, where a dielectric is placed at different distances from the point [174, 175]. An interesting result is that the streamer across the dielectric propagates roughly two times faster than the streamer through the bulk gas, which is argon in this case. Measurements in air were performed by Morales *et al.* [176]; here the influence of pulse rise time and UV radiation was studied. The propagation velocities along a surface in air were measured by Deng *et al.* [177]. Their values are lower than the ones found for argon, the much slower voltage pulse rise time is probably an important factor here. A very interesting aspect of the surface discharge is that surfaces of birefringent material give information of the local electric field strength and the surface charge density. Bismuth germanium oxide crystals are generally used for this purpose, the method is described in detail by Gegot *et al.* [178]. Tanaka *et al.* [179] have used this method to obtain 2D profiles of charge and potential on the surface. They found that the horizontal component of the electric field reaches a maximum on the tip of the streamer and is derived to be 15–30 kV cm^{-1}, that is, equivalent to ~2–3 eV electron energy. This low value is in agreement with the idea that the surface makes it easier for the streamer tip to propagate. Numerical calculations are very difficult in this area because quantitative data on the aforementioned fundamental processes are not available and because the surface streamer process is inherently 3D. 3D gas phase streamer simulations are now becoming available [74, 82, 83, 90], so surface streamer simulations might follow soon.

Packed bed barrier discharges mostly use beads of dielectric material to get a large surface area for enhancement of chemical reactions. These beads can be covered with a catalyst [180]. It is usually assumed that the beads are covered with a surface discharge. This situation is more complex than the planar surface discharge and it appears that no detailed studies of the plasma properties and the plasma chemical kinetics are available.

1.4.4 Applications of Barrier Discharges

The coaxial electrode configuration of Figure 1.7d was first described by Werner Siemens in 1857 [181]. This opened the way for stable and large-scale production of ozone. In 1886, De Meritens discovered that ozone could destroy microorganisms, and in 1906, the city of Nice had already introduced ozone treatment for drinking water. Nowadays, ozone technology is mature, a lot of information can be found in books [182–184] and in a fully devoted journal [185]. Practically all this ozone is made by barrier discharges.

An important aspect of an industrial ozone generator is its yield, usually expressed in grams per kilowatt-hour. A calculation based on the enthalpy of formation shows that a 100% efficient reactor would produce 1.22 kg kWh^{-1}. Probably, the best laboratory result for ozone production in air is still 180 g kWh^{-1} as reported by Masuda *et al.* [158]. For a commercial equipment, where a lot of emphasis is put on lifetime and reliability of the equipment, the production is usually in the range of 1–50 g kWh^{-1}. A pilot plant for flue gas cleaning based on pulsed corona achieved 45 and 60 g kWh^{-1} [186] for pulses with high and low energy, respectively. This demonstrates again that corona and barrier discharges seem to perform equally in terms of chemical yield, although there are indications that the efficiency of nanosecond-pulsed corona discharges can be higher [1]. A new and interesting method for ozone production is the use of xenon excimer lamps [11]. Such lamps very efficiently produce light with a wavelength of 172 nm. This light dissociates oxygen, and as claimed by Salvermoser: "A 172 nm VUV ozone generator operating in a cold environment with ambient air as a feed gas could easily produce 2 wt% of ozone with a wall plug ozone yield of 150 g/kWh without any NO_x-contamination present" [11]. The VUV light source can be a barrier or a corona discharge.

A DBD or surface discharge can have a (small) influence on gas flows. This can be used to improve the flow around an airplane wing, for flame ignition, or for cooling of electrical components. The main advantage of such flow control over conventional methods is that no moving parts are needed and that the size of the device can be kept very small [187, 188].

Occasionally, specific barrier configurations are developed for fundamental investigations. A good example is the single filament electrode geometry developed by Wagner and Kozlov for their cross-correlation spectroscopy [189, 190]. Two

Table 1.2 Barrier discharge applications.

Discharge type	Application	References
Ozone generator	Gas and water cleaning	[184]
Surface barrier	Control aerodynamics	[193, 194]
Surface barrier	Textile modifications	[16, 116, 156]
Atmospheric pressure plasma jet	Microbial decontamination	[163]
Plasma needle	Sterilization/wound healing	[160, 195]
Excimer lamps	Bacteria removal/ozone generation	[11, 12]
Corona with barrier	Phenol/dye removal from water	[7, 196, 197]
Microdischarges	Plasma display panels	[18, 159]
(Packed bed) barrier	Soot removal diesel engines	[198]
Barrier glow discharge	Thin-film deposition solar cells	[17, 19]
Barrier with packed bed catalyst	Chemical conversions	[13–15]

spherical electrodes are used with a gap distance of 1 mm, and one or both are covered with a glass layer. The single filament that develops where the electrodes are closest has such a stable position that the spectra can be recorded by accumulating photons from many pulses. Liu *et al.* have developed a barrier discharge configuration inside a cavity [191]. This enables detection of HO_2 and OH radicals with high sensitivity with the cavity ring down method [192]. Such diagnostics are very important for the development of good models of the chemical activity of the discharge.

A huge amount of literature is available on what can be done with barrier discharges. Only a few examples are mentioned in Table 1.2; in the references given, much more information can be found. At the moment of writing, fast-rising topics are the surface barrier discharges and the plasma jets.

1.5 Gliding Arcs

Arc discharges are thermal discharges and are not considered in this chapter. However, gliding arcs have properties of both thermal and nonthermal plasma conditions. They are highly reactive and often have a high selectivity for chemical processes. The main reason why it is used is because it can provide a plasma with useful properties both from thermal plasmas (large electron densities, currents, and power) and nonthermal plasmas (low gas temperature).

A gliding arc is usually generated between two diverging electrodes typical in a gas flow. The discharge ignites at the shortest distance between the electrodes (few millimeters). Typical breakdown voltages are a few kilovolts. The formation of a hot quasi-thermal plasma corresponds with a decrease in voltage and strong increase in current. Owing to the gas flow (or in absence of the gas flow due

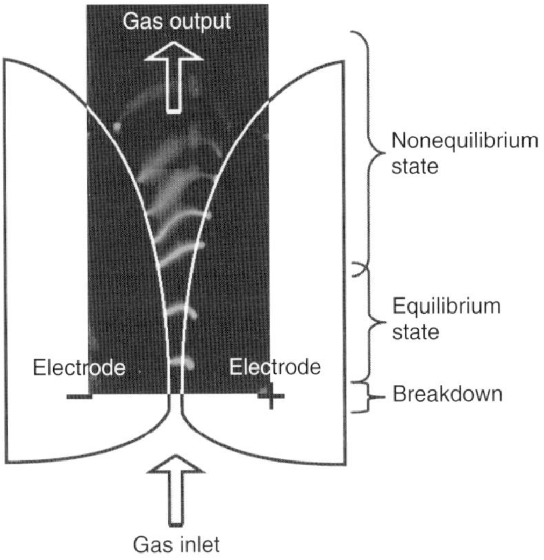

Figure 1.10 Nice example of the superposition of several short exposures of the different stages of a gliding arc discharge. The extension of the discharge channel is clearly visible in the nonthermal stage (upper part). (Source: Taken with permission from Fridman [29].)

to thermal buoyancy) the discharge moves upward and the length of the plasma column increases (see Figure 1.10). This increasing length causes an increase of the heat losses in the column, which exceeds the input energy of the power supply. The quasi-thermal plasma converts into a nonthermal plasma corresponding with a decrease in current and increase in voltage due to the increasing resistivity of the plasma. Eventually, the plasma extinguishes as the power supply cannot maintain such a long plasma column. At this point the recombination of the plasma starts and a reignition of the discharge occurs at the minimum distance between the electrodes. This causes the self-pulsing nature of the gliding arc discharges, which is always clearly visible in current–voltage waveforms and typically occurs on 10 ms timescales.

The exact plasma properties strongly depend on the input power, which can range from about 100 W up to the order of 40 kW. The above description assumes that the transition of a thermal to a nonthermal plasma occurs, although for low powers (and thus low currents) and low flow fields, the gliding arc is nonthermal during its entire lifetime. For the above power range, the gas temperature can range from 2500 up to 10 000 K in the initial "quasi-thermal" state of the plasma. For higher powers, the plasma cools down during the quasi-thermal to nonthermal transition to gas temperatures of 1000–2000 K or even lower, while the electron temperature remains in the order of 1 eV. Note that the largest part of the power is consumed in the nonequilibrium stage of the discharge. Measurements and estimates of the electron density indicate 10^{17}–10^{18} m^{-3} [199]. The chemical efficiency of these discharges is due to the two phases. In the thermal phase, the molecules introduced in the discharge are strongly dissociated. The fast transition from a thermal to nonthermal discharge allows a fast recombination of the dissociation products to molecules, which are required for the application. As the flow is rather large, it allows large throughputs with residence times of the reactants in the order of milliseconds. This is the reason why these discharges are typically the workhorses of plasma chemists. Gliding arc discharge has been studied for combustion [200], gas cleaning [200], production of syngas [200], and water treatment [201, 202]. Gliding arcs can also be magnetically stabilized [203]. More details about the physics of gliding discharges can be found in the review papers by Fridman *et al.* [200].

1.6 Concluding Remarks

As in any review of limited length, this work is not complete. We have attempted to include an overview of (recent) work on nonthermal plasmas but realize that it is far from complete. We also had to limit ourselves to the most used nonthermal plasma types. Therefore we have not discussed some recent developments such as plasma jets and plasma bullets in detail. Furthermore, because we have focused on atmospheric pressure plasmas, we have said nothing about microwave-driven plasmas and other nonthermal plasmas that are primarily used at lower pressures.

Some parts of this book deal also with discharges in and in contact with liquids. The main types of discharges are, however, identical to ordinary gas discharges except for some peculiar properties of the so-called direct streamer discharges in liquids [204].

References

1. van Heesch, E.J.M., Winands, G.J.J., and Pemen, A.J.M. (2008) Evaluation of pulsed streamer corona experiments to determine the O* radical yield. *J. Phys. D: Appl. Phys.*, **41** (23), 234015.
2. Heil, B.G., Czarnetzki, U., Brinkmann, R.P., and Mussenbrock, T. (2008) On the possibility of making a geometrically symmetric RF-CCP discharge electrically asymmetric. *J. Phys. D: Appl. Phys.*, **41**, 165202, DOI: 10.1088/0022-3727/41/16/165202.
3. Ebert, U., Nijdam, S., Li, C., Luque, A., Briels, T., and van Veldhuizen, E. (2010) Review of recent results on streamer discharges and discussion of their relevance for sprites and lightning. *J. Geophys. Res.: Space Phys.*, **115**, A00E43.
4. van Veldhuizen, E.M. (2000) *Electrical Discharges for Environmental Purposes: Fundamentals and Applications*, Nova Science Publishers, New York.
5. Winands, G.J.J., Liu, Z., Pemen, A.J.M., van Heesch, E.J.M., and Yan, K. (2008) Analysis of streamer properties in air as function of pulse and reactor parameters by iccd photography. *J. Phys. D: Appl. Phys.*, **41**, 234001.
6. Grabowski, L., van Veldhuizen, E., Pemen, A., and Rutgers, W. (2006) Corona above water reactor for systematic study of aqueous phenol degradation. *Plasma Chem. Plasma Process.*, **26**, 3–17, doi: 10.1007/s11090-005-8721-8
7. Malik, M.A. (2010) Water purification by plasmas: which reactors are most energy efficient? *Plasma Chem. Plasma Process.*, **30**, 21–31.

8. Braun, D., Kuechler, U., and Pietsch, G. (1991) Microdischarges in air-fed ozonizers. *J. Phys. D: Appl. Phys.*, **24**, 564–572.
9. Eliasson, B., Hirth, M., and Kogelschatz, U. (1986) Ozone synthesis from oxygen in dielectric barrier discharges. *J. Phys. D: Appl. Phys.*, **20**, 1421–1437.
10. Šimor, M., Ráhel, J., Vojtek, P., Cernák, M., and Brablec, A. (2002) Atmospheric-pressure diffuse coplanar surface discharge for surface treatments. *Appl. Phys. Lett.*, **81**, 2716–2718.
11. Salvermoser, M., Murnick, D., and Kogelschatz, U. (2008) Influence of water vapor on photochemical ozone generation with efficient 172 nm xenon excimer lamps. *Ozone Sci. Eng.*, **30**, 228–237.
12. Erofeev, M.V., Schitz, D.V., Skakun, V.S., Sosnin, E.A., and Tarasenko, V.F. (2010) Compact dielectric barrier discharge excilamps. *Phys. Scr.*, **82**, 045403.
13. Horvath, G., Mason, N.J., Polachova, L., Zahoran, M., Moravsky, L., and Matejcik, S. (2010) Packed bed DBD discharge experiments in admixtures of N2 and CH4. *Plasma Chem. Plasma Process.*, **30**, 565–577.
14. Mok, Y.S., Kang, H.C., Lee, H.J., Koh, D.J., and Shin, D.N. (2010) Effect of nonthermal plasma on the methanation of carbon monoxide over nickel catalyst. *Plasma Chem. Plasma Process.*, **30**, 437–447.
15. Huang, L., Xing-wang, Z., Chen, L., and Le-cheng, L. (2011) Direct oxidation of methane to methanol over cu-based catalyst in an ac dielectric barrier discharge. *Plasma Chem. Plasma Process.*, **31**, 67–77, DOI: 10.1007/s11090-010-9272-1.
16. Cernak, M., Cernakova, L., Hudec, I., Kovacik, D., and Zahoranov, A. (2009) Diffuse coplanar surface barrier discharge and its applications for in-line processing of low-added-value materials. *Eur. Phys. J. Appl. Phys.*, **47**, 22806.
17. Starostin, S., Premkumar, P.A., van Veldhuizen, M.C.E., de Vries, H., Paffen, R., and van de Sanden, M. (2009) On the formation mechanisms of the diffuse atmospheric pressure dielectric barrier discharge in cvd processes of thin silica-like films. *Plasma Sources Sci. Technol.*, **18**, 045021.
18. Klages, C.P., Hinze, A., Lachmann, K., Berger, C., Borris, J., Eichler, M., von Hausen, M., Zänker, A., and Thomas, M. (2007) Surface technology with cold microplasmas. *Plasma Processes Polym.*, **4**, 208–218.
19. Sarra-Bournet, C., Gherardi, N., Glénat, H., Laroche, G., and Massines, F. (2010) Effect of C2H4/N2 ratio in an atmospheric pressure dielectric barrier discharge on the plasma deposition of hydrogenated amorphous carbon-nitride films (a-C:N:H). *Plasma Chem. Plasma Process.*, **30**, 213–239.
20. Caruana, D. (2010) Plasmas for aerodynamic control. *Plasma Phys. Control. Fusion*, **52**, 124045.
21. Starikovskii, A., Nikipelov, A., Nudnova, M., and Roupassov, D. (2009) Sdbd plasma actuator with nanosecond pulse-periodic discharge. *Plasma Sources Sci. Technol.*, **18**, 034015.
22. Unfer, T. and Boeuf, J. (2009) Modelling of a nanosecond surface discharge actuator. *J. Phys. D: Appl. Phys.*, **42**, 194017.
23. Foest, R., Kindel, E., Lange, H., Ohl, A., Stieber, M., and Weltmann, K.D. (2007) Rf capillary jet – a tool for localized surface treatment. *Contrib. Plasma Phys.*, **47**, 119–128.
24. Laroussi, M. and Lu, X. (2005) Room-temperature atmospheric pressure plasma plume for biomedical applications. *Appl. Phys. Lett.*, **87**, 113902.
25. Pai, D.Z., Stancu, G.D., Lacoste, D.A., and Laux, C.O. (2009) Nanosecond repetitively pulsed discharges in air at atmospheric pressure-the glow regime. *Plasma Sources Sci. Technol.*, **18** (4), 045030.
26. Baars-Hibbe, L., Sichler, P., Schrader, C., Lucas, N., Gericke, K.H., and Büttgenbach, S. (2005) High frequency glow discharges at atmospheric pressure with micro-structured electrode

arrays. *J. Phys. D: Appl. Phys.*, **38**, 510–517.

27. Becker, K., Schoenbach, K., and Eden, J. (2006) Microplasmas and applications. *J. Phys. D: Appl. Phys.*, **39**, R55–R70.
28. Hrycak, B., Jasinski, M., and Mizeraczyk, J. (2010) Spectroscopic investigations of microwave microplasmas in various gases at atmospheric pressure. *Eur. Phys. J. D*, **60**, 609–619.
29. Fridman, A. (2008) *Plasma Chemistry*, Cambridge University Press.
30. Bruggeman, P., Walsh, J.L., Schram, D.C., Leys, C., and Kong, M.G. (2009) Time dependent optical emission spectroscopy of sub-microsecond pulsed plasmas in air with water cathode. *Plasma Sources Sci. Technol.*, **18** (4), 045023.
31. Li, C., Brok, W.J.M., Ebert, U., and van der Mullen, J.J.A.M. (2007) Deviations from the local field approximation in negative streamer heads. *J. Appl. Phys.*, **101** (12), 123305.
32. Li, C., Ebert, U., and Hundsdorfer, W. (2010) Spatially hybrid computations for streamer discharges with generic features of pulled fronts: I. planar fronts. *J. Comput. Phys.*, **229** (1), 200–220.
33. Kossyi, I.A., Kostinsky, A.Y., Matveyev, A.A., and Silakov, V.P. (1992) Kinetic scheme of the non-equilibrium discharge in nitrogen-oxygen mixtures. *Plasma Sources Sci. Technol.*, **1**, 207.
34. Morrow, R. and Lowke, J.J. (1997) Streamer propagation in air. *J. Phys. D: Appl. Phys.*, **30**, 614.
35. Liu, N. and Pasko, V.P. (2004) Effects of photoionization on propagation and branching of positive and negative streamers in sprites. *J. Geophys. Res.*, **109**, 1.
36. Dujko, S., Ebert, U., White, R., and Petrovic, Z. (2011) Boltzmann equation analysis of electron transport in a N_2-O_2 streamer discharge. *Jap. J. Appl. Phys.* **50**, 08JC01, DOI: 10.1143/JJAP.50.08JC01.
37. Nikonov, V., Bartnikas, R., and Wertheimer, M. (2001) Surface charge and photoionization effects in short air gaps undergoing discharges at atmospheric pressure. *J. Phys. D: Appl. Phys.*, **34**, 2979–2986.
38. Auday, G., Guillot, P., and Galy, J. (2000) Secondary emission of dielectrics used in plasma display panels. *J. Appl. Phys.*, **88**, 4871–4874.
39. Josepson, R., Laan, M., Aarik, J., and Kasikov, A. (2008) Photoinduced field-assisted electron emission from dielectric-coated electrodes into gases. *J. Phys. D: Appl. Phys.*, **42**, 135209.
40. Fridman, A., Chirokov, A., and Gutsol, A. (2005) Non-thermal atmospheric pressure discharges. *J. Phys. D: Appl. Phys.*, **38**, R1.
41. Chang, J., Lawless, P., and Yamamoto, T. (1991) Corona discharge processes. *IEEE Trans. Plasma Sci.*, **19** (6), 1152–1166.
42. Ono, R. and Oda, T. (2007) Ozone production process in pulsed positive dielectric barrier discharge. *J. Phys. D: Appl. Phys.*, **40** (1), 176.
43. Kim, H. (2004) Nonthermal plasma processing for air-pollution control: a historical review, current issues, and future prospects. *Plasma Process. Polym.*, **1** (2), 91–110.
44. Namihira, T., Wang, D., Katsuki, S., Hackam, R., and Akiyama, H. (2003) Propagation velocity of pulsed streamer discharges in atmospheric air. *IEEE Trans. Plasma Sci.*, **31** (5), 1091.
45. Penetrante, B.M. (1993) *Non- Thermal Plasma Techniques for Pollution Control*, NATO ASI Series, Vol. **G34**, Springer-Verlag, pp. 65–89.
46. Urashima, K., Chang, J.S., Park, J.Y., Lee, D.C., Chakrabarti, A., and Ito, T. (1998) Reduction of nox from natural gas combustion flue gases by corona discharge radical injection techniques [thermal power plant emissions control]. *IEEE Trans. Ind. Appl.*, **34** (5), 934–939.
47. Penetrante, B.M., Hsiao, M.C., Merritt, B.T., Vogtlin, G.E., Wallman, P.H., Neiger, M., Wolf, O., Hammer, T., and Broer, S. (1996) Pulsed corona and dielectric-barrier discharge processing of NO in N2. *Appl. Phys. Lett.*, **68** (26), 3719–3721.

48. Itikawa, Y. and Mason, N. (2005) Cross sections for electron collisions with water molecules. *J. Phys. Chem. Ref. Data*, **34** (1), 1–22.
49. Bruggeman, P. and Schram, D.C. (2010) On OH production in water containing atmospheric pressure plasmas. *Plasma Sources Sci. Technol.*, **19** (4), 045025.
50. Rehbein, N. and Cooray, V. (2001) Nox production in spark and corona discharges. *J. Electrostat.*, **51–52**, 333–339.
51. Eliasson, B. and Kogelschatz, U. (1991) Modelling and applications of silent discharge plasmas. *IEEE Trans. Plasma Sci.*, **19**, 309–323.
52. Aleksandrov, N.L. and Bazelyan, E.M. (1999) Ionization processes in spark discharge plasmas. *Plasma Sources Sci. Technol.*, **8**, 285.
53. Laux, C., Spence, T., Kruger, C., and Zare, R. (2003) Optical diagnostics of atmospheric pressure air plasmas. *Plasma Sources Sci. Technol.*, **12** (2), 125–138.
54. Kozlov, K.V., Wagner, H.E., Brandenburg, R., and Michel, P. (2001) Spatio-temporally resolved spectroscopic diagnostics of the barrier discharge in air at atmospheric pressure. *J. Phys. D: Appl. Phys.*, **34** (21), 3164.
55. Liu, N., Pasko, V.P., Burkhardt, D.H., Frey, H.U., Mende, S.B., Su, H.T., Chen, A.B., Hsu, R.R., Lee, L.C., Fukunishi, H. *et al.* (2006) Comparison of results from sprite streamer modeling with spectrophotometric measurements by ISUAL instrument on FORMOSAT-2 satellite. *Geophys. Res. Lett.*, **33**, 01101.
56. Bruggeman, P., Iza, F., Guns, P., Lauwers, D., Kong, M.G., Gonzalvo, Y.A., Leys, C., and Schram, D.C. (2010) Electronic quenching of OH(A) by water in atmospheric pressure plasmas and its influence on the gas temperature determination by OH(A − X) emission. *Plasma Sources Sci. Technol.*, **19** (1), 015016.
57. Niemi, K., Gathen, V., and Döbele, H. (2005) Absolute atomic oxygen density measurements by two-photon absorption laser-induced fluorescence spectroscopy in an RF-excited atmospheric pressure plasma jet. *Plasma Sources Sci. Technol.*, **14**, 375.
58. Stancu, G.D., Kaddouri, F., Lacoste, D.A., and Laux, C.O. (2010) Atmospheric pressure plasma diagnostics by OES, CRDS and TALIF. *J. Phys. D: Appl. Phys.*, **43** (12), 124002.
59. Muraoka, K. and Kono, A. (2011) Laser Thomson scattering for low-temperature plasmas. *J. Phys. D: Appl. Phys.*, **44** (4), 043001.
60. Palomares, J.M., Iordanova, E.I., Gamero, A., Sola, A., and van der Mullen, J.J.A.M. (2010) Atmospheric microwave-induced plasmas in Ar/H2 mixtures studied with a combination of passive and active spectroscopic methods. *J. Phys. D: Appl. Phys.*, **43** (39), 395202.
61. Stoffels, E., Gonzalvo, Y.A., Whitmore, T.D., Seymour, D.L., and Rees, J.A. (2007) Mass spectrometric detection of short-living radicals produced by a plasma needle. *Plasma Sources Sci. Technol.*, **16** (3), 549.
62. Bruggeman, P., Iza, F., Lauwers, D., and Gonzalvo, Y.A. (2010) Mass spectrometry study of positive and negative ions in a capacitively coupled atmospheric pressure RF excited glow discharge in He-water mixtures. *J. Phys. D: Appl. Phys.*, **43** (1), 012003.
63. Clements, J.S., Mizuno, A., Finney, W.C., and Davis, R.H. (1989) Combined removal of SO2, NOx, and fly ash from simulated flue gas using pulsed streamer corona. *IEEE Trans. Ind. Appl.*, **25** (1), 62–69.
64. Winands, G.J.J., Yan, K., Pemen, A.J.M., Nair, S.A., Liu, Z., and van Heesch, E.J.M. (2006) An industrial streamer corona plasma system for gas cleaning. *IEEE Trans. Plasma Sci.*, **34** (5, Part 4), 2426–2433.
65. Kogelschatz, U. (2004) Atmospheric-pressure plasma technology. *Plasma Phys. Control. Fusion*, **46** (12B), B63.
66. Eichwald, O., Ducasse, O., Dubois, D., Abahazem, A., Merbahi, N., Benhenni, M., and Yousfi, M. (2008) Experimental analysis and modelling of positive

streamer in air: towards an estimation of O and N radical production. *J. Phys. D: Appl. Phys.*, **41** (23), 234002.

67. Morrow, R. (1997) The theory of positive glow corona. *J. Phys. D: Appl. Phys.*, **30** (22), 3099.
68. Shang, K. and Wu, Y. (2010) Effect of electrode configuration and corona polarity on NO removal by pulse corona plasma. Power and Energy Engineering Conference (APPEEC), 2010 Asia-Pacific, pp. 1–4.
69. Yan, K., van Heesch, E.J.M., Pemen, A.J.M., and Huijbrechts, P.A.H.J. (2001) From chemical kinetics to streamer corona reactor and voltage pulse generator. *Plasma Chem. Plasma Process.*, **21**, 107–137.
70. Ratushnaya, V., Luque, A., and Ebert, U. (2012) Electrodynamic characterization of long positive streamers in air. *J. Phys. D: Appl. Phys.*, in preparation.
71. Ebert, U., Brau, F., Derks, G., Hundsdorfer, W., Kao, C.Y., Li, C., Luque, A., Meulenbroek, B., Nijdam, S., Ratushnaya, V., Schäfer, L., and Tanveer, S. (2011) Multiple scales in streamer discharges, with an emphasis on moving boundary approximations. *Nonlinearity*, **24**, C1.
72. Babich, L.P. (2003) *High-Energy Phenomena in Electric Discharges in Dense Gases*, Futurepast Inc, Arlington, VA.
73. Moss, G.D., Pasko, V.P., Liu, N., and Veronis, G. (2006) Monte carlo model for analysis of thermal runaway electrons in streamer tips in transient luminous events and streamer zones of lightning leaders. *J. Geophys. Res.*, **111** (A2), A02307.
74. Li, C., Ebert, U., and Hundsdorfer, W. (2009) 3D hybrid computations for streamer discharges and production of run-away electrons. *J. Phys. D: Appl. Phys.*, **42** (202003), 202–203.
75. Chanrion, O. and Neubert, T. (2010) Production of runaway electrons by negative streamer discharges. *J. Geophys. Res.*, **115**, A00E32.
76. Nguyen, C.V., van Deursen, A.P.J., and Ebert, U. (2008) Multiple X-ray bursts from long discharges in air. *J. Phys. D: Appl. Phys.*, **41**, 234012.
77. Briels, T.M.P., Kos, J., van Veldhuizen, E.M., and Ebert, U. (2006) Circuit dependence of the diameter of pulsed positive streamers in air. *J. Phys. D: Appl. Phys.*, **39**, 5201.
78. Briels, T.M.P., Kos, J., Winands, G.J.J., van Veldhuizen, E.M., and Ebert, U. (2008) Positive and negative streamers in ambient air: measuring diameter, velocity and dissipated energy. *J. Phys. D: Appl. Phys.*, **41**, 234004.
79. Naidis, G.V. (2009) Positive and negative streamers in air: velocity-diameter relation. *Phys. Rev. E*, **79** (5), 057401.
80. Luque, A. and Ebert, U. (2010) Sprites in varying air density: charge conservation, glowing negative trails and changing velocity. *Geophys. Res. Lett*, **31**, L06806.
81. Luque, A., Ratushnaya, V., and Ebert, U. (2008) Positive and negative streamers in ambient air: modeling evolution and velocities. *J. Phys. D: Appl. Phys.*, **41**, 234005.
82. Luque, A. and Ebert, U. (2012) Density models for streamer discharges: beyond cylindrical symmetry and homogeneous media. *J. Comput. Phys.*, **231**, 904–918, DOI: 10.1016/j.jcp.2011.04.019.
83. Li, C., Ebert, U., and Hundsdorfer, W. (2012) Spatially hybrid computations for streamer discharges: II. fully 3D simulations. *J. Comput. Phys.*, **231**, 1020–1050, DOI: 10.1016/j.jcp.2011.07.023.
84. Mathew, D., Bastiaens, H.M.J., Boller, K.J., and Peters, P.J.M. (2007) Effect of preionization, fluorine concentration, and current density on the discharge uniformity in F2 excimer laser gas mixtures. *J. Appl. Phys.*, **102** (3), 033305.
85. Meek, J.M. (1940) A theory of spark discharge. *Phys. Rev.*, **57** (8), 722–728.
86. Raether, H. (1939) Die entwicklung der elektronenlawine in den funkenkanal. *Z. Phys. A: Hadrons Nuclei*, **112** (7), 464–489.
87. Montijn, C. and Ebert, U. (2006) Diffusion correction to the Raether and Meek criterion for the avalanche-to-streamer transition. *J. Phys. D: Appl. Phys.*, **39**, 2979.

88. Pancheshnyi, S. (2005) Role of electronegative gas admixtures in streamer start, propagation and branching phenomena. *Plasma Sources Sci. Technol.*, **14**, 645.
89. Wormeester, G., Pancheshnyi, S., Luque, A., Nijdam, S., and Ebert, U. (2010) Probing photo-ionization: simulations of positive streamers in varying N_2:O_2-mixtures. *J. Phys. D: Appl. Phys.*, **43**, 505201, DOI: 10.1088/0022-3727/43/50/505201.
90. Luque, A., Ebert, U., Montijn, C., and Hundsdorfer, W. (2007) Photoionisation in negative streamers: fast computations and two propagation modes. *Appl. Phys. Lett.*, **90**, 081501, DOI: 10.1063/1.2435934.
91. Briels, T.M.P., van Veldhuizen, E.M., and Ebert, U. (2008) Positive streamers in air and nitrogen of varying density: experiments on similarity laws. *J. Phys. D: Appl. Phys.*, **41**, 234008.
92. Briels, T.M.P., van Veldhuizen, E.M., and Ebert, U. (2008) Positive streamers in ambient air and a N2:O2-mixture (99.8: 0.2). *IEEE Trans. Plasma Sci.*, **36**, 906–907.
93. Nijdam, S., Miermans, K., van Veldhuizen, E., and Ebert, U. (2011) A peculiar streamer morphology created by a complex voltage pulse. *IEEE Trans. Plasma. Sci.*, **39**, 2216–2217, DOI: 10.1109/TPS.2011.2158661.
94. Briels, T.M.P., van Veldhuizen, E.M., and Ebert, U. (2005) Branching of positive discharge streamers in air at varying pressures. *IEEE Trans. Plasma Sci.*, **33**, 264.
95. Ebert, U., Montijn, C., Briels, T.M.P., Hundsdorfer, W., Meulenbroek, B., Rocco, A., and van Veldhuizen, E.M. (2006) The multiscale nature of streamers. *Plasma Sources Sci. Technol.*, **15**, S118.
96. Marode, E. (1975) The mechanism of spark breakdown in air at atmospheric pressure between a positive point and a plane. I. experimental: nature of the streamer track. *J. Appl. Phys.*, **46** (5), 2005–2015.
97. Sigmond, R.S. (1984) The residual streamer channel: return strokes and secondary streamers. *J. Appl. Phys.*, **56** (5), 1355–1370.
98. Ono, R. and Oda, T. (2003) Formation and structure of primary and secondary streamers in positive pulsed corona discharge-effect of oxygen concentration and applied voltage. *J. Phys. D: Appl. Phys.*, **36** (16), 1952–1958.
99. Liu, N. (2010) Model of sprite luminous trail caused by increasing streamer current. *Geophys. Res. Lett*, **37**, L04102.
100. Nijdam, S., Moerman, J.S., Briels, T.M.P., van Veldhuizen, E.M., and Ebert, U. (2008) Stereo-photography of streamers in air. *Appl. Phys. Lett.*, **92**, 101502.
101. Nijdam, S., Geurts, C.G.C., van Veldhuizen, E.M., and Ebert, U. (2009) Reconnection and merging of positive streamers in air. *J. Phys. D: Appl. Phys.*, **42** (4), 045201.
102. Arrayás, M., Ebert, U., and Hundsdorfer, W. (2002) Spontaneous branching of anode-directed streamers between planar electrodes. *Phys. Rev. Lett.*, **88** (17), 174502.
103. Montijn, C., Ebert, U., and Hundsdorfer, W. (2006) Numerical convergence of the branching time of negative streamers. *Phys. Rev. E*, **73** (6), 65401.
104. Luque, A. and Ebert, U. (2011) Electron density fluctuations accelerate the branching of streamer discharges in air. *Phys. Rev. E.*, **84**, 046411, DOI: 10.1103/PhysRevE.84.046411.
105. Loeb, L.B. and Meek, J.M. (1940) The mechanism of spark discharge in air at atmospheric pressure. I. *J. Appl. Phys.*, **11** (6), 438–447.
106. Nijdam, S., van de Wetering, F.M.J.H., Blanc, R., van Veldhuizen, E.M., and Ebert, U. (2010) Probing photo-ionization: experiments on positive streamers in pure gases and mixtures. *J. Phys. D: Appl. Phys.*, **43**, 145204.
107. Wormeester, G., Nijdam, S., and Ebert, U. (2011) Feather-like structures in positive streamers. *Jap. J. Appl. Phys.*, **50**, 08JA01, DOI: 10.1143/JJAP.50.08JA01.

108. van Veldhuizen, E.M. and Rutgers, W.R. (2002) Pulsed positive corona streamer propagation and branching. *J. Phys. D: Appl. Phys.*, **35**, 2169.
109. Babaeva, N.Y. and Kushner, M.J. (2008) Streamer branching: the role of inhomogeneities and bubbles. *IEEE Trans. Plasma Sci.*, **36** (4), 892–893.
110. Niemeyer, L., Pietronero, L., and Wiesmann, H.J. (1984) Fractal dimension of dielectric breakdown. *Phys. Rev. Lett.*, **52**, 1033.
111. Pasko, V.P., Inan, U.S., and Bell, T.F. (1998) Spatial structure of sprites. *Geophys. Res. Lett.*, **25**, 2123–2126.
112. Akyuz, M., Larsson, A., Cooray, V., and Strandberg, G. (2003) 3D simulations of streamer branching in air. *J. Electrostat.*, **59**, 115.
113. Raizer, Y.P. (1991) *Gas Discharge Physics*, Springer-Verlag, Berlin.
114. Šijacic, D.D. and Ebert, U. (2002) Transition from townsend to glow discharge: subcritical, mixed, or supercritical characteristics. *Phys. Rev. E*, **66** (6), 066410.
115. von Engel, A., Seeliger, R., and Steinbeck, M. (1933) Über die Glimmentladung bei hohen Drucken. *Z. Phys.*, **85**, 144.
116. Massines, F. and Gouda, G. (1998) A comparison of polypropylene-surface treatment by filamentary, homogeneous and glow discharges in helium at atmospheric pressure. *J. Phys. D: Appl. Phys.*, **31**, 3411–3420.
117. Kunhardt, E. (2000) Generation of large – volume, atmospheric – pressure, nonequilibrium plasmas. *IEEE Trans. Plasma Sci.*, **28** (1), 189–200.
118. Barinov, Y., Kaplan, V., Rozhdestvenskii, V., and Shkolnik, S. (1998) Determination of the electron density in a discharge with nonmetallic liquid electrodes in atmospheric-pressure air from the absorption of microwave probe radiation. *Tech. Phys. Lett.*, **24** (12), 929–931.
119. Lu, X.P. and Laroussi, M. (2008) Electron density and temperature measurement of an atmospheric pressure plasma by millimeter wave interferometer. *Appl. Phys. Lett.*, **92** (5), 051501.
120. Bruggeman, P., Liu, J., Degroote, J., Kong, M.G., Vierendeels, J., and Leys, C. (2008) Dc excited glow discharges in atmospheric pressure air in pin-to-water electrode systems. *J. Phys. D: Appl. Phys.*, **41** (21), 215201.
121. Staack, D., Farouk, B., Gutsol, A.F., and Fridman, A. (2007) Spatially resolved temperature measurements of atmospheric-pressure normal glow microplasmas in air. *IEEE Trans. Plasma Sci.*, **35** (5, Part 2), 1448–1455.
122. Iza, F., Lee, J.K., and Kong, M.G. (2007) Electron kinetics in radio-frequency atmospheric-pressure microplasmas. *Phys. Rev. Lett.*, **99** (7), 075004.
123. Aldea, E., Peeters, P., de Vries, H., and van de Sanden, M. (2005) Atmospheric glow stabilization. do we need pre-ionization? *Surf. Coat. Technol.*, **200**, 46–50.
124. Laroussi, M., Alexeff, I., Richardson, J., and Dyer, F. (2002) The resistive barrier discharge. *IEEE Trans. Plasma Sci.*, **30** (1, Part 1), 158–159.
125. Andre, P., Barinov, Y., Faure, G., Kaplan, V., Lefort, A., Shkol'nik, S., and Vacher, D. (2001) Experimental study of discharge with liquid non-metallic (tap-water) electrodes in air at atmospheric pressure. *J. Phys. D: Appl. Phys.*, **34** (24), 3456–3465.
126. Lu, X. and Laroussi, M. (2005) Atmospheric pressure glow discharge in air using a water electrode. *IEEE Trans. Plasma Sci.*, **33** (2, Part 1), 272–273.
127. Bruggeman, P., Ribezl, E., Maslani, A., Degroote, J., Malesevic, A., Rego, R., Vierendeels, J., and Leys, C. (2008) Characteristics of atmospheric pressure air discharges with a liquid cathode and a metal anode. *Plasma Sources Sci. Technol.*, **17** (2), 025012.
128. Gherardi, N. and Massines, F. (2001) Mechanisms controlling the transition from glow silent discharge to streamer discharge in nitrogen. *IEEE Trans. Plasma Sci.*, **29** (3), 536–544.
129. Massines, F., Rabehi, A., Decomps, P., Gadri, R., Segur, P., and Mayoux, C. (1998) Experimental and theoretical study of a glow discharge at atmospheric pressure controlled by

dielectric barrier. *J. Appl. Phys.*, **83** (6), 2950–2957.
130. Gherardi, N., Gouda, G., Gat, E., Ricard, A., and Massines, F. (2000) Transition from glow silent discharge to micro-discharges in nitrogen gas. *Plasma Sources Sci. Technol.*, **9**, 340.
131. Raizer, Y.P., Ebert, U., and Šijacic, D. (2004) Dependence of the transition from Townsend to glow discharge on secondary emission. *Phys. Rev. E*, **70** (1), 17401.
132. Šijacic, D.D., Ebert, U., and Rafatov, I. (2004) Period doubling cascade in glow discharges: local versus global differential conductivity. *Phys. Rev. E*, **70** (5), 056220.
133. Šijacic, D.D., Ebert, U., and Rafatov, I. (2005) Oscillations in dc driven barrier discharges: numerical solutions, stability analysis, and phase diagram. *Phys. Rev. E*, **71** (6), 066402.
134. Rafatov, I.R., Šijacic, D.D., and Ebert, U. (2007) Spatiotemporal patterns in a dc semiconductor-gas-discharge system: stability analysis and full numerical solutions. *Phys. Rev. E*, **76** (3), 036206.
135. Raizer, Y.P., Gurevich, E.L., and Mokrov, M.S. (2006) Self-sustained oscillations in a low-current discharge with a semiconductor serving as a cathode and ballast resistor: II. Theory. *Tech. Phys.*, **51** (2), 185–197.
136. Shi, J., Deng, X., Hall, R., Punnett, J., and Kong, M. (2003) Three modes in a radio frequency atmospheric pressure glow discharge. *J. Appl. Phys.*, **94** (10), 6303–6310.
137. Shi, J. and Kong, M. (2005) Mechanisms of the alpha and gamma modes in radio-frequency atmospheric glow discharges. *J. Appl. Phys.*, **97** (2), 023306.
138. Goossens, O., Callebaut, T., Akishe v, Y., Napartovich, A., Trushkin, N., and Leys, C. (2002) The DC glow discharge at atmospheric pressure. *IEEE Trans. Plasma Sci.*, **30** (1, Part 1), 176–177.
139. Akishev, Y., Goossens, O., Callebaut, T., Leys, C., Napartovich, A., and Trushkin, N. (2001) The influence of electrode geometry and gas flow on corona-to-glow and glow-to-spark threshold currents in air. *J. Phys. D: Appl. Phys.*, **34** (18), 2875–2882.
140. Machala, Z., Janda, M., Hensel, K., Jedlovsky, I., Lestinska, L., Foltin, V., Martisovits, V., and Morvova, M. (2007) Emission spectroscopy of atmospheric pressure plasmas for bio-medical and environmental applications. *J. Mol. Spectrosc.*, **243** (2), 194–201.
141. Schulz-von der Gathen, V., Schaper, L., Knake, N., Reuter, S., Niemi, K., Gans, T., and Winter, J. (2008) Spatially resolved diagnostics on a microscale atmospheric pressure plasma jet. *J. Phys. D: Appl. Phys.*, **41** (19), 194004.
142. Waskoenig, J., Niemi, K., Knake, N., Graham, L.M., Reuter, S., Schulz-von der Gathen, V., and Gans, T. (2010) Atomic oxygen formation in a radio-frequency driven micro-atmospheric pressure plasma jet. *Plasma Sources Sci. Technol.*, **19** (4), 045018.
143. Niemi, K., Reuter, S., Graham, L.M., Waskoenig, J., and Gans, T. (2009) Diagnostic based modeling for determining absolute atomic oxygen densities in atmospheric pressure helium-oxygen plasmas. *Appl. Phys. Lett.*, **95** (15), 151504.
144. Liu, D.X., Rong, M.Z., Wang, X.H., Iza, F., Kong, M.G., and Bruggeman, P. (2010) Main species and physicochemical processes in cold atmospheric-pressure He + O-2 plasmas. *Plasma Process. Polym.*, **7** (9–10), 846–865.
145. Liu, D.X., Bruggeman, P., Iza, F., Rong, M.Z., and Kong, M.G. (2010) Global model of low-temperature atmospheric-pressure He + H2O plasmas. *Plasma Sources Sci. Technol.*, **19** (2), 025018.
146. Akishev, Y., Grushin, M., Kochetov, I., Karalnik, V., Napartovich, A., and Trushkin, N. (2005) Negative corona, glow and spark discharges in ambient air and transitions between them. *Plasma Sources Sci. Technol.*, **14** (2), S18–S25.
147. Korolev, Y.D., Frants, O.B., Landl, N.V., Geyman, V.G., and Matveev, I.B. (2007) Glow-to-spark transitions in a plasma system for ignition and combustion

control. *IEEE Trans. Plasma Sci.*, **35** (6, Part 1), 1651–1657.

148. Lewis, D. and Woolsey, G. (1981) Spark discharges in iodine vapour. *J. Phys. D: Appl. Phys.*, **14** (8), 1445–1158.

149. Takaki, K., Kitamura, D., and Fujiwara, T. (2000) Characteristics of a high-current transient glow discharge in dry air. *J. Phys. D: Appl. Phys.*, **33** (11), 1369–1375.

150. Chalmers, I. and Duffy, H. (1971) Observations of arc-forming stages of spark breakdown using an image intensifier and converter. *J. Phys. D: Appl. Phys.*, **4** (9), 1302–1305.

151. Bruggeman, P., Guns, P., Degroote, J., Vierendeels, J., and Leys, C. (2008) Influence of the water surface on the glow-to-spark transition in a metal-pin-to-water electrode system. *Plasma Sources Sci. Technol.*, **17** (4), 045014.

152. Chalmers, I. (1971) Transient glow discharge in nitrogen and dry air. *J. Phys. D: Appl. Phys.*, **4** (8), 1147.

153. Suleebka, P., Barrault, M., and Craggs, J. (1975) Constriction of a high-pressure glow-discharge in hydrogen. *J. Phys. D: Appl. Phys.*, **8** (18), 2190–2297.

154. Pai, D.Z., Lacoste, D.A., and Laux, C.O. (2010) Transitions between corona, glow, and spark regimes of nanosecond repetitively pulsed discharges in air at atmospheric pressure. *J. Appl. Phys.*, **107** (9), 093303.

155. Morshuis, P. (2005) Degradation of solid dielectrics due to internal partial discharge. *IEEE Trans. Dielectrics Electr. Insul.*, **12**, 905–913.

156. Onsuratoom, S., Rujiravanit, R., Sreethawong, T., Tokura, S., and Chavadej, S. (2010) Silver loading on dbd plasma-modified woven pet surface for antimicrobial property improvement. *Plasma Chem. Plasma Process.*, **30**, 191–206.

157. Stollenwerk, L. (2010) Interaction of current filaments in a dielectric barrier discharge system. *Plasma Phys. Control. Fusion*, **52**, 124017.

158. Masuda, S., Akutsu, K., Kuroda, M., Awatsu, Y., and Shibuya, Y. (1988) A ceramic-based ozonizer using high-frequency discharge. *IEEE Trans. Ind. Appl.*, **24**, 223–231.

159. Hagelaar, G., Klein, M., Snijkers, R., and Kroesen, G. (2001) Energy loss mechanisms in the microdischarges in plasma display panels. *J. Appl. Phys.*, **89**, 2033–2039.

160. Stoffels, E., Sakiyama, Y., and Graves, D. (2008) Cold atmospheric plasma: charged species and their interactions with cells and tissues. *IEEE Trans. Plasma Sci.*, **36**, 1441–1457.

161. Stoffels, E., Flikweert, A., Stoffels, W., and Kroesen, G. (2002) Plasma needle: a non-destructive atmospheric plasma source for fine surface treatment of (bio)materials. *Plasma Sources Sci. Technol.*, **11**, 383–388.

162. Jiang, N., Ji, A., and Cao, Z. (2010) Atmospheric pressure plasma jets beyond ground electrode as charge overflow in a dielectric barrier discharge setup. *J. Appl. Phys.*, **108**, 033302.

163. Ehlbeck, J., Schnabel, U., Polak, M., Winter, J., von Woedtke, T., Brandenburg, R., von dem Hagen, T., and Weltmann, K.D. (2011) Low temperature atmospheric pressure plasma sources for microbial decontamination (topical review). *J. Phys. D: Appl. Phys.*, **44**, 013002.

164. Lee, H., Park, G., Seo, Y., Im, Y., Shim, S., and Lee, H. (2011) Modelling of atmospheric pressure plasmas for biomedical applications. *J. Phys. D: Appl. Phys.*, **44**, 053001, DOI: 10.1088/0022-3727/44/5/053001.

165. Jarrige, J., Laroussi, M., and Karakas, E. (2010) Formation and dynamics of plasma bullets in a non-thermal plasma jet: influence of the high-voltage parameters on the plume characteristics. *Plasma Sources Sci. Technol.*, **19**, 065005.

166. Dhali, S.K. and Williams, P.F. (1987) Two-dimensional studies of streamers in gases. *J. Appl. Phys.*, **62** (12), 4696–4707.

167. Gallimberti, I. (1988) Impulse corona simulation for flue gas treatment. *Pure Appl. Chem.*, **60**, 663–674.

168. Babaeva, N.Y. and Naidis, G.V. (1996) Two-dimensional modelling of positive streamer dynamics in non-uniform

electric fields in air. *J. Phys. D: Appl. Phys.*, **29**, 2423.

169. Vitello, P.A., Penetrante, B.M., and Bardsley, J.N. (1994) Simulation of negative-streamer dynamics in nitrogen. *Phys. Rev. E*, **49** (6), 5574–5598.

170. Balcon, N., Aanesland, A., and Boswell, R. (2007) Pulsed rf discharges, glow and filamentary mode at atmospheric pressure in argon. *Plasma Sources Sci. Technol.*, **16** (2), 217.

171. Dong, L., Qi, Y., Zhao, Z., and Li, Y. (2008) Electron density of an individual microdischarge channel in patterns in a dielectric barrier discharge at atmospheric pressure. *Plasma Sources Sci. Technol.*, **17** (1), 015015.

172. Zhu, X.M., Pu, Y.K., Balcon, N., and Boswell, R. (2009) Measurement of the electron density in atmospheric-pressure low-temperature argon discharges by line-ratio method of optical emission spectroscopy. *J. Phys. D: Appl. Phys.*, **42** (14), 142003.

173. Tan, B., Allen, N., and Rodrigo, H. (2007) Progression of positive corona on cylindrical insulating surfaces. I. Influence of dielectric material. *IEEE Trans. Dielectrics Electr. Insul.*, **14**, 111–118.

174. Sobota, A., van Veldhuizen, E.M., and Stoffels, W.W. (2008) Discharge ignition near a dielectric. *IEEE Trans. Plasma Sci.*, **36**, 912.

175. Sobota, A., Lebouvier, A., Kramer, N., Stoffels, W., Manders, F., and Haverlag, M. (2009) Speed of streamers in argon over a flat surface of a dielectric. *J. Phys. D: Appl. Phys.*, **42**, 015211.

176. Morales, K., Krile, J., Neuber, A., and Krompholz, H. (2007) Dielectric surface flashover at atmospheric conditions with unipolar pulsed voltage excitation. *IEEE Trans. Dielectrics Electr. Insul.*, **14**, 774–782.

177. Deng, J., Matsuoka, S., Kumada, A., and Hidaka, K. (2010) The influence of residual charge on surface discharge propagation. *J. Phys. D: Appl. Phys.*, **43**, 495203.

178. Gegot, F., Callegari, T., Aillerie, M., and Boeuf, J. (2008) Experimental protocol and critical assessment of the Pockels method for the measurement of surface charging in a dielectric barrier discharge. *J. Phys. D: Appl. Phys.*, **41**, 135204.

179. Tanaka, D., Matsuoka, S., Kumada, A., and Hidaka, K. (2009) Two-dimensional potential and charge distributions of positive surface streamer. *J. Phys. D: Appl. Phys.*, **42**, 075204.

180. Suttikul, T., Sreethawong, T., Sekiguchi, H., and Chavadej, S. (2011) Ethylene epoxidation over alumina- and silica-supported silver catalysts in low-temperature ac dielectric barrier discharge. *Plasma Chem. Plasma Process.*, **31**, 273–290, DOI: 10.1007/s11090-010-9280-1.

181. Rubin, M. (2001) The history of ozone. The schönbein period, 1839–1868. *Bull. Hist. Chem.*, **26**, 40–56.

182. Rice, R. and Netzer, A. (1984) *Handbook of Ozone Technologies and Applications*, Ozone for Drinking Water Treatment, Vol. **II**, Butterworth Publishers, Boston.

183. Vosmaer, A. (1916) *Ozone: Its Manufacture, Properties and Uses*, Van Nostrand, New York.

184. Gottschalk, C., Libra, J., and Saupe, A. (2010) *Ozonation of Water and Waste Water: A Practical Guide to Understanding Ozone and its Applications*, Wiley-VCH Verlag GmbH.

185. Loeb, B. (2010) Editorial. *Ozone: Sci. Eng.*, **32**, 381–382.

186. Winands, G.J.J., Liu, Z., Pemen, A.J.M., van Heesch, E.J.M., Yan, K., and van Veldhuizen, E.M. (2006) Temporal development and chemical efficiency of positive streamers in a large scale wire-plate reactor as a function of voltage waveform parameters. *J. Phys. D: Appl. Phys.*, **39**, 3010.

187. Moreau, E. (2007) Airflow control by non-thermal plasma actuators. *J. Phys. D: Appl. Phys.*, **40** (3), 605.

188. Starikovskii, A., Anikin, N., Kosarev, I., Mintoussov, E., Nudnova, M., Rakitin, A., Roupassov, D., Starikovskaia, S., and Zhukov, V. (2008) Nanosecond-pulsed discharges for plasma-assisted combustion and aerodynamics. *J. Propul. Power*, **24** (6), 1182.

189. Hoder, T., Brandenburg, R., Basner, R., Weltmann, K.D., Kozlov, K., and Wagner, H.E. (2010) A comparative study of three different types of barrier discharges in air at atmospheric pressure by cross-correlation spectroscopy. *J. Phys. D: Appl. Phys.*, **43**, 124009.

190. Kloc, P., Wagner, H.E., Trunec, D., Navrátil, Z., and Fedoseev, G. (2010) An investigation of dielectric barrier discharge in ar and ar/nh3 mixture using cross-correlation spectroscopy. *J. Phys. D: Appl. Phys.*, **43**, 345205.

191. Liu, Z.W., Xu, Y., Yang, X.F., Zhu, A.M., Zhao, G.L., and Wang, W.G. (2008) Determination of the ho2 radical in dielectric barrier discharge plasmas using near-infrared cavity ring-down spectroscopy. *J. Phys. D: Appl. Phys.*, **41**, 045203.

192. Zhao, G., Zhu, A., Wu, J., Liu, Z., and Xu, Y. (2010) Measurement of oh radicals in dielectric barrier discharge plasmas by cavity ring-down spectroscopy. *Plasma Sci. Technol.*, **12**, 166–171.

193. Dong, B., Bauchire, J.M., Pouvesle, J., Magnier, P., and Hong, D. (2008) Experimental study of a dbd surface discharge for the active control of subsonic airflow. *J. Phys. D: Appl. Phys.*, **41**, 155201.

194. Opaits, D., Roupassov, D., Starikovskaia, S., Starikovskii, A., Zavialov, I.N., and Saddoughi, S. (2005) Plasma control of boundary layer using low-temperature non-equilibrium plasma of gas discharge. 43rd AIAA Aerospace Sciences Meeting and Exhibit, Reno, Nevada, paper AIAA 2005-1180.

195. Stoffels, E., Kieft, I., Sladek, R., van den Bedem, L., van der Laan, E., and Steinbuch, M. (2006) Plasma needle for in vivo medical treatment: recent developments and perspectives. *Plasma Sources Sci. Technol.*, **15**, S169–S180.

196. Pokryvailo, A., Wolf, M., Yankelevich, Y., Wald, S., E.M. van Veldhuizen, Grabowski, L., Rutgers, W., Reiser, M., Eckhardt, T., Glocker, B., Kempenaers, P., and Welleman, A. (2006) High-power pulsed corona for treatment of pollutants in heterogeneous media. *IEEE Trans. Plasma Sci.*, **34**, 1731–1743.

197. Grabowski, L., van Veldhuizen, E., Pemen, A., and Rutgers, W. (2007) Breakdown of methylene blue and methyl orange by pulsed corona discharge. *Plasma Sources Sci. Technol.*, **16**, 226–232.

198. Yao, S. (2009) Plasma reactors for diesel particulate matter removal. *Recent Patents Chem. Eng.*, **2**, 67–75.

199. Kalra, C., Gutsol, A., and Fridman, A. (2005) Gliding arc discharges as a source of intermediate plasma for methane partial oxidation. *IEEE Trans. Plasma Sci.*, **33** (1, Part 1), 32–41.

200. Fridman, A., Gutsol, A., Gangoli, S., Ju, Y., and Ombrellol, T. (2008) Characteristics of gliding arc and its application in combustion enhancement. *J. Propul. Power*, **24** (6), 1216–1228.

201. Benstaali, B., Boubert, P., Cheron, B., Addou, A., and Brisset, J. (2002) Density and rotational temperature measurements of the OH degrees and NO degrees radicals produced by a gliding arc in humid air. *Plasma Chem. Plasma Process.*, **22** (4), 553–571.

202. Burlica, R., Kirkpatrick, M., and Locke, B. (2006) Formation of reactive species in gliding arc discharges with liquid water. *J. Electrostat.*, **64** (1), 35–43.

203. Gangoli, S.P., Gutsol, A.F., and Fridman, A.A. (2010) A non-equilibrium plasma source: magnetically stabilized gliding arc discharge: I. Design and diagnostics. *Plasma Sources Sci. Technol.*, **19** (6), 065003.

204. Bruggeman, P. and Leys, C. (2009) Non-thermal plasmas in and in contact with liquids. *J. Phys. D: Appl. Phys.*, **42** (5), 053001.

2
Catalysts Used in Plasma-Assisted Catalytic Processes: Preparation, Activation, and Regeneration

Vasile I. Parvulescu

2.1
Introduction

The catalytic processes that found a beneficial contribution in using plasma are nowadays quite numerous. In concordance with this enlargement in applications, the types of catalysts that have been used in these processes are more and more diverse. They include simple and mixed metal oxides with semiconductor or isolator properties, zeolites, metal-supported catalysts with different reducibility properties, oxide-supported catalysts, and so on.

From the viewpoint of plasma physics, the presence of catalytic or noncatalytic pellets would significantly enhance the electric field, especially around the contact points between pellets and pellets/electrodes [1], and in accordance, the preparation of the catalysts for such applications should done considering these particularities. In addition, simulation studies reported that the way these catalysts are used exerts a high importance and, for example, a packed bed reactor could achieve a higher electric field compared with a nonpacked one [2–4]. The enhancement of electric field also depends on the contact angle and the dielectric constant of the packing pellets [1]. However, we have to accept that in the presence of a catalyst inside a plasma reactor, plasma can induce an enhancement of the temperature, which will finally generate a thermal catalysis as well. Since the energy efficiency is improved in this way, this is not a negative aspect.

In addition to these aspects, the presence of catalysts can increase the effective reaction time because of the adsorption of contaminants and intermediates on the catalyst surface, resulting in a shift of the reaction selectivity, a preferential consumption of active species by surface reactions (oxidation of adsorbed hydrocarbons) in comparison to undesired gas-phase reactions, and the formation of new catalytically active sites by the energy impact from the discharge, for example, as observed for silica and Y zeolites [5, 6].

As it is very well documented in this book, the first step of such kind of plasma-assisted processes consists in the generation of ozone and/or different types of radicals, free electrons, and so on. Thus, in plasma, O atoms are generated via electron-impact reactions, $O_2 \Downarrow e \longrightarrow O \Downarrow O \Downarrow e$. Hence, the higher the oxygen

Plasma Chemistry and Catalysis in Gases and Liquids, First Edition.
Edited by Vasile I. Parvulescu, Monica Magureanu, and Petr Lukes.

content in the feeding gas, the more O atoms will be formed. To be effective, the catalyst should be able to interact with these very reactive species generating superficial active structures. This step is a surface reaction. After that they will be able to either generate a fast total oxidation with mineralization or form new chemical bonds with the synthesis of new molecules. This step is also a surface reaction. The final step would be the complete desorption of the reaction products with liberation of the surface for a new interaction.

Plasma-assisted catalytic reactions are starting to become a mature domain. There are already an impressive number of publications in this field. Accordingly, the number of the catalysts that have been investigated is also very large and not always in a logical correlation with the type of application. On the basis of the analysis of the state of the art, this chapter first provides an inventory of the methods that can be used for the preparation of these catalysts. The second objective is to associate the structural and textural properties of the generated catalysts with the preparation methodologies. The third objective is to discuss the modalities in which the materials can be shaped in different forms with the aim of being utilized in plasma reactors. The stability of these materials and the methodologies of regeneration are also analyzed.

2.2
Specific Features Generated by Plasma-Assisted Catalytic Applications

Working under plasma conditions may exert a certain influence on the catalyst itself, and this depends on the type of plasma in which the catalyst is used [7]. Besides the chemical composition, the way a catalyst is activated for a certain process is crucial to provide an efficient catalyst. There are many reports showing that heterogeneous and homogeneous nonequilibrium plasma chemistry offer a range of potential advantages compared to conventional thermal activation and synthesis methods. Changing plasma characteristics can eventually result in the enhanced production of new active species, thus increasing the oxidizing power of the plasma discharge. Changes in the catalyst structure/texture can also result during the catalytic process. Therefore, to avoid any interference and to investigate the influence of plasma discharge on the catalytic performance, the catalyst should be pretreated in plasma and compared with an untreated catalyst. The literature has already reported changes in the catalysts structure, metal dispersion, metal reduction degree, surface area, and so on due to the plasma effect. On the other hand, changes of plasma properties resulting from the introduction of catalyst material have also been observed. It has been reported that discharge types can even change. Accordingly, it was reported that microdischarges are formed within the catalyst pores [7].

As mentioned, several changes of the catalysts can be generated by the plasma discharge [8]. First, discharges may enhance the dispersion of active catalytic components [9]. Nonthermal plasma experiments showed changes in the stability and catalytic activity of the exposed catalyst materials. Second, the oxidation state of the material can be influenced when exposed to plasma discharge [9–13]. To support this, a Mn_2O_3 catalyst was exposed to a nonthermal plasma (energy density of 756 $J\,l^{-1}$) for 40 h. After this experiment, Mn_3O_4 was detected. This lower-valent manganese oxide is known to have a larger oxidation capability [8]. Similarly, Wallis *et al.* [11] reported that due to plasma catalyst interactions, less parent Ti–O bonds are found on TiO_2 surfaces. Pribytkov *et al.* [12] and Jun *et al.* [13] both agreed that in hybrid plasma catalyst configurations, new types of active sites with unusual and valuable catalytic properties may be formed. Similar conclusions were made by Roland *et al.* [10], who observed the formation of a stable Al–O–O* paramagnetic species (lifetime>14 days). These were formed during the application of Al_2O_3 directly in the discharge zone at the interior of the pores by direct plasma processes (electrons, UV, plasma species such as OH, OD, ...). Finally, it was postulated that plasma exposure could even result in a specific surface area enhancement or in a change of catalytic structure [9].

A single-stage plasma catalysis reactor is a reactor in which the catalyst is directly introduced in the discharge region [1]. The enhancement of the changes induced in the catalysts as well as of the catalytic effect is directly connected to the electric field, which depends, among other properties, on the solid curvature [1–4].

For instance, the experimental results of Futamura *et al.* [14] showed that the energy efficiency for converting C_3H_8 via C_3H_8 steam reforming obtained with a packed bed reactor (PBR) packed with $BaTiO_3$ pellets is 6.5–8.2 times higher than that achieved with a dielectric barrier discharge (DBD) reactor, depending on the specific energy density.

2.3 Chemical Composition and Texture

Data reported in the literature demonstrate that a very large variety of materials have been used in connection with plasma for plasma-catalytic investigations. Among these, oxide supports (TiO_2, Al_2O_3, and SiO_2) and various zeolites [11, 16–22], supported oxides and intimate mixed oxides [9, 23–29], as well as supported metals [30–33] are the most frequent catalysts.

Figure 2.1 depicts the elements considered in the preparation of these catalysts. Almost 30% of the elements of the periodic table have been used in these preparations.

Preparation of these catalysts followed typical or adapted rules used for such a purpose. In accordance, this chapter discusses the basic principles of preparation of texture-controlled materials such as zeolites and the semiconductor and isolator oxides, techniques to synthesize metal-supported oxides with controlled metal particle size and metal oxidation state, and methodologies to support metal oxides. Special attention is also devoted to the synthesis of perovskites and plated electrodes.

Plasma sputtering processing is a very versatile and efficient alternate chemical technique of catalyst synthesis. This technique allows the deposition of very hard reducible metals even in zero oxidation state. It is also a technique that allows

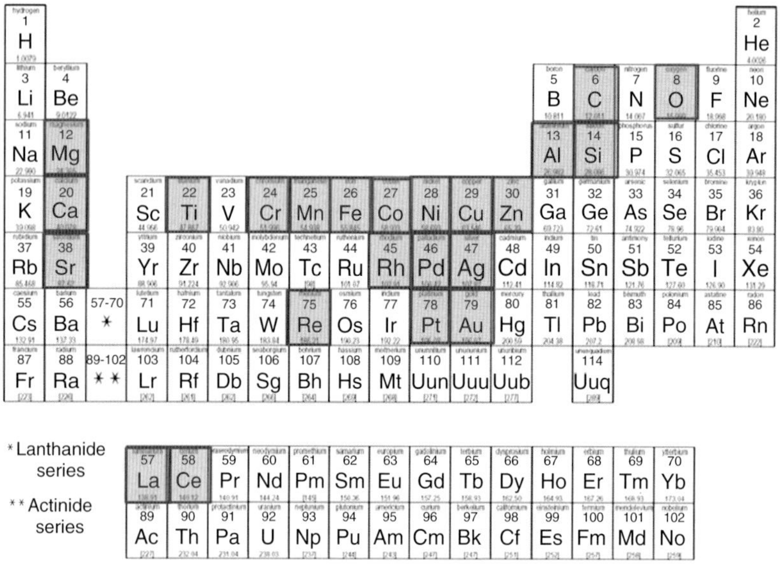

Figure 2.1 Elements used as building blocks of catalysts reported in connection with plasma.

the control of the layer deposition from ultrathin to micrometer range and the morphology of the different metals and oxides. In this context, this chapter also analyzes the parameters controlling the preparation of the catalysts using this methodology.

Besides the composition and phase structure, the catalysts are also characterized by textural properties. The main properties defining the texture are surface area, pore volume, pore size, pore size distribution, particle size, and particle shape. Obviously, these properties exert in a catalyzed reaction an influence that is in many cases comparable with that of the chemical composition. Electron paramagnetic resonance spectroscopic studies with nonporous and porous materials offer a good example in this sense [34]. Thus, γ-alumina samples have shown the formation of a paramagnetic site due to the application of a nonthermal plasma at ambient temperature. The fact that this effect is only observed for γ-Al_2O_3 and not for α-Al_2O_3 was easily interpreted by their dramatically different specific surface areas.

It is now clear that the preparation method controls not only the chemical composition and the phase structure but also the texture and morphology of the catalysts. Generally, methods such as *precipitation* lead to small-surface-area materials but with a high crystallinity, while *sol–gel* and *template-assisted syntheses* lead to large-surface-area amorphous materials. *Hydrothermal syntheses* may produce large-surface-area crystalline materials. The specific features of the various preparation methods are discussed in a direct correlation with the synthesis of materials already used in plasma-assisted catalytic applications.

2.4
Methodologies Used for the Preparation of Catalysts for Plasma-Assisted Catalytic Reactions

2.4.1
Oxides and Oxide Supports

Support materials such as γ-Al_2O_3, SiO_2, TiO_2, or ZrO_2 are commonly used as supplied with an eventual additional forming process, such as grinding to 500–850 mm in size [35], pastillation, granulation, and so on. There are also many examples in which the supports are prepared in laboratory using less-conventional techniques.

2.4.1.1 Al_2O_3

Alumina is a support largely used in industry with numerous practical applications, and currently, it is also being investigated under plasma conditions [36]. Industrial synthesis of aluminum oxide is done either by the *neutralization* of solutions of aluminum salts in the presence of a strong base or by the *hydrolysis of aluminum alkoxides* [37]. Heat treatments of the dried hydroxides have led to different kinds of transition aluminas such as γ, η, χ, θ, δ, κ, and α depending on the precursor, calcination temperature, composition (most of commercial alumina still contain alkali in their structure), and presence of additives (Scheme 2.1) [38]. γ, η, and α are the most important representatives of this class, the first two exhibiting spinel structures.

Flame hydrolysis is another route that can be used for the synthesis of alumina [39, 40]. The term "*flame hydrolysis*" or high-temperature hydrolysis describes a process in which a gaseous mixture of a metal chloride precursor, hydrogen (the term is also extended to carbon monoxide), and air is made to react in a continuously operated flame reactor [40]. The reaction occurs at temperatures higher than 1473 K, and the flow type of reactor allows a high productivity of nano/microscale metal oxides.

Spray pyrolysis is also a process by which pyrogenic oxides may be generated. Spray pyrolysis occurs via a droplet-to-particle process, whereby droplets containing

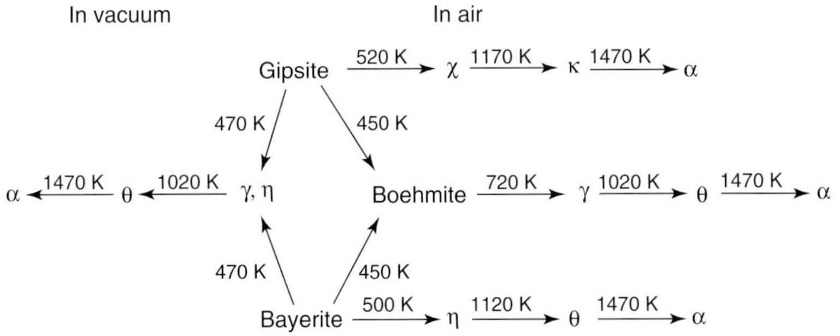

Scheme 2.1 Synthesis of transition alumina by thermal treatment. (Source: With permission from Ref. [38].)

precursors are mechanically formed by liquid atomization and then pyrolyzed in flames [40]. Pyrogenic alumina is characterized by a crystalline structure mostly consisting of γ- and δ-forms instead of the stable α-form. The primary particle size of pyrogenic alumina is in the range of 13 nm, corresponding to a specific surface area of about 100 $m^2\ g^{-1}$.

Other techniques allowing the preparation of nanoscale metal oxides via gas to particle conversion processes are the *decomposition of suitable precursors on hot walls*, *plasma reactors*, and *laser ablation* [40]. However, these are relatively high energy consumption methodologies and, in addition, result in low productivities. Accordingly, they have only minor technical relevance in manufacturing commercial quantities of nanoscale particles.

2.4.1.2 SiO_2

Silica can also be prepared following different routes. *Sol–gel processes* (silica gel), *precipitation* (precipitated silica, although this is sometimes also called *silica gel*), or *flame hydrolysis* (fumed silica) are currently used for the preparation of this support [41].

Precipitation starts from a solution of sodium silicate that is acidified with sulfuric acid [42]. The resulting silicic acids immediately undergo polycondensation and further growth to colloidal silica particles. In addition to pH, the process is controlled by several other factors such as degree of polymerization of silicon in the silicate precursor, silica concentration, type and concentration of electrolyte, temperature, and so on. As a function of these parameters, the product results as a silica hydrogel or silica precipitates. The silica hydrogel represents a coherent system composed of a three-dimensional network of agglomerated spheres with sodium sulfate solution as the dispersing liquid.

The porous structure of silica can be controlled more easily using the *sol–gel methodology*. Typically in this case, the precursor is a silicon alkoxide that undergoes a hydrolysis "catalyzed" either by an inorganic or organic acid or a base. The reaction takes place using less water than that required by the stoichiometric reaction. This allows in the second step [43] the polymerization of the resulting units via the condensation of the hydroxyl (or OR) groups (Scheme 2.2). Finally, all the OR groups are removed with the formation of the gel.

Aging of the gel is a very slow process and may take days. The product is a transparent gel that after a kind of drying and calcination at about 823 K generates a very high surface area microporous amorphous silica support. The washing step may influence the pore structure of the resulting material, with silica gels being more sensitive to the conditions of washing than precipitated silicas. Pore structure is also influenced by the hydrolysis route, the base one leading to larger pores than the acidic. This procedure also allows the control of hydro/liophilicity of the resulting solid. If instead of the OR groups the precursor contains covalent Si–R unities, these cannot hydrolyze, leading to a liophilic surface.

Crystalline structures can be achieved if one combines the sol–gel process with hydrothermal treatment in the presence of a structure-directing agent (SDR) [44, 45]. Examples of SDRs are tetraethylamine, tetrapropylamine, tetrabutylamine,

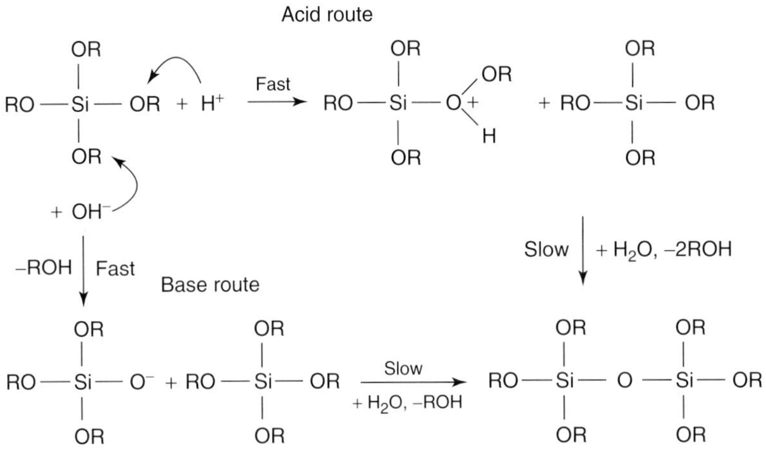

Scheme 2.2 Acid and base sol–gel routes in the synthesis of silica.

n-propylamine, 1,2-diaminoethane, and so on. The precursors of these syntheses are either high-purity reactive oxide powders or soluble silicates. The variables in these syntheses are temperature and alkalinity. The hydrogel, formed in the first sol–gel step, is usually kept for a certain period below the crystallization temperature (aging) followed by a hydrothermal treatment at elevated temperatures where the crystallization occurs. The crystalline product is commonly known as *silicalite*.

Flame hydrolysis is also being used to produce silica at industrial scale [46]. Silicon tetrachloride is the precursor for the production of pyrogenic silica. It is vaporized, mixed with dry air and hydrogen, and then fed into the flame reactor where during the combustion, hydrogen and oxygen form water, which quantitatively hydrolyzes the $SiCl_4$, resulting in nano/microparticles of SiO_2 [40]. The particle size can be adjusted by the flame parameters in order to generate tailored particles with specific properties. Any other vaporizable silicon-containing precursor, such as methyltrichlorosilane or trichlorosilane can be decomposed using the flame hydrolysis process.

Quartz is a crystalline form of silica and has also been used in plasma-assisted application [5]. *Hydrothermal* reactions have been reported to produce quartz from micron- to submicron-sized powders [47–49]. This reaction starts with the hydrothermal dissolution and reprecipitation of silica under conditions designed to encourage rapid nucleation of quartz. In a typical reaction, a dry high-surface-area amorphous silica precursor is added to NaOH (at pH $\sim$ 8) and heated at 473–573 K for several hours.

2.4.1.3 **TiO_2**

Titanium dioxide is another oxide frequently reported in plasma-catalytic applications [11, 50, 51]. Owing to similarities in properties with silicon, it can be prepared following the routes described above for silica. However, both alkoxides and titanium tetrachloride are more reactive than the corresponding silicon compounds, and this requires more precautions in the synthesis. In accordance, the *sol–gel processes* use as catalysts only organic acids and bases [52].

Pyrogenic titania is obtained from titanium chloride in flame of either hydrogen or CO [53]. Another difference as compared to silica is the fact that titania can exist in three different transition states: anatase, crystallizing in the tetragonal system (stable for temperatures smaller than 873 K); brookite, crystallizing in the orthorhombic system (with a maximum stability around 1073 K); and rutile, crystallizing in primitive tetragonal system (it starts to be formed at temperatures higher than 1188 K). In order to obtain one of these states, the design of the system should present a careful control of the flame temperature. This methodology also allows doping of titania with different elements contained in volatile chlorides.

The precipitation uses the sulfuric acid route and leads to the thermodynamically stable rutile modification. However, the purity of the product is limited since sulfur is still present.

2.4.1.4 ZrO_2

Although not investigated in plasma-catalytic applications, zirconium oxide is another potential support for such investigations. It is industrially produced using technologies very similar to those reported for silica and titania. The commercial product predominantly consists of the thermodynamically stable monoclinic ZrO_2 phase and, to a lesser degree, the metastable tetragonal phase. Starting compounds used for the formation of hydrous zirconia are $ZrOCl_2$, $ZrO(NO_3)_2$, and zirconium alkoxides [54–56]. *Precipitating* agents are aqueous ammonia and solutions of either KOH or NaOH. Hydrolysis of zirconium alkoxides in alcohol solutions using *sol–gel* is another alternative. ZrO_2 can also be obtained by *flame hydrolysis*. Temperatures higher than 1443 K favor the conversion from a monoclinic to tetragonal structure, and above 2643 K, to the cubic one.

2.4.2 Zeolites

Zeolites are a very important class of porous materials, and the principles of their design and synthesis were very well described in the literature [57–61]. The literature also contains many reports of using such materials in plasma-assisted catalytic processes [50, 62].

Zeolites are microporous, aluminosilicate materials with a very regular pore structure of molecular dimensions. This property is related to the term "*molecular sieve*" that confers a particular property to these materials, that is, the ability to selectively sort molecules based primarily on a size exclusion process.

The preparation of zeolites occurs via *hydrothermal synthesis* starting from a mixture of silicon and aluminum compounds, alkali metal cations (M), with or without SDRs, and water in a supersaturated solution. Syntheses may use different sources of silicon or aluminum. Thus, colloidal silica, sodium silicate, pyrogenic silica, or silicon alkoxides such as tetramethyl and tetraethyl orthosilicate can be used as silicon source compounds, while aluminate salts, aluminum alkoxides, pseudo-boehmite, and so on, can be used as aluminum source.

The general chemical composition of zeolites accounts to the formula

$$x\text{SiO}_2 \cdot y\text{Al}_2\text{O}_3 \cdot y\text{M}_2\text{O}_\text{n} \cdot z\text{SDR} \cdot u\text{H}_2\text{O}$$

The type of resulting zeolite is dependent on many parameters such as the nature of the precursors, temperature, and pH (directly correlated with the concentration of the OH^- anions). The OH^- anion species control the depolymerization and/or hydrolysis of the amorphous aluminosilicate particles, and thus the nucleation process. They also influence the Si : Al ratio of the crystalline product.

The role of the SDR is also very complex. It participates in all the synthesis steps: gelation, precursor formation, nucleation, and crystal growth [63–65]. The SDR organizes aluminosilicate oligomers into a particular geometry and, as a result, provides precursor species for nucleation and growth of the template zeolite structure. The steric properties of the SDR determine both the chemical composition (i.e., the Si/Al ratio) and the topology. Table 2.1 presents the structure of various zeolites in direct relation with the preparation conditions and the nature of the SDR [66–69]. Neutral molecules as well as cations or ion pairs are able to fulfill these structure- and composition-directing functions of SDR.

Metal-containing molecular sieves represent a class of materials in which part of the constituent species is partially substituted by transition metals. Aluminum phosphate (APO) and silicon-aluminum phosphate (SAPO) materials are well known not only for intrinsic properties but also for the capacity to incorporate such species. Of this large class, Mn-APO-5 and Mn-SAPO-11 have been investigated in plasma-assisted catalytic oxidation [70]. They have been synthesized using specific protocols by *hydrothermal method*. In an identical way, other transition metals can be inserted as well.

Mn-SAPO-11 was obtained following the procedures for SAPO-11 using a gel composition in molar oxide ratios of 0.08 MnO : 1.0 Pr_2NH : 0.1 SiO_2 : Al_2O_3 : P_2O_5 : 42 H_2O. A reaction mixture corresponding to this composition was prepared by combining water and aluminum isopropoxide, to which was added a concentrated solution of orthophosphoric acid under vigorous stirring and manganese nitrate. To this was added fumed silica and then, di-*n*-propylamine. The final mixture was stirred until homogeneous and then sealed in a pressure vessel and heated at 423 K at autogenous pressure for five days [71].

2.4.2.1 **Metal-Containing Molecular Sieves**

Mn-APO-5 was prepared via a similar methodology [71]. Using a gel composition of 1.05 TEA : 0.08 MnO : Al_2O_3 : P_2O_5 : 40 H_2O the typical synthesis protocol started from phosphoric acid that was mixed with distilled water at room temperature followed by the addition of the required amount of $Mn(NO_3)_2$ and alumina under vigorous stirring. Then, triethylamine was added dropwise, with the temperature being controlled. The crystallization was carried out in an autoclave at 473 K for two days.

Table 2.1 Structure of various zeolites in direct relation with the preparation conditions and the nature of the SDR.

Zeolite	Structure	Cavity window (Å)	Member ring	SDR	Reaction conditions
Zeolite A	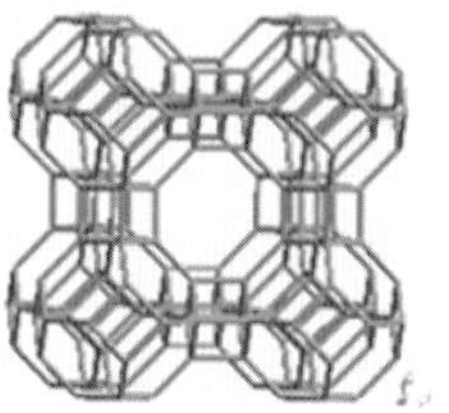	4.1	8	–	$SiO_2/Al_2O_3 = 1$ $Na_2O/SiO_2 = 2.16$ $H_2O/Na_2O = 4.0$ Aging at 298 K for 24 h 3 h at 363 K
Faujasite Zeolite X Zeolite Y	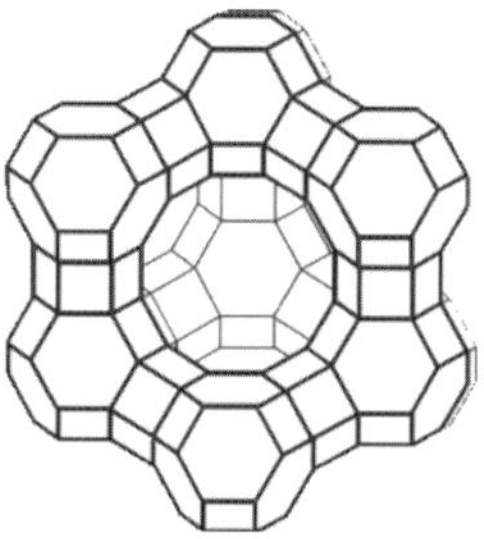	7.4	12	–	X: $SiO_2/Al_2O_3 = 3.2$ $Na_2O/SiO_2 = 1.33$, $H_2O/Na_2O = 25–65$. 19 h at room temperature, 6 h 373 K Y: $SiO_2/Al_2O_3 = 10$ $Na_2O/SiO_2 = 0.7$ $H_2O/Na_2O = 40$ 6 h at 368 K
Zeolite L	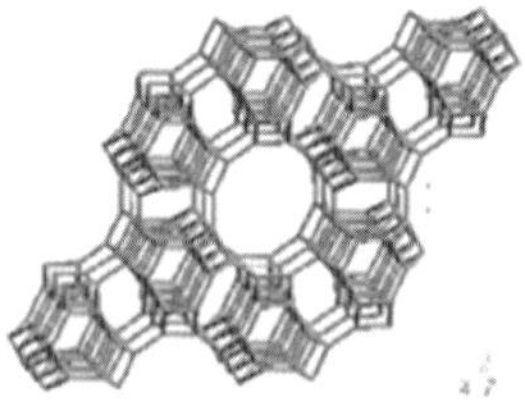	7.7	12	–	$SiO_2/Al_2O_3 = 15–30$, $(Na_2O + K_2O)/SiO_2 = 0.40$, $H_2O/(Na_2O + K_2O) = 25–40$, a two-step crystallization at 293 and 373 K with 72 h for each step
Mordenite	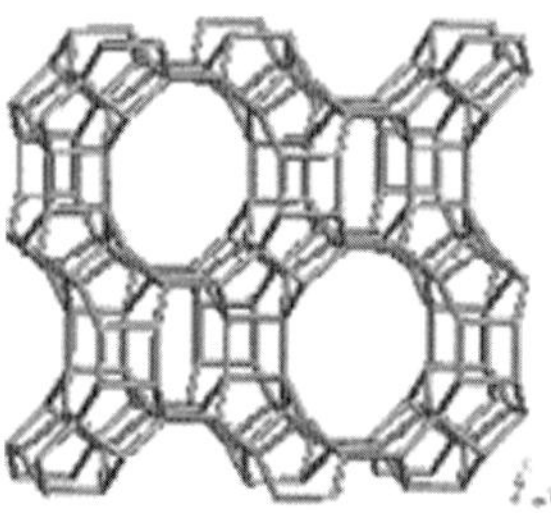	7 × 6.5	12	–	$SiO_2/Al_2O_3 = 12$ $Na_2O/SiO_2 = 0.147$ $H_2O/Na_2O = 74$ 24 h at 473 K

Table 2.1 *(continued)*.

Zeolite	Structure	Cavity window (Å)	Member ring	SDR	Reaction conditions
MFI ZSM-5	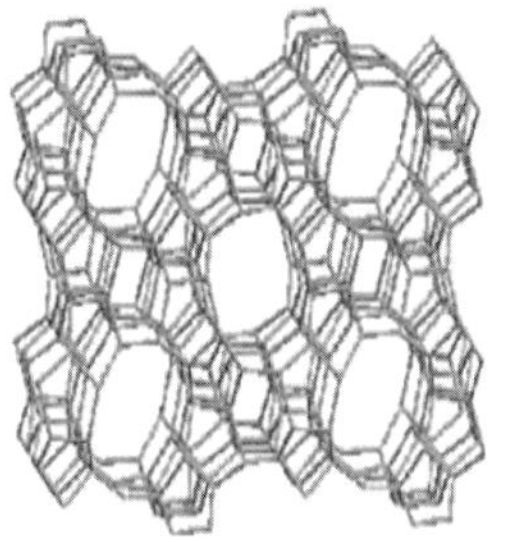	5.5 × 5.1	10	Et_3PrN^+ Pr_4N^+ Pr_4N^+ +*iso*-$PrNH_2$ 1,6 hexane diamine	$SiO_2/Al_2O_3 > 15$ $Na_2O/SiO_2 < 0.8$ $H_2O/Na_2O = 160$ 60 h at 413–443 K
Beta	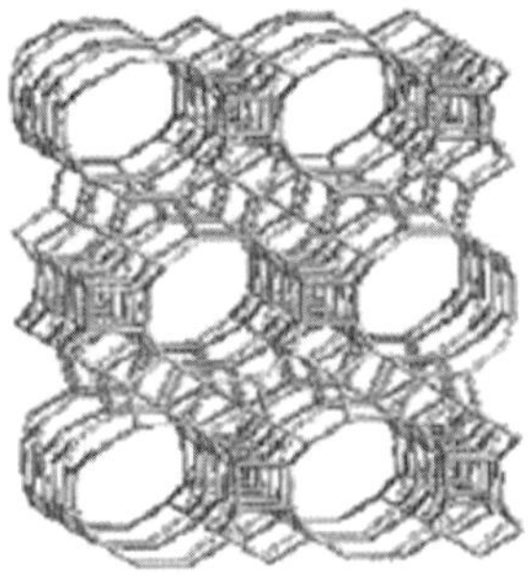	6.6 × 6.7	12	Et_4NOH Et_4NOH+ 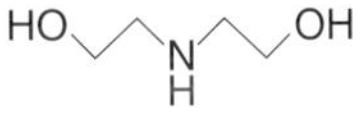	$SiO_2/Al_2O_3 < 19$ $Na_2O/SiO_2 < 0.25$ $H_2O/Na_2O = 160$ 60 h at 368–443 K

2.4.3
Active Oxides

Except for supports, there are numerous examples where individual oxides are used as heterogeneous catalysts in plasma-assisted applications. MnO_2 was recently reported for the destruction of environmental pollutants, using a multistage packed bed discharge reactor, operated with an industrial scale flow rate of 300 l min^{-1} [72]. It is worth mentioning that NO_x (which are common by-products in plasma processing) have not been detected in these studies.

Commercial synthesis involves electrodeposition of manganese dioxide onto a titanium anode from a hot aqueous solution of H_2SO_4 and $MnSO_4$.

MnO_2 can also be prepared using chemical tools [73]. The most common starts from natural manganese dioxide and converts it using dinitrogen tetroxide and water to manganese (II) nitrate solution. Then, the evaporation of the water leaves the crystalline nitrate salt. At temperatures of 673 K, the salt decomposes, releasing N_2O_4 (that is reused) and leaving a residue of purified manganese dioxide.

MnO_2 nanomaterials of different crystallographic types and crystal morphologies have been selectively synthesized from the same manganese (II) nitrate via a hydrothermal route [74]. Thus β-MnO_2 nano/microstructures, including one-dimensional nanowires, nanorods, and nanoneedles, as well as 2D hexagram-like and dendritelike hierarchical forms, were obtained by simple hydrothermal decomposition of $Mn(NO_3)_2$ solution under controlled reaction conditions.

The addition of phosphate to manganese oxides also has a beneficial effect in plasma-catalytic reactions [75]. The catalysts prepared following two different procedures confirmed such a behavior. In the first route, the catalyst was prepared via the reaction of MnO_2 with concentrated solutions of H_3PO_4. Then the suspension with dimethylaminoalcohol was hydrothermally treated for 48 h at 443 K in an autoclave. In the second approach, thc *hydrothermal* treatment was carried out under microwave conditions using, in addition, hexadecyltrimethylammonium bromide and tetramethylammonium hydroxide as SDRs. The final catalysts were obtained after the materials were calcined at 823 K.

2.4.4 Mixed Oxides

2.4.4.1 Intimate Mixed Oxides

Intimate mixed oxides can be synthesized using various methodologies: solid state reaction of mixed oxides at high temperatures, coprecipitation, sol–gel, spray pyrolysis, and reactive grinding of the single oxides.

Solid state reaction of mixed oxides at high temperatures is currently used for plasma catalysts application.

$Li_2Si_2O_5$ is one sinterable additive of the $Ca_{0.7}Sr_{0.3}TiO_3$ for dielectric barrier discharge plasma [76]. It was prepared by a conventional solid state reaction using powders of Li_2CO_3 and amorphous SiO_2. They were mixed by ball milling and calcined at 1283 K for 10 h, and then remilled with zirconia balls.

Glassy catalysts have also been investigated in plasma-assisted reactions. SiO_2–B_2O_3–SiB_4 is one of these [77]. Preparation of such thermally stable Si-borosilicate glasses requires the reaction between silicon and aluminoborosilicate glass by *melting* at temperatures over 1623 K [78].

2.4.4.2 Perovskites

Perovskites are mixed oxides with structure, ABO_3 (A – usually a lanthanide in dodecahedral coordination and B – transition metal for catalytic purposes, in octahedral coordination), that may accommodate almost all elements of the periodic table [79, 80]. In addition, the perovskite lattice can accommodate multiple cationic substitutions with only small changes, and as a consequence, several properties of the solid, such as sintering and catalytic performance, can be modified. This compositional flexibility induces interesting and useful properties, but only a few base formulations are known for their activity in total oxidation reactions, namely, those based on Fe, Co, Ni, and Mn at B-site with La (partially substituted by other lanthanide or alkaline earth metals) at A-site [81].

They can be prepared using the same routes as for the mixed oxides. To make more clear the differences among these preparation routes, the synthesis of perovskites is discussed considering the same chemical composition, that is, for $LaCoO_3$ perovskites.

Solid state reaction of mixed oxides occurs at high temperatures [82]. The perovskite is prepared by a first short step of hand grinding of La_2O_3 and Co_3O_4 in order to obtain a homogeneous mixture of the single oxides. The powder obtained is calcined for 10 h at 1273 K (with a reasonable ramp rate = 20 K min^{-1}) for perovskite crystallization.

The coprecipitation method involves the simultaneous precipitation of the nitrate precursors $La(NO_3)_3{\cdot}6H_2O$ and $Co(NO_3)_2{\cdot}6H_2O$ [82]. They are first dissolved in distilled water. The precipitation is carried out in the presence of a concentrated base solution (NaOH, KOH, etc.) that is quickly added to the precursors under vigorous stirring until pH = 10.5. The precipitate obtained is filtered and washed with distilled water until the pH of the precipitate is around 7. Then the compound is dried at 353 K for one night and calcined at 973 K under air (12 h, at a very small ramp rate = 2 K min^{-1}).

These methods suffer from the drawback of a very small surface area of the resulting materials, typically less than 1 m^2 g^{-1}. Usually, the catalytic applications require more large surface areas. An increase of this parameter of more than 20 times can be achieved by using the following methodologies.

The *sol–gel route* starts from $La(NO_3)_3.6H_2O$ and $Co(NO_3)_2.6H_2O$ as precursors and citric acid monohydrate as the complexating molecule [83]. Other polyacids can also be used in this approach [84]. Typical procedures use aqueous solutions with a cation ratio La : Co of 1 : 1 in an excess of complexating molecule. The slurry is stirred for about 5 h, and water is slowly evaporated at 313 K in a rotary evaporator. The drying process is completed by heating the powder under vacuum at 333 K under a pressure of 200 kPa for 16 h. The powder obtained is calcined at 973 K for 6 h in static air. The *sol–gel route* has also been used for the synthesis of $La_{0.8}$ $K_{0.2}MnO_3$ starting from stoichiometric quantities of citrates and nitrates of Mn and La [85].

The synthesis of the same perovskites can also be carried out via an adapted sol–gel route [84]. This starts from an ethanolic solution of cobalt nitrate that is added dropwise to an ethanolic solution of lanthanum nitrate until a La : Co ratio of 1 : 1 is reached. Ethylene glycol is added to this mixture as complexing agent. Hydrolysis of these complexes is carried out using water, in a large Ln^{3+} : H_2O excess. The gelification was done at room temperature under a small vacuum. Drying is carried out using the same conditions, but by this route, the ABO_3 structure can be achieved at a lower calcination temperature, that is, 333 K.

Spray pyrolysis [86, 87] involves the uniform nebulization of nitrate solutions containing $La(NO_3)_3{\cdot}6H_2O$ and $Co(NO_3)_2{\cdot}6H_2O$ prepared as a 0.1 M liquid solution of precursors. Two online furnaces at 250 and 873 K evaporate the solvent (distilled water) with the dissolved nitrates and produce an initially amorphous perovskite powder. The collected material is subsequently annealed at 873 K for 4 h, thus producing a crystalline perovskite with rhombohedral symmetry. Using this technique,

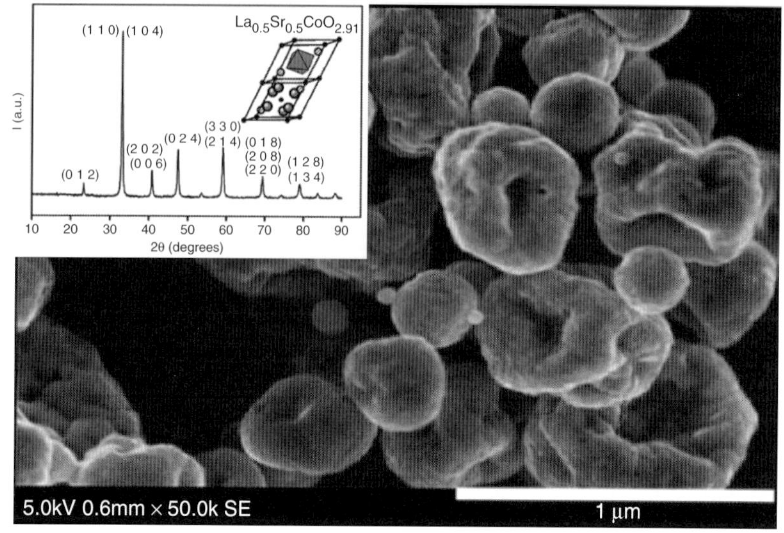

Figure 2.2 $La_{0.5}Sr_{0.5}CoO_3$ perovskite: SEM micrograph and XRD. (Source: Printed with permission from Ref. [87].)

lanthanum-substituted perovskites can be prepared as well. No separate individual oxides have been identified in the analysis of these materials (Figure 2.2) [88].

Reactive grinding of the single oxides is a method that requires several steps [89]. The oxide precursors are ground for 4 h under O_2, in a laboratory grinder (at a very high rotation speed = 1040 rpm). At the end of this step, the single oxide conversion into perovskite is complete. In the second step of grinding, an additive (ZnO) is added to the perovskite. After the second grinding step, the compound obtained (perovskite + ZnO) is washed repeatedly with dilute NH_4NO_3 to free the sample from any trace of additive. This step results in an increase of the specific surface area developed by the solid. Then the precipitate is dried at 353 K for one night and calcined at 773 K for 4 h.

Coprecipitation coupled with reactive grinding is done in three steps [82]. The first step consists in the coprecipitation of the nitrates of lanthanum and cobalt, following the procedure described for coprecipitation. Then, treatment of the precipitate obtained follows the steps indicated for the reactive grinding methodology.

The changes in the nature of the rare-earth or transitional metal does not cause any change in the discussed preparation protocols [90].

In addition to the rare-earth-based perovskites, the nonporous ferroelectric perovskite $BaTiO_3$ with a high dielectric constant ($\varepsilon_r \approx 3000$) received a great interest for plasma-assisted catalytic reactions [10]. $BaTiO_3$ exhibits very specific chemical and physical properties, such as oxygen transport, ferroelectric, pyroelectric, piezoelectric, and dielectric behavior [91]. These properties mainly depend on the crystal size, crystal structure, shape, stoichiometry, homogeneity, and the surface and interface properties.

Preparation of $BaTiO_3$ follows methodologies similar to those reported for rare-earth-based perovskites. Initially, the synthesis of these materials used classical methods such as *conventional solid state reaction* and *coprecipitation* [92]. *Sol–gel* using metal alkoxide precursors was then largely reported [93, 94]. Instead of an alkoxide, barium organic salts such as barium acetate can also be used [95]. The synthesis of this material can also be carried out via an alkoxide–hydroxide *sol-precipitation method* [96]. It involves the hydrolysis and condensation of barium hydroxide octahydrate and titanium (IV) isopropoxide in an alcoholic solution. Monodisperse nanoparticles are also prepared via sol–gel protocols using both the surfactant [97] and β-diketones chelating synthesis routes [98].

Hydrothermal synthesis uses $TiCl_4$ diluted in hydrochloric acid that is mixed with $BaCl_2{\cdot}2H_2O$ dissolved in deionized water. The white homogeneous colloidal barium titanium slurry obtained after precipitation with NaOH aqueous solution is then autoclavized at predetermined temperature for a certain time [99]. Microwave-hydrothermal treatment of the slurry can improve the synthesis of $BaTiO_3$ powders [100]. The effects of water density on polymorph of $BaTiO_3$ particles synthesized hydrothermally under sub- and supercritical conditions have also been studied [101]. Experiments were performed within the temperature range of 573–693 K and the pressure of 20–40 MPa using a flow reaction system.

Ball-milling-assisted hydrothermal synthesis is another method used to prepare $BaTiO_3$ powder from nano- and submicron-sized TiO_2 powders [102]. These were dispersed in distilled water with $Ba(OH)_2{\cdot}8H_2O$, and after adjusting the pH to 14, the suspension was reacted hydrothermally at 358–473 K for 8 h with a milling media of 5 mm ZrO_2 balls.

Ignition method involves a combustion process of BaO_2-TiO_2-black carbon (as fuel) exothermic mixture, *attrition milling* of reaction products, and subsequent heat treatment, with the temperature ranging from 973 to 1273 K. The product is a tetragonal submicron barium titanate particle perovskite [103].

2.4.5 Supported Oxides

Metal oxides are known to have ozone decomposing properties. Oxides of transition metals such as Mn, Co, Ni, and Ag can be used as ozone degrading catalyst. These compounds are often supported on materials such as γ-Al_2O_3, TiO_2, SiO_2, zeolites, activated carbon, or combinations of these [28, 104]. In addition, Heisig *et al.* [104] found that combinations of metal oxides with various organic and inorganic additives are effective in enhancing the physical stability and activity of the catalysts.

The preparation of the supported oxides can be carried out using various procedures: *impregnation, coprecipitation, coprecipitation-impregnation, mechanical mixing, sol–gel,* and so on.

Impregnation involves bringing the solution into the pore space of the support [105]. Accordingly, the precursor is solved in a small volume of solution and is expected to be retained on the support after drying. From the methodological point of view, there are two alternatives, one in which the volume of the impregnating solution coincides with the pore volume, known as *capillary impregnation*, and a second in which the volume of the impregnating solution is much in excess compared with the pore volume, that is, the *diffusional impregnation*.

The literature reports both impregnation methodologies for the preparation of catalysts for plasma applications. MnO_x/Al_2O_3 with different Mn contents were prepared using manganese acetate as the precursor and capillary impregnation as the methodology [106]. For the preparation of the catalyst, Al_2O_3 pellets were impregnated in the $Mn(CH_3COO)_2$ aqueous solution of desired concentration, followed by evaporation to dryness in a rotary evaporator at 343 K. The resulting samples were then dried and calcined in air at 773 K to obtain the desired catalyst.

MnO_2-coated alumina or other oxides such as TiO_2 and SiO_2 cylinders [107, 108] were prepared by *wet impregnation* method using the nitrate salts. An alternative method is impregnation with aqueous $HMnO_4$, drying, and calcination [109]. Various characterization techniques suggest that Lewis acid sites, a high surface concentration of MnO_2, and redox properties are important in achieving high catalytic performance at low temperatures.

High loadings of nickel-oxide-based catalysts (8% w) were prepared by the *wet impregnation* method using a concentrated $Ni(NO_3)_2{\cdot}6H_2O$ solution and γ-Al_2O_3 as the support [110, 111]. To ensure sufficient contact of the inner pores of alumina with the solution, ultrasonication is a very efficient tool. After filtration and drying, the activation of the catalysts is carried out at 823 K, that is, at the stability limit of the phase γ of the support. Low loadings of Ni/γ-Al_2O_3 catalysts have also been investigated in plasma-assisted catalytic reaction [112]. They were prepared following a very similar impregnation methodology.

Silver was also found to be an active species in various plasma-catalyzed oxidation reactions [35]. Many of these used γ-Al_2O_3- or TiO_2-supported silver oxide. Typically, such catalysts are prepared by loading the silver on the γ-Al_2O_3 or TiO_2 chips by an *impregnation* method using $AgNO_3$ as precursor [35]. After the evaporation of water, the catalyst was heated at a quite high temperature (873 K) to generate the active catalyst.

Incipient wetness impregnation technique has also been used for the preparation of In/Al_2O_3 plasma-activated catalysts from aqueous solutions of indium(III) nitrate pentahydrate [113].

CuO-ZnO/Al_2O_3 is one of the most widely used catalysts in the industry and has found plasma-catalytic applications too [114, 115]. It is also an interesting case study for the preparation of the catalysts. The classic route comprises *the coprecipitation* of metal nitrates [$(Cu(NO_3)_2$, $Zn(NO_3)_2$, $Al(NO_3)_3$] with a solution of Na_2CO_3 as a precipitant. During the precipitation process, pH (±0.1 unit), temperature, and aging time should be strictly controlled [116] in order to generate a strained copper lattice rather than bulk copper [117]. Strained copper particles in close interaction with the ZnO account for the catalytic behavior. The hydroxy carbonate route leads to the formation of several mixed metal hydroxy carbonates, including aurichalcite [$(Cu, Zn)_5(CO3)_2(OH)_6$], zincian-malachite [$(Cu, Zn)_2(OH)_2CO_3$], and a Cu–Zn hydrotalcite-like phase [$(Cu, Zn)_6$-$Al_2(OH)_{16}CO_3 \cdot 4H_2O$] [118]. All these phases decompose to give well-dispersed oxidic phases. The presence of residual carbonate in the calcined sample plays an important role, supposedly due to subsequent formation of a copper suboxide species, increasing the chemical activity of catalysts. Studies on the effect of the nature of precipitant (Na_2CO_3, $NaHCO_3/Na_2CO_3$ buffer, and $NaHCO_3$) showed that it has little importance [119].

Other methods used for the preparation of this supported mixed oxide are *separate package, complex package, dry mixing, wet mixing, coprecipitation-sedimentation*, and *coprecipitation-impregnation* [120]. They include mixing of the copper, zinc and aluminum salts, and a reducing agent (citric acid, urea, or aminoacetic acid); ball milling the precursors at a certain milling rate; and calcining. Composite catalysts were also prepared by *mixing* the freshly precipitated catalyst precursors, whereas hybrid catalysts are prepared by *mechanical mixing* of the two calcined catalyst components by grinding. However, these catalysts usually exhibit much lower activity, as compared to those prepared using the conventional method [121].

Sol–gel synthesis of mixed oxide followed by the impregnation of alumina support or by slurry coating has been reported as well [122–124]. Various gelation initiators including propylene oxide were used for this purpose [125].

Other plasma-assisted catalytic processes using mixed oxide catalysts prepared using the same methodologies are $CuO/ZnO/MgO/Al_2O_3$ [24] and $CuOMnO_2/TiO_2$ [28].

The preparation of the catalysts is highly dependent on the chemical composition. Thus, replacing zinc oxide with manganese oxide in $CuO\text{-}MnO_x/Al_2O_3$ limits the preparation routes to impregnation of alumina with the corresponding nitrate salts [126]. This catalyst was reported to be active under both plasma [27] and separate ozonization conditions [127]. The role of manganese in this catalyst is to improve the dispersivity of CuO and to inhibit the formation of the $CuAl_2O_4$ spinel phase. Other catalysts such as $Fe_2O_3\text{-}MnO_x/Al_2O_3$ can also be prepared via the same procedure.

2.4.5.1 Metal Oxides on Metal Foams and Metal Textiles

Nickel foam is a type of metallic foam reported to be effective in plasma-assisted catalytic applications [128]. Metal foam describes a wide variety of porous metallic materials, or a mixture of polyurethane foams with a metallic coating. They can be prepared using various methods such as (i) coating a resin foam via *impregnation* with Ni salt (oxalate, acetate, basic carbonate) followed by a reductive calcination [129], (ii) pasting fine graphite powders into pores of thin polyurethane foam film to supply electron-conducting seeds for nickel *deposition by electroless plating* reaction and removing the remaining polyurethane foam by organic solvent treatment and graphite particles by ultrasonic cleaning [130], or (iii) making a mixture of a solution of a metal salt, soluble polymer, and a suitable solvent that is converted into a gel body. Then the gel body is converted to inorganic foam by heating [131].

Supported CuO, Co_3O_4, Mn_2O_3, and Fe_2O_3 catalysts for plasma-assisted catalytic reactions can be prepared by *impregnation* of the above foams with the corresponding metal salts followed by drying and calcining at temperatures of 873 K [132].

Another chemical alternative is the *dipping* of nickel foam into the solution of the metal salt. This alternative can also be used for the deposition of noble metals using acidified solutions of metal halides [128].

Sintered metal fiber filters made of stainless steel (Cr 16–18; Ni 10–13; Mo 2–2.5; C $<$ 0.01; Fe balance) have been used as support for metal oxides for plasma applications as well [26, 133]. This material consists of thin uniform metal fibers with diameter of ~2 mm. For the deposition of MnO_x and CoO_x, the stainless steel filter was first oxidized at 873 K, followed by *wetness impregnation* with Co and Mn nitrate aqueous solutions of desired concentrations.

2.4.6
Metal Catalysts

2.4.6.1 Embedded Nanoparticles

Silver colloids are one example of embedded nanoparticles for plasma-assisted catalytic application [134]. They were prepared by reducing $AgNO_3$ in an aqueous solution. In a typical preparation, 0.294 g sodium citrate was added to 50 cm^3 of 10 mM $AgNO_3$ aqueous solution in an ice bath. $NaBH_4$ (0.019 g) was added to the solution at once with strong stirring, and a black powder formed in solution. The solution was then filtered and dried at room temperature under vacuum.

The embedding of these colloids was carried out using a *sol–gel methodology* according to which aluminum butoxide ($Al(OBu)_3$) was dissolved in isobutanol, which was subsequently added to an isobutanol solution of Pluronic 84($(EO)_{19}$ $(PO)_{39}(EO)_{19}$), which acted as the surfactant. The resulting mixture was then heated to 243 K for 6 h. Water was added, and the reaction was maintained at 353 K for 10 h followed by 373 K for 20 h. The Ag colloid dissolved in water was added to this mixture under vigorous stirring. The samples were preserved at room temperature for 48 h, dried under vacuum at 383 K, and then calcined at 773 K with a slope of 0.5 $K\,min^{-1}$. This preparation yielded samples with loadings of silver between 1% w and 5% w. Changing the molar ratio alkoxide : alcohol : water led to modification of both phase distribution and texture of these catalysts (Figure 2.3) [134].

Gold nanoparticles embedded in SBA-15 catalysts exhibit a high stability in plasma-assisted reactions because of the methodology used for their preparation and their stabilization in the silica matrix [33]. Gold nanoparticles confined in the walls of mesoporous silica were synthesized by dissolving Pluronic P123 ($EO_{20}PO_{70}EO_{20}$) in HCl solution; subsequently, a mixture of tetraethyl orthosilicate and 1,4 bis(triethoxysilyl)propane tetrasulfide was added, followed by the addition of predesignated amounts of aqueous $HAuCl_4$ solution (Figure 2.4) [135]. The gel was aged at 373 K and then calcined at 773 K.

2.4.6.2 Catalysts Prepared via Electroplating

Electroplating is a technique primarily used for depositing a layer of material to a surface [137]. This occurs via a plating process in which metal ions in a solution are moved by an electric field to coat an electrode. The process uses electrical current to reduce cations of the desired material from a solution.

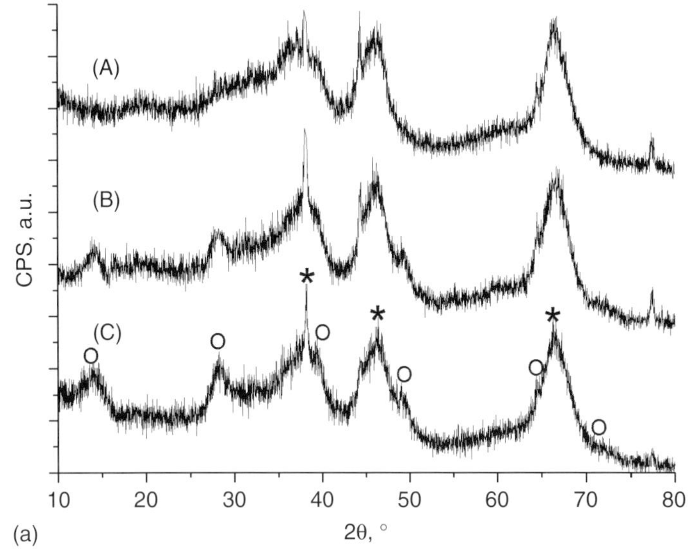

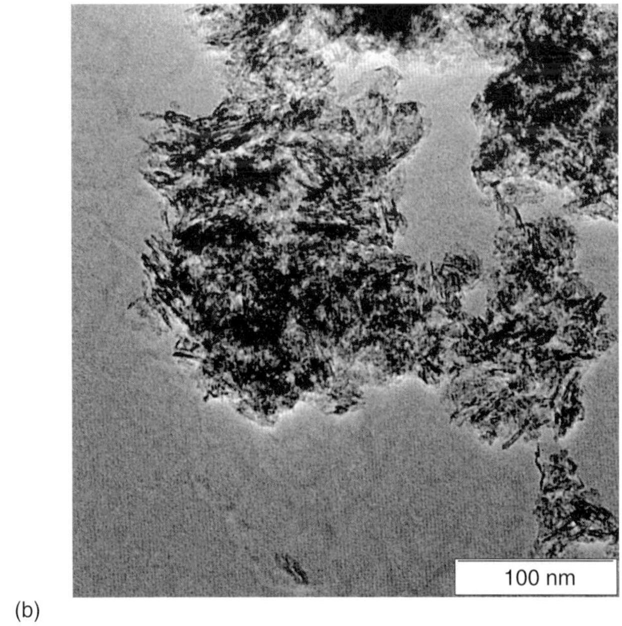

Figure 2.3 (a) XRD patterns of the alumina embedded silver colloids: (A) 3% w AlAg (molar ratio alkoxide : alcohol : water = 1 : 10: 25), (B) 3% w AlAg (molar ratio alkoxide : alcohol : water = 1 : 5 : 10), (C) 5% w AlAg (molar ratio alkoxide : alcohol : water = 1 : 5 : 10) (o-γ-Al_2O_3 phase, * – Ag^0 phase) and (b) TEM picture of 3% w AlAg (molar ratio alkoxide : alcohol : water = 1 : 5 : 10). (Source: With permission from Ref. [134].)

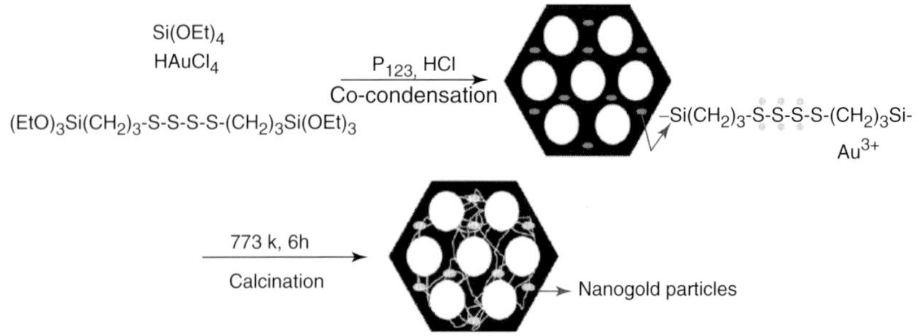

Figure 2.4 Graphical representation of the preparation processes for mesoporous silica catalysts with gold nanoparticles in the walls. (Source: With Permission from Ref. [136].)

This technique is provides a very simple and efficient instrument to modify the electrodes usually used in generating plasma, and thus to introduce the catalyst directly in the plasma region. There are several examples in the literature presenting such catalysts. Thus, various catalytically active metals, such as iron and nickel, were electroplated onto a copper rod at a thickness of 100 μm [138]. In the experimental setup, the exposed end of the supporting metal rod was used as the electrical contact for the inner metal electrode to the high-voltage alternating current power source.

2.4.6.3 **Catalysts Prepared via Chemical Vapor Infiltration**

Chemical vapor infiltration (CVI) is a variant of *chemical vapor deposition. Chemical vapor deposition* implies deposition of the active species onto a surface, *whereas CVI* implies deposition within a body. CVI originated in efforts to densify porous graphite bodies by infiltration with carbon. The technique has developed commercially so that half of the carbon composites currently produced is made by this method. The earliest report of CVI for ceramics was a 1964 patent for infiltrating fibrous alumina with chromium carbides [139, 140].

According to this technique, the reactant gases diffuse into an isothermal porous preform, made of long continuous fibers, and form a deposition. When the infiltrated species is a metal, the precursor is either a metal alkoxide or a metal chloride that suffers a pyrolytic decomposition [141]. The setup requires an adequate furnace. Following this procedure, Rh and Rh/Pt were inserted in a carbon electrode, resulting in a catalyst for environmental plasma applications [142]. The feasibility of the CVI of rhenium was also demonstrated. Statistical analyses showed that the amount of rhenium deposited increased for higher chlorine flow rates and lower preform temperatures, 1023 versus 1173 K, for the chlorination and sublimation processes, respectively [143].

2.4.6.4 **Metal Wires**

Molybdenum and iron wires were used as catalysts for plasma synthesis of ammonia and hydrazine in the radio frequency nitrogen–hydrogen plasma, and it was reported that the ammonia yield increases with increasing the electron work function of the metal used as catalyst [144, 145].

Typically, materials wire-containing refractory metals having melting points higher than 2473 K, such as molybdenum, are prepared by electron beam melting and pouring the molten refractory material into a molder or extruder [146]. In some cases graphite is used as lubricant. This requires an additional wet hydrogen annealing in order to remove it [147].

For more oxidizable metals having melting points lower than 2473 K, such as iron, it is necessary that the entire process be carried out in a reductive atmosphere [148]. The initial rate of reduction is greater the higher the temperature (623–1173 K). Reduction proceeds in two stages, first to FeO and then to Fe. The micro- and ultrapore structures of the resulting wires also depend on the reduction temperature and heating rate.

Nanowires of these two categories of metals can be obtained using specific routes. Metallic molybdenum (Mo(0)) wires with diameters ranging from 15 nm to 1.0 μm and lengths of up to 500 μm were prepared in a two-step procedure [149]. Molybdenum oxide wires were electrodeposited selectively at step edges and then reduced in hydrogen gas at 773 K to yield Mo(0). The hemicylindrical wires prepared by this technique were self-uniform. Another method to prepare molybdenum nanowires consists in their encapsulation inside double-walled carbon nanotubes [150]. Using this procedure, wires having inner diameters ranging from 0.6 to 0.8 nm can be obtained. These individual structures form spontaneously within the hollow core of tubes in the absence of any reducing agent.

2.4.6.5 **Supported Metals**

Reduced metals have been reported in plasma-assisted catalytic applications too. They can be prepared following different methodologies. The most facile is the *reduction* of supported metal oxides prepared following the above-mentioned methods using a reducing agent such as hydrogen. For this purpose, the samples are introduced in a flow of H_2 in an inert gas at temperatures superior to those thermodynamically favoring the reduction of the corresponding oxides to metals.

An example is Ni/Al_2O_3 [151, 152]. However, for plasma applications where ozone and atomic oxygen species always exist, the use of reducible species is always questionable.

Another aspect that should be considered when reduction is applied is the stability of the support. Both the reduction temperature and the reductant may influence the support stability. If this treatment does not intend to generate changes of the support as well, it should be carried out with all the precautions.

Stable supported nanoparticle catalysts can be prepared following the approach proposed for Ni/Mg amesite [153, 154]. Amesites are phylloaluminosilicates with septechlorite structure. Like hydrotalcites, they exhibit lamellar structures (Scheme 2.3) that can be synthesized by careful coprecipitation. Following the same methodology, nickel and magnesium can be inserted in the inter layer of these structures, resulting in Ni–Mg amesite with the formula $(Ni_{1.8}Mg_2Al_{2.2})[Al_{2.2}Si_{1.8}O_{10}](OH)_8$. After treatment at 873–1073 K in

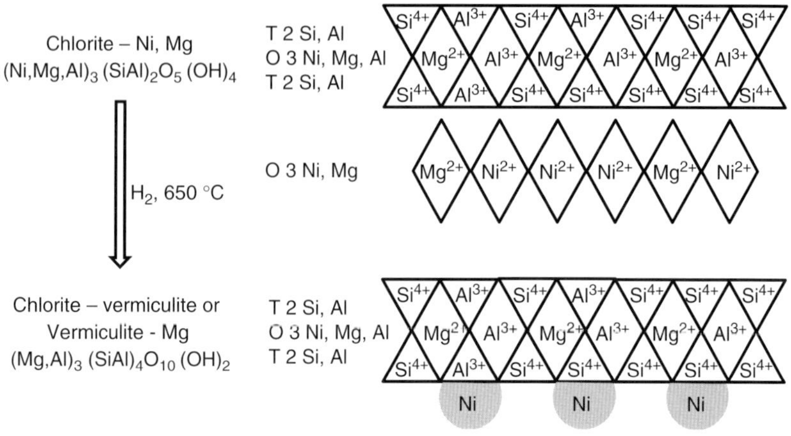

Scheme 2.3 Synthesis of metallic nanoparticles via amesite route. (Source: With permission from Ref. [151].)

hydrogen, the samples contain dispersed metallic nickel particles supported on Mg-chlorite-vermiculite. It is important to note that none of the catalysts obtained contained SiO_2. The authors reported that the samples were stable in inert gas and hydrogen atmospheres at 1123 K, as well as in hydrogen plus steam at 20 bar and 923 K. Thus, one can consider that Ni-containing amesitelike compounds are, indeed, suitable catalysts for the methane steam reforming process.

The size of the nanoparticles resulting following this methodology ranges in between 5 and 10 nm. Other nanoparticles such as cobalt can also be obtained using this procedure. They have been used in plasma-assisted catalytic methane oxidation [154].

2.4.6.6 **Supported Noble Metals**

The most common procedures to prepare supported noble metals are *impregnation* of the support with the corresponding salts when catalysts with a large dispersion of metal particles are obtained, and the *deposition–precipitation* method when, on the contrary, a narrow particle size-distribution is achieved.

Obviously, noble metals such as Pd, Rh, or Pt are prepared using the impregnation techniques. As mentioned above, the supported particles prepared in this way ranged between 4 and 30 nm [50].

Catalysis on gold is strongly dependent on the size of metal particles [155]. Therefore the use of the *deposition–precipitation* method is highly preferred. This is also the case of catalysts used in plasma-assisted applications, where the Au catalyst was supported on the TiO_2 pellets by deposition–precipitation method [50]. The method consists of fast precipitation of $HAuCl_4$ in the presence of NaOH and the support. The addition of NaOH raises the pH of $HAuCl_4$ and allows hydrolysis of $HAuCl_4$ to take place to form nanogold hydroxide that is attached to the support [136]. Using this procedure, diameters of Au particles on the TiO_2 pellets in the range of 3–7 nm can be obtained.

2.5 Catalysts Forming

The catalysts used in plasma reactors are usually not in a powder form. They are typically shaped in different forms with diameters between 2 and 5 mm [106]. Not only the effectiveness of these catalysts but also the stability and mechanical resistance are enhanced if the constituent powder is shaped. Both the form and the size of the shaped materials are important since they are catalyzing a chemical process. To achieve an optimum balance between the minimization of the pore diffusion effects in the catalyst particles, which typically requires small particles, and the pressure drop across the catalyst bed, which requires large particles, it is necessary to establish the correct shape.

To achieve this goal, there are several techniques that are extensively used both at the laboratory and industrial level: tableting, granulation, extrudation, and pelletization.

2.5.1 Tableting

Figure 2.5a shows typical images of tablets. Tableting (pilling) produces very regular shaped bodies (tablets or pills) in the form of cylinders [156]. This form is achieved by cold isostatic pressing of the material powders. The first condition of this process is the use of powders with homogeneous size and shape, preferably spheres. Obviously, the materials shaped in this form have high mechanical strength.

The deformation behavior of a solid has a major influence in the tableting process, as the response of the material to applied stress is the main factor controlling the properties of the resulting tablets [157]. The elastic component of the response to the applied pressure is also important in this process since it is stored in the pressed material. This is directly related to the stability because when the pressure is released it relaxes producing a collapse of the tablet.

Not all materials can be tableted by simple pressing. Raising the pressure is acceptable till a certain value. A further increase of the pressure can generate changes in both the structure and texture of the materials. Zeolites, and even more mesoporous materials, are very sensitive to this parameter, and an increase of the tableting pressure causes a collapse of the pores.

To avoid very high pressures, this technique frequently uses lubricants as adjuvants. Their selection depends on both the nature of the materials and on the applications of the shaped catalysts. The chemical composition of the lubricants includes a variety of compounds from mineral structures such as graphite or talc to organic compounds such as polyvinyl alcohol, polyethylene, waxes or greases,

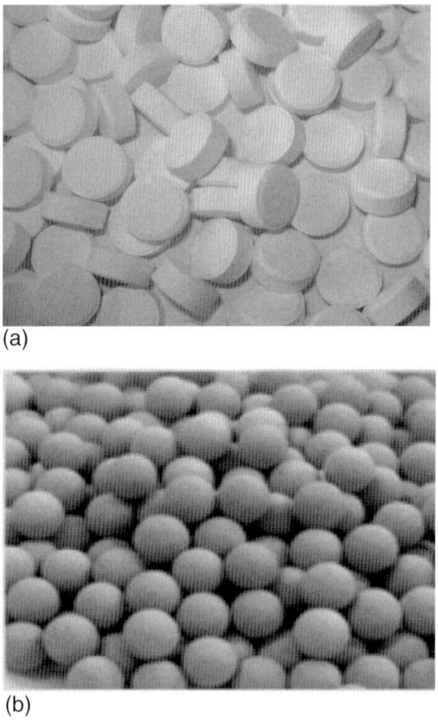

Figure 2.5 (a) Tablet and (b) ball-formed catalysts.

or stearin. They are added in a low content (0.5–1%) under sonic homogenization. Organic compounds can be removed from the tablet by burning under air atmosphere or even under plasma, while inorganic ones remain inserted in the catalyst tablet. For thermally labile materials, burning can induce important changes of the parent catalyst because the temperature developed during this step can generate phase transformations. This is especially the case of γ-Al_2O_3, which at high temperatures is transformed to the form α. The inorganic lubricants do not introduce the problems related to combustion but they act as diluent for the catalytic active phase and also induce restrictions in the liquid and gas transport through the catalyst.

The tableting method using thermally removable lubricants is very feasible for perovskites, especially when they are prepared using *solid state* routes that involve high reaction temperatures. However, in other cases, the lubricant is necessary to be selected from the category of inorganic oxides. In one example, appropriate quantities of $CaTiO_3$, $SrTiO_3$, and $Li_2Si_2O_5$ were wet mixed with ZrO_2 balls and ethanol for 16 h. The mixtures were dried at 85 °C and then ground by an agate mortar. The specimens were uniaxially pressed at 20 MPa, subjected to cold isostatic pressing under a pressure of 200 MPa, and sintered at 1473 K for 2 h in air [76, 110, 111].

2.5.2 Spherudizing

The oil drop method is the most described method for producing very regular beads of spherical resistant particles (Figure 2.5b). The classic example to produce this shape refers to alumina and consists of dropping an aqueous acidic alumina material gel that becomes spherical while falling through a water-immiscible liquid and coagulates under basic pH conditions [158, 159]. This process involves in the first step the preparation of a suspension of aluminum hydroxide (as boehmite) whose viscosity of the slurry is carefully controlled to allow shaping by drop formation through nozzles. The droplets of slurry are formed in air at the nozzle tips and fall through the air into a column containing an upper phase of a water-immiscible liquid, most frequently an oil [160]. The surface tension of the hydrocarbon forces the droplet to become spherical. The spheres then pass into a lower ammoniacal phase that provokes acidic alumina charge annulation to zero point charge and thus solidification. The process may also use pore-forming agents in order to increase the mesoporosity/macroporosity of the spherical beads. They are organic compounds such as oils, fats, glycerides, and so on. The alumina particles obtained with this methodology have diameter of 1–5 mm.

There are also methodologies other than batchwise to produce spherical granules. Fluidized bed can also be used to prepare such granules [161]. In this case, the feed suspension prepared from water, fresh bauxite, recycled bauxite dust, and auxiliary agents such as polyvinyl alcohol and ammonium citrate is atomized in a fluid bed at 823 K. This method ensures a fast estimation and control of the granule diameter.

Spherical granules can also be produced by dispersing pyrogenically produced aluminum oxide in water, followed by a *spray dry* step [162]. *Spray dry* technique uses slurries containing small particles of the catalyst to be formed, in a mixture with matrix components that bind the catalyst particles and impart the attrition resistance. The mixture is sprayed through a nozzle into a chamber into which hot gas is flowing cocurrently, leading to a fast evaporation and the formation of spherical granules.

In all these cases, the granules should be calcined to provide the stability required by plasma-assisted catalytic applications.

2.5.3 Pelletization

Pelletization is a process comprising dry mixing of the catalyst powder with a binder in particulate form and then moistening the mixture with a granulating liquid, that is, an aqueous solution of a water-soluble inorganic salt [163]. Small particles are agitated and stick to each other and thus grow to larger aggregates. The process is effected in a drum inclined to the horizontal, by introducing the powder at one end and removing the pelletized at the other end, whereas it is forcibly transported inside the drum during a desired residence time with execution of a rocking movement with substantial avoidance of shear forces [164].

Figure 2.6 Pellet-formed catalysts.

The organic binder can be selected from starch and cellulose derivatives such as polyacrylates, polymaleates, or polyvinyl pyrrolidone. The nature and properties of the binder are strongly correlated with the manner the liquid is delivered to the growing pellets. Thus it can be delivered by pouring, spraying, or melting, if a solid binder is used with a relatively low melting point.

Figure 2.6 depicts the typical shapes obtained using this forming procedure. Granulation results in less dense and strong particles compared to tableting or extrusion. Also, the shape and size of the particles is more irregular.

Pellets are also currently used in plasma-assisted catalytic reactions. In order to be used in these applications, the pellets should be dried and calcined. As an example, pellets of CuO-ZnO-Al_2O_3 catalysts 5.4 mm long and 5.2 mm in diameter were reported to be effective when they were packed inside the reactor [114].

2.5.4 Extrusion

The extrusion forming method is one of the most used processes for shaping catalysts, mostly in the case of zeolites. It involves the preparation of a paste of the particles including several additives that it is pressed through a die sieve. The form of the die can lead to the formation of a variety of different shapes, from the simplest case of cylinders to the more complicated shapes such as rings, stars,

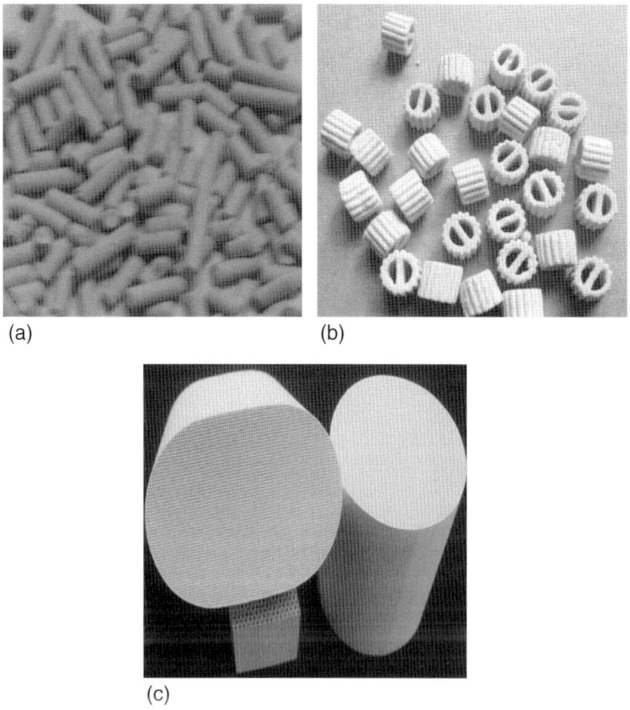

Figure 2.7 Example of extrudates: (a) cylinders, (b) miniliths [165], and (c) honeycombs.

trilobates, starrings, and so on. Monolithic honeycombs can also be produced by using a suitable die. After collecting from the die, the extrudates are typically cut off, which leads to extrudate catalysts that are all approximately identical in size.

Figure 2.7 give few examples of commercial extrudates.

Preparation of the paste that should behave as a viscoelastic fluid is a decisive step in shaping the materials as extrudates. In addition to the size of the powder and the water content, the peptizing agents and the binders are the most important parameters. The role of the peptizing agents is to stabilize the sol and to hydrolyze oxobridges between particles, which are aggregated by bond formation, and thus further assist in deagglomeration [157]. Nitric acid, formic acid, or acetic acid solutions are often used in this scope because in addition to the properties mentioned above they have the advantage of being decomposed tracelessly in the later calcination step. The role of the binders is to enhance the mechanical strength, and for such a purpose, inorganic oxide sols or clays are usually selected. Plasticizers (as additives to improve the rheological behavior of the paste), lubricants (to help the extrusion process), and porogens (organic compounds added with the purpose of creating porosity during the calcination step) are also currently used to prepare extrudates.

There are also many examples of using extrudates for plasma applications. Thus, Aerolyst was selected, as in plasma catalyst, as an extrudate with cylindrical shape and average diameter and length of 2.7–3 mm and 4 mm, respectively [28]. $CuO-MnO_2/TiO_2$ was also selected for postplasma catalysis as an extruded catalyst material cylindrically shaped with an average diameter and length of 1.5 and 4 mm, respectively [28].

2.5.5 Foams

Forming foams is a methodology applied mostly for ceramics, and it is also of interest for plasma applications. Ceramic foam articles of controlled permeability and structural uniformity are typically prepared by a process composed of an open-celled organic polymer foam material possessing a predetermined permeability and resilience, that is, impregnated with an aqueous slurry of a thixotropic ceramic composition (Figure 2.8a) [166]. The resulting uniformly impregnated foam structure is then dried to volatilize moisture and heated to remove the organic foam component. Metal foams are prepared using the same concept, but the deposition of the metal is carried out using specific tools such as chemical vapor deposition, physical vapor deposition, or thermal decomposition [167].

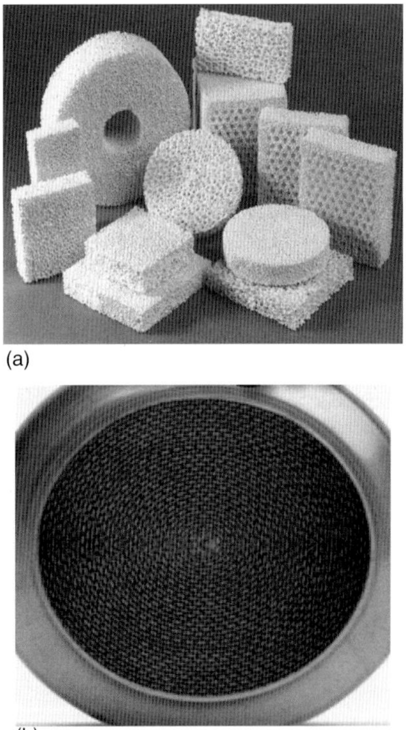

(a)

(b)

Figure 2.8 Example of ceramic foams (a) and metal textile and (b) catalysts.

2.5.6
Metal Textile Catalysts

Metal textiles are obtained using typical metallurgic procedures (Figure 2.8b) [168]. Among the various materials and textures, they can also find applications in plasma-assisted catalytic reactions as supports for the active phases.

2.6
Regeneration of the Catalysts Used in Plasma Assisted Reactions

The stability of the catalysts under working conditions is one of the most important properties governing their application in practice. There are two factors that can induce deactivation under plasma working conditions. The first one refers to structural (phase transformation) and textural (pore collapse) modifications induced either by punctual overheating or by solid state reactions. This kind of transformation is irreversible and is due to an incorrect activation/pretreatment of the catalysts used in such applications.

The second one refers to the formation of coke or soot. This generates a real concern in the catalytic processes of hydrocarbons, and even more in many environmental applications. Thus, soot is typically deposited on the catalysts used for elimination of noxes released by cars and trucks. However, this represents a reversible process, and there are tools that can be used to remove both coke and soot. On the basis of this very important concern, the preparation of the catalysts with a poorer population of centers that can induce the deactivation of the catalysts as a result of the coke formation and the regeneration of the catalysts are extremely important topics and can be solved by using preparation tools also.

Regeneration of the catalysts by plasma has also been reported. The regeneration of a $CuO\text{-}ZnO\text{-}Al_2O_3/\gamma\text{-}Al_2O_3$ catalyst is one example [169]. It was thus reported that the regeneration with oxygen may allow a complete combustion of coke. Under conditions of partial coke combustion, the removal of the fraction deposited on the metal oxide makes it possible to recover the initial activity of the catalyst, which subsequently undergoes fast deactivation again. Kinetic evaluation of such a process can differentiate between the different states (i.e., free of coke and covered by coke) [170]. It is worth to take into consideration that in this type of regeneration an overheating that can induce an irreversible deactivation is again possible.

A very important feature of working with plasma was recently reported [6]. The mineralization of organic compounds in plasma in the absence of the catalysts is always generating condensed polyaromatic compounds (named either coke or soot). When the catalyst is introduced directly to plasma, a selection of a proper catalyst can reduce the radicalic reaction route, namely, that responsible for the formation of these compounds. This has a double advantage, that is, the elimination of coke and an increased stability of the catalyst, and relates this process to the selection of the catalyst.

2.7 Plasma Produced Catalysts and Supports

Fullerene molecules (Figure 2.9a) consist of carbon atoms and come in many forms; the most abundant forms are carbon 60 (which has a soccer ball shape) (Figure 2.9a), carbon 70 (which has more of a rugby ball shape) (Figure 2.9c), and carbon 84 (spherical). Unlike diamond and graphite, that is, the more familiar forms of carbon, the carbon atoms are located at the corner of the polyhedral structure consisting of pentagons and hexagons [171, 172]. Following the discovery of fullerenes, another new carbon structure known as *carbon nanotubes* was reported (Figure 2.9d) [173, 174]. Nanotubes are seamless cylinders of hexagonal carbon networks (tubes) that are formed either as a single-walled molecule or as multiwalled molecules.

A tremendous growth in the field of carbon nanomaterials has led to the emergence of carbon nanotubes, fullerenes, mesoporous carbon, and more recently, graphene or a combination of these [175]. Some of these materials have found applications and have shown considerable potential as catalysts [176–178]. The high temperatures and hydrocarbon precursors involved in their synthesis usually yield highly inert graphitic surfaces, making them very attractive for oxidation reactions even under plasma conditions [141]. There are also examples in which the coated electrode also had a graphitic overcoat to increase the surface area [141].

Plasma is a very efficient tool to prepare such materials. Coevaporation of different elements with carbon leads to the formation of functionalized fullerenes such as heterofullerenes and endohedrals [179–181].

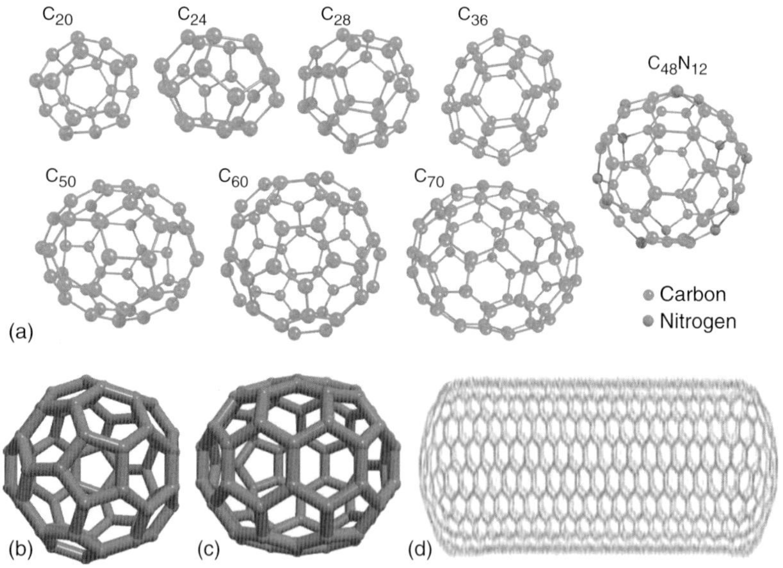

Figure 2.9 Fullerenes (a), C60 (b), C70 (c), and carbon nanotubes (d).

Many elements such as Fe, Co, Ni, rare-earth metal atoms, or B were reported to significantly influence the process of carbon arc plasma formation of fullerenes and carbon nanotubes [182]. The concentration of C60 is usually lowered in the presence of these catalysts, while the nanotube content increases distinctly. Not only single-walled nanotubes (SWCNTs) but also multi-walled nanotubes (MWCNTs) were produced in this way. Generally, Fe and Co were reported to exhibit higher activity than Cu [183]. However, this is still controversial, other reports indicating Cu as more active than Co and Fe [184].

The experimental setup is based on the arcing procedure [181] the reactor configuration including both alternating current (AC) or direct current (DC) feeding modes. The homogeneous graphite electrodes containing one of the above elements are arced in He atmosphere under pressures up to 60 kPa.

A similar approach can also be used for the preparation of other valuable catalysts and supports. Doping titania with nitrogen, carbon, or sulfur was reported to enhance the visible-light sensitivity of TiO_2, thereby increasing the performance of both photovoltaic and photocatalytic devices. Usually, this is generated using chemical methodologies and urea as reactive species [185]. Plasmas may offer an alternative to these techniques. Thus using inductively coupled radio frequency plasmas it was recently demonstrated that nitrogen-doped TiO_2 films could be obtained from a wide range of nitrogen precursors [186]. These treatments resulted in anatase-phased materials with as high as 34% nitrogen content. The nitrogen-doped TiO_2 films produced via plasma treatments displayed colors ranging from gray to brown to blue to black, paralleling the N/Ti ratios of the films.

N-doped TiO_2 was also prepared by using liquid-phase nonthermal plasma technology [187]. Nonthermal plasma produced in water solutions forms the basis of an innovative advanced oxidation technology of water treatment [188, 189] and is frequently used to remove chemical and biological wastes under liquid state even in a large flux of these. In the particular case of the decomposition of the organic compound containing nitrogen, this process is made in the benefit of the N-doping TiO_2. The analysis of the resulting material showed that the crystal structure remained unchanged as anatase after plasma treating at 13.5 W for 40 min.

Microporous Ni-doped TiO_2 film photocatalysts can also be prepared by plasma electrolytic oxidation [190]. The effects of Ni doping on the structure, composition, and optical absorption properties of the film catalysts indicated that Ni changes the phase composition and the lattice parameters (interplanar crystal spacing and cell volume) of the films. The results showed that the film catalyst is composed of anatase and rutile TiO_2 with microporous structure. After the addition of nickel, the optical absorption range of TiO_2 film gradually expands and shifts to the red with increasing the metal dosages.

Another very creative method offered by plasma to prepare heterogeneous catalysts is plasma etching. Ag-coated catalysts using the electrospun nanofiber template is one example [191]. This method offers a facile strategy to fabricate three-dimensional hierarchically porous Ag films, with clean surfaces. The films are built of Ag porous nanotubes and are homogeneous in macrosize but rough and porous in nanoscale. Each nanotube block is micro/nanostructured with evenly distributed nanopores on the tube walls. According to the authors, each film architecture (i.e., the shape, arrangement, distribution density of porous nanotubes, and the number and size of nanopores) can be easily controlled by the nanofiber template configuration, Ag coating, and plasma etching conditions. Such hierarchically porous films could also be very useful as sensors and nanodevices.

2.7.1 Sputtering

Plasma sputtering is an alternate way to chemical techniques. High frequency (100 MHz) plasma is initiated in argon gas at low pressure (10^{-1}–10 Pa) with a small input power. The metal atom source is a bare metal which is negatively biased (300–350 V) with respect to the plasma potential (V = 100 V) so that the Ar^+ ions are attracted and gain sufficient energy to induce sputtering [192].

The literature presents many examples of catalysts prepared using this technique. Pd, Pt–Rh, and Pt, Rh, thin films were deposited on various amorphous substrates: amorphous carbon membrane, SiO_2 native oxide of Si(100) wafers, SiC, and Si_3N_4 thick layers on Si(100). Rh was also deposited on cordierite monolith for testing in methane partial oxidation reaction. It was also possible to coat inside monolith channels because metal atoms are sputtered with relatively large energy, that is, ~2 eV. This allows the sputtered atoms to diffuse along the channel inner surface [193].

Thin films of transition metals such as Pd, Pt, Rh, and corresponding alloys are known to grow via clusters, mainly because the cohesive energy is often larger than the interaction energy with the support. After completing the maximum possible cluster density, clusters start to coalesce and form meandering structures, which reach percolation [193].

Working under aqueous atmosphere, oxide catalysts and catalyst supports can be prepared using the same approach [194, 195]. Thin-film oxides and supported metal-clusters-based catalysts were also obtained using this technique, including nickel oxide on titania [196], tantalum oxide on titania [197], vanadium oxide on titania [198], or gold on titania [199].

2.8 Conclusions

The literature reports an impressive number of heterogeneous catalysts for plasma-assisted applications, including actually all the categories of materials that have been utilized until now in typical heterogeneous catalytic applications. Accordingly, the preparation of these catalysts followed multiple methodologies.

Therefore a review of the preparation methodologies allows a discussion of all the basic methods used for the preparation of materials. In this chapter, we tried to address only the basic principles of these methods with exemplification on the catalysts that have been really tested in such reactions.

Plasma applications also require shaped catalysts, and a large variety of formed materials has also been reported until now.

In spite of this very important effort, there is no insight demonstrating the advantage of any specific preparation methodology. Although the important features of plasma are well known to the physicists, the cooperation with chemistry to give answer to this problem is still missing or at the beginning and requires a special attention in the future. In many cases, the positive results appear to be the effect of chance and not the result of a rational preparation design. Also, it is missing a correlation between the shape of the catalysts and their efficiency in a plasma-assisted catalytic reaction.

In conclusion, the new preparations of plasma-active catalysts should use the known methodologies to provide model materials that incorporate particularities already observed in plasma applications.

References

1. Chen, H.L., Lee, H.M., Chen, S.H., Chao, Y., and Chang, M.B. (2008) Review of plasma catalysis on hydrocarbon reforming for hydrogen production – Interaction, integration, and prospects. *Appl. Catal., B Environ.*, **85**, 1.
2. Takuma, T. (1991) Field behavior at a triple junction in composite dielectric arrangements. *IEEE Trans. Electr. Insul.*, **26**, 500.
3. Chang, J.S., Kostov, K.G., Urashima, K., Yamamoto, T., Okayasu, Y., Kato, T., Iwaizumi, T., and Yoshimura, K. (2000) Removal of NF_3 from semiconductor process flue gases by tandem packed bed plasma and adsorbent hybrid systems. *IEEE Trans. Ind. Appl.*, **36**, 1251.
4. Kang, S.K., Park, J.M., Kim, Y., and Hong, S.H. (2003) Numerical study on influecnes of barrier arrangement on dielectric barrier discharge characteristics. *IEEE Trans. Plasma Sci.*, **31**, 504.
5. Holzer, F., Roland, U., and Kopinke, F.D. (2002) Combination of non-thermal plasma and heterogeneous catalysis for oxidation of volatile organic compounds. Part 1. Accessibility of the intra-particle volume. *Appl. Catal. Environ.*, **38**, 163.
6. Magureanu, M., Piroi, D., Mandache, N.B., Pârvulescu, V.I., Pârvulescu, V., Cojocaru, B., Cadigan, C., Richards, R., Daly, H., and Hardacre, C. (2011) In situ study of ozone and hybrid plasma Ag–Al catalysts for the oxidation of toluene: evidence of the nature of the active sites. *Appl. Catal., B Environ.*, **104**, 84.
7. Badyal, J.P.S. (1996) Catalysis and plasma chemistry at solid surfaces. *Top. Catal.*, **3**, 255.
8. Van Durme, J., Dewulf, J., Leys, C., and Van Langenhove, H. (2008) Combining non-thermal plasma with heterogeneous catalysis in waste gas treatment: a review. *Appl. Catal., B Environ.*, **78**, 324.
9. Guo, Y.F., Ye, D.Q., Chen, K.F., He, J.C., and Chen, W.L. (2006) Toluene decomposition using a wire-plate dielectric barrier discharge reactor with manganese oxide catalyst in situ. *J. Mol. Catal. A: Chem.*, **245**, 93.
10. Roland, U., Holzer, F., and Kopinke, F.D. (2005) Combination of non-thermal plasma and heterogeneous catalysis for oxidation of volatile organic compounds. Part 2. Ozone decomposition and deactivation of g-Al_2O_3. *Appl. Catal., B Environ.*, **58**, 217.
11. Wallis, A.E., Whitehead, J.C., and Zhang, K. (2007) Plasma-assisted catalysis for the destruction of CFC-12 in

atmospheric pressure gas streams using TiO_2. *Catal. Lett.*, **113**, 29.
12. Pribytkov, A.S., Baeva, G.N., Telegina, N.S., Tarasov, A.L., Stakheev, A.Y., Tel'nov, A.V., and Golubeva, V.N. (2006) Effect of electron irradiation on the catalytic properties of supported Pd catalysts. *Kinet. Catal.*, **47**, 765.
13. Jun, J., Kim, J.C., Shin, J.H., Lee, K.W., and Baek, Y.S. (2004) Effect of electron beam irradiation on CO_2 reforming of methane over Ni/Al_2O_3 catalysts. *Radiat. Phys. Chem.*, **71**, 1095.
14. Futamura, S., Kabashima, H., and Einaga, H. (2004) Steam reforming of aliphatic hydrocarbons with nonthermal plasma. *IEEE Trans. Ind. Appl.*, **40**, 1476.
15. Chen, H.L., Lee, H.M., Chen, S.H., and Chang, M.B. (2008) Review of packed-bed plasma reactor for ozone generation and air pollution control. *Ind. Eng. Chem. Res.*, **47**, 2122.
16. Malik, M.A., Minamitani, Y., and Schoenbach, K.H. (2005) Comparison of catalytic activity of aluminum oxide and silica gel for decomposition of volatile organic compounds (VOCs) in a plasma catalytic reactor. *IEEE Trans. Plasma Sci.*, **33**, 50.
17. Oh, S.M., Kim, H.H., Ogata, A., Einaga, H., Futamura, S., and Park, D.W. (2005) Effect of zeolite in surface discharge plasma on the decomposition of toluene. *Catal. Lett.*, **99**, 101.
18. Pekarek, S., Pospisil, M., and Krysa, J. (2006) Non-thermal plasma and TiO_2-assisted n-heptane decomposition. *Plasma Process. Polym.*, **3**, 308.
19. Lu, B., Zhang, X., Yu, X., Feng, T., and Yao, S. (2006) Catalytic oxidation of benzene using DBD corona discharges. *J. Hazard. Mater.*, **137**, 633.
20. Intriago, L., Diaz, E., Ordonez, S., and Vega, A. (2006) Combustion of trichloroethylene and dichloromethane over protonic zeolites: Influence of adsorption properties on the catalytic performance. *Microporous Mesoporous Mater.*, **91**, 161.
21. Morent, R., Dewulf, J., Steenhaut, N., Leys, C., and Van Langenhove, H. (2006) Hybrid plasma-catalyst system for the removal of trichloroethylene in air. *J. Adv.Oxid. Technol.*, **9**, 53.
22. Han, S.B. and Oda, T. (2007) Decomposition mechanism of trichloroethylene based on by-product distribution in the hybrid barrier discharge plasma process. *Plasma Sources Sci. Technol.*, **16**, 413.
23. Chae, J.O., Demidiouk, V., Yeulash, M., Choi, I.C., and Jung, T.G. (2004) Experimental study for indoor air control by plasma-catalyst hybrid system. *IEEE Trans.Plasma Sci.*, **32**, 493.
24. Chang, M.B. and Lee, H.M. (2004) Abatement of perfluorocarbons with combined plasma catalysis in atmospheric-pressure environment. *Catal. Today*, **89**, 109.
25. Delagrange, S., Pinard, L., and Tatibouet, J.M. (2006) Combination of a non-thermal plasma and a catalyst for toluene removal from air: manganese based oxide catalysts. *Appl. Catal., B Environ.*, **68**, 92.
26. Subrahmanyam, C., Magureanu, M., Renken, A., and Minsker-Kiwi, L. (2006) Catalytic abatement of volatile organic compounds assisted by non-thermal plasma. Part 1. A novel dielectric barrier discharge reactor containing catalytic electrode. *Appl. Catal., B Environ.*, **65**, 150.
27. Grossmannova, H., Neirynck, D., and Leys, C. (2006) Atmospheric discharge combined with Cu-Mn/Al2O3 catalyst unit for the removal of toluene. *Czech.J. Phys.*, **56**, B1156.
28. Van Durme, J., Dewulf, J., Sysmans, W., Leys, C., and Van Langenhove, H. (2007) Efficient toluene abatement in indoor air by a plasma catalytic hybrid system. *Appl. Catal., B Environ.*, **74**, 161.
29. Subrahmanyam, C., Renken, A., and Kiwi-Minsker, L. (2007) Novel catalytic dielectric barrier discharge reactor for gas-phase abatement of isopropanol. *Plasma Chem. Plasma Process.*, **27**, 13.
30. Kim, H.H., Ogata, A., and Futamura, S. (2005) Atmospheric plasma-driven catalysis for the low temperature decomposition of dilute aromatic compounds. *J. Phys. D: Appl. Phys.*, **38**, 1292.

31. Kim, H.H., Ogata, A., and Futamura, S. (2006) Effect of different catalysts on the decomposition of VOCs using flow-type plasma-driven catalysis. *IEEE Trans. Plasma Sci.*, **34**, 984.
32. Ding, H.X., Zhu, A.M., Lu, F.G., Xu, Y., Zhang, J., and Yang, X.F. (2006) Low-temperature plasma-catalytic oxidation of formaldehyde in atmospheric pressure gas streams. *J. Phys. D: Appl. Phys.*, **39**, 3603.
33. Magureanu, M., Mandache, N.B., Hu, J., Richards, R., Florea, M., and Pârvulescu, V.I. (2007) Plasma-assisted catalysis total oxidation of trichloroethylene over gold nano-particles embedded in SBA-15 catalysts. *Appl. Catal., B Environ.*, **76**, 275.
34. Roland, U., Holzer, F., Poppl, A., and Kopinke, F.D. (2005) Combination of non-thermal plasma and heterogeneous catalysis for oxidation of volatile organic compounds. Part 3. Electron paramagnetic resonance (EPR) studies of plasma-treated porous alumina. *Appl. Catal., B Environ.*, **58**, 227.
35. Harling, A.M., Demidyuk, V., Fischer, S.J., and Whitehead, J.C. (2008) Plasma-catalysis destruction of aromatics for environmental clean-up: effect of temperature and configuration. *Appl. Catal., B Environ.*, **82**, 180.
36. Wallis, A.E., Whitehead, J.C., and Zhang, K. (2007) The removal of dichloromethane from atmospheric pressure air streams using plasma-assisted catalysis. *Appl. Catal., B Environ.*, **72**, 282.
37. Tanabe, K., Misono, M., Ono, Y., Hattori, H. (1989) *New Solid Acids and Bases, Studies in Surface Science and Catalysis*, vol. 51, Elsevier, Amsteram, p. 78.
38. Foger, K. (1984) Dispersed metal catalysts. *Catal. Sci. Technol.*, **6**, 231.
39. Kodas, T. and Hampden-Smith, M.J. (1999) *Aerosol Processing of Materials*, Wiley-VCH Verlag GmbH, New York, p. 712.
40. Kerner, D. and Rochnia, M. (2008) in *Handbook of Heterogeneous Catalysis*, vol. 1 (eds G. Ertl, H. Knozinger, F. Schuth, and J. Weitkamp), Wiley-VCH Verlag GmbH, Weinheim, p. 286.
41. Schuth, F., Hesse, M., and Unger, K.K. (2008) in *Handbook of Heterogeneous Catalysis*, vol. 1 (eds G. Ertl, H. Knozinger, F. Schuth, and J. Weitkamp), Wiley-VCH Verlag GmbH, Weinheim, p. 111.
42. Iler, R.K. (1979) *The Chemistry of Silica and Silicates*, John Wiley & Sons, Inc., New York.
43. Brinker, J.F. and Scherer, G.W. (1990) *Sol-Gel Science: The Physics and Chemistry of Sol-Gel Processing*, Academic Press, Boston.
44. Occelli, M.L. and Robson, H.E. (1989) *Zeolite Synthesis*, Ameican Chemical Society, Washington, DC.
45. Byrappa, K. and Yoshimura, M. (2001) *Handbook of Hydrothermal Technology, A Technology for Crystal Growth and Materials Processing*, Noyes Publications, New Jersey, p. 315.
46. Kloepfer, H. (1942) DE Patent 762723, to Degussa.
47. Bacsa, R.R. and Gratzel, M. (1996) Rutile formation in hydrothermally crystallized nanosized titania. *J. Am. Ceram. Soc.*, **79**, 2185.
48. Cheng, H., Ma, J., Zhao, Z., and Qi, L. (1995) Hydrothermal preparation of uniform nanosize rutile and anatase particles. *Chem. Mater.*, **7**, 663.
49. Bertone, J.F., Cizeron, J., Wahi, R.K., Bosworth, J.K., and Colvin, V.L. (2003) Hydrothermal synthesis of quartz nanocrystals. *Nano Lett.*, **3**, 655.
50. Kim, H.H., Ogata, A., and Futamura, S. (2008) Oxygen partial pressure-dependent behavior of various catalysts for the total oxidation of VOCs using cycled system of adsorption and oxygen plasma. *Appl. Catal., B Environ.*, **79**, 356.
51. Marouf-Khelifa, K., Abdelmalek, F., Khelifa, A., and Addou, A. (2008) TiO_2-assisted degradation of a perfluorinated surfactant in aqueous solutions treated by gliding arc discharge. *Chemosphere*, **70**, 1995.
52. Radu, D.C., Pârvulescu, V., Câmpeanu, V., Bartha, E., Jonas, A., Grange, P., and Pârvulescu, V.I. (2003) Chemoselective oxidation of

2-thiomcthyl-4,6-dimethyl-pyrimidine and 2-thiobenzyl-4,6-dimethyl-pyrimidine over titania-silica catalysts. *Appl. Catal., A Gen.*, **242**, 77.
53. Seicaru, O., Pârvulescu, V.I., Simion, G., Dumitru, I., Moreh, E., and Anger, I. (1996) Romanian Patent 111355.
54. Parvulescu, V.I., Parvulescu, V., Endruschat, U., and Poncelet, G. (2001) Preparation and characterization of mesoporous zirconium oxide (I). *Appl. Catal., A Gen.*, **214**, 273.
55. Parvulescu, V.I., Parvulescu, V., Endruschat, U., Lehmann, C.W., Grange, P., Poncelet, G., and Bönnemann, H. (2001) Preparation and characterization of mesoporous zirconium oxide. Part 2. *Microporous Mesoporous Mater*, **44–45**, 221.
56. Jentoft, F.C. (2008) in *Handbook of Heterogeneous Catalysis*, vol. 1 (eds G. Ertl, H. Knozinger, F. Schuth, and J. Weitkamp), Wiley-VCH Verlag GmbH, Weinheim, p. 266.
57. Breck, D.W. (1974) *Zeolite Molecular Sieves*, John Wiley & Sons, Inc., New York, London.
58. Barrer, H.R.M. (1982) *Hydrothermal Chemistry of Zeolites*, Academic Press Inc., London.
59. Jacobs, P.A. and Martens, J.A. (1987) *Synthesis of High-Silica Aluminosilicate Zeolites*, Studies in Surface Science and Catalysis, Vol. 33, Elsevier, Amsterdam.
60. Dyer, A. (1988) *An Introduction to Zeolite Molecular Sieves*, John Wiley & Sons, Inc., New York, London.
61. Van Bekkum, H., Flanigen, E.M., and Jansen, J.C. (1991) *Studies in Surface Science and Catalysis*, vol. 58, Elsevier, Amsterdam.
62. Fan, H.Y., Shi, C., Li, X.S., Yang, X.F., Xu, Y., and Zhu, A.M. (2009) Low-temperature NOx selective reduction by hydrocarbons on H-mordenite catalysts in dielectric barrier discharge plasma. *Plasma Chem Plasma Process.*, **29**, 43.
63. Feijen, E.J.P., Martens, J.A., and Jacobs, P.A. (1994) Zeolites and their mechanism of synthesis. *Stud. Surf. Sci. Catal.*, **84**, 3.
64. Feijen, E.J.P., Martens, J.A., Jacobs, P.A. (1999) in *Preparation of Solid Catalysts* (eds G. Ertl, H. Knoezinger, and J. Weitkamp), Wiley-VCH, Weinheim, p. 262.
65. Livage, J. (1994) in *Advanced Zeolite Science and Applications*, Studies in Surface Science and Catalysis, Vol. 85 (eds J.C. Jansen, M. Stocker, H.G. Karge, and J. Weitkamp), Elsevier, Amsterdam, p. 1.
66. Domine Berges, M. and Domine, D. (1968) Fr Patent 136937, to 7Air Liquide SA.
67. Bibby, D.M., Milestone, N.B., and Aldridge, L.P. (1980) Ammonium-tetraalkyl ammonium systems in the synthesis of zeolites. *Nature*, **285**, 30.
68. Lohse, U., Altrichter, B., Donath, R., Fricke, R., Jancke, K., Parlitz, B., and Schreier, E. (1996) Synthesis of zeolite beta. Part 1. Using tetraethylammonium hydroxide/bromide with addition of chelates as templating agents. *J. Chem. Soc., Faraday Trans.*, **92**, 159.
69. Xiong, G., Yu, Y., Feng, Z.C., Xin, Q., Xiao, F.S., and Li, C. (2001) UV Raman spectroscopic study on the synthesis mechanism of zeolite X. *Microporous Mesoporous Mater.*, **42**, 317.
70. Magureanu, M., Mandache, N.B., Eloy, P., Gaigneaux, E.M., and Parvulescu, V.I. (2005) Plasma-assisted catalysis for volatile organic compounds abatement. *Appl. Catal., B Environ.*, **61**, 12.
71. Lok, B.M., Messina, C.A., Patton, R.L., Gajek, R.T., Cannan, T.R., Flanigen, E.M. (1984) US Patent 4440871, to Union Carbide Corporation.
72. Harling, A.M., Glover, D.J., Whitehead, J.C., and Zhang, K. (2008) Industrial scale destruction of environmental pollutants using a novel plasma reactor. *Ind. Eng. Chem. Res.*, **47**, 5856.
73. Preisler, E. (1980) Moderne verfahren der großchemie: braunstein. *Chem. unserer Zeit*, **14**, 137.
74. Cheng, F., Zhao, J., Song, W., Li, C., Ma, H., Chen, J., and Shen, P. (2006) Facile controlled synthesis of MnO_2 nanostructures of novel shapes and

their application in batteries. *Inorg. Chem.*, **45**, 2038.

75. Magureanu, M., Mandache, N.B., Gaigneaux, E., Paun, C., and Parvulescu, V.I. (2006) Toluene oxidation in a plasma-catalytic system. *J. Appl. Phys.*, **99**, 123301.
76. Li, R., Tang, Q., Yin, S., and Sato, T. (2006) Plasma catalysis for CO2 decomposition by using different dielectric materials. *Fuel Process. Technol.*, **87**, 617.
77. Afonina, N.E., Gromov, V.G., and Kovalev, V.L. (2002) Investigation of the influence of different heterogeneous recombination mechanisms on the heat fluxes to a catalytic surface in dissociated carbon dioxide. *Fluid Dyn.*, **37**, 117.
78. Karpichenko, E.A. and Efimenko, L.P. (1996) Preparation and thermal stability of Si-borosilicate glass composites. *Inorg. Mater.*, **32**, 552.
79. Tascon, J.M.D., Fierro, J.L.G., and Tejuca, L.J. (1992) in *Properties and Applications of Perovskite Type Oxides* (eds L.J. Tejuca and J.L.G. Fierro), Marcel Dekker, New York, Chapter 8, p. 171.
80. Pena, M.A. and Fierro, J.L.G. (2001) Chemical structures and performance of perovskite oxides. *Chem. Rev.*, **101**, 1981.
81. Alifanti, M., Florea, M., Filotti, G., Kuncser, V., Cortes-Corberan, V., and Parvulescu, V.I. (2006) In situ structural changes during toluene complete oxidation on supported $EuCoO_3$ monitored with ^{151}Eu Mössbauer spectroscopy. *Catal. Today*, **117**, 329.
82. Royer, S., Berube, F., and Kaliaguine, S. (2005) Oxygen storage capacity of $La_{1-x}A'_xBO_3$ perovskites (with A′ = Sr, Ce; B = Co, Mn)-relation with catalytic activity in the CH_4 oxidation reaction. *Appl. Catal., A Gen.*, **282**, 273.
83. Alifanti, M., Florea, M., and Parvulescu, V.I. (2007) Ceria based oxides as supports for $LaCoO_3$ perovskite; catalysts for total oxidation of VOC. *Appl. Catal., B Environ.*, **70**, 400.
84. Alifanti, M., Bueno, G., Parvulescu, V., Parvulescu, V.I., and Cortes Corberan, V. (2009) Oxidation of ethane on high specific surface $SmCoO_3$ and $PrCoO_3$ perovskites. *Catal. Today*, **143**, 309.
85. Lin, H., Huang, Z., Shangguan, W., and Peng, X. (2007) Temperature-programmed oxidation of diesel particulate matter in a hybrid catalysis–plasma reactor. *Proc. Combust. Inst.*, **31**, 3335.
86. López-Navarrete, E., Caballero, A., Orera, V.M., Lázaro, F.J., and Ocaña, M. (2003) Oxidation state and localization of chromium ions in Cr-doped cassiterite and Cr-doped malayaite. *Acta Mater.*, **51**, 2371.
87. Hueso, J.L., Cotrino, J., Caballeroa, A., Espinósa, J.P., and González-Elipe, A.R. (2007) Plasma catalysis with perovskite-type catalysts for the removal of NO and CH_4 from combustion exhausts. *J. Catal.*, **247**, 288.
88. Hueso, J.L., Caballero, A., Cotrino, J., and González-Elipe, A.R. (2007) Plasma catalysis over lanthanum substituted perovskites. *Catal. Commun.*, **8**, 1739.
89. Royer, S., Van Neste, A., Davidson, R., McIntyre, S., and Kaliaguine, S. (2004) Methane oxidation over nanocrystalline $LaCo_{1-x}Fe_xO_3$: resistance to SO_2 poisoning. *Ind.Eng. Chem. Res.*, **43**, 5670.
90. Levasseur, B. and Kaliaguine, S. (2008) Effect of the rare earth in the perovskite-type mixed oxides $AMnO_3$ (A = Y, La, Pr, Sm, Dy) as catalysts in methanol oxidation. *J. Solid State Chem.*, **181**, 2953.
91. Chandler, C., Roger, D.C., and Smith, M.H. (1993) Chemical aspects of solution routes to perovskite-phase mixed-metal oxides from metal-organic precursors. *Chem. Rev.*, **93**, 1205.
92. Phule, P. and Risbud, R.H. (1988) Sol-gel synthesis and characterization of barium titanate ($BaTi_4O_9$ and $BaTiO_3$) powders. *Mater. Res. Soc. Symp. Proc.*, **121**, 275.
93. Livage, J., Henry, M., and Sanchez, C. (1990) Sol-gel chemistry of transition metal oxides. *Prog. Solid State Chem.*, **18**, 942.
94. Yogo, T., Kikuta, K.I., Yamada, S., and Hirano, S.I. (1994) Synthesis of barium titanate/polymer composites from

metal alkoxides. *J. Sol-Gel Sci. Technol.*, **2**, 175.
95. Cernea, M., Monnereau, O., Llewellyn, P., Tortet, L., and Galassi, C. (2006) Sol–gel synthesis and characterization of Ce doped-$BaTiO_3$. *J. Eur. Ceram. Soc.*, **26**, 3241.
96. Yoon, S., Dornseiffer, J., Schneller, T., Hennings, D., Iwaya, S., Pithan, C., and Waser, R. (2010) Percolative $BaTiO_3$–Ni composite nanopowders from alkoxide-mediated synthesis. *J. Eur. Ceram. Soc.*, **30**, 561.
97. O'Brien, S., Murray, C.B., and Brus, L.E. (2001) Synthesis and characterization of nanocrystals of barium titanate, toward a generalized synthesis of oxide nanoparticles. *J. Am. Chem. Soc.*, **123**, 12085.
98. Pramanik, N.C., Seok, S.I., and Ahn, B.Y. (2006) Wet-chemical synthesis of crystalline $BaTiO_3$ from stable chelated titanium complex: Formation mechanism and dispersibility in organic solvents. *J. Colloid Interface Sci.*, **300**, 569.
99. Chen, C., Wei, Y., Jiao, X., and Chen, D. (2008) Hydrothermal synthesis of $BaTiO_3$: crystal phase and the Ba^{2+} ions leaching behavior in aqueous medium. *Mater. Chem. Phys.*, **110**, 186.
100. Khollam, Y.B., Deshpande, A.S., Patil, A.J., Potdar, H.S., Deshpande, S.B., and Date, S.K. (2001) Microwave-hydrothermal synthesis of equi-axed and submicron-sized $BaTiO_3$ powders. *Mater. Chem. Phys.*, **71**, 304.
101. Hakuta, Y., Ura, H., Hayashi, H., and Arai, K. (2005) Effect of water density on polymorph of $BaTiO_3$ nanoparticles synthesized under sub and supercritical water conditions. *Mater. Lett.*, **59**, 1387.
102. Hotta, Y., Tsunekawa, K., Isobe, T., Sato, K., and Watari, K. (2008) Synthesis of $BaTiO_3$ powders by a ball milling-assisted hydrothermal reaction. *Mater. Sci. Eng., A*, **475**, 12.
103. Won, H.I., Nersisyan, H.H., and Won, C.W. (2007) Low temperature solid-phase synthesis of tetragonal $BaTiO_3$ powders and its characterization. *Mater. Lett.*, **61**, 1492.
104. Heisig, C., Zhang, W., and Oyama, S.T. (1997) Decomposition of ozone using carbon-supported metal oxide catalysts. *Appl. Catal., B Environ.*, **14**, 117.
105. Marceau, E., Carrier, X., Che, M., Clause, O., and Marcily, C. (2008) in *Handbook of Heterogeneous Catalysis*, vol. 1 (eds G. Ertl, H. Knozinger, F. Schuth, and J. Weitkamp), Wiley-VCH Verlag GmbH, Weinheim, p. 473.
106. Fan, X., Zhu, T.L., Wang, M.Y., and Li, X.M. (2009) Removal of low-concentration BTX in air using a combined plasma catalysis system. *Chemosphere*, **75**, 1301.
107. Jarrige, J. and Vervisch, P. (2009) Plasma-enhanced catalysis of propane and isopropyl alcohol at ambient temperature on a MnO_2-based catalyst. *Appl. Catal., B Environ.*, **90**, 74.
108. Smirniotis, P.G., Sreekanth, P.M., Pena, D.A., and Jenkins, R.G. (2006) Manganese oxide catalysts supported on TiO_2, Al_2O_3, and SiO_2: a comparison for low-temperature SCR of NO with NH_3. *Ind. Eng. Chem. Res.*, **45**, 6436.
109. Kotowski, W., Lach, J., Pyzikowski, J., Kuszka, W., and Magnuszewska, Z. (1980) Brit. GB Patent 2025252.
110. Chao, Y., Huang, C.T., Lee, H.M., and Chang, M.B. (2008) Hydrogen production via partial oxidation of methane with plasma-assisted catalysis. *Int. J. Hydrogen Energy*, **33**, 664.
111. Chao, Y., Lee, H.M., Chen, S.H., and Chang, M.B. (2009) Onboard motorcycle plasma-assisted catalysis system – Role of plasma and operating strategy. *Int. J. Hydrogen Energy*, **34**, 6271.
112. Bromberg, L., Cohn, D.R., Rabinovici, A., O'Brien, C., and Hochgreb, S. (1998) Plasma reforming of methane. *Energy Fuels*, **12**, 11.
113. Tran, D.N., Aardahl, C.L., Rappe, K.G., Park, P.W., and Boyer, C.L. (2004) Reduction of NO_x by plasma-facilitated catalysis over In-doped g-alumina. *Appl. Catal., B Environ.*, **48**, 155.
114. Yu, S.J. and Chang, M.B. (2001) Oxidative conversion of PFC via plasma

processing with dielectric barrier discharges. *Plasma Chem. Plasma Process.*, **21**, 311.
115. Chang, M.B. and Yu, S.J. (2001) An atmospheric-pressure plasma process for C_2F_6 removal. *Environ. Sci. Technol.*, **35**, 1587.
116. Kiener, C., Kurtz, M., Wilmer, H., Hoffmann, C., Schmidt, H.W., Grunwaldt, J.D., Muhler, M., and Schüth, F. (2003) High-throughput screening under demanding conditions: Cu/ZnO catalysts in high pressure methanol synthesis as an example. *J. Catal.*, **216**, 110.
117. Günter, M.M., Ressler, T., Bems, B., Büscher, C., Genger, T., Hinrichsen, O., Muhler, M., and Schlögl, R. (2001) Implication of the microstructure of binary Cu/ZnO catalysts for their catalytic activity in methanol synthesis. *Catal. Lett.*, **71**, 37.
118. Baltes, C., Vukojevi, S., and Schüth, F. (2008) Correlations between synthesis, precursor, and catalyst structure and activity of a large set of $CuO/ZnO/Al_2O_3$ catalysts for methanol synthesis. *J. Catal.*, **258**, 334.
119. Palgunadi, J., Yati, I., and Jung, K.D. (2010) Catalytic activity of Cu-Zn-Al-Mn admixed with gamma-alumina for the synthesis of DME from syngas: manganese effect or just method of preparation? *React. Kinet. Mech. Catal.*, **101**, 117.
120. Lei, H., Nie, R., Wang, J., Hou, Z., Fei, J., and Zheng, X. (2010) Chin. Patent 101850254.
121. Naik, S.P., Du, H., Wan, H., Bui, V., Miller, J.D., and Zmierczak, W.W. (2008) A comparative study of ZnO-CuO-Al_2O_3/SiO_2-Al_2O_3 composite and hybrid catalysts for direct synthesis of dimethyl ether from syngas. *Ind. Eng. Chem. Res.*, **47**, 9791.
122. Kaddouri, A. and Mazzocchia, C. (2009) Comparative study of the physico-chemical properties of nanocrystalline CuO-ZnO-Al_2O_3 prepared from different precursors: hydrogen production by vaporeforming of bio-ethanol. *Catal. Lett.*, **131**, 234.
123. Jun, K.W., Min, K.S., Song, S.L., and Jeong, S.H. (2009) WO 2009-KR4128, to Hyundai Heavy Industries Co., Ltd.
124. Phan, X.K., Bakhtiary, H.D., Myrstad, R., Thormann, J., Pfeifer, P., Venvik, H.J., and Holmen, A. (2010) Preparation and performance of a catalyst-coated stacked foil microreactor for the methanol synthesis. *Ind. Eng. Chem. Res.*, **49**, 10934.
125. Guo, Y., Meyer-Zaika, W., Muhler, M., Vukojevic, S., and Epple, M. (2006) Cu/Zn/Al xerogels and aerogels prepared by a sol-gel reaction as catalysts for methanol synthesis. *Eur. J. Inorg. Chem.*, 4774.
126. Lopes, R.J.G., Silva, A.M.T., and Quinta-Ferreira, R.M. (2007) Screening of catalysts and effect of temperature for kinetic degradation studies of aromatic compounds during wet oxidation. *Appl. Catal., B Environ.*, **73**, 193.
127. Martins, R.C. and Quinta-Ferreira, R.M. (2009) Screening of ceria-based and commercial ceramic catalysts for catalytic ozonation of simulated olive mill wastewaters. *Ind. Eng. Chem. Res.*, **48**, 1196.
128. Cheng, Y., Liu, Y., Cao, D., Wang, G., and Gao, Y. (2011) Effects of acetone on electrooxidation of 2-propanol in alkaline medium on the Pd/Ni-foam electrode. *J. Power Sources*, **196**, 3124.
129. Morikawa, T. and Eguchi, H. (1987) Jpn. Kokai Tokkyo Koho 62263974 to Osaka Prefecture.
130. Shin, J.H. and Kim, K.W. (1995) Preparation of thin nickel foam for nickel-metal hydride battery. *Han'guk Pyomyon Konghak Hoechi*, **28**, 83.
131. Burrell, A.K., McCleskey, T.M., Jia, Q., Bauer, E., Blackmore, K.J., Mueller, A.H.,and Dirmyer, M. (2010) US Patent Appl. Publ. 516011000.
132. Huang, H., Ye, D., and Guan, X. (2008) The simultaneous catalytic removal of VOCs and O_3 in a post-plasma. *Catal. Today*, **139**, 43.
133. Magureanu, M., Mandache, N.B., Parvulescu, V.I., Subrahmanyam, C., Renken, A., and Kiwi-Minsker, L. (2007) Improved performance of non-thermal plasma reactor during

decomposition of trichloroethylene: optimization of the reactor geometry and introduction of catalytic electrode. *Appl. Catal., B Environ.*, **74**, 270.
134. Parvulescu, V.I., Cojocaru, B., Parvulescu, V., Richards, R., Li, Z., Cadigan, C., Granger, P., Miquel, P., and Hardacre, C. (2010) Sol–gel-entrapped nano silver catalysts-correlation between active silver species and catalytic behavior. *J. Catal.*, **272**, 92.
135. Hu, J., Chen, L., Zhu, K., Suchopar, A., and Richards, R. (2007) Aerobic oxidation of alcohols catalyzed by gold nano-particles confined in the walls of mesoporous silica. *Catal. Today*, **122**, 277.
136. Kah, J.C.Y., Phonthammachai, N., Wan, R.C.Y., Song, J., White, T., Mhaisalkar, S., Ahmad, I., Sheppard, C., and Olivo, M. (2008) Synthesis of gold nanoshells based on the deposition precipitation process. *Gold Bull.*, **41**, 23.
137. Degarmo, E.P., Black, J.T., and Kohser, R.A. (2003) *Materials and Processes in Manufacturing*, 9th edn, Wiley-VCH Verlag GmbH, Weinheim.
138. Xing, Y., Liu, Z., Couttenye, R.A., Willis, W.S., Suib, S.L., Fanson, P.T., Hirata, H., and Ibe, M. (2007) Generation of hydrogen and light hydrocarbons for automotive exhaust gas purification: conversion of n-hexane in a PACT (plasma and catalysis integrated technologies) reactor. *J. Catal.*, **250**, 67.
139. Jenkin, W.C. (1964) US Patent 3,160,517.
140. Besmann, T.M., Matlin, W.M., and Stinton, D.P. (1995) Chemical Vapor Infiltration Process Modelling and Optimization, DOE Scientific and Technical Information, *http://www.osti.gov/bridge/purl.cover.jsp.*
141. Sarantopoulos, C., Puzenat, E., Guillard, C., Herrmann, J.M., Gleizes, A., and Maury, F. (2009) Microfibrous TiO_2 supported photocatalysts prepared by metal-organic chemical vapor infiltration for indoor air and waste water purification. *Appl. Catal., B Environ.*, **91**, 225.
142. Kirkpatrick, M.J., Finney, W.C., and Locke, B.R. (2003) Chlorinated organic compound removal by gas phase pulsed streamer corona electrical discharge with reticulated vitreous carbon electrodes. *Plasmas Polym.*, **8**, 165.
143. King, H.C., Renier, M.C., Ellzey, K.E., and Lackey, W.J. (2003) Chemical vapor infiltration of rhenium. *Chem. Vap. Deposition*, **9**, 59.
144. Uyama, H., Nakamura, T., Tanaka, S., and Matsumoto, O. (1993) Catalytic effect of iron wires on the syntheses of ammonia and hydrazine in a radio-frequency discharge. *Plasma Chem. Plasma Process.*, **13**, 117.
145. Tanaka, S., Uyama, H., and Matsumoto, O. (1994) Synergistic effects of catalysts and plasmas on the synthesis of ammonia and hydrazine. *Plasma Chem. Plasma Process.*, **14**, 491.
146. Tetsuya, T. (1964) The Electron Beam Melting and Brittle Fracture of Metal Molybdenum. *J. Japan Inst. Metals.*, **28**, 670.
147. Rowe, C.E.D., and Hinch, G.R. (1986) "Sintered Molybdenum Alloy Process," US Patent 4,622,068, to Murex Limited, GB.
148. Chufarov, G.I. and Averbukh, B.D. (1934) Reduction of iron oxides by gaseous reducing agents. II. The rate of reduction of magnetic iron oxide by hydrogen. *Zh. Fizich. Khim.*, **5**, 1292.
149. Zach, M.P., Ng, K.H., and Penner, R.M. (2000) Molybdenum nanowires by electrodeposition. *Science*, **290**, 2120.
150. Muramatsu, H., Hayashi, T., Kim, Y.A., Shimamoto, D., Endo, M., Terrones, M., and Dresselhaus, M.S. (2008) Synthesis and isolation of molybdenum atomic wires. *Nano Lett.*, **8**, 237.
151. Heintze, M. and Pietruszka, B. (2004) Plasma catalytic conversion of methane into syngas: the combined effect of discharge activation and catalysis. *Catal. Today*, **89**, 21.
152. Pietruszka, B. and Heintze, M. (2004) Methane conversion at low temperature: the combined application of catalysis and non-equilibrium plasma. *Catal. Today*, **90**, 151.

153. Khassin, A.A., Yurieva, T.M., Demeshkina, M.P., Kustova, G.N., Itenberg, I.S., Kaichev, V.V., Plyasova, L.M., Anufrienko, V.F., Molina, I.Y., Larina, T.V., Baronskaya, N.A., and Parmon, V.N. (2003) Characterization of the nickel-amesite-chlorite-vermiculite system. 1. Silicon Binding in Ni-Mg-Al phylloaluminosilicates. *Phys. Chem. Chem. Phys.*, **5**, 4025.
154. Khassin, A.A., Pietruszka, B.L., Heintze, M., and Parmon, V.N. (2004) Methane oxidation in a dielectric barrier discharge. The iimpact of discharge power and discharge gap filling. *React. Kinet. Catal. Lett.*, **82**, 111.
155. Neatu, F., Li, Z., Richards, R., Toullec, P.Y., Genet, J.P., Dumbuya, K., Gottfried, M.J., Steinrueck, H.P., Parvulescu, V.I., and Michelet, V. (2008) Heterogeneous gold catalysts for efficient access to functionalized lactones. *Chem. Eur. J.*, **14**, 9412.
156. Stiles, A.B. (1983) *Catalyst Mannufacture, Laboratory and Commercial Preparations*, Marcel Dekker, Inc., New York.
157. Schuth, F. and Hesse, M. (2008) in *Handbook of Heterogeneous Catalysis*, vol. 1 (eds G. Ertl, H. Knozinger, F. Schuth, and J. Weitkamp), Wiley-VCH Verlag GmbH, Weinheim, p. 676.
158. Hoekstruas, J. (1952) Patent 2,620,314, to UOP.
159. Hoekstruas, J. (1954) Patent 2,666,749, to UOP.
160. Page, J.F.L. (1999) in *Preparation of Solid Catalysts* (eds G. Ertl, H. Knozinger, and J. Weitkamp), VCH Verlag Gesellchaft, Weinheim, p. 579.
161. Lunghofer, E.P., Mortensen, S., and Ward, A.P. (1984) US Patent 4440866, to Niro Atomizer, DK.
162. Meyer, J., Neugebauer, P., and Steigerwald, M. (2004) US Patent 6743269, to Degussa AG, De.
163. Townend, J., Gradwell, A.J., and Withenshaw, J.D. (1995) US Patent 5433881, to Warwick International G. L., GB.
164. Gergely, G. and Tritthart, W. (1998) US Patent 5831123.
165. *https://www.e-catalysts.com/supportsearch/tutorials/minilith.htm* (2011).
166. Yarwood, J.C., Dore, J.E., and Preuss, R.K. (1978) US Patent 4075303, to Swiss Aluminium Ltd., CH.
167. Queheillalt, D.T., Hass, D.D., Sypeck, D.J., and Wadley, H.N.G. (2001) Synthesis of open-cell metal foams by templated directed vapor deposition. *J. Mater. Res.*, **16**, 1028.
168. Kruse, G.E. and Macdonald, R.E. (1984) US Patent 4471495, to Whiting and Davis Company, Inc., USA.
169. Sierra, I., Erena, J., Aguayo, A.T., Arandes, J.M., and Bilbao, J. (2010) Regeneration of $CuO-ZnO-Al_2O_3/\gamma-Al_2O_3$ catalyst in the direct synthesis of dimethyl ether. *Appl. Catal., B Environ.*, **94**, 108.
170. Sierra, I., Erena, J., Aguayo, A.T., Olazar, M., and Bilbao, J. (2010) Deactivation kinetics for direct dimethyl ether synthesis on a $CuO-ZnO-Al_2O_3/\gamma-Al_2O_3$ catalyst. *Ind. Eng. Chem. Res.*, **49**, 481.
171. Kroto, H.W., Heath, J.R., O'Brien, S.C., Curl, R.F., and Smalley, R.E. (1985) C60: buckminsterfullerene. *Nature*, **318**, 162.
172. Kratschmer, W., Lamb, L.D., Fostiropoulos, K., and Huffman, D.R. (1990) Solid C60: a new form of carbon. *Nature*, **347**, 354.
173. Ijima, S. (1991) Helical microtubules of graphitic carbon. *Nature*, **354**, 56.
174. Iijima, S. and Ichihashi, T. (1993) Single-shell carbon nanotubes of 1-nm diameter. *Nature*, **363**, 603.
175. Nasibulin, A.G., Pikhitsa, P.V., Jiang, H., Brown, D.P., Krasheninnikov, A.V., Anisimov, A.S., Queipo, P., Moisala, A., Gonzalez, D., Lientschnig, G., Hassanien, A., Shandakov, S.D., Lolli, G., Resasco, D.E., Choi, M., Tománek, D., and Kauppinen, E.I. (2007) A novel hybrid carbon material. *Nat. Nanotechnol.*, **2**, 156.
176. Keller, N., Maksimova, N.I., Roddatis, V.V., Schur, M., Mestl, G., Butenko, Y.V., Kuznetsov, V.L., and Schlogl, R. (2002) The catalytic use of onion-like carbon materials for styrene synthesis

by oxidative dehydrogenation of ethylbenzene. *Angew. Chem. Int. Ed. Engl.*, **41**, 1885.

177. Kuang, Y., Islam, N.M., Nabae, Y., Hayakawa, T., and Kakimoto, M.A. (2010) Selective aerobic oxidation of benzylic alcohols catalyzed by carbon-based catalysts: a nonmetallic oxidation system. *Angew. Chem. Int. Ed.*, **49**, 436.
178. Jagadeesan, D. and Eswaramoorthy, M. (2010) Functionalized carbon nanomaterials derived from carbohydrates. *Chem. Asian J.*, **5**, 232.
179. Huczko, A. (1997) Heterohedral fullerenes and nanotubes: formation and characteristics. *Fullerene Sci. Technol.*, **5**, 1091.
180. Sliwa, W. (1996) Metallofullerenes. *Transition Met. Chem.*, **21**, 583.
181. Lange, H., Huczko, A., Sioda, M., Pacheco, M., Razafinimanana, M., and Gleizes, A. (2002) Influence of gadolinium on carbon arc plasma and formation of fullerenes and nanotubes. *Plasma Chem. Plasma Process.*, **22**, 523.
182. Huczko, A., Lange, H., and Sogabe, T. (2000) Influence of Fe and Co/Ni on carbon arc plasma and formation of fullerenes and nanotubes. *J. Phys. Chem. A*, **104**, 10708.
183. Ivanov, V., Nagy, J.B., Lambin, P., Lucas, A., Zhang, X.B., Zhang, X.F., Bernaerts, D., Van Tendeloo, G., Amelinckx, S., and Van Landuyt, J. (1994) The study of carbon nanotubules produced by catalytic method. *Chem. Phys. Lett.*, **223**, 329.
184. Tian, Y.J., Zhang, Y.L., Yü, Q., Wang, X.Z., Hu, Z., Zhang, Y.F., and Xie, K.C. (2004) Effect of catalysis on coal to nanotube in thermal plasma. *Catal. Today*, **89**, 233.
185. Cojocaru, B., Neatu, S., Parvulescu, V.I., Somoghi, V., Petrea, N., Epure, G., Alvaro, M., and Garcia, H. (2009) Synergism of activated carbon and undoped and N-doped TiO2 in the photocatalytic degradation of chemical warfare soman, VX and yperite. *ChemSusChem*, **2**, 427.
186. Pulsipher, D.J., Martin, I.T., and Fisher, E.R. (2010) Controlled nitrogen doping and film colorimetrics in porous TiO2 materials using plasma processing. *ACS Appl. Mater. Interfaces*, **2**, 1743.
187. Cheng, H.H., Chen, S.S., Cheng, Y.W., Tseng, W.L., and Wang, Y.H. (2010) Liquid-phase non-thermal plasma-prepared N-doped $TiO_{(2)}$ for azo dye degradation with the catalyst separation system by ceramic membranes. *Water Sci. Technol.*, **62**, 1060.
188. Lukes, P. and Locke, B.R. (2005) Degradation of substituted phenols in a hybrid gas-liquid electrical discharge reactor. *Ind. Eng. Chem. Res.*, **44**, 2921.
189. Sunka, P., Babicky, V., Clupek, M., Lukes, P., Simek, M., Schmidt, J., and Cernak, M. (1999) Generation of chemically active species by electrical discharges in water. *Plasma Sources Sci. Technol.*, **8**, 258.
190. Yao, Z., Jia, F., Tian, S., Li, C., Jiang, Z., and Bai, X. (2010) Microporous Ni-doped TiO_2 film photocatalyst by plasma electrolytic oxidation. *ACS Appl. Mater. Interfaces*, **2**, 2617.
191. He, H., Cai, W., Lin, Y., Dai, Z., and Chen, B. (2010) Surface decoration of ZnO nanorod arrays by electrophoresis in the Au colloidal solution prepared by laser ablation in water. *Langmuir*, **26**, 8925.
192. Laure, C., Brault, P., Thomman, A.L., Boswell, R., Rousseau, B., and Estrade-Szwarckopf, H. (1996) Plasma assisted evaporation of palladium. *Plasma Sources Sci. Technol.*, **5**, 510.
193. Brault, P., Thomann, A.L., Rozenbaum, J.P., Cormier, J.M., Lefaucbeux, P., Andreazza, C., and Andreazza, P. (2001) The use of plasmas in catalysis: catalyst preparation and hydrogen production. *Ann. Chim. Sci. Mat.*, 26, 69.
194. Dumitriu, D., Bally, A.R., Ballif, C., Parvulescu, V.I., Schmid, P.E., Sanjines, R., and Levy, F. (1998) Reactive sputtering as a tool for preparing photocatalysts, in *Preparation of Catalysts VII*, vol. 118 (eds P.A.J.B. Delmon, R. Maggi, J.A. Martens, P. Grange, and G. Poncelet), Elsevier Science B.V, Amsterdam, p. 485.

195. Dumitriu, D., Bally, A.R., Ballif, C., Hones, P., Schmid, P.E., Sanjines, R., Levy, F., and Parvulescu, V.I. (2000) Photocatalytic degradation of phenol by TiO_2 thin films prepared by sputtering. *Appl. Catal., B Environ.*, **25**, 83.

196. Visinescu, C.M., Sanjines, R., Lévy, F., and Parvulescu, V.I. (2005) Photocatalytic degradation of acetone by Ni-doped titania thin films prepared by dc reactive sputtering. *Appl. Catal., B Environ.*, **60**, 155.

197. Visinescu, C.M., Sanjines, R., Lévy, F., Marcu, V., and Parvulescu, V.I. (2005) Tantalum doped titania photocatalysts: preparation by dc reactive sputtering and catalytic behavior. *J. Photochem. Photobiol.*, **174**, 106.

198. Neatu, S., Sacaliuc-Parvulescu, E., Levy, F., and Parvulescu, V.I. (2009) Photocatalytic decomposition of acetone over dc-magnetron sputtering supported vanadia/TiO_2 catalysts. *Catal. Today*, **142**, 165.

199. Cojocaru, B., Neaţu, S., Sacaliuc-Parvulescu, E., Lévy, F., Parvulescu, V.I., and Garcia, H. (2011) Influence of gold particle size on the photocatalytic activity for acetone oxidation of Au/TiO_2 catalysts prepared by dc-magnetron sputtering. *Appl. Catal., B Environ.*, **107**, 140.

3
NO_x Abatement by Plasma Catalysis

Gérald Djéga-Mariadassou, François Baudin, Ahmed Khacef, and Patrick Da Costa

3.1
Introduction

3.1.1
Why Nonthermal Plasma-Assisted Catalytic NO_x Remediation?

In atmospheric chemistry and air pollution, nitrogen oxides refer specifically to NO_x (NO and NO_2). They are precursors of ozone and remain the most serious hazards to human health among the regulated compounds. Their selective reduction to N_2 is still a matter of research. Among the existing pollution sources, motor vehicles are seen as the major contributors to air pollution by NO_x, unburned hydrocarbons (UHCs), and fine particulate matter (PM). *No stable and sufficiently active catalyst* for selective catalytic reduction (SCR) of NO_x emission from automotive exhaust gases in lean (i.e., oxygen-rich) conditions has been designed yet.

In order to understand the role of plasma in deNO_x catalysis, a *deNO_x reaction model* is required. A *three-function catalyst model* for hydrocarbon (HC) SCR of NO_x has been previously defined, in the *absence* of plasma, by Djéga-Mariadassou and coworkers [1–4] and is first summarized (Section 3.2).

Nonthermal plasmas (NTPs), also referred to as *nonequilibrium plasmas* or *cold plasmas,* appear as potent technologies for the removal of air pollutants such as NO_x, sulfur oxides (SO_x), methane, and volatile organic compounds (VOCs) in the exhaust gas stream of stationary or mobile sources, under atmospheric pressure. Many applications of NTPs have been developed [5], and treatments for diesel and gasoline engine exhausts under lean conditions have been proposed [6–10].

NTPs for environmental applications may be produced by a variety of electrical discharge devices (corona discharge, surface discharge, dielectric barrier discharge (DBD), and DC discharge) [11–14] or by electron beam irradiation [15]. The electron beam technique, which is very efficient for removal of NO_x and VOCs, has been used first. However, pulsed discharges are much more suited than e-beams for some industrial and domestic applications because of their high selectivity, moderate operating conditions (atmospheric pressure and room temperature), and relatively low maintenance requirements, resulting in relatively low energy costs

Plasma Chemistry and Catalysis in Gases and Liquids, First Edition.
Edited by Vasile I. Parvulescu, Monica Magureanu, and Petr Lukes.

of the pollutant treatment. DBD processing is a very mature technology, first investigated by Siemens in the 1850s for ozone generation. It is now routinely used in different industrial and fundamental applications such as water purification, polymer treatment, UV light generation, biological and medical treatment, pollution control, and exhaust cleaning from CO, NO_x, SO_x, and VOCs.

In the past years, these various thermal plasmas and NTPs such as DBD, corona, gliding arc, microwave, glow discharge, and pulsed discharge have also been widely investigated for methane conversion to higher HCs [16–18], and methane reforming [19, 20]. Alumina was already used as the catalyst in the plasma-catalytic process and for NO_x abatement [21]. Although various studies have been performed, only a few of them deal with plasma technology as a way to oxidize methane to CO_2 from emissions of combined heat powers (CHPs). However, recently, plasma-assisted catalysis was studied by Da Costa *et al.* [22, 23] using noble metals, and by Hueso *et al.* [24] using lanthanum-substituted Perovskites and silica, as catalysts for methane oxidation.

3.2 General deNO_x Model over Supported Metal Cations and Role of NTP Reactor: "Plasma-Assisted Catalytic deNO_x Reaction"

It will be shown that an NTP reactor can play the role of two of the three functions defined in the model, by coupling the plasma reactor to the catalytic-deNO_x one.

The *three-function catalyst model* for HC SCR of NO_x has been described [1–4] based on experimental evidence for each function, during temperature-programmed surface reactions (TPSRs). It has been verified during stationary experiments. The model has been mainly written for reductants that are very often inactive at the temperature "T_{NO}," where NO is able to dissociate and reduce. It has been shown that T_{NO} can be predetermined by the adsorption of NO at room temperature and its subsequent temperature-programmed desorption (TPD). It is verified in Section 3.5.

The release of N_2 occurs within function F3 (Figure 3.1). It involves the dissociation of NO (via a dinitrosyl-adsorbed intermediate) followed by subsequent formation of N_2 and scavenging of the adsorbed oxygen species (O_{ads}) left from NO dissociation. If the initial HC in the feed is not oxidized to CO_2/H_2O by scavenging O_{ads} at T_{NO}, the removal of adsorbed oxygen has to be done by the oxidation of activated reductants ($C_xH_yO_z$ designed as "oxygenates" in the chapter). This reaction corresponds to "a supported homogeneous catalytic process" involving a surface transition metal complex. The corresponding catalytic sequence of elementary steps occurs in the coordinative sphere of the metal cation.

Function F2 (Figure 3.1) has to turn over simultaneously to function F3. It has to deliver the active reductant species $C_xH_yO_z$ to function F3, at the T_{NO} temperature where function F3 cycle turns over. Function F2 is the mild oxidation of HC (or any initial oxygenate [25] by NO_2, through organic nitrogen-containing intermediates (RNO_x). The important feature is that these RNO_x species are *quite thermally*

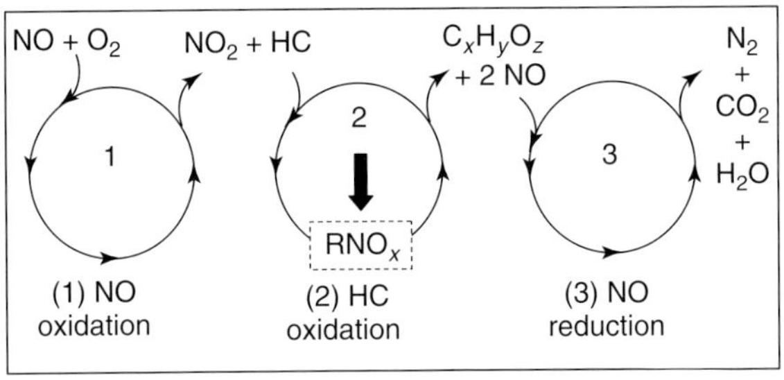

Function F1	Function F2	Function F3
NO oxidation to NO_2	HC mild oxidation to oxygenates by NO_2	2NO reduction to N_2 assisted by $C_xH_yO_z$ total oxidation over a metal cation
NO_2	$HC + NO_2 = C_xH_yO_z$	$(2x-z+y/2)NO + C_xH_yO_z = (x-z/2+y/4)N_2 + xCO_2 + y/2\ H_2O$

Figure 3.1 The three-function model for designing deNO$_x$ catalysts in the presence of a nonreactive hydrocarbon as reductant [2–4].

instable: they decompose with temperature to $C_xH_yO_z$ + NO (not to N_2), according to the following global equation:

$$\mathrm{HC}\ (\text{or } C_xH_yO_z) + \mathbf{NO_2} = C_{x'}H_{y'}O_{z'} + \mathbf{NO} \tag{3.1}$$

It is therefore obvious that functions F2 and F3 have to turn over simultaneously. Nevertheless, at the molecular level, these two functions are not in the same catalytic cycle.

Function F1 has also to turn over simultaneously with functions F2 and F3, as it has to provide NO_2 to function F2 (Eq. (3.1)). The oxidation of NO to NO_2 is therefore the first function of any efficient catalyst.

The concept of "simultaneous turn over" between catalytic functions in multi-functional catalysis is widely accepted, and this aspect of the proposed mechanism is a "normal" behavior in steady state.

Figure 3.1 summarizes the model and the corresponding global reactions for each function.

The three-function model is now extended, in this chapter, to other cases where reductants can be more easily oxidized to CO_2/H_2O. Figures 3.2–3.6 describe the different cases.

- $T_{HC} \ll T_{NO}$ (Figure 3.2). The temperature T_{HC} of the reductant oxidation to CO_2/H_2O is too low compared to that of NO activation T_{NO}. In this case, as NO is not activated, deNO$_x$ is not observed at T_{HC}; furthermore, there will be a lack of reductant at T_{NO} and a very low deNO$_x$ process will be generally observed on the catalyst at this temperature. The unique global reaction observed around T_{HC} is

$$C_nH_m + \left(n + \frac{1}{4}m\right) O_2 = n\,CO_2 + \frac{1}{2}m\,H_2O \tag{3.2}$$

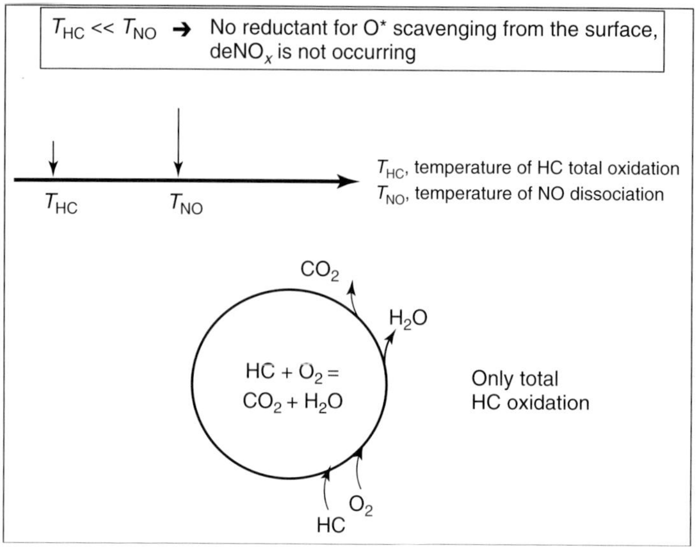

Figure 3.2 The three-function model depending on the temperature of reductant oxidation (to CO_2/H_2O) compared to that of activation/dissociation of NO.

- $T_{HC} = T_{NO}$ (Figure 3.3). The oxidation of HC is occurring in the temperature range where NO is activated/reduced (around T_{NO}) and the deNO$_x$ reaction takes place according to the global equation:

$$C_nH_m + \left(2n + \frac{1}{2}m\right) NO = \left(n + \frac{1}{4}m\right) N_2 + n\,CO_2 + \frac{1}{2}m\,H_2O \tag{3.3}$$

- The as-defined function F3 of the model turns over. It will be shown hereafter that competitive oxidation of HC to CO_2/H_2O by O_2 can occur simultaneously (not reported in Figure 3.3).
- $T_{HC} = T_{NO}$ (Figure 3.4). The oxidation of HC is still occurring at T_{NO}. If functions F1 and F2 turn over simultaneously (depending on the catalyst design and composition), oxygenates can be delivered to function F3. In this case, a new deNO$_x$ catalytic cycle will be kinetically associated to the HC-deNO$_x$ catalytic cycle of function F3 (Figure 3.4). It is the $C_xH_yO_z$-deNO$_x$ one, according to the global equation:

$$C_xH_yO_z + \left(2x + \frac{1}{2}y - z\right) NO = \left(x + \frac{1}{4}y - \frac{1}{2}z\right) N_2 + x\,CO_2 + \frac{1}{2}y\,H_2O \tag{3.4}$$

- Similarly, it will be shown that competitive oxidation of $C_xH_yO_z$ to CO_2/H_2O can occur simultaneously (not reported in Figure 3.4).
- $T_{HC} = T_{NO}$ (Figure 3.5). *Effect of the competitive total oxidations of HC and oxygenates on the deNO$_x$ process.* In addition to the (b) and (c) cases, two competitive catalytic cycles can turn over, *kinetically coupled* with the HC- and $C_xH_yO_z$-deNO$_x$

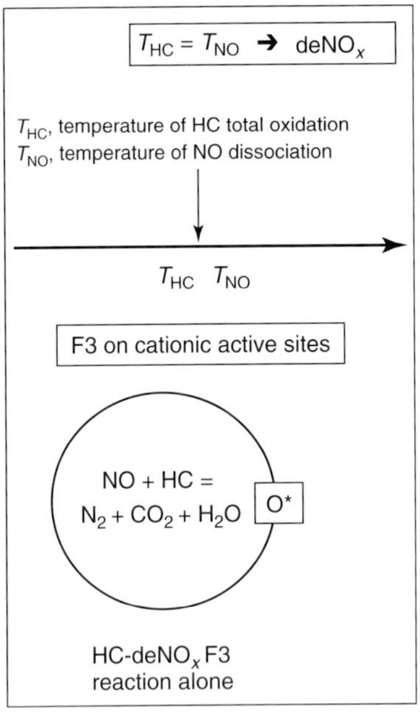

Figure 3.3 Function F3 turns over according to Eq. (3.3) when $T_{HC} = T_{NO}$.

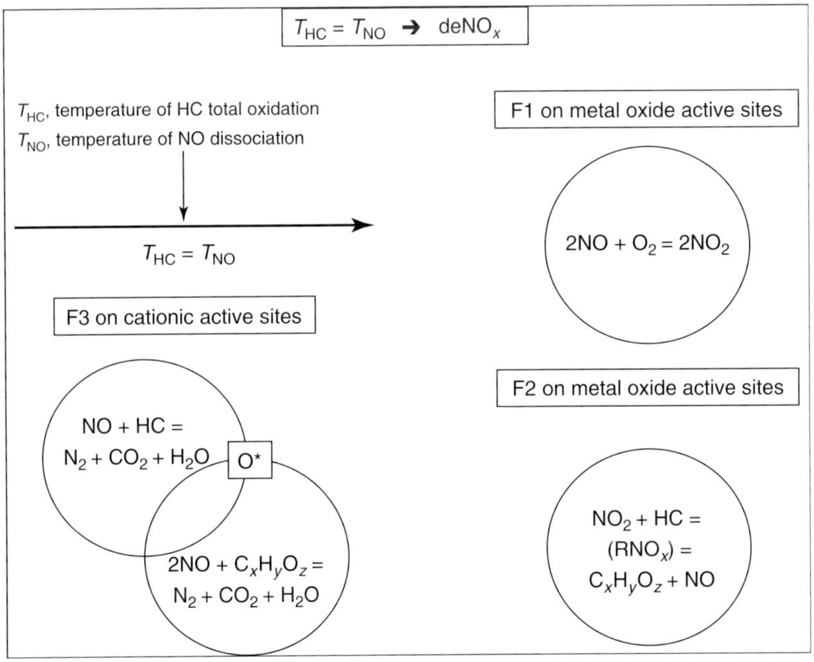

Figure 3.4 $T_{HC} = T_{C_xH_yO_z} = T_{NO}$: kinetic coupling of HC and $C_xH_yO_z$ function F3 cycles (4).

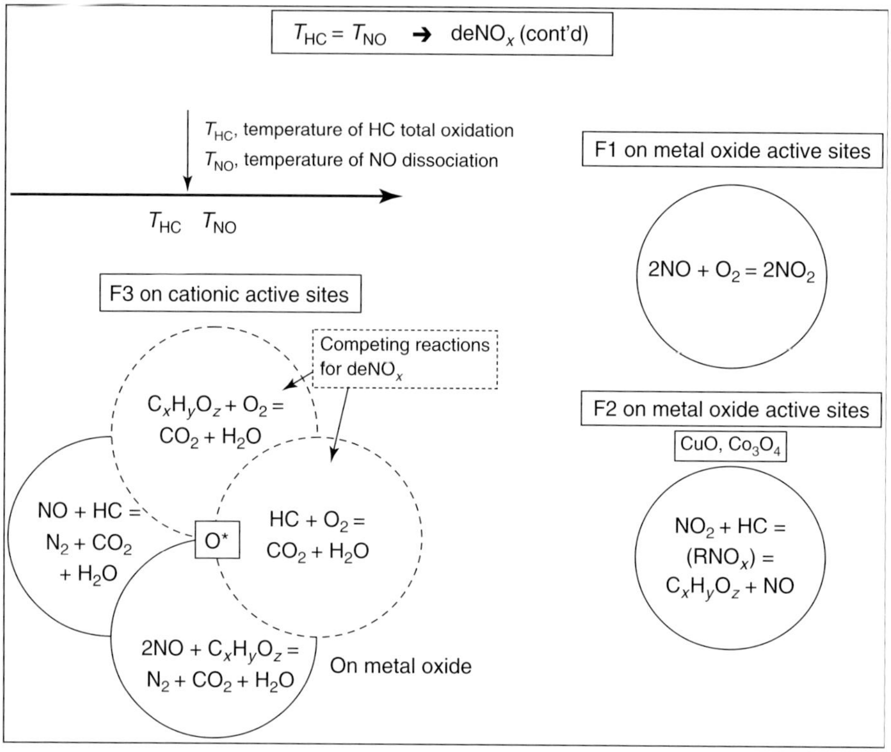

Figure 3.5 Competitive oxidation of HC and $C_xH_yO_z$ to CO_2/H_2O by O_2.

cycles of Figure 3.4; they are the two oxidations of HC and $C_xH_yO_z$ to CO_2/H_2O that can turn over simultaneously (not reported in Figure 3.4).

- Furthermore, depending on the catalyst chemical composition and design, two other parallel catalytic cycles (not reported in Figure 3.5) can turn over: HC and $C_xH_yO_z$ oxidation on different catalytic sites, such as other oxides active in oxidation (Co_3O_4, $CeZrO_2$, CuO, etc.). These direct reductant oxidations to CO_2 and H_2O clearly and drastically lead to a strong consumption of reductants and therefore to a strong decrease of NO_x conversion.
- *It is shown in Section 3.3 that an NTP reactor can produce stabilized oxygenates that can be transferred to the deNO$_x$ reactor, in the whole range of reaction temperature.*
- $T_{HC} \gg T_{NO}$ (Figure 3.6). In this case, the reductant cannot scavenge the adsorbed oxygen species left by NO dissociation (O_{ads}) and here is the great interest of the present model: the catalyst has to be designed in order to proceed to the mild oxidation of HC to oxygenates, at T_{NO}. The three functions are represented in Figure 3.6. Furthermore, as seen in Figure 3.5, the direct global oxidation of $C_xH_yO_z$ can turn over simultaneously, decreasing the concentration of the oxygenate species for the $deNO_x$ function F3 cycle. The consumption of oxygenate on another oxide site, active in its oxidation, is not reported in Figure 3.6.

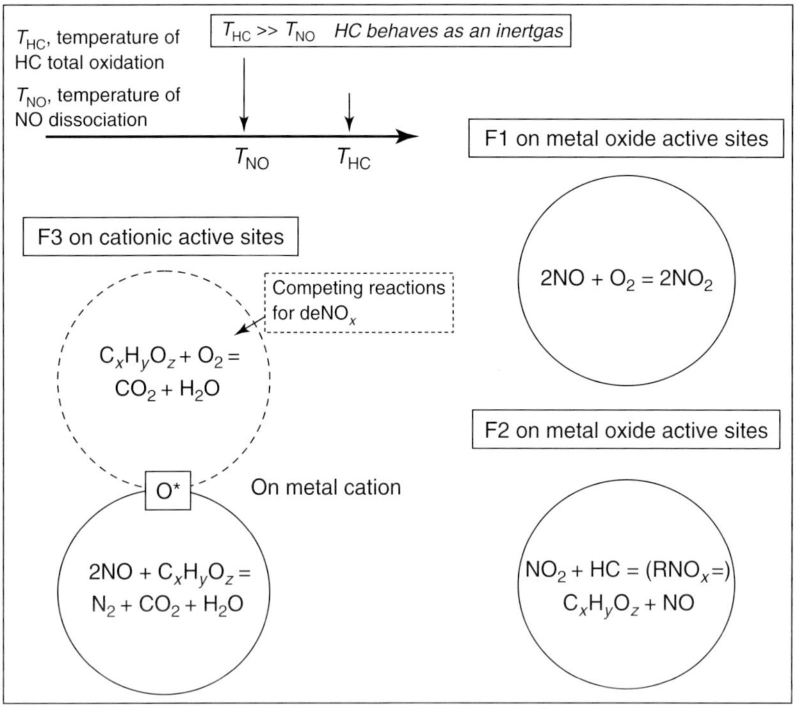

Figure 3.6 The function F3 for $T_{HC} \gg T_{NO}$. $C_xH_yO_z$ are the efficient reductants. Oxygenates are produced owing to function F1 and function F2.

The nonthermal-assisted deNO$_x$ catalytic device is able to *continuously* deliver to the deNO$_x$ reactor efficient reductants working at T_{NO}.

The present model has to be used to optimize catalysts. More particularly, the coupling of plasma and deNO$_x$ reactors has to decrease, as far as possible, the competitive direct oxidations of reductants by O_2. Furthermore, active sites have to be created, corresponding to the three functions that have to turn over simultaneously at T_{NO}, owing to active species delivered by the plasma reactor.

As it is very difficult to find *the best* design of the catalyst to simultaneously initiate the three functions by itself, *an external device can be developed to substitute functions F1 and F2, providing the catalyst the good oxygenated species, in the full range of temperature.*

There are different ways of developing such an idea. *One of them is the NTP-assisted deNO$_x$ reaction.* One of the earlier works, by Penetrante and coworkers [6], has shown that the coupling of an NTP reactor, in front of the alumina catalyst, was able to produce a significant deNO$_x$ of the feed. They did not give, in their paper, any interpretation of the observed results, but published the composition of the stabilized gas mixture at the outlet of the plasma reactor, just before the catalytic reactor containing alumina. The gas mixture contained more oxygenated compounds.

Hoard and Balmer [7] and Doraï and Kushner [26] have also found $C_xH_yO_z$ (CH_2O, CH_2O_2, NO_2, and RNO_x (CH_3ONO_2), which are "intermediates" needed for functions F1 and F2 and, furthermore, for the third function itself.

3.3
About the Nonthermal Plasma for NO_x Remediation

The conventional NTP reactors that are widely used for various environmental applications are subdivided according to the type of discharge mode (pulse, DC, AC, radio frequency (RF), microwave), presence of a dielectric barrier or catalyst, and geometry (cylinder, plate). It is important to note that the chemical potential of each discharge mode differs enormously from one discharge to another. Roughly speaking, the efficiency of a plasma discharge to remove pollutant from gas stream mainly depends on its ability to produce a large amount of active species (O atoms, OH radicals) in the plasma volume. It has been well established that DBD and corona discharge fulfill that condition [5, 27]. DBDs are attractive for plasma remediation because of their ability to operate in a stable manner at atmospheric pressure, with high average power and with low voltage (a few kilovolts to ten kilovolts) compared to coronas or electron beams. DBDs are compact, efficient plasma sources commonly used as ozonizers. Detailed characteristics of NTP reactors may be found in [27–29], and only a brief description of the DBD reactor that we used is given hereafter. The successful application of DBDs to plasma remediation of pollutants from mobile sources such as diesel exhausts will largely depend on the ability to meet the goals of obtaining high efficiency of remediation (low electron volt per molecule) and low production of undesirable compounds.

In the DBD reactor, the electrode configuration is characterized by the presence of at least one dielectric barrier in the current path, preventing the formation of a conducting channel, in addition to the gas gap used for discharge initiation. At atmospheric pressure, in most gases the discharge is maintained by a large number of short-lived localized current filaments called *microdischarges* [30–32]. The NTP conditions in these microdischarges can be influenced and optimized for different applications. Besides this multifilaments mode with a seemingly random distribution of microdischarges, regularly patterned discharges and apparently homogeneous glow discharges have been obtained in such electrode configurations.

For the vehicle exhaust systems, it was established that the observed chemistry in the plasma includes the conversion of NO to NO_2 as well as the partial oxidation of HCs. The presence of the UHCs in the exhaust is very important for the plasma-catalytic deNO_x process for multiple reasons [6, 8, 13, 33–36].

First, UHCs enhance the gas-phase oxidation of NO to NO_2 and lower the energy cost for this oxidation.

Secondly, their partial oxidation leads to production of chemical species such as aldehydes and alcohols useful for the catalytic reduction of NO_x. For some catalysts, the partially oxygenated HCs are much more effective compared to original HCs in reducing NO_x to N_2.

Thirdly, UHCs prevent the oxidation of SO_2, thus making the plasma-catalytic process tolerant to the sulfur content of the fuel.

The need for an effective after-treatment process capable of reducing NO_x under lean conditions has produced a large interest in plasma-assisted catalytic reduction [37–41]. From the literature, especially the *Society of Automotive Engineers* (SAE) published papers, catalysts such as γ-Al_2O_3 and NaY zeolite have shown high activity when combined with NTP.

3.3.1 The Nanosecond Pulsed DBD Reactor Coupled with a Catalytic deNO$_x$ Reactor: a Laboratory Scale Device Easily Scaled Up at Pilot Level

DBD reactors in cylindrical configuration combined with heterogeneous catalysts have been used. This combination can be either single stage or two stage (Figure 3.7). In the two-stage system, the catalyst materials are usually placed downstream from the NTP reactor (PPC: *postplasma catalysis configuration*). In the single-stage system, the catalysts are placed directly in the NTP reactor (IPC: *in-plasma catalysis* configuration). When the IPC is operated, short lifetime reactive species, in the range of tens of nanoseconds, as radicals and excited states ($O(^3P)$, $O(^1D)$, and OH), created by the NTP can be used efficiently by the catalyst. On the contrary, when the catalyst is placed postplasma (PPC), long lifetime reactive species, in the range of milliseconds, such as O_3, H_2O_2, NO, and NO_2, could be used.

As part of our studies, PPC configuration was used for deNO$_x$ experiments, while both the PPC and IPC configurations were tested in VOC oxidation experiments. It has been stated that the performance of the NTP for the removal of VOC can be improved by the introduction of the catalytic material into the discharge zone (IPC) [42–44].

The plasma DBD reactor consisted of a tungsten wire (0.9 mm diameter) centered in a quartz tube (inner and outer diameters of 11 and 13 mm, respectively). A brass

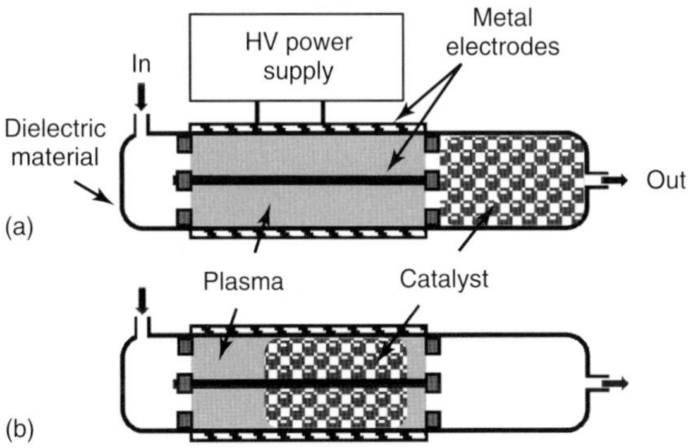

Figure 3.7 Schematic overview of the plasma–catalyst configurations: (a) PPC and (b) IPC.

Figure 3.8 The plasma DBD reactor followed by the $deNO_x$ one in an open furnace.

mesh covered the dielectric tube and formed the outer electrode. The length of the outer electrode can be adjusted and then determines the active volume of the plasma reactor. The DBD reactor was placed inside a tubular furnace, and the gas mixture temperature could be adjusted from room temperature up to 500 °C (Figure 3.8).

For $deNO_x$ studies, the plasma reactor was driven by a cable transformer powered by high voltage (HV) ceramic capacitors disposed in Blumlein-like configuration. Details of the HV pulse generator are given in Ref. [45]. This pulse generator is capable of delivering HV up to 30 kV into 80 ns full width at half maximum (FWHM) pulses and short rate rise time (~kV ns^{-1}) at repetition rate up to 200 Hz. The fast voltage rise time allows achieving significant overvoltage at breakdown. The energy deposition in the plasma reactor (E_d) is defined as the discharge power to the gas flow rate ratio and is given by the following formula:

$$E_d\,(J\,l^{-1}) = \frac{E_p \cdot f}{Q}$$

where E_p is the discharge pulse energy (J per pulse), f the pulse repetition frequency (Hz), and Q the gas flow (l s^{-1}) rate at standard conditions. An example of the time behavior of the voltage–current, discharge power, and discharge pulse energy is given in Figure 3.9. For the experiments presented here, the input energy density was adjusted to about 36 J l^{-1}, whatever the temperature.

The $deNO_x$ experiments have been conducted to investigate gas mixtures with more and more complex compositions. The goal is to reach synthetic gas exhausts simulating diesel and lean-burn gasoline engine exhausts. Typically, the N_2-based mixture consists of O_2, H_2O, NO, and HCs (C_3H_6, C_3H_8, n-C_{10}, and toluene). The gas mixtures were prepared in a gas-handling system and their composition was controlled using calibrated high-precision mass flow controllers. Maximum concentration of different gas components is O_2 (10%), H_2O (10%), NO (500 ppm), NO_2 (500 ppm), C_3H_6 (2000 ppmC), C_3H_8 (150 ppm), n-C_{10} (1100 ppmC), toluene (450 ppmC), and N_2 as balance.

The reactor outflow was analyzed using a set of specific detectors (Figure 3.10). A NO_x Chemiluminescence Analyzer (Eco Physics CLD 700 AL) allowed the simultaneous detection of NO, NO_2, and NO_x. An Infrared (IR) analyzer (Ultramat 6) was used to monitor N_2O and a flame ionization detector (FID) detector to follow the total concentration of HC. Gas chromatography–mass spectrometry (GC–MS) analyses were performed online

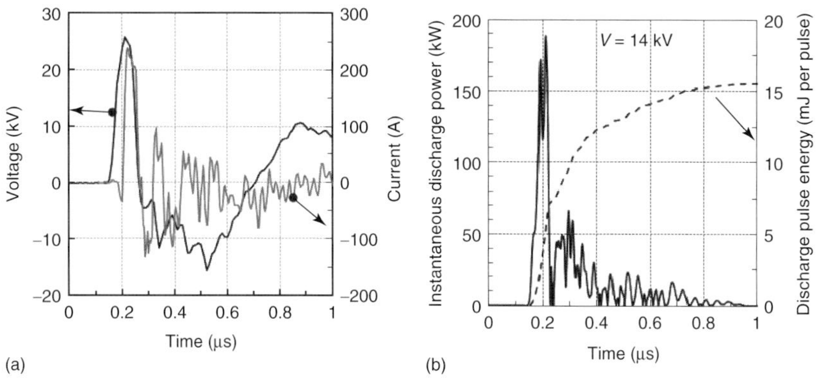

Figure 3.9 Typical waveforms of (a) pulse voltage–current and (b) instantaneous discharge power and discharge pulse energy.

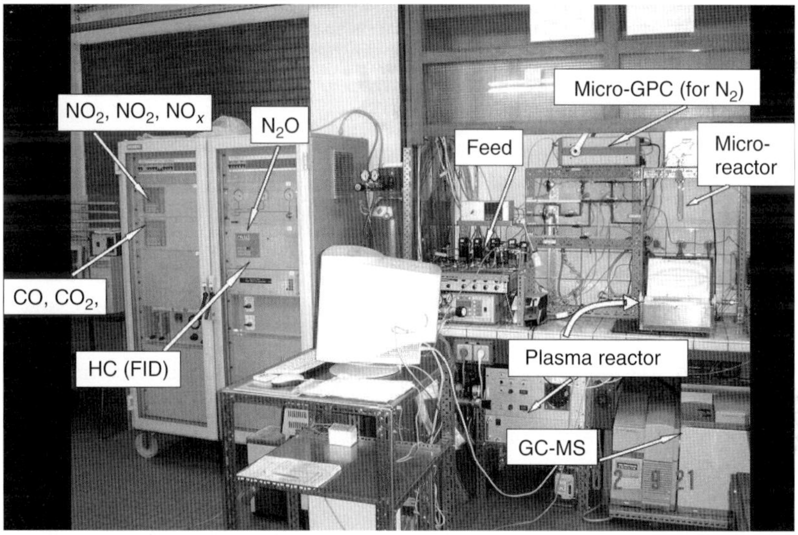

Figure 3.10 Experimental device for a complete study of deNO_x process: plasma DBD reactor, GC/MS, micro-gas phase chomatograph (GPC) (for N_2), specific detectors, and standard microreactor.

using an Agilent device (GC 6890-MS 5973 N) equipped with a CP-PoraBOND *Q* capillary column with temperature programming from 50 to 280 °C. The NIST spectral Agilent database was used to identify the detected products. This device also allowed the analysis of various gases such as H_2, O_2, N_2, CO, and CO_2, in a scale ranging from parts per million to percentage. Some additional chemical analysis were performed using Fourier transform infrared absorption spectrometer (FTIR-Nicolet Magna 550 II) equipped with a heated 10 m multiple pass absorption cell.

3.3.2
Nonthermal Plasma Chemistry and Kinetics

In these plasmas, the chemistry is initiated by high-energy free electrons (3–6 eV) and the entire gas stream remains at room temperature. Oxidation is the dominant process for systems containing dilute concentrations of pollutants (NO_x or VOCs) in gas mixtures of N_2, O_2, and H_2O, particularly when the O_2 concentration is 5% or higher. In addition, ozone (mainly produced through three-body reaction of O with O_2) can play an important role in the oxidation processes, particularly at low temperature. Even though the electrons are short lived under atmospheric pressure conditions and rarely collide with a pollutant molecule, they undergo many collisions with the dominant background gas molecules (electron-impact dissociation and ionization) to produce radicals (N, O, OH, etc.) that, in turn, lead to the chemical conversion of the NO_x or VOC molecules.

During plasma discharge in typical exhausts containing N_2, O_2, and H_2O with NO (the major form of NO_x), primary radicals (N, OH, and O) are created by electron-impact reactions and rapidly consumed by reactions in the remediation pathway. They are then regenerated during the next current pulse. The secondary radicals (HO_2, NO_3, and O_3) are not formed by direct electron-impact events but rather by reactions involving the primary radicals. They are produced and consumed on longer time scales than the primary radicals and their densities are more sensitive to plasma conditions. These radicals initiate chemical reactions, leading to chemical conversion of pollutant molecules. Production of oxidizing agents predicted by a self-consistent 0D model in N_2–O_2–H_2O mixture at 25 °C and 1 bar [46] is illustrated in Figure 3.11. O and OH maximum densities are achieved in some tens of nanoseconds, whereas the density of ozone slowly increases to reach a maximum about 1 ms after the discharge.

Much effort has already been expended on the modeling of the removal by plasma of pollutant species from vehicle exhaust streams, particularly the direct removal of NO_x. Attention has now turned to studying the chemistry occurring when an HC is added to the mixture, promoting the conversion of NO to NO_2. Detailed kinetic schemes and discussions of the mechanisms involved in gas-phase chemistry in the plasma-processing O_2, NO, N_2, and HC mixtures were studied extensively [47–49].

In an HC-free mixture, the consumption of NO_x mainly occurs through the oxidation channel. This is due to the fact that the rates of producing O and OH are higher compared to N because of the lower dissociation energy of H_2O and O_2 compared to N_2. UHCs, often present in diesel exhausts, significantly influence NO_x chemistry during plasma remediation by oxidizing NO into NO_2. Globally, the plasma chemistry shows that the atomic oxygen produced in the discharge is the initiator of the HC chemistry. These reactions produce HC radical intermediates such as RO_2, OH, and HO_2. After the initiation of the kinetics, OH radicals rather than O atoms become the main HC-consuming species. Although diesel exhausts are humid, with a typical water content of few percents, the production of OH radicals by electron-impact dissociation of water is slow [50], and the main source

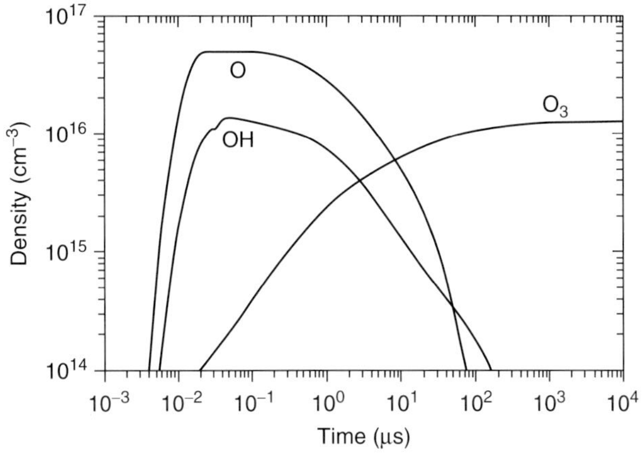

Figure 3.11 Time evolution of O, OH, and O_3 densities as predicted by a 0D self-consistent model of the photo-triggered discharge for the N_2–O_2–H_2O (77.6–20–2.4) mixture at 25 °C, 1 bar, and specific energy deposition of about 100 J l^{-1} [46].

of OH comes from the HC oxidation chemistry. The HO_2 radical and, to a lesser extent, peroxy radicals, RO_2, are responsible for the conversion of NO to NO_2[10]. The RO radicals then go on to produce the aldehydes [49]. The reaction scheme recycles the OH and HO_2 radicals in a chain mechanism. Niessen *et al.* [51] used ethene (C_2H_4) as a model UHC and showed that the W-value (energy required to remove one molecule) for NO improved from 60 eV without C_2H_4 to 10 eV with 2000 ppm of C_2H_4. In experiments by Khacef *et al.* [34], NO removal improved from 50% at 90 J l^{-1} in the absence of propene (C_3H_6) to 90% when adding 500 ppm C_3H_6, with most of the NO being converted to NO_2.

In addition to NO_2, plasma processing of a mixture containing O_2–NO–H_2O–HC–N_2 led to the production of CO, CO_2, aldehydes (CH_3CHO, CH_2O), alcohols (CH_3OH), nitrate and nitrite compounds of RNO_x type (CH_3ONO, CH_3ONO_2), nitromethane (CH_3NO_2), formic acid (CH_2O_2), propylene oxide (C_3H_6O), and to some extent of acids (HNO_2, HNO_3). Formations of such molecules are predicted by kinetic models [33, 36, 47, 49, 50] and detected by gas chromatography and FTIR spectroscopy as well [34, 52, 53]. Examples of FTIR and GC/MS results obtained at the exit of the pulsed DBD in a dry gas mixture are shown in Figures 3.12 and 3.13. Other examples are extensively described in Section 3.5.

In C_3H_6–O_2–NO–N_2 mixture, the maximum efficiency for oxidation of NO to NO_2 increases as the C1/NO_x ratio is increased. A C1/NO_x ratio of 4 is required to get 80% oxidation efficiency. In the absence of propene, the oxidation efficiency is very low (54%) even at high values of electrical energy density input to the plasma. When the gas mixture was heated, the NO to NO_2 oxidation reaction is counteracted at high temperature by the reduction reaction, as shown in Figure 3.14. The evolution of the FID signal (total HC concentration) as the

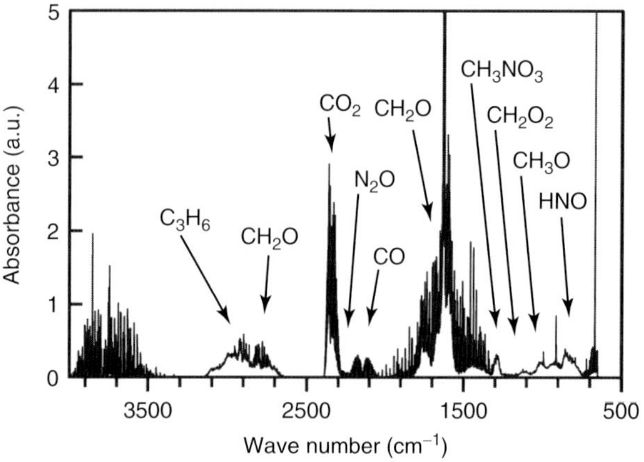

Figure 3.12 Typical FTIR spectrum of the products from pulsed DBD processing (27 J l^{-1} input energy density, room temperature) of O_2 (10%), NO (500 ppm), C_3H_6 (1500 ppmC), and N_2.

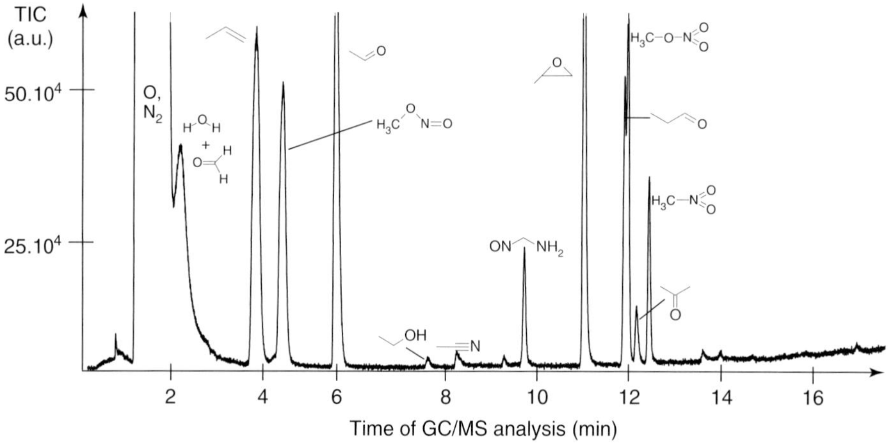

Figure 3.13 Typical chromatogram plot of the gas phase from pulsed DBD processing (36 J l^{-1}, 150 °C) of O_2 (8%), NO (300 ppm), C_3H_6 (450 ppmC), and N_2.

temperature increases shows that some amount of HC species remains in the gas phase even at temperatures as high as 400 °C.

3.3.3
Plasma Energy Deposition and Energy Cost

One of the most important problems of applied NTP for applications such as emission control of industrial and automotive exhaust gases is the minimization of the energy consumption to be competitive with other technologies. The energy cost

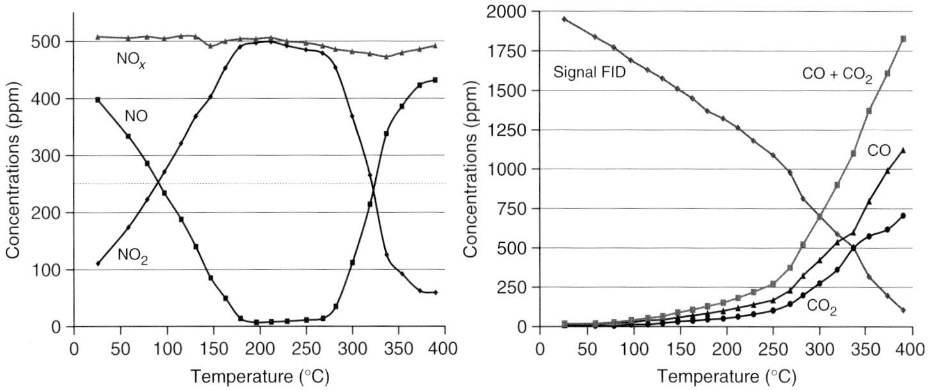

Figure 3.14 Species concentrations as a function of the gas temperature at the exit of HV-pulsed DBD at 36 J l^{-1}. Gas mixture: C_3H_6 (2000 ppmC), NO (500 ppm), O_2 (8% vol.), and N_2.

of the plasma-chemical processes are closely related to its mechanisms and the same plasma processes in different discharge systems or under different conditions result in entirely different expenses of energy. Usually, the plasma-processing literature presents the fraction of NO_x removed as a function of parameters such as the specific energy deposited in the discharge, the residence time of the gas in the reactor, or the applied voltage to the plasma reactor [37–39]. The composition of the output gaseous components was recorded as a function of energy density, which is the parameter commonly used to evaluate the NO_x conversion efficiency in the plasma. However, if this parameter is important to characterize the electrical energy consumption of the process, we should take into account the way to achieve the chosen energy value [54]. It means that for a specific energy deposited in given plasma reactor, the plasma chemistry strongly depends on the type of the discharge (e.g., pulsed or AC voltage) and their HV parameters (amplitude, rise time, duration, and frequency).

Figure 3.15a shows an example of NO concentrations and C_3H_6 conversion rate measured at the exit of an HV-pulsed wire-cylinder DBD reactor for two discharge regimes: (i) low energy per pulse and high repetition rate and (ii) high energy per pulse and low repetition rate. These results emphasize that the NO removal efficiency is optimum for a low deposited energy per pulse and a high HV pulse repetition frequency. For example, plasma processing at about 27 J l^{-1} (35 mJ per pulse and 200 Hz) shows that more than 90% of NO molecules and 36% of C_3H_6 molecules were converted, respectively. Those conversions are only around 29 and 15%, respectively, for 195 mJ per pulse and 30 Hz case. Whatever the pulsed discharge regime used in the O_2–NO–C_3H_6–N_2 mixture, the main products of the process are NO_2, CO, and CO_2. Formaldehyde (CH_2O) and a large variety of by-products such as methyl nitrate (CH_3ONO_2), formic acid (CH_2O_2), and nitrous acid (HONO) are observed. However, marked differences in final concentrations of these species were observed according to the regime used, as shown in Figure 3.15b

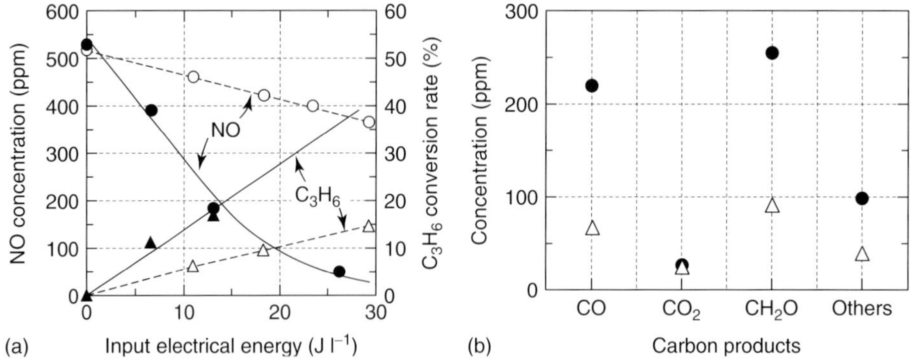

Figure 3.15 (a) NO concentration and C_3H_6 conversion rate and (b) carbon product concentrations at the outlet of an HV-pulsed wire-cylinder DBD reactor for two excitation regimes: $f = 200$ Hz and $E_p = 35$ mJ (filled symbols); $f = 40$ Hz and $E_p = 195$ mJ (open symbols). Gas mixture: O_2 (10%), NO (500 ppm), C_3H_6 (1500 ppmC), N_2 at 1 bar and 25 °C.

with respect to the carbon products. "Others" represent by-products (formic acid, methyl nitrate, etc.) not quantified in that study.

It appears from these experiments that if the energy is deposited in small fractions more often, the energy utilization could be more efficient. The energy cost of one removed toxic molecule or W-value (electron volts per molecule) mainly depends on the efficiency of the energy transfer from the power source to the plasma reactor, configuration of electrodes, and efficiency of chemical reactions [55]. These results are consistent with the modeling of Doraï and Kushner [56]. In that model, the energy cost to remove one NO_x molecule decreases from 240 eV for a single pulse (58 J l^{-1}) to 185 eV when the same energy was distributed over 20 pulses. Figure 3.16 shows the change in the NO molecule removal energy cost with temperature in the presence of C_3H_6. At the energy density of 27 J l^{-1}, the cost of

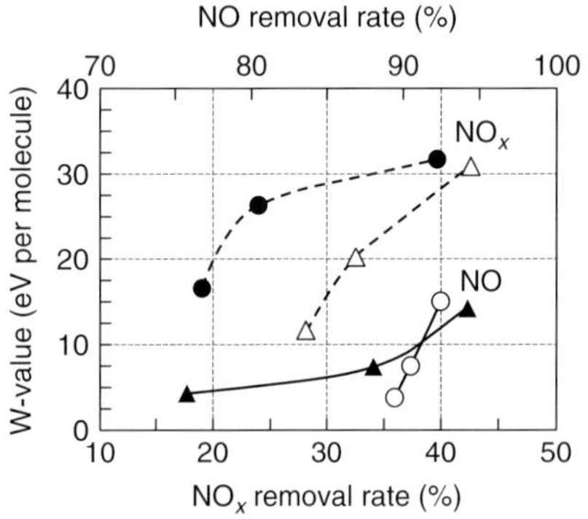

Figure 3.16 Energy cost versus removal rate for NO and NO_x for HV-pulsed DBD reactor at 100 °C (filled symbols) and 260 °C (open symbols). Gas mixture: O_2 (10%), H_2O (10%), NO (500 ppm), C_3H_6 (1500 ppmC), and N_2.

15 eV per NO with efficiency of 95% was achieved at the temperature of 100 °C. For the same efficiency, the energy cost at higher gas temperature (260 °C) was larger. These results could be compared with previous data obtained by Bröer *et al.* [57] in synthetic gas mixture containing higher O_2 concentration (18%), C_2H_4 instead of C_3H_6, and at higher frequency (1 kHz). These authors found that the maximum NO removal rate was 92% with a W-value around 100 eV.

As a conclusion for pulsed DBD, the optimization of the energy deposition leads to lowering the energy cost of NO_x conversion, which is one of the determinant parameters in pollution control processes [58]. Pulses with short rise time and high amplitude voltage have been found to be advantageous because of their ability to rapidly produce high E/N (electric field/number density) [59]. It has also been found that it is more efficient with respect to remediation to deposit a given amount of energy in a large number of shorter duration pulses.

3.4 Special Application of NTP to Catalytic Oxidation of Methane on Alumina-Supported Noble Metal Catalysts

Catalytic processes for total oxidation of methane have been studied as an alternative to conventional thermal combustion [60–62]. The main application of total catalytic oxidation is the elimination of methane emissions from natural gas, which decreases the greenhouse gas effect [63]. Although the several advantages of natural gas combustion, the emission of unburned methane (natural gas is typically 90–95% of methane) is a negative point, because methane is a potent greenhouse gas, which is recognized as contributing more to global atmosphere warming than CO_2 at equivalent emission rates; moreover, its lifetime is quite long [63]. As methane emissions are now being taken into account in present and future regulations, the emissions must be necessarily reduced [64].

Several studies have been performed to design catalytic materials that present the highest activity at the lowest temperature and the best resistance to poisons present in exhaust gases [65–67]. The most commonly used catalysts for complete methane oxidation are platinum (Pt)- and palladium (Pd)-based catalysts. There are several studies in which the Pd-based catalyst and various supports have been tested, such as alumina, Ta_2O_3, TiO_2, CeO_2, and ZrO_2 [68]. Fewer studies have been performed with respect to Pt catalysts supported on oxides supports compared to Pd catalysts [63]. The future of catalytic oxidation consists of the association of platinum and palladium in bimetallic catalytic systems in order to combine the high activity of Pd catalysts to the high resistance of Pt ones against sulfur poisoning. Recently, rhodium has been also used as a promoter on palladium-based catalysts [69].

In this section, we only paid attention to results of experiments with a DBD for total methane oxidation in the presence or absence of a catalyst. We also focused on the effect of water in the feed to reach typical industrial conditions. The same DBD device was used for this study and for NO_x removal studies.

Table 3.1 Temperature of n% methane conversion, T_n, under different energy depositions in the presence of DBD alone for: 150 ppm NO, 8% O_2, 7% CO_2, and 1000 ppm CH_4 in N_2 as balance.

Energy deposition ($J\,l^{-1}$)	n% methane conversion temperature (T_n for n% conversion) (°C)			
	T_{10}	T_{30}	T_{50}	T_{80}
36	348	430	470	–
58	330	410	450	–
80	315	385	430	480

3.4.1
Effect of DBD on the Methane Oxidation in Combined Heat Power (CHP) Conditions

In our experiments, two different reaction mixtures were considered. The reactants are the following: 150 ppm NO, 8% O_2, 7% CO_2, 1000 ppm CH_4 in N_2 as balance and 150 ppm NO, 8% O_2, 7% CO_2, 1000 ppm CH_4, 4% H_2O in N_2 as balance, which is representative of CHP conditions. Furthermore, 150 ppm of NO was chosen in order to reproduce a gas composition close to those observed in exhaust gases of CHP. The total gas flow was maintained at 0.25 $l\,min^{-1}$ NTP. More details are discussed elsewhere [23].

3.4.1.1 Effect of Dielectric Material on Methane Oxidation

Table 3.1 presents the effect of the energy deposition on the steady-state methane conversion as a function of temperature for the following mixture: 150 ppm NO, 8% O_2, 7% CO_2, 1000 ppm CH_4 in N_2 as balance. The methane conversion increases with the energy deposition. One can see in Table 3.1 that 50% of methane is converted at 470, 450, and 430 °C for $E_d = 36$, 58, and 80 $J\,l^{-1}$, respectively. The effect of NO was also investigated as a function of energy deposition. The NO_x concentration increased on increasing the temperature and the energy deposition. Only above 400 °C, NO_x were formed. Thus, about 10 ppm of NO_x were formed with an $E_d = 36\,J\,l^{-1}$ at 450 °C. This energy deposition is the best to minimize the NO_x formation during the methane oxidation process.

3.4.1.2 Effect of Water on Methane Conversion as a Function of Energy Deposition

The influence of water on methane oxidation in the presence of plasma DBD was also studied as a function of energy deposition in gas-phase reaction. For this study, 36 Jl^{-1} as energy has been chosen. The results are presented in Figure 3.17. The oxidation of methane is enhanced in the presence of wet feed only at high temperatures (above 400 °C). Indeed, at 450 °C, 40% of methane oxidation is achieved in the absence of water, whereas 48% is converted in the presence of water. With higher energy deposition (58 $J\,l^{-1}$), the influence of water

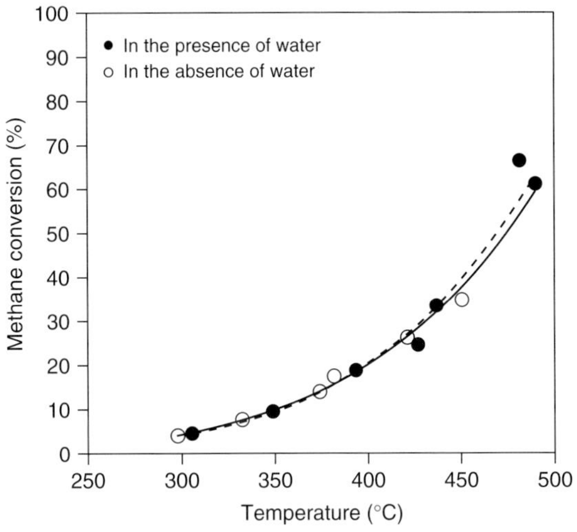

Figure 3.17 Effect of H_2O in the feed on the CH_4 conversion in the absence of a catalyst with an energy deposition of 36 J l^{-1}.

is more evident and becomes significant at 450 °C [23]. Thus, the presence of water increased the total oxidation of methane by promoting the CO_2 formation. It could be due to the high affinity of water to electrons, in the presence of energy deposition [70, 71]. In our conditions, even at high temperatures, in the presence of water the amount of NO_x formed is not significant.

3.4.2 Effect of Catalyst Composition on Methane Conversion as a Function of Energy Deposition

3.4.2.1 Effect of the Support on Plasma-Catalytic Oxidation of Methane

The alumina catalyst was placed after the discharge zone of plasma to improve the selectivity and the efficiency of the plasma process. Alumina alone is only active at high temperatures (higher than 600 °C) [23]. The methane conversion was followed as a function of temperature and energy deposition, in the presence of plasma + alumina. As reported in Table 3.2, the effect of energy deposition on methane conversion is evident only at high temperatures for a plasma-catalytic system. Table 3.2 also indicates that the methane conversion increases with the energy deposition, in the presence of plasma + alumina. The temperature found for 50% methane conversion decreases with increasing E_d. These results are in agreement with those presented in Table 3.1, obtained in the absence of alumina. For a fixed energy deposition, the presence of alumina leads to a decrease in the temperature of conversion of about 10–20 °C. This better activity in the presence of plasma + alumina could be explained by the better reactivity of the activated species, induced by the plasma. In fact, the plasma created reactive species from methane, and

Table 3.2 Temperature of *n*% methane conversion, T_n, under different energy depositions as a function of the catalyst for: 150 ppm NO, 8% O_2, 7% CO_2, and 1000 ppm CH_4 in N_2 as balance.

Catalyst	GHSV (h^{-1})	Energy deposition ($J\,l^{-1}$)	*n*% methane conversion temperature (°C)		
			T_{30}	T_{50}	T_{80}
None	–	36	430	470	–
None	–	58	410	450	–
Support (Al_2O_3)	20 000	0	470	–	–
		36	412	448	–
		58	398	435	470
Pd(0.5)/Al_2O_3	40 000	0	315	330	362
		36	315	330	362
		58	315	330	362
Pd(0.36)/Al_2O_3	100 000	0	–	–	–
		36	422	460	–
		58	395	438	485
Pt(0.36)/Al_2O_3	20 000	0	462	–	–
		36	412	448	–
		58	388	426	462

then these species reacted with the catalyst, leading to higher methane conversion [24].

3.4.2.2 Effect of the Noble Metals on Plasma-Catalytic Oxidation of Methane in the Absence of Water in the Feed

In these experiments, the total flow rate was maintained constant at 250 ml min^{-1}, and the space velocity gas hourly space velocity (GHSV) was varied depending on the catalyst to have a significant activity; more details are given elsewhere [24].

Palladium-Based Catalysts Palladium-based catalysts are well known to be the most active catalysts for the catalytic total oxidation of methane [63]. Experiments were performed with a Pd(0.5)/Al_2O_3 catalyst using a GHSV = 40 000 h^{-1} to minimize the influence of the catalyst in the catalytic plasma-assisted system (Table 3.2). Fifty percent of methane is converted at 330 °C for $E_d = 0$ and 36 J l^{-1}. Thus, there is no influence of the energy deposition in the presence of Pd(0.5)/Al_2O_3. In the studied experimental conditions, the palladium-based catalysts are highly active in the temperature range from 350 to 500 °C (in which the plasma activates the methane). Thus, for palladium-based catalysts, there is no influence of plasma on the methane total oxidation. Let us note that for all the experiments, only CO_2 was observed as a product, and that in the operating conditions, only about 20 ppm of NO_x was formed at 500 °C, whereas 100% methane is converted ($E_d = 36\ J\ l^{-1}$).

Platinum-Based Catalysts A platinum-based catalyst, Pt(0.36)/Al_2O_3, has been prepared and studied in the absence of water with a GHSV 20 000 h^{-1}. The reaction of methane oxidation was carried out in the absence of plasma and in the presence of an energy deposition ranging from 36 to 58 $J\ l^{-1}$ (Table 3.2). One can see that in the absence of plasma, the Pt(0.36)/Al_2O_3 catalyst oxidized the methane above 350 °C. These results are in agreement with the literature [63]. Indeed, the platinum-based catalysts are less active than the palladium ones for methane total oxidation in the absence of water [63]. In the presence of energy deposition (36 $J\ l^{-1}$), the methane conversion increases with the energy deposition. In Table 3.2, one can see that 30% of methane is converted at 462, 412, and 388 °C for $E_d = 0$, 36 and 58 $J\ l^{-1}$, respectively. Thus, the influence of plasma is clearly evident in the presence of a platinum-based catalyst. The catalytic plasma-assisted system is then efficient for energy deposition of 36 $J\ l^{-1}$ and of course for higher energy deposition.

3.4.2.3 Influence of Water on the Plasma-Assisted Catalytic Methane Oxidation in CHP Conditions

Table 3.3 presents the results of methane oxidation in the presence of water as functions of energy deposition and catalyst.

Table 3.3 Temperature of *n*% conversion, T_n, under different energy depositions in the presence of a catalyst, in the presence (W) of water, or in the absence (W/O) of water in the feed.

Energy deposition ($J\ l^{-1}$)		Catalysts			
		Pt(0.5)/Al_2O_3		Pt(0.36)/Al_2O_3	
		W/O H_2O	W H_2O	W/O H_2O	W H_2O
0	T_{10}	290	340	425	475
	T_{30}	321	368	475	–
	T_{50}	340	390	–	–
36	T_{10}	350	295	350	350
	T_{30}	310	358	421	421
	T_{50}	335	375	452	452
58	T_{10}	295	295	327	327
	T_{30}	310	352	389	390
	T_{50}	335	372	441	442

Reaction mixture: W/O H_2O : 150 ppm NO, 8% O_2, 7% CO_2, and 1000 ppm CH_4 in N_2 as balance and W H_2O:150 ppm NO, 8% O_2, 7% CO_2, 1000 ppm CH_4, and 4% H_2O in N_2 as balance.

Influence of Wet Mixture on Support The activity of alumina-based catalysts for methane oxidation was presented elsewhere [23]. In the presence of water, the methane oxidation was totally inhibited. The alumina alone was inactive in the studied range of temperature. The one effect observed was the effect of energy deposition, as already reported in dry conditions [23].

Influence of Water on Platinum- and Palladium-Based Catalysts Since numerous authors mentioned a strong inhibition of water vapor on the rate of methane oxidation, which could be responsible for the observed deactivation of Pd catalysts [72–74], we decided to confirm the effect of water on the noble-metal-based catalysts in CHP conditions. As reported in the literature [72–74], the methane oxidation rate decreases with the addition of water in the feed. Indeed, on Pt(0.36)/Al_2O_3, 10% of methane conversion is obtained at 425 and 475 °C in the absence of plasma in dry and wet conditions, respectively (Table 3.3). In addition, 50% of methane is converted at 340 °C in the absence of water and at 385 °C in the presence of water on Pd(0.5)/Al_2O_3 (GHSV = 40 000 h^{-1}). The loss of activity on water addition was then consistent with previous studies on Pd/Al_2O_3, indicating the strong inhibition by water vapor [72, 74, 75].

Coupled Plasma–Pt(*X*)/Al_2O_3 or Plasma–Pd(*X*)/Al_2O_3 for Methane Oxidation in the Presence of Water in the Feed As already reported in the absence of plasma (Table 3.2), the platinum catalyst is inactive for methane conversion in the presence of water. With plasma addition, one can conclude the same. In fact, in Table 3.3, for $E_d = 36\ \mathrm{J\,l^{-1}}$, whatever the experimental conditions, 50% of methane is converted at the same temperature, at 452 °C. The enhancement of methane oxidation is only due to the increase in energy deposition, as already reported [57]. We can then conclude that systems with plasma DBD coupled with platinum-based catalysts are not efficient systems for methane oxidation in wet exhaust conditions. On the contrary, a difference in reactivity is observed in plasma coupled with palladium-based catalysts in the presence of water (Table 3.3). As already reported in the absence of plasma and in literature [72–75], a loss of activity was observed in the presence of water. However, for plasma–palladium systems, the influence of energy deposition becomes significant. Indeed, in comparison to catalytic systems alone, an increase in activity is observed in wet conditions, whereas no significant synergetic effect was observed in dry conditions. For Pd(0.5)/Al_2O_3 (GHSV = 40 000 h^{-1}), 50% of methane conversion is obtained at 390, 375, and 372 °C for E_d = 0, 36, and 58 $\mathrm{J\,l^{-1}}$, respectively. For both palladium catalysts, in wet conditions, the methane conversion increases on increasing the energy deposition. The negative effect of water on Pd catalysts, which can be explained by adsorption competition [72] or by the change of active phase [73, 74, 76], is partially drawn aside by the positive effect of water on plasma systems, which generates radicals, accelerating the CO_2 formation [70, 71]. In wet CHP conditions, the most efficient system for methane oxidation is the plasma + Pd(*X*)/Al_2O_3 system.

3.4.3 Conclusions

The methane total oxidation was studied as a function of temperature, by changing the energy deposition (E_d). At low-energy deposition, $E_d = 36\ J\,l^{-1}$, methane was converted from 300 °C. The methane conversion increased on increasing the energy deposition. However, for high temperatures, than for the high methane oxidation efficiency, NO_x were formed during the plasma process. Therefore, to minimize NO_x formation and to obtain the highest methane conversion, we carried out experiments at temperatures around 400 °C, but in the presence of low-energy deposition, $E_d = 36\ J\,l^{-1}$. The alumina alone, used as a catalyst, was active at 425 °C and above. The activity of this catalyst is then very poor. However, no NO_x were formed in the absence of plasma with the catalytic system alone. Finally, the plasma-catalytic oxidation of methane was studied in the presence of alumina. For the same energy deposition, the methane oxidation is favored in the presence of plasma + alumina system, in comparison to plasma or alumina alone. This better activity in the presence of plasma + alumina could be explained by the better reactivity of the activated species, induced by the plasma. In fact, the plasma created reactive species from methane, and then these species reacted with the catalyst, placed after the discharge, leading to higher methane conversion. We can conclude that the combination of plasma and alumina led to a higher activity for methane total oxidation than alumina or plasma used alone. However, the temperatures of total oxidation remain very high. To decrease the temperature reaction, the alumina catalyst, which had not presented a high activity, could be replaced by a typical methane oxidation catalyst. Thus, the methane oxidation reaction was studied in CHP experimental conditions using catalytic systems such as palladium or platinum catalysts, using plasma DBD alone with energy deposition of $36\ J\,l^{-1}$, or using plasma and catalyst systems. In the absence of water, palladium catalysts exhibit a superior catalytic activity in methane oxidation. Furthermore, the plasma-alumina-supported palladium system is as efficient as $Pd(X)/Al_2O_3$ alone. There is then no influence of plasma on methane oxidation for these systems, in the absence of water. In the presence of water in the feed, the methane oxidation rate decreases with the addition of water in the feed. Indeed, on noble-metal-based catalysts, the same methane conversion is obtained at temperatures 50 °C higher in wet conditions in comparison to dry mixtures. The loss of activity on water addition was then consistent with previous studies on Pd/Al_2O_3, indicating a strong inhibition by water vapor and a modification of surface active phase PdO. In the same experimental conditions, the platinum-based catalysts are inactive. For DBD plasma alone, it was found that the methane oxidation rate increases with the water addition and energy deposition. Finally, plasma–catalyst systems were studied in wet conditions. In the presence of plasma, the methane oxidation rate observed for plasma–$Pt(X)/Al_2O_3$ system only corresponds to the influence of energy deposition. Indeed, the presence of water creates radicals, which leads to a higher rate of methane oxidation. Moreover, in the presence of plasma, for both palladium catalysts, in wet conditions, the rate of methane conversion is enhanced significantly compared to the catalytic system alone. In these latter experiments, the negative effect of water on palladium catalysts is partially drawn aside by the positive effect of water on plasma systems. In wet CHP conditions, the most efficient system for methane oxidation is the plasma–Pd(*X*)/Al_2O_3 system, depending on the desired reaction temperature. These coupled plasma/catalysis systems could be a good alternative to decrease the amount of noble metal in the combustion of methane and more generally in HC oxidation processes.

3.5 NTP-Assisted Catalytic NO_x Remediation from Lean Model Exhausts Gases

3.5.1 Consumption of Oxygenates and RNO_x from Plasma during the Reduction of NO_x According to the Function F3: Plasma-Assisted Propene-deNO_x in the Presence of $Ce_{0.68}Zr_{0.32}O_2$

3.5.1.1 Conversion of NO_x and Total HC versus Temperature (Light-Off Plot)

In order to check the efficiency and the role of plasma on the deNO_x process by providing oxygenates and nitrocompounds to the catalyst from low temperature, the conversion of NO_x and total HC versus temperature has been conducted in stationary conditions, by successive isotherms. Results are reported in Figure 3.18; these plots correspond to the light-off of the catalyst for both NO_x and reductants (HC, RNO_x or $C_xH_yO_z$). The feed is a mixture of C_3H_6 (2000 ppmC), NO (500 ppm), O_2 (8% vol.), and N_2 (complement). When the plasma is ON an energy density of 36 J l^{-1} is provided. The NTP reactor is located before the catalyst reactor.

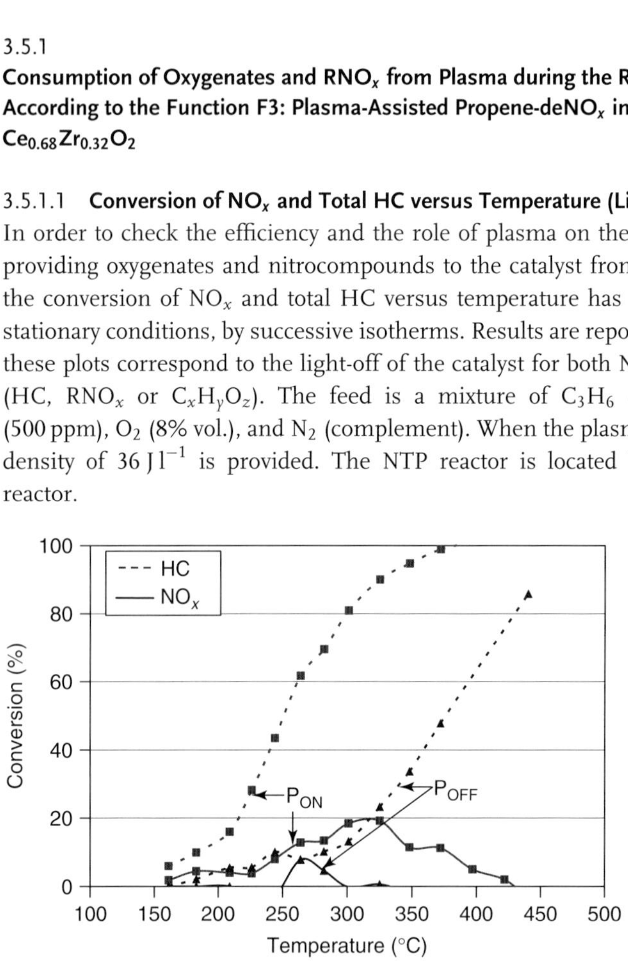

Figure 3.18 NO_x and total HC conversion versus temperature. Feed: C_3H_6 (2000 ppmC) NO (500 ppm), O_2 (8% vol.), and N_2 (complement); catalyst: $Ce_{0.68}Zr_{0.32}O_2$. GHSV: 93000 h^{-1}; P_{ON}: plasma on; P_{OFF}: plasma off.

- **deNO_x**. As in the case of three-way catalysis [1, 4], the support alone is active in NO reduction: 8% at 265 °C (Figure 3.18, full line, Poff), Ce^{4+} being the active site for the function F3.
  In the case of plasma–$Ce_{0.68}Zr_{0.32}O_2$ coupling, NO_x conversion is rising from 10% at 250 °C to 20% at 325 °C. Let us note that both conversion and temperature window are rising.
- **HC conversion**. "HC" stands for hydrocarbon, oxygenates, and nitrocompounds as detected by the FID specific detector. Figure 3.18 emphasizes the very strong effect of plasma on HC total oxidation to CO_2 (and H_2O, not measured): this very high conversion shows: (i) the simultaneity between deNO_x and the reducers oxidation in the same range of temperature from 150 °C; (ii) the simultaneity between the lack of HC from 325 °C and the decrease in deNO_x above this temperature; and (iii) the kinetic coupling of HC-deNO_x and HC oxidation catalytic cycles in accordance with the function F3 of the model (Figure 3.5).

3.5.1.2 GC/MS Analysis

A strong propene conversion to oxygenates and nitrocompounds in plasma alone is observed at 265 °C (Figure 3.19). Table 3.4 and Figure 3.19a report the major compounds detected at the outlet of the plasma reactor in the absence of catalyst.

Figure 3.19b shows that only propene, propane, methyl nitrite, acetaldehyde, acetonitrile, and methyl nitrate are detected in the presence of $Ce_{0.68}Zr_{0.32}O_2$ at 265 °C.

- *Drastic consumption of HC species in the presence of a catalyst at* 265 °C, *corresponding to* 15% *deNO_x*.
- Comparison between Figures 3.19a,b shows that oxygenates from the plasma reactor such as acetaldehyde, methylformate, propylene oxide, propanal, acetone, and propenol have been consumed. Only traces of acetaldehyde are detected.

Table 3.4 Main species detected at the outlet of the plasma reactor (36 J l^{-1}) at 265 °C from C_3H_6 (2000 ppmC), NO (500 ppm), O_2 (8% vol.), and N_2.

HC and oxygenates	N-containing compounds
Propene	Methyl nitrite
Acetaldehyde	Acetonitrile
Methyl	Nitroso
Formate	Methylamine
Propylene	Methyl nitrate
Oxide	Nitromethane
Propenal	Ethyl nitrate
Propenol	–

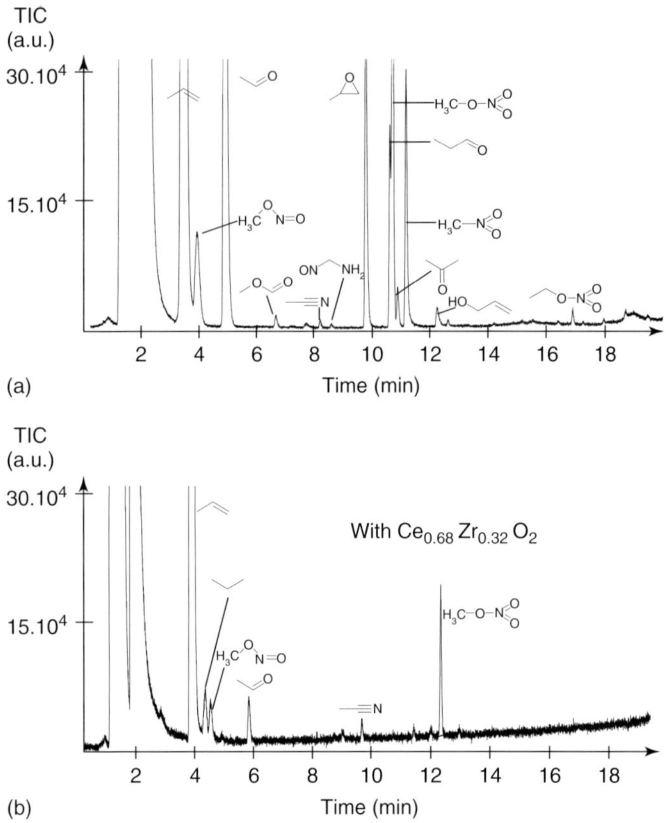

Figure 3.19 GC/MS of species from C_3H_6 (2000 ppmC), NO (500 ppm), O_2 (8% vol.), and N_2, in the presence of plasma (36 J l^{-1}) (a) without catalyst and (b) with $Ce_{0.68}Zr_{0.32}O_2$ at 265 °C.

Furthermore, the nitro methane (RNO_x type in the model) has completely disappeared, in concordance with both functions F2 and F3 of the $deNO_x$ model.

- As a conclusion to this first case, oxygenates $C_xH_yO_z$ coming from the initial propene (HC) in the plasma reactor are probably the reductant species of $deNO_x$ reaction during coupling of the "plasma – $Ce_{0.68}Zr_{0.32}O_2$" reactors.

3.5.2
The NTP is Able to Significantly Increase the $deNO_x$ Activity, Extend the Operating Temperature Window while Decreasing the Reaction Temperature

Propene being a common reductant, plasma-assisted propene-$deNO_x$ over alumina will be used for the sake of demonstration.

3.5.2.1 TPD of NO for Prediction of the $deNO_x$ Temperature over Alumina without Plasma

As shown in Section 3.3 on the three-function model, the theory is based on the temperature at which NO is activated and can either desorb or dissociate to N_{ads} and O_{ads} [2–4] (Figure 3.20).

One of the most interesting results of Penetrante and coworkers [6] is the coupling of an NTP reactor with a $deNO_x$ one *containing alumina alone*. The explanation of their results is completely in accordance with the concept of O_{ads} scavenging by HCs or oxygenates that are able to suffer their total oxidation at the temperature of NO activation (desorption/dissociation).

It has been shown by Djéga-Mariadassou [3] that a TPD of preadsorbed NO at room temperature can lead to the temperature of $deNO_x$ process.

The NO adsorption has been made by flowing NO (150 ppm), O_2 (8% vol.), and N_2 (complement). It is followed by TPD/10 °C min^{-1} in a flowing O_2-N_2 mixture. (Let us note that adsorption/TPD is still better for the prediction of $deNO_x$ temperature without adding O_2 that provokes the simultaneous NO oxidation to NO_2, a parallel reaction as there is no oxygen as reagent in the $deNO_x$ function F3 cycle.)

Figure 3.20 shows the two domains where $deNO_x$ is able to occur: the first NO desorption is observed at 255 °C and is referred to as the expected "low-temperature" $deNO_x$, in contrast to the "high-temperature" one, occurring at 485 °C. The low-temperature NO_2 comes from the NO oxidation to NO_2 on alumina, whereas the NO_2 peak observed at high temperature comes from the surface nitrate species dissociation and NO_2 desorption.

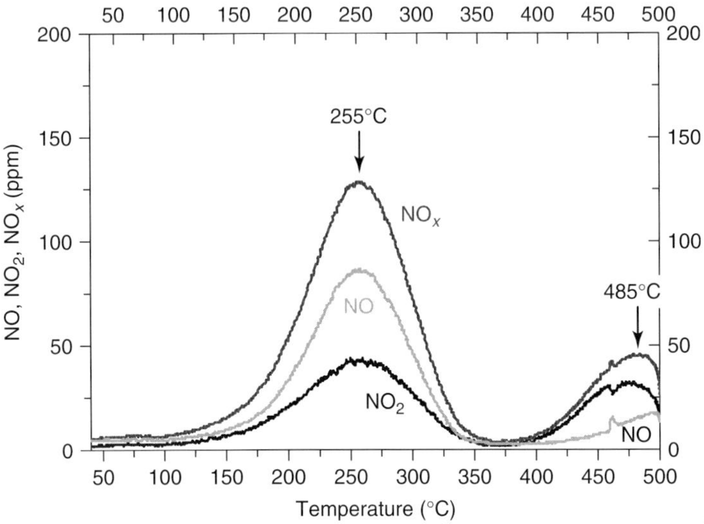

Figure 3.20 NO, NO_2, and NO_x (NO + NO_2) evolution during TPD of NO in the presence of O_2, as described in the text, over alumina precalcined for 2 h at 500 °C.

This figure shows that two domains of deNO_x can exist over this alumina, 125–350 and 400–500 °C, as long as O_{ads} left by NO dissociation is able to be scavenged by a reductant that oxidized to $CO_2 + H_2O$.

3.5.2.2 **Coupling of a NTP Reactor with a Catalyst (Alumina) Reactor for Catalytic-Assisted deNO_x**

The experiment has been carried out by flowing a C_3H_6–NO–O_2–N_2 mixture in the two successive reactors (Figure 3.7). Figure 3.21 shows the conversion of global HC detected by a specific FID detector (without discrimination of the nature of compounds), HC being defined as propene, oxygenates $C_xH_yO_z$, and nitrogen-containing species. Conversion of NO_x versus temperature is also reported.

- *In the absence of plasma,* Al_2O_3 only begins to be active in deNO_x at high temperatures, starting at 350 °C, with a NO_x conversion of about 10% above 425 °C (Figure 3.21).
- Furthermore, HCs are oxidized to CO_2/H_2O from 270 °C, that is, before NO activation on alumina.

  In accordance with the model, the deNO_x process is poor (low turnover of the function F3 catalytic cycle) because of the lack of reducer at high temperature: the four kinetically coupled catalytic cycles of the function F3 model turn over, leading to a high competition of oxidation of propene and oxygenates for scavenging O_{ads} from NO and O_{ads} from O_2 on the same active site.
- *In the presence of plasma,* NO_x conversion starts at a low temperature, *simultaneously* with the total oxidation of HC to CO_2/H_2O: the function F3 of deNO_x model turns over. The NO_x conversion is higher than 20% between 200 and 360 °C, 40% being obtained at 300 °C.

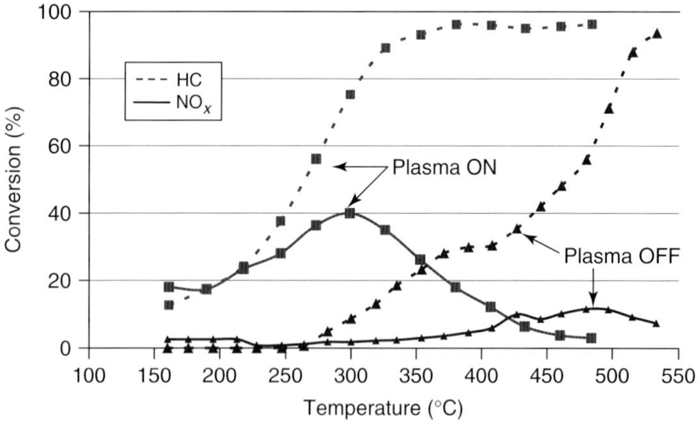

Figure 3.21 NO_x and global HC conversions versus temperature. Successive isotherms. Feed: NO (500 ppm), C_3H_6 (2000 ppmC), O_2 (8% vol.), and N_2; catalyst: Al_2O_3; GHSV: 54 000 h^{-1}.

This behavior clearly demonstrates the role of the NTP as a promoter of the deNO$_x$ reaction. Comparing Figure 3.21 (deNO$_x$) to Figure 3.20 (TPD of NO), the NTP delivers active reducers (oxygenates) to alumina and activates the NO reduction at the temperature of the first NO desorption peak observed in Figure 3.20.

As a conclusion to this section, it can be seen that the NTP is activating the low-temperature deNO$_x$ function of alumina while providing a wide operating temperature window for deNO$_x$ reaction.

To get a larger broad temperature window for deNO$_x$ process, a "composite" catalyst system can be used, as seen in next section.

3.5.3
Concept of a "Composite" Catalyst Able to Extend the deNO$_x$ Operating Temperature Window

In a review of the selective reduction of NO_x with HCs under lean-burn conditions, Burch *et al.* [77] have concluded that the main problems are related to complex real exhaust emissions and temperature range, which is wider for an engine test than for a laboratory one.

A "unit" catalyst (with a single cation or oxide) generally works in a narrow range of temperature. Its activity in the deNO$_x$ process is limited by the reductant oxidation and concentration, as well as by the temperature of NO activation (dissociation/reduction).

To extend the temperature range of the catalytic material, a *composite catalyst* can be designed associating unit catalysts working in different temperature ranges. These unit catalysts have to be selected according to the nature of reductants: propene, propane, *n*-decane, and toluene are studied in this chapter. A narrow overlap of unit catalysts will be useful to get a continuous deNO$_x$ activity of the composite catalyst. The light-off of unit catalysts should extend the range of deNO$_x$ activity.

- *Scheme of the composite catalyst.*
  A composite catalyst device is presented in Figure 3.22. The unit catalysts have to turn over successively in the temperature range. Cat 1 is active at high temperatures and should be the first in the composite device [53]; Cat 3 turns over at low temperatures and will be the last one for avoiding the reductant oxidation necessary for Cat 1 to work at high temperatures.

As a consequence, catalytic cycles of unit catalysts will be classified from three to one in the temperature scale.

The unit catalyst working at high temperatures (around 400 °C) will be alumina (as found in the absence of plasma in Section 5.3.2). $Ag/Ce_{0.68}Zr_{0.2}O_2$ has been selected for its activity at low temperatures [53]. Finally $Rh{-}Pd/Ce_{0.68}Zr_{0.32}O_2$ [78] has been found to be working at intermediate temperatures. The composite catalyst used for the next section is summarized in Table 3.5.

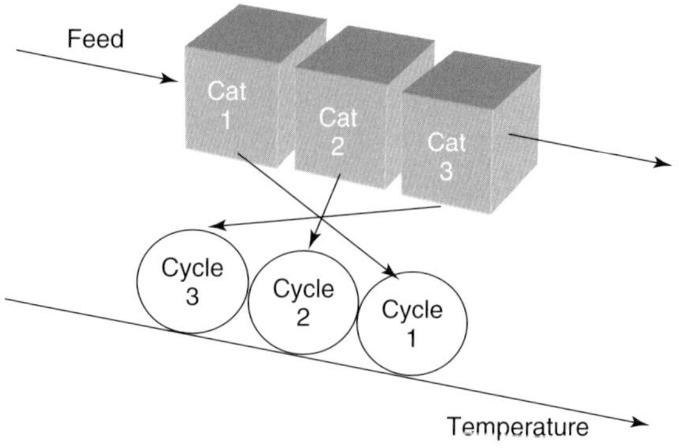

Figure 3.22 Scheme of a composite catalyst. "Cat" 1–3 are three unit catalysts. Cycles 1–3 represent the turn over of each catalyst unit versus temperature.

Table 3.5 Design of the composite catalyst.

C_1	C_2	C_3
Al_2O_3	$Rh-Pd/Ce_{0.68}Zr_{0.32}O_2$	$Ag/Ce_{0.68}Zr_{0.2}O_2$

3.5.4 Propene-deNO_x on the "Al_2O_3 /// $Rh-Pd/Ce_{0.68}Zr_{0.32}O_2$ /// $Ag/Ce_{0.68}Zr_{0.32}O_2$" Composite Catalyst

3.5.4.1 NO_x and C_3H_6 Global Conversion versus Temperature

Experiments were conducted with successive isotherms in stationary conditions (Figure 3.23). The composite catalyst was precalcined in flowing synthetic air, with a GHSV = 54000 h^{-1}. *Propene* was the reductant in the following feed composition: C_3H_6 (2000 ppmC) NO (500 ppm), O_2 (8% vol.), and N_2 (balance).

- *In the absence of plasma,* two conversion peaks are observed (Figure 3.23) with 24 and 18% conversion, at 270 and 485 °C, respectively. Considering previous studies on unit catalysts, the first peak can be attributed to $Rh-Pd/Ce_{0,68}Zr_{0,32}O_2$ /// $Ag/Ce_{0,68}Zr_{0,32}O_2$ and the second peak to Al_2O_3.
- *In the presence of plasma,* NO_x conversion higher than 50% is observed in the 240–275 °C range of temperature.

In both cases, global propene is oxidized from 150 °C, simultaneously with the two deNO_x peaks. Nevertheless, for the second peak at high temperature, the propene oxidation – due to the deNO_x function F3 of the composite catalyst – is competing with that due to the oxidation by dioxygen (see the function F3 model with four kinetically coupled catalytic cycles in Section 3.2).

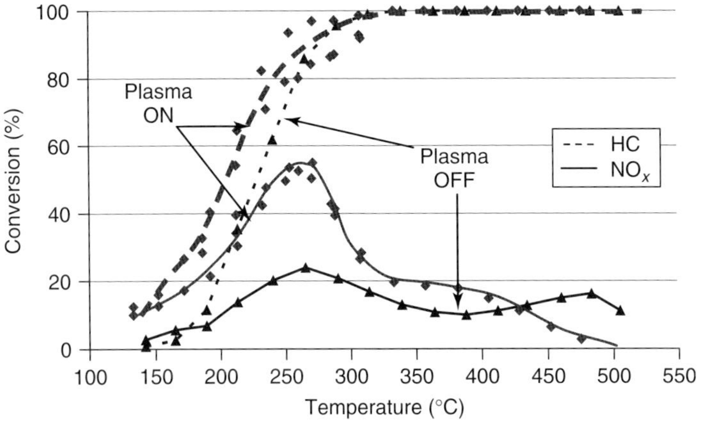

Figure 3.23 NO_x and total HC conversion versus temperature. Feed: C_3H_6 (2000 ppmC) NO (500 ppm), O_2 (8% vol.), and N_2 (complement). Composite catalyst: Al_2O_3 /// Rh–Pd/$Ce_{0.68}Zr_{0.32}O_2$ /// Ag/$Ce_{0.68}Zr_{0.32}O_2$; GHSV: 93000 h^{-1}. With or without plasma (E_d = 36 $J\,l^{-1}$).

The NTP leads to a 30% higher deNO_x activity of the composite catalyst, the range of temperature being relatively broader.

3.5.4.2 GC/MS Analysis of Gas Compounds at the Outlet of the Catalyst Reactor

In the presence of plasma, oxygenates ($C_xH_yO_z$) are due to the plasma interaction with propene, whereas without plasma, oxygenates (ex-RNO_x) are delivered by the catalyst and responsible for the deNO_x reaction only at high temperatures (Figure 3.24a,b).

- *In the presence of plasma at* 120 °C, the Figure 3.24a shows the presence of propene, methyl nitrite, acetaldehyde, methyl formate, acetonitrile, nitrosomethylamine, propenal, propanal, methyl nitrate, acetone, nitromethane, and methyl propenal.

3.5.5 NTP Assisted Catalytic deNO_x Reaction in the Presence of a Multireductant Feed: NO (500 ppm), Decane (1100 ppmC), Toluene (450 ppmC), Propene (400 ppmC), and Propane (150 ppmC), O_2 (8% vol), Ar (Balance)

3.5.5.1 Conversion of NO_x and Global HC versus Temperature

Without plasma, NO_x conversion is about 40% at 250 °C. The promoter effect of NTP leads to 50–55% in the 230–280 °C temperature range (Figure 3.25). For the sake of deNO_x evidence, nitrogen has been measured using a micro-gas chromatograph (μ-GC). Let us note that the NO_x consumption quantitatively corresponds to the formation of N_2 above 200 °C. The composite catalyst is able to proceed to the deNO_x reaction in the presence of the complex gas mixture.

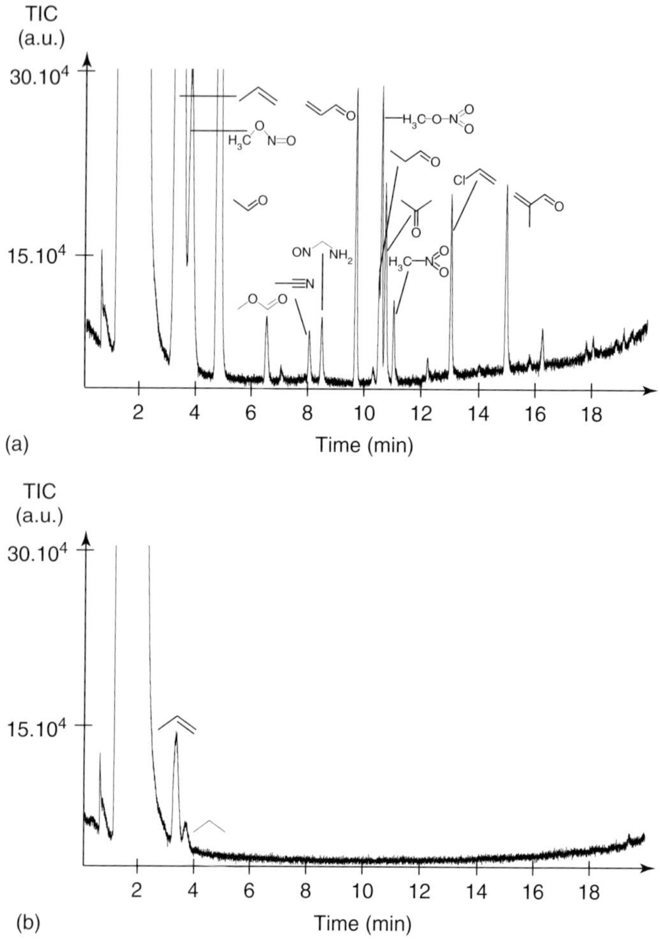

Figure 3.24 GC/MS of products from C_3H_6 (2000 ppmC), NO (500 ppm), O_2 (8% vol.), and N_2, *in the presence of plasma* (36 J l^{-1}), with composite catalyst: Al_2O_3 /// Rh–Pd/$Ce_{0.68}Zr_{0.32}O_2$ /// Ag/$Ce_{0.68}Zr_{0.32}O_2$. (a) $T = 120\,^{\circ}C$ and (b) $T = 260\,^{\circ}C$.

3.5.5.2 GC/MS Analysis of Products at the Outlet of Associated Reactors

- *In the absence of a catalyst,* Figure 3.26 shows the numerous species formed in the plasma and reported in Table 3.6. Baudin [53] has shown this result to be in agreement with the GC/MS of pure HCs of the gas mixture.
- *In the presence of the composite catalyst* (Figure 3.27), the GC/MS analysis only detected chloromethane, propene, chloroethane, methyl nitrate, benzene, toluene, styrene, benzaldehyde, acetophenone, and decane. It has to be noted that all $C_xH_yO_z$ and nonaromatic RNO_x have been completely consumed: *these species produced by plasma are the reductants of the deNO_x reaction.*

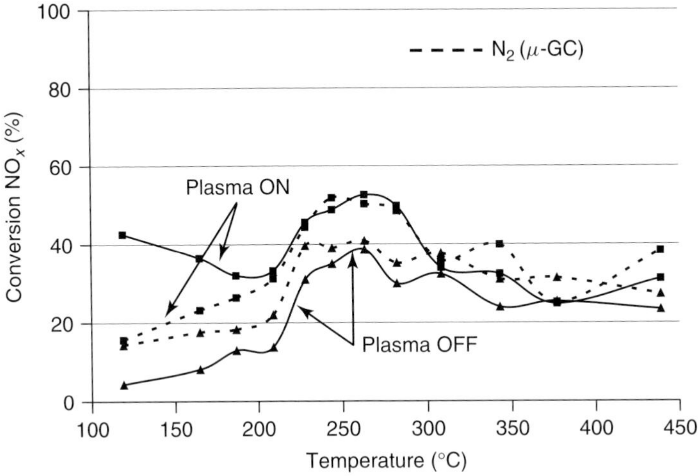

Figure 3.25 NO_x conversion and N_2 formation versus temperature. Feed: NO (500 ppm), decane (1100 ppmC), toluene (450 ppmC), propene (400 ppmC), propane (150 ppmC), O_2 (8% vol.), and Ar. Composite catalyst: Al_2O_3/// Rh–Pd/$Ce_{0.68}Zr_{0.32}O_2$ /// Ag/$Ce_{0.68}Zr_{0.32}O_2$ calcined for 2 h at 500 °C, in flowing synthetic air (GHSV: 93 000 h^{-1}); plasma $E_d = 36\ J\,l^{-1}$.

Table 3.6 List of compounds detected by GC/MS (Figure 3.26) classified as HC-$C_xH_yO_z$ and N-containing compounds.

Methanol	Methyl nitrite
Propane	Nitromethane
Methyl formate	Nitrosomethylamine
Ethanol	Methyl nitrate
Acetonitrile	Nitromethane
Propylene oxide	Ethane nitrile
Propanal	C_4–RNO_x
Acetone	
$C_3H_yO_z$	
$C_4H_yO_z$	
$C_5H_yO_z$	
Toluene	
$C_6H_yO_z$	
$C_7H_yO_z$	
Styrene	
Benzaldehyde	
Decane	

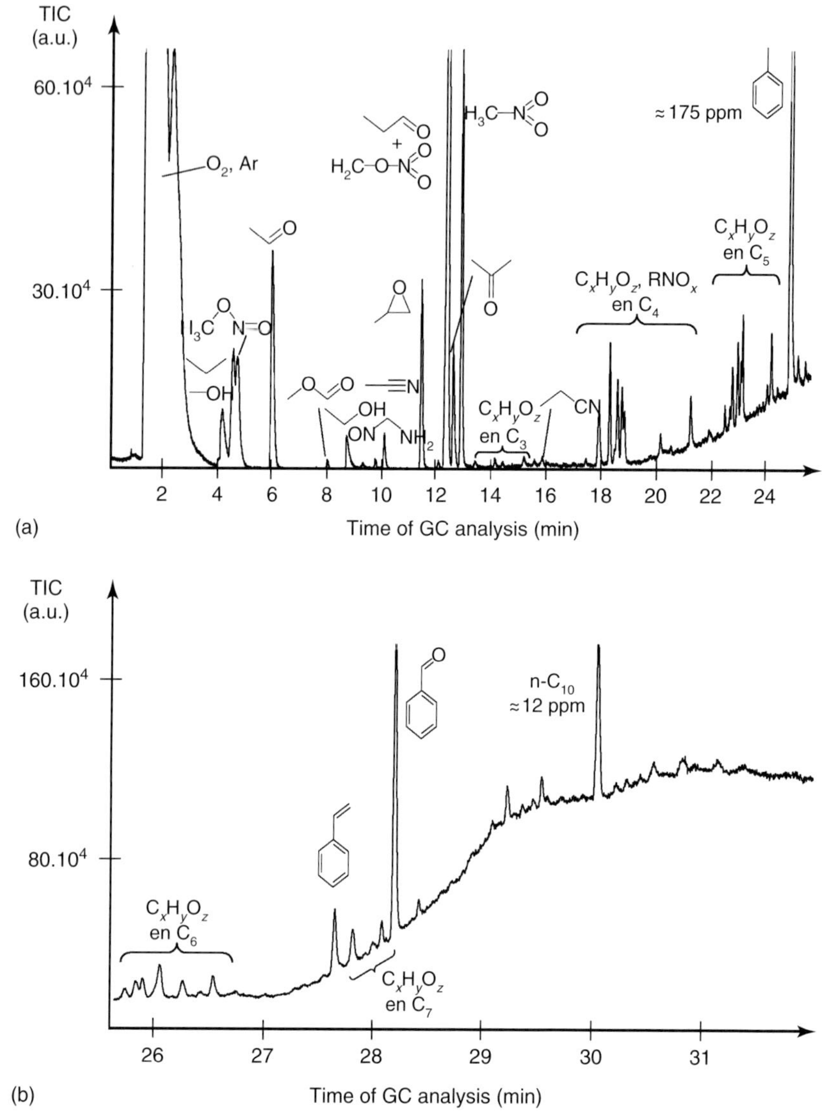

Figure 3.26 GC/MS analysis of species at the outlet of plasma reactor (36 J l^{-1}), in the absence of a catalyst. (Total ion current (TIC) vs time of analysis, in two separated parts (a,b) for the sake of presentation.) Feed: NO (500 ppm), decane (1100 ppmC), toluene (450 ppmC), propene (400 ppmC), propane (150 ppmC), O_2 (8% vol.), and Ar; plasma $E_d = 36$ J l^{-1}; $T = 235\,^{\circ}C$.

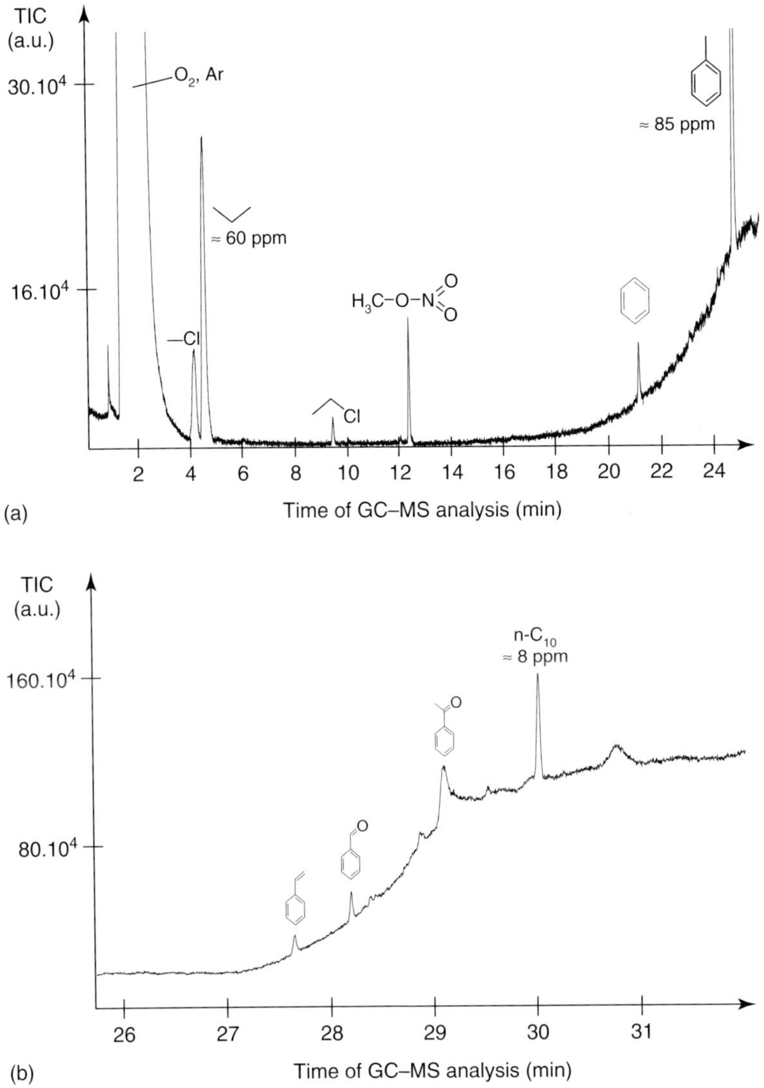

Figure 3.27 GC/MS analysis of species at the outlet of the plasma reactor (36 $J\,l^{-1}$), *in the presence of a catalyst*. (Total ion current (TIC) vs time of analysis, in two separated parts (a,b) for the sake of presentation.) Feed: NO (500 ppm), decane (1100 ppmC), toluene (450 ppmC), propene (400 ppmC), propane (150 ppmC), O_2 (8% vol.), and Ar. Composite catalyst: Al_2O_3/// Rh−Pd/$Ce_{0.68}Zr_{0.32}O_2$ /// Ag/$Ce_{0.68}Zr_{0.32}O_2$ calcined for 2 h at 500 °C, in flowing synthetic air (GHSV: 93000 h^{-1}); plasma $E_d = 36\ J\,l^{-1}$; $T = 230\,°C$.

3.6 Conclusions

Associated with the NTP theory, the three-function catalyst model developed in this chapter concerns supported and well-dispersed metal cations on oxides.

Different cases of deNO_x processes have been derived from the general model.

- They all depend on the temperature (T_{NO}) of NO dissociation/reduction to O_{ads} and N_2 release in the gas phase.
- The different cases are mainly linked to the relative position of T_{NO} in regard to the temperature (T_{HC}) of the oxidation (to CO_2/H_2O) of reductants (HC or $C_xH_yO_z$).
- More particularly, when the temperature of the reductant is higher than that of T_{NO} (very often occurring experimentally), the three functions have to turn over simultaneously at T_{NO}: (i) oxidation of NO to NO_2; (ii) mild oxidation of HC to $C_xH_yO_z$ (oxygenates); and (iii) NO reduction assisted by the oxidation of reductants to CO_2/H_2O.
- The catalyst has to simultaneously produce , by itself, all these reactions.

The NTP reactor has been shown to play several roles.

- The NTP plays the role of two of the three functions defined in the catalysis model when coupling the plasma reactor to the catalytic deNO_x device (as far the catalyst has been designed in order to have the three functions!):
  - the conversion of NO to NO_2;
  - the production of oxygenated compounds $C_xH_yO_z$ and organic nitrogen-containing intermediates RNO_x that are able to decompose to oxygenates.
- It has been shown that in addition to NO_2, plasma processing of a mixture containing O_2–NO–H_2O–HC–N_2 leads to the production of aldehydes (CH_3CHO, CH_2O), alcohols (CH_3OH), nitrate and intermediate organic nitroso compounds of R–NO_x type (CH_3ONO, CH_3ONO_2), nitro methane (CH_3NO_2), formic acid (CH_2O_2), propylene oxide (C_3H_6O), and to some extent of acids (HNO_2, HNO_3). Formations of such molecules are predicted by kinetic models of NTP.
- The NTP provides these active species to the catalyst and allows the extension of the temperature windows of deNO_x reaction from room temperature.
- Furthermore, it has been found that UHCs of gas exhausts of diesel engines prevent the oxidation of SO_2, thus making the plasma-catalytic process tolerant to the sulfur content of the fuel.
- DBD processing being a very mature technology, it is now routinely used in industrial applications and could be adapted to exhaust gas deNO_x treatments.

The coupling of NTP-catalytic reactor has been illustrated comparing three kinds of data:

- The light-off curves of catalytic reactor NTP off and on;
- The NTP activity on reaction mixtures;
- The NTP-catalytic reactor coupling.

Some deNO_x experiments have been selected to investigate gas mixtures with more and more complex compositions. The goal was to produce synthetic gas exhausts simulating diesel and lean-burn gasoline engine exhausts. Typically, the N_2-based mixture consists of O_2, H_2O, NO, and HCs (C_3H_6, C_3H_8, n-C_{10}, and toluene). In all cases, the plasma reactor has been shown to deliver to the catalyst reactor a rich mixture of NO_2, oxygenates, and intermediate organic nitroso compounds, *completely consumed* during the deNO_x catalytic process and leading to efficient NO_x abatements.

"Composite" catalysts have been shown to extend the deNO_x operating temperature window in the presence of NTP coupling.

The particular case of plasma-catalytic oxidation of methane was studied in the presence of alumina. This better activity in the presence of plasma + alumina could be explained by the better reactivity of the activated species induced by the plasma. The combination of plasma and alumina led to a higher activity for methane total oxidation than alumina or plasma used alone. In the presence of plasma, the methane oxidation rate observed for the plasma–Pt(X)/Al_2O_3 system only corresponds to the influence of energy deposition. Indeed, the presence of water creates radicals, which leads to a higher rate of methane oxidation. Moreover, in the presence of plasma, for both palladium catalysts, in wet conditions, the rate of methane conversion is enhanced significantly compared to the catalytic system alone. In these latter experiments, the negative effect of water on palladium catalysts is partially drawn aside by the positive effect of water on plasma systems. In wet CHP conditions, the most efficient system for methane oxidation is the plasma–Pd(X)/Al_2O_3 system, depending on the desired reaction temperature. These coupled plasma/catalysis systems could be a good alternative to decrease the amount of noble metal in the combustion of methane and more generally in the HC oxidation processes.

Acknowledgments

The authors greatly acknowledge all engineers from industries (PSA Peugeot-Citroën, Renault, GDF Suez, ADEME) who worked with the "Laboratoire Réactivité de Surface," and "Laboratoire GREMI" as well as *Michel Boudart, Jean Michel Pouvesle,* and *Jean Marie Cormier* for many fruitful discussions on deNO_x and plasma physics and chemistry.

References

1. Djéga-Mariadassou, G., Fajardie, F., Tempère, J.F., Manoli, J.M., Touret, O., and Blanchard, G. (2000) A general model for both three-way and deNO_x catalysis: dissociative or associative nitric oxide adsorption, and its assisted decomposition in the presence of a reductant. *J. Mol. Catal. A: Chem.*, **161**, 179–189.
2. Djéga-Mariadassou, G. and Boudart, M. (2003) Classical kinetics of catalytic reactions. *J. Catal.*, **216**, 89–97.
3. Djéga-Mariadassou, G. (2004) From three-way to DeNO_x catalysis: a general model. *Catal. Today*, **90**, 27–34.
4. Djéga-Mariadassou, G., Berger, M., Gorce, O., Park, J.W., Pernot, H.,

Potvin, C., Thomas, C., and Da Costa, P. (2007) in *Past and Present in deNO$_x$ Catalysis*, Studies in Surface Science and Catalysis, Vol. **171**, 1st edn (eds P. Granger and V.I. Pârvulescu), Elsevier, The Netherlands, pp. 145–173.
5. Penetrante, B.M. and Schultheis, S.E. (eds) (1993) *Non-thermal Plasma Techniques for Pollution Control (Part A and B)*, Springer, New York.
6. Penetrante, B.M., Brusasco, R.M., Merrit, B.T., Pitz, W.J., Vogtlin, G.E., Kung, M.C., Kung, H.H., Wan, C.Z., and Voss, K.E. (1998) Plasma-Assisted Catalytic Reduction of NOx. SAE Technical Paper 982508, pp. 57–66.
7. Hoard, J. and Balmer, M.L. (1998) Analysis of Plasma-Catalysis for Diesel NOx Oxidation. SAE Technical Paper 982429, pp. 13–19.
8. Balmer, M.L., Tonkin, R., Yoon, S., Kolwaite, A., Barlow, S., Maupin, G., and Hoard, J. (1999) NOx Destruction Behaviour of Select Materials when Combined with a Non-Thermal Plasma. SAE Technical Paper 1999-01-3640, pp. 67–73.
9. Higashi, M., Uchida, S., Suzuki, N., and Fujii, K. (1992) Soot elimination and NOx and SOx reduction in diesel-engine exhaust by a combination of discharge plasma and oil dynamics. *IEEE Trans. Plasma Sci.*, **20** (1), 1–12.
10. Penetrante, B.M., Brusasco, R.M., Merrit, B.T., Pitz, W.J., and Vogtlin, G.E. (1999) Feasability of Plasma Aftertreatment for Simultaneous Control of NOx and Particulates. SAE Technical Paper 1999-01-3637, pp. 45–50.
11. Masuda, S. and Nakao, H. (1990) Control of NOx by positive and negative pulsed corona discharges. *IEEE Trans. Ind. Appl.*, **26** (2), 374–383.
12. Dhali, S.K. and Sardja, I. (1991) Dielectric barrier discharge for processing of SO_2/NOx. *J. Appl. Phys.*, **69** (9), 6319–6324.
13. Penetrante, B.M., Hsiao, M.C., Bardsley, J.N., Merrit, B.T., Vogtlin, G.E., Wallman, P.H., Kuthi, A., Burkhart, C.P., and Bayless, J.R. (1996) Electron beam and pulsed corona processing of volatile organic compounds in gas streams. *Pure Appl. Chem.*, **68** (5), 1083–1087.
14. Evans, D., Rosocha, L.A., Anderson, G.K., Coogan, J.J., and Kushner, M.J. (1993) Plasma remediation of trichloroethylene in silent discharge plasmas. *J. Appl. Phys.*, **74** (9), 5378–5386.
15. Frank, N.W. (1995) Introduction and historical review of electron beam processing for environmental pollution control. *Radiat. Phys. Chem.*, **45** (6), 989–1002.
16. Okumoto, M. and Mizuno, A. (2001) Conversion of methane for higher hydrocarbon fuel synthesis using pulsed discharge plasma method. *Catal. Today*, **71** (1–2), 211–217.
17. Jiang, T., Li, Y., Liu, C.J., Xu, G., Eliasson, B., and Xue, B. (2002) Plasma methane conversion using dielectric-barrier discharges with zeolite A. *Catal. Today*, **72** (3–4), 229–235.
18. Cho, W., Baek, Y., Moon, S.K., and Kim, Y.C. (2002) Oxidative coupling of methane with microwave and RF plasma catalytic reaction over transitional metals loaded on ZSM-5. *Catal. Today*, **74** (3–4), 207–223.
19. Heintze, M. and Pietruszka, B. (2004) Plasma catalytic conversion of methane into syngas: the combined effect of discharge activation and catalysis. *Catal. Today*, **89** (1–2), 21–25.
20. Pietruszka, B., Anklam, M., and Heintze, M. (2004) Plasma-assisted partial oxidation of methane to synthesis gas in a dielectric barrier discharge. *Appl. Catal., A*, **261** (1), 19–24.
21. Penetrante, B.M., Brusasco, R.M., Meritt, B.T., Pitz, W.J., Volgtlin, G.E., Kung, M.C., Kung, H.H., Won, C.Z., and Voss, K.E. (1998) Plasma-Assisted Catalytic Reduction of NOx. SAE Technical Paper 982508, pp. 57–66.
22. Marques, R., Da Costa, S., and Da Costa, P. (2008) Plasma-assisted catalytic oxidation of methane: on the influence of plasma energy deposition and feed composition. *Appl. Catal., B*, **82** (1–2), 50–57.
23. Da Costa, P., Marques, R., and Da Costa, S. (2008) Plasma catalytic oxidation of methane on

alumina-supported noble metal catalysts. *Appl. Catal., B*, **84** (1–2), 214–222.
24. Hueso, J.L., Caballero, A., Cotrino, J., and Gonzalez-Elipe, A.R. (2007) Plasma catalysis over lanthanum substituted perovskites. *Catal. Commun.*, **8** (11), 1739–1742.
25. Baudin, F., Da Costa, P., Thomas, C., Calvo, S., Lendresse, Y., Schneider, S., and Djéga-Mariadassou, G. (2004) NO_x reduction over $CeZrO_2$ supported iridium catalyst in the presence of propanol. *Top. Catal.*, **30-31**, 97–101.
26. Doraï, R. and Kushner, M.J. (1999) SAE Technical Paper 01-3683.
27. Van Veldhuizen, E.M. (ed.) (2000) *Electrical Discharge for Environment Purpose. Fundamentals and Applications*, Nova Science Publishers Inc., New York.
28. Kim, H.H. (2004) Non-thermal plasma processing for air-pollution control: A historical review, current issues, and future prospects. *Plasma Processes Polym.*, **1** (2), 91–110.
29. Chang, J.S., Lawless, P.L., and Yamamoto, T. (1991) *IEEE Trans. Plasma Sci.*, **19**, 1152.
30. Eliasson, B. and Kogelschatz, U. (1991) Modeling and applications of silent discharge plasmas. *IEEE Trans. Plasma Sci.*, **19** (2), 309–323.
31. Eliasson, B. and Kogelschatz, U. (1991) Non-equilibrium volume plasma chemical processing. *IEEE Trans. Plasma Sci.*, **19** (6), 1063–1077.
32. Falkenstein, Z. and Coogan, J.J. (1997) Microdischarge behavior in the silent discharge of nitrogen-oxygen and water-air mixtures. *J. Phys. D: Appl. Phys.*, **30** (5), 817–825.
33. Shin, H.H. and Yoon, W.S. (2000) Effect of Hydrocarbons on the Promotion of NO-NO2 Conversion in Non-Thermal Plasma deNOx Treatment. SAE Technical Paper 2000-01-2969, pp. 103–110.
34. Khacef, A., Cormier, J.M., and Pouvesle, J.M. (2002) NO_x remediation in oxygen-rich exhaust gas using atmospheric pressure non-thermal plasma generated by a pulsed nanosecond dielectric barrier discharge. *J. Phys. D: Appl. Phys.*, **35** (13), 1491–1498.
35. Dorai, R. and Kushner, M.J. (1999) Effect of Propene on the Remediation of NOx from Engine Exhausts. SAE Technical Paper 1999-01-3683, pp. 81–87.
36. Filimonova, E.A., Kim, Y.H., Hong, S.H., and Song, Y.H. (2002) Multiparametric investigation on NOx removal from simulated diesel exhaust with hydrocarbons by pulsed corona discharge. *J. Phys. D: Appl. Phys.*, **35** (21), 2795–2807.
37. Balmer, M.L., Fisher, G, and Hoard, J. (eds) (1999) *Non-Thermal Plasma for Exhaust Emission Control*, SAE Special Publication SP 1483, SAE, Warrendale.
38. Servati, H. and Hoard, J. (eds) (1998) *Plasma Exhaust Aftertreatment*, SAE Special Publication SP 1395, SAE, Warrendale.
39. Balmer,M.L., Fisher, G., and Hoardn, J. (eds) (2000) *Non-Thermal Plasma*, SAE Special Publication SP 1566, SAE, Warrendale.
40. Lampert, J.K. (2000) An Assessment of the Plasma Assisted Catalytic Reactor (PACR) Approach to Lean NOx Abatement: The Relative Reducibility of NO and NO2 Using #2 Diesel Fuel as the Reductant. SAE Technical Paper 2000-01-2962, pp. 67–72.
41. Balmer, M.L., Tonkyn, R.G., Kim, A.Y., Yoon, I.S., Jimenez, D., Orlando, T., Barlow, S.E., and Hoard, J. (1998) Diesel NOx Reduction on Surfaces in Plasma. SAE Technical Paper 982511, pp. 73–78.
42. Quoc An, H.T., Pham Huu, T., Le Van, T., Cormier, J.M., and Khacef, A. (2011) Application of atmospheric non-thermal plasma-catalysis hybrid system for air pollution control: toluene removal. *Catal. Today*, doi: 10.1016/j.cattod. 2010.10.005
43. Van Durme, J., Dewulf, J., Leys, C., and Van Langenhove, H. (2008) Combining non-thermal plasma with heterogeneous catalysis in waste gas treatment: a review. *Appl. Catal., B Environ.*, **78**, 324–333.
44. Blin-Simiand, N., Tardiveau, P., Risacher, A., Jorand, F., and Pasquiers, S. (2005) Removal of 2-heptanone by dielectric barrier discharges – the effect of a catalyst support. *Plasma Process. Plasma Polym.*, **2** (3), 256–262.

45. Khacef, A., Viladrosa, V., Cachoncinlle, C., Robert, E., and Pouvesle, J.M. (1997) High repetition rate compact source of nanosecond pulses of 5-100 keV x-ray photons. *Rev. Sci. Instrum.*, **68** (6), 2292–2297.
46. Pasquiers, S. (2004) Removal of pollutants by plasma catalytic processes. *Eur. Phys. J. Appl. Phys.*, **28** (3), 319–324.
47. Lombardi, G., Blin-Simiand, N., Jorand, F., Magne, L., Pasquiers, S., Postel, C., and Vacher, J.R. (2007) Production and reactivity of the hydroxyl radical in homogeneous high pressure plasmas of atmospheric gases containing traces of light olefins. *Plasma Chem. Plasma Process.*, **27** (4), 414–445.
48. Dorai, R. and Kushner, M.J. (2003) Consequences of unburned hydrocarbons on microstreamer dynamics and chemistry during plasma remediation of NOx using dielectric barrier discharges. *J. Phys. D: Appl. Phys.*, **36** (9), 1075–1083.
49. Martin, A.R., Shawcross, J.T., and Whitehead, J.C. (2004) *J. Phys. D: Appl. Phys.*, **37**, 42.
50. Dorai, R. and Kushner, M.J. (2000) Consequences of propene and propane on plasma remediation of NO_X. *J. Appl. Phys.*, **88** (6), 3739–3747.
51. Niessen, W., Wolf, O., Schruft, R., and Neiger, M. (1998) The influence of ethene on the conversion of NOx in a dielectric barrier discharge. *J. Phys. D: Appl. Phys.*, **31** (5), 542–550.
52. Gorce, O., Jurado, H., Thomas, C., Djéga-Mariadassou, G., Khacef, A., Cormier, J.M., Pouvesle, J.M., Blanchard, G., Calvo, S., and Lendresse, Y. (2001) Non-Thermal Assisted Catalytic NOx Remediation from a Lean Model Exhaust. SAE Technical Paper 2001-01-3508, pp. 47–51.
53. Baudin, F. (2004) Catalyse deNOx assistée par plasma non thermique. PhD Thesis. Université Pierre et Marie Curie, France.
54. Khacef, A., Cormier, J.M., and Pouvesle, J.M. (2006) Energy deposition effect on the NOx remediation in oxidative media using atmospheric non-thermal plasmas. *Eur. Phys. J. Appl. Phys.*, **33**, 195–198.
55. Puchkarev, V., Roth, G., and Gundensen, M. (1998) Plasma processing of diesel exhaust by pulsed corona discharge. SAE Technical Paper 982516, pp. 107–111.
56. Dorai, R. and Kushner, M.J. (2001) Effect of multiple pulses on the plasma chemistry during the remediation of NOx using dielectric barrier discharges. *J. Phys. D: Appl. Phys.*, **34** (4), 574–583.
57. Bröer, S., Hammer, T., and Kishimoto, T. (1997) Proceedings of the 12th International Conference of Gas Discharges and Their Applications, Greifswald, September 8–12, 1997.
58. Khacef, A., Cormier, J.M., and Pouvesle, J.M. (2005) Non-thermal plasma NOx remediation: from binary gas mixture to lean-burn gasoline and diesel engine exhaust. *J. Adv. Oxid. Technol.*, **8** (2), 150–157.
59. Vitello, P.A., Penetrante, B.M., and Bardsley, J.N. (1994) Simulation of negative streamer dynamics in nitrogen. *Phys. Rev. E*, **49** (6), 5575–5598.
60. Prasad, R., Kennedy, L.A., and Ruckenstein, E. (1984) Catalytic combustion. *Catal. Rev. Sci. Eng.*, **26**, 1–58.
61. Pfefferle, L.D. and Pfefferle, W.C. (1987) Catalysis in combustion. *Catal. Rev. Sci. Eng.*, **29**, 219–267.
62. Zwinkels, M.F.M., Jaras, S.G., Menon, P.G., and Griffin, T.A. (1993) Catalytic materials for high-temperature catalytic combustion. *Catal. Rev. Sci. Eng.*, **35** (3), 319–358.
63. Gélin, P. and Primet, M. (2002) Complete oxidation of methane at low temperature over noble metal based catalysts: a review. *Appl. Catal., B*, **39** (1), 1–37.
64. Barry, A.A.L., Van Setten, M.M., and Moulijn, J.A. (2001) Science and technology of catalytic diesel particulate filters. *Catal. Rev.*, **43** (4), 489–564.
65. Oh, S.E., Mitchell, P.J., and Siewert, R.M. (1991) Methane oxidation over alumina-supported noble metal catalysts with and without cerium additives. *J. Catal.*, **132** (2), 287–301.
66. Burch, R. and Loader, P.K. (1994) Investigation of Pt/Al_2O_3 and Pd/Al_2O_3 catalysts for the combustion of methane

at low concentrations. *Appl. Catal., B,* **5** (1–2), 149–164.

67. Ma, L., Trimm, D.L., and Jiang, C. (1996) The design and testing of an autothermal reactor for the conversion of light hydrocarbons to hydrogen I. The kinetics of the catalytic oxidation of light hydrocarbons. *Appl. Catal., A,* **138** (2), 275–283.
68. Farrauto, R.J., Lampert, J.K., Hobson, M.C., and Waterman, E.M. (1995) Thermal decomposition and reformation of PdO catalysts; support effects. *Appl. Catal., B,* **6** (3), 263–270.
69. Maione, A., André, F., and Ruiz, P. (2007) The effect of Rh addition on Pd/γ-Al_2O_3 catalysts deposited on FeCrAlloy fibers for total combustion of methane. *Appl. Catal., A,* **333** (1), 1–10.
70. Lowke, J.J. and Morrow, R. (1995) Theoretical analysis of removal of oxides of sulphur and nitrogen in pulsed operation of electrostatic precipitators, IEEE Trans. *Plasma Sci.,* **23**, 661–671.
71. Ono, R. and Oda, T. (2000) Measurement of hydroxyl radicals in an atmospheric pressure discharge plasma by using laser-induced fluorescence, IEEE Trans. *Ind. Appl.,* **36** (1–2), 82–86.
72. Van Giezen, J.C., van den Berg, F.R., Kleinen, J.L., van Dillen, A.J., and Geus, J.W. (1999) The effect of water on the activity of supported palladium catalysts in the catalytic combustion of methane. *Catal. Today,* **47** (1–4), 287–293.
73. Burch, R., Urbano, F.J., and Loader, P.K. (1995) Methane combustion over palladium catalysts: The effect of carbon dioxide and water on activity. *Appl. Catal.,* A **123** (1–2), 173–184.
74. Roth, D., Gélin, P., Primet, M. and Tena, E. (2000) Catalytic behaviour of Cl-free and Cl-containing Pd/Al_2O_3 catalysts in the total oxidation of methane at low temperature *Appl. Catal.,* A **203** (1), 37–45.
75. Ketteler, G., Ogletree D.F., Bluhm H., Liu, H., Hebenstreit, E.L.D. and Salmeron, M. (2005) In situ spectroscopic study of the oxidation and reduction of Pd(111), *JACS,* **127** (51), 18269–18273.
76. Cullis, C.F., and Willatt, B.M. (1984) The inhibition of hydrocarbon oxidation over supported precious metal catalysts. *J. Catal.,* **86**(1), 187–200.
77. Burch, R., Breen, J.P., and Meunier, F.C. (2002) A review of the selective reduction of NO_x with hydrocarbons under lean-burn conditions with non-zeolitic oxide and platinum group metal catalysts. *Appl. Catal., B,* **39**, 283–303.
78. Gorce, O. (2000) Design d'un catalyseur de deNO_x à partir d'un modèle généralisé de réduction des oxydes d'azote PhD Thesis. University Pierre et Marie Curie (Paris 6), France.

4
VOC Removal from Air by Plasma-Assisted Catalysis-Experimental Work

Monica Magureanu

4.1
Introduction

Volatile organic compounds (VOCs) refer to organic chemical compounds that have a sufficiently low boiling point to allow evaporation in air without decomposition (usually below 250 °C) and that can affect the environment and human health. Depending on the definitions in different countries, only substances boiling above 50 °C, or even methane, are included.

4.1.1
Sources of VOC Emission in the Atmosphere

VOC emissions in the atmosphere result from both natural sources and human activities. Natural or biogenic sources include vegetation, forest fires, and animals. Many factors influence VOC emission by plants, such as temperature, light, stage of plant development, air composition, and the presence of air pollutants, moisture, mechanical stress, and injury [1, 2]. Biogenic VOCs include isoprene and monoterpenes, carbonyl compounds, various alkanes, alkenes, organic acids, alcohols, esters, ethers, and so on [1, 3].

Man-made or anthropogenic sources can be divided into four main categories, including transportation (emission from cars, trucks, and buses, as well as nonroad vehicles such as aircraft, ships, and farming and construction equipment), the use of organic solvents and solvent-containing products (paints, inks, glues, adhesives, antifreezing agents, etc.), production processes and storage (chemical industry, production of pharmaceuticals, manufacturing of paper and certain food products, etc.), and combustion processes (combustion of various fuels such as coal, wood, oil, and gas) [4].

Plasma Chemistry and Catalysis in Gases and Liquids, First Edition.
Edited by Vasile I. Parvulescu, Monica Magureanu, and Petr Lukes.

4.1.2
Environmental and Health Problems Related to VOCs

VOCs are the precursors of photochemical smog, which appears as a result of chemical reactions in the presence of sunlight and nitrogen oxides in the atmosphere. The chemistry involved in the formation of photochemical smog is very complex [5, 6]. Briefly, it proceeds through photodissociation of nitrogen dioxide (NO_2), forming atomic oxygen, followed by the generation of other oxidizers such as ozone and hydroxyl radicals by reactions with hydrocarbons and water molecules. These strong oxidizers react with VOCs, generating hydrocarbon radicals. Further oxidation leads to the formation of aldehydes and aldehyde peroxyacids, and by reactions with NO_2, peroxyacyl nitrates are formed [7]. These are only a few of the components of smog, which cause environmental and health impacts, leading to irritation of the eyes and respiratory tract, damage to lung tissue and reduction in lung function, and so on.

VOCs are also recognized as precursors to ground-level or tropospheric ozone, a key component of photochemical smog. Ozone can be transported by wind currents and affect regions far from original sources, damaging forest ecosystems and agricultural crops in rural areas [8, 9]. Ozone is also a pollutant of concern because it is associated with extensive health effects, most notably related to the respiratory system [10, 11].

Another environmental problem created by VOC emissions is global warming. Some VOCs, especially halogenated compounds, act as potent greenhouse gases, with global warming potential (GWP) exceeding that of CO_2 by orders of magnitude.

The extent and nature of the health effects produced directly by VOCs depend on many factors including type of VOC, level of exposure, duration of exposure, and exposure pathway (ingestion, respiration, and dermal absorption). The toxicity of VOCs varies largely as a function of the nature of the organic compound; while some of them have no known health effect, others are highly toxic.

Exposure to VOCs may result in a spectrum of illnesses ranging from mild, such as irritation, to very severe effects, including cancer. Eye and respiratory tract irritation are several of the most frequent acute symptoms that appear during or soon after exposure to some VOCs [12, 13]. Other acute symptoms experienced by people exposed to VOCs are headaches, nausea, dizziness, fatigue, visual disorders, allergic skin reactions, and memory impairment.

Many VOCs are not acutely toxic but have chronic effects. Industrial workers who handle these compounds in their workplace experience a high risk of developing symptoms from long-term exposure to VOCs. In general, long-term exposure to some VOCs, even at low concentrations, may cause damage to the liver, kidneys, and central nervous system. Exposure to VOCs, such as polycyclic aromatic hydrocarbons and aldehydes, has been reported to elevate the risk of cardiovascular disease by affecting atherogenesis, thrombosis, or blood pressure regulation [14]. Many organic compounds cause cancer in animals, and some are suspected of causing cancer in humans (e.g., formaldehyde, trichloroethylene (TCE), dichloroethylene, perchloroethylene, chloroform) or are known to cause cancer in humans, such as benzene. Scientific evidence, particularly epidemiologic evidence, regarding the contribution of environmental and occupational exposures to various types of cancer is reviewed in [15, 16].

Research findings have demonstrated that some air pollutants, including VOCs, occur more frequently and at a higher concentration in indoor air than in outdoor air. Since people spend most of their time indoors nowadays, the quality of indoor air is a matter of utmost importance. The nature of the VOCs that are ubiquitous in indoor environment and the evidence of adverse health effects associated with exposure to some of these compounds have been recently reviewed in [17, 18].

4.1.3 Techniques for VOC Removal

The technology of choice for a given case depends on many parameters such as the type of pollutants and their concentration, the gas flow rate, and the degree of removal legally required [19]. It depends also on the recovery value of the VOCs [20]. Recovery is useful in the case of single-VOC exhaust streams, and if the cost of recovery is less than the cost of purchasing new VOCs, which typically implies relatively concentrated exhaust streams. In this situation, nondestructive methods represent a good choice. If the VOC stream has no significant recovery value, as in case of mixtures of VOCs or for toxic compounds, then destruction is more appropriate.

Current available techniques for VOC removal are thermal oxidation, catalytic oxidation, photocatalysis, adsorption, absorption, biofiltration, condensation, and membrane separation. Each of these methods has advantages as well as drawbacks. A detailed description of the most widely used techniques for pollution control is provided in [21]. In the following, a brief description of each method and its operation principle and conditions are given, as well as a short summary of the range of applicability. Plasma treatment and plasma catalysis are also included, since during the past years significant research efforts were devoted to this nonthermal approach to VOC oxidation and some practical applications have emerged.

4.1.3.1 Thermal Oxidation

Thermal oxidation of VOCs operates at high temperatures, above the autoignition temperature of the treated VOCs, in the presence of oxygen. Usually, the temperature range is 700–900 °C and, if maintained for a sufficient time, provides almost complete oxidation (over 95%) of the VOCs to carbon dioxide and water. Another advantage of thermal oxidation is its low sensitivity to the type of VOCs. However, such high temperatures require significant amounts of energy, usually supplied as natural gas, leading to high operation costs. Therefore, thermal oxidation is mostly used for relatively high VOC concentrations and is impractical at low VOC concentrations. Fuel consumption (and hence operating costs) of thermal oxidation can be reduced by heat recovery using recuperative heat exchangers or employing ceramic beds. In this case, the high equipment cost and size and the reduced operating costs have to be balanced. The control of the operating temperature is very important for this technology, since the formation of NO_x, dioxins, and furan can occur, depending on the gas temperature. Effluent gas scrubbing may be needed to control the acid vapor, in case of treatment of halogenated VOCs.

4.1.3.2 **Catalytic Oxidation**

Catalytic oxidation of VOCs operates at substantially lower temperatures than thermal oxidation, which may be in some cases as low as 250–400 °C. Typical catalysts for VOC oxidation are precious metals (such as Pt, Pd) supported on ceramic or metal monoliths (honeycombs) or on ceramic pellets, base metals supported on ceramic pellets, or metal oxides. With proper selection of the catalyst and operating conditions, the catalytic process also provides quite good destruction efficiency to the VOC, above 95%. Advantages of this technology are less NO_x formation, given low operating temperatures, and less formation of partial oxidation products, such as carbon monoxide and aldehydes. The main drawback is catalyst poisoning and the sensitivity of the catalysts to high temperatures, leading to deactivation. Especially, noble metal catalysts are sensitive to contaminants in the feed stream (such as metal or metal oxide dust) and can be easily poisoned. Heavy hydrocarbons or particulates can also deposit on the catalysts and cause their deactivation by masking the active sites. However, in some cases, catalyst lifetime can be extended by regeneration techniques. Unfortunately, in many cases regenerating the catalyst is not economical; the poisoning is irreversible because catalyst lifetime can also be limited by failure of the support structure (or agglomeration of finely dispersed particles). In the case of precious metals, recycling is more common than regeneration. Another disadvantage of catalytic oxidation is its sensitivity to the type of VOCs, and therefore several different catalysts may be required for the effective oxidation of a gas containing a mixture of VOCs. As in the case of thermal oxidation, treatment of halogenated VOCs leads to the formation of acid vapor, which can be removed by off-gas scrubbing.

4.1.3.3 **Photocatalysis**

For such kind of applications, the photocatalytic technique typically uses semiconductor pristine metal oxides (mainly TiO_2, but also ZnO, WO_3, Fe_2O_3, etc.) or metal-doped metal oxides as catalysts, activated in the presence of UV radiation. Under illumination by photons having larger energy than the bandgap of the semiconductor, highly reactive electron–hole pairs are produced, which leads to reduction and oxidation, respectively, of the VOCs adsorbed on the photocatalyst surface. The most widely used photocatalyst for environmental applications is TiO_2 because of its physical and chemical stability, lower cost, nontoxicity, and resistance to corrosion. Surface modification of TiO_2 by doping with various metal ions has been reported to be beneficial for photocatalytic reactions. Photocatalysis can be operated at room temperature, and it is a nonselective process having a broad activity toward various contaminants. The disadvantages of this method are related mainly to the relatively low efficiency of the lamps and longer residence time requirements.

4.1.3.4 **Adsorption**

In the adsorption process, the VOCs are physically adsorbed on the surface of an adsorbent material. The method is suitable for low VOC concentrations, where removal efficiency can exceed 90–95%. Adsorption is a useful method for the recovery of VOCs with intermediate molecular weights, typically about 45–130. Smaller compounds do not adsorb well because of their high volatility. Larger compounds are strongly adsorbed and difficult to remove during adsorbent regeneration. The most common adsorbent is activated carbon, but other materials such as zeolites, alumina, silica gel, and polymers have been used as well in some processes. Temperature control is needed, since adsorption is most effective at relatively low temperatures, so that cooling of hot exhaust gas streams may be necessary. In addition, the humidity of the gas stream should be kept low in order to have higher adsorption capacity. Absorbents require regeneration or disposal after a certain period. In any case, they need to be replaced regularly in order to prevent VOC escape after saturation.

4.1.3.5 **Absorption**

In the absorption process, a soluble gaseous pollutant is removed from the gas stream by dissolution in a solvent liquid. Absorption takes place if the partial pressure of the soluble gas in the gas mixture exceeds the vapor pressure of the solute gas in the liquid film in contact with the gas. Some absorption systems use water as the primary absorbing liquid, while others use a low-volatility organic liquid. The absorbing liquid containing the absorbed pollutant can be disposed of, or the pollutant can be separated from the liquid and recovered by distillation or stripping (desorption), while the absorbing liquid is regenerated and recycled. For VOC absorption, organic liquids that give the best solubility values are preferred. A problem of this process consists in the choice of the solvent, so that it does not become a source of VOC pollution itself when it does not have sufficiently low vapor pressure. Another condition for the choice of absorbing liquid is its stability in contact with the carrier gas.

4.1.3.6 **Biofiltration**

Biofiltration is an oxidation process based on passing the gas stream through an active microorganism bed (e.g., fungi, bacteria). VOCs provide a food source for the microorganisms. Through biotransformation of the VOCs, end products are formed, including carbon dioxide, water, nitrogen, and mineral salts. This method is generally used for the treatment of low concentrations of VOCs. Biofiltration is a low-temperature process and implies relatively low operation costs. However, this method requires a large facility because of the long gas residence times needed. For successful biofiltration, the design of the biofilter should ensure a proper environment for the microorganisms, including a rather strict control of temperature, humidity, pH, oxygen supply, absence of toxic materials, and inorganic nutrient supply for the microorganisms.

4.1.3.7 Condensation

Condensation is a nondestructive method for VOC recovery. This technology is based on reducing the temperature of the gas stream to the level necessary for the condensation of VOCs, thus enabling the recovery of the VOCs in the liquid phase. It is suitable especially for high VOC concentration and low gas flow rate. The optimum temperature range is highly dependent on the vapor pressure of the treated VOC. Condensation is a flexible technique, having the ability to respond to changes in the flow rate and concentration of VOCs. The liquid produced via condensation may require treatment for water removal or may require additional separation (typically distillation) if multiple VOCs are recovered. Especially when low boiling point compounds have to be treated, a very low condenser temperature is required, which will lead to the condensation of large amounts of water. A disadvantage of condensation is aerosol formation, which is entrained and must be removed by filters. Another problem of condensation appears when solvents are used, which become solid at room temperature and will block the condenser. Then, the condenser must be operated at a higher temperature, limiting the efficiency. In this case, an additional treatment step could be used.

4.1.3.8 Membrane Separation

Membrane separation is best suited to treat relatively low-flow gas streams containing low to moderate VOC concentrations. The technology utilizes a polymeric membrane that is more permeable to condensable organic vapor than it is to noncondensable gases. Thus, the gas stream is separated into a permeate that contains concentrated VOCs and a treated stream that is depleted of VOCs. Since the method concentrates the VOC, it can be used with a condenser to recover the VOC if it has sufficient value. Drawbacks of this technique are the inability to handle fluctuations in VOC concentrations and the membrane sensitivity to moisture.

4.1.3.9 Plasma and Plasma Catalysis

The main advantage of nonthermal plasma is the ability to use the energy introduced in the discharge in a selective manner, using it to generate energetic electrons, while the background gas remains close to room temperature. The high-energy electrons react with the background gas molecules, generating chemically active species (radicals, ions, excited species, etc.). These reactive species can subsequently react with the pollutant molecules and decompose them. In the presence of oxygen, strong oxidizers are formed, such as atomic oxygen, hydroxyl radicals, ozone, and so on, which lead to VOC oxidation. Therefore, a highly reactive environment can be created in nonthermal plasma without spending energy on heating the entire gas stream. The main drawback of nonthermal plasma is the formation of undesired reaction by-products, since plasma reactions are rather nonselective. A solution to this problem is the combination of plasma and heterogeneous catalysis. In this way, the high efficiency of nonthermal activation provided by plasma combined with the high selectivity offered by catalysis can lead to a synergetic effect. Commercial solutions using plasma are currently available for very low pollutant concentrations; odor control appliances have been developed by PlasmaClean, United Kingdom, or by Airtec, Germany, which is based on nonthermal plasma followed by an activated carbon.

4.2
Plasma-Catalytic Hybrid Systems for VOC Decomposition

4.2.1
Nonthermal Plasma Reactors

In nonthermal plasmas or nonequilibrium plasmas the electron mean energy is much higher than the mean energy of ions and neutral gas particles. The electron energy can reach 1–10 eV, while the gas temperature remains close to room temperature [22]. There are numerous ways to generate nonthermal plasma at atmospheric pressure and ambient temperature, such as corona discharges, dielectric barrier discharges (DBDs), nonthermal arcs, and microwave and radio-frequency discharges. The plasma reactors employed in a large number of experiments on VOC decomposition are based on corona discharges and DBDs. As their operating principles are addressed in detail in Chapter 1, only a very brief description is provided here.

Corona discharges are generated in strongly inhomogeneous electric fields associated with thin wires, needles, or sharp edges of an electrode. The most well-known corona geometries are pin-to-plate, wire-cylinder, and wire-plane, which are illustrated in Figure 4.1. The discharge can be operated with constant voltage (DC corona) or pulsed voltage. Characteristic parameters of corona discharges are presented in [23, 24].

The DBD is produced in an arrangement consisting of two electrodes, with a dielectric layer covering one of them or sometimes both electrodes. The most common discharge configurations are the planar electrode configuration and the coaxial configuration, which are illustrated in Figure 4.2. Sometimes, geometries

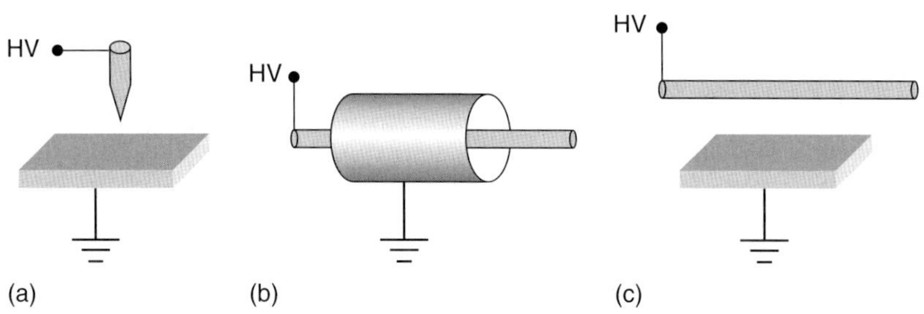

Figure 4.1 Geometries of corona discharges: (a) pin-to-plate; (b) wire-cylinder; and (c) wire-plane. HV, high voltage.

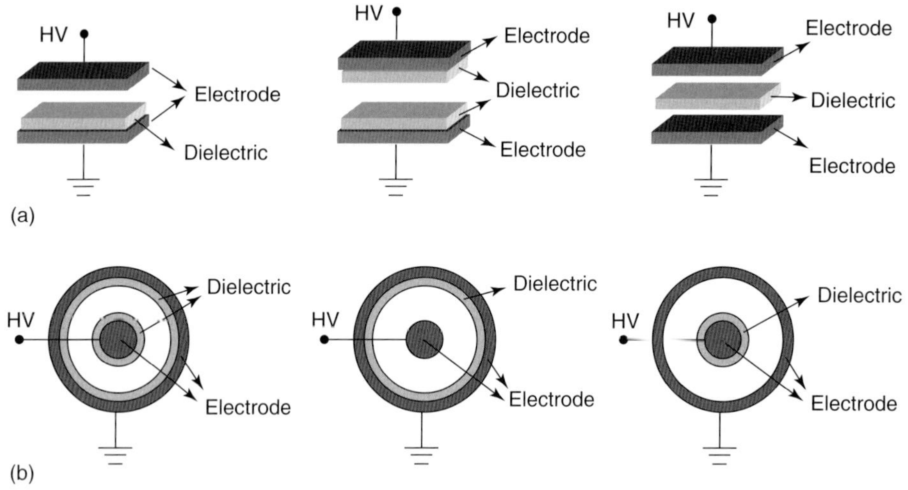

Figure 4.2 Geometries of dielectric-barrier discharges: (a) planar configuration and (b) coaxial configuration.

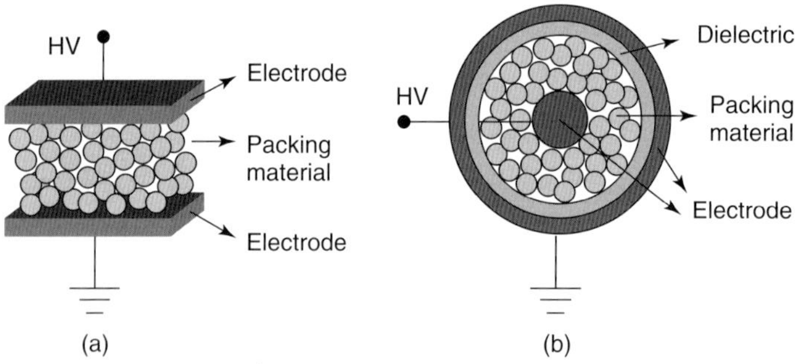

Figure 4.3 Geometries of packed-bed discharges: (a) planar configuration and (b) coaxial configuration.

combining needles or wires and planar or cylindrical electrodes covered by dielectric are also used. DBDs can be operated with AC voltage or pulsed voltage. The most important parameters characteristic of a DBD are summarized in [22, 24].

Packed-bed reactors are also widely used in VOC decomposition experiments. In this configuration, the discharge gap is filled with dielectric material. Glass, $BaTiO_3$, Al_2O_3, TiO_2, and zeolite pellets are some of the most usual packing materials. Common configurations for packed-bed reactors are illustrated in Figure 4.3.

Especially with ferroelectric packed-bed reactors, wide reactor discharge gaps and high reactor volumes at moderate breakdown voltages can be obtained. The local electric field at the contact points between pellets can be significantly enhanced because of the very high dielectric constant [25–28]. It was reported that the introduction of ferroelectric pellets into the discharge zone shifts the electron distribution toward higher energies [26]. The amplification of the electric field and, consequently, the increased production of chemically active species in the discharge zone are considered as the main advantage of ferroelectric packed-bed reactors, responsible for the higher decomposition efficiency of VOCs [26, 28].

4.2.2
Considerations on Process Selectivity

The desirable process in order to generate the least harmful pollutants is total oxidation of the VOCs to carbon dioxide and water. However, it was found that a major drawback of VOC decomposition by nonthermal plasma is the rather low selectivity to CO_2 and the formation of unwanted reaction by-products.

The scaling parameter generally used in plasma and plasma-catalytic decomposition of VOCs is the *specific input energy* (SIE), defined as the ratio between the average power dissipated in the electrical discharge and the total gas flow rate. The important parameters that characterize the efficiency of VOC oxidation are the conversion and the selectivity toward CO_2. *Conversion* is defined as the ratio between the amount of VOC decomposed in the process and the amount of VOC in the influent gas (Eq. (4.1)). The *selectivity toward CO_2* is defined as the percentage of the total amount of VOC converted, which is transformed into CO_2 (Eq. (4.2)).

$$\text{conv}\,(\%) = \left(1 - \frac{[\text{VOC}]_{\text{out}}}{[\text{VOC}]_{\text{in}}}\right) \times 100 \tag{4.1}$$

$$S_{CO_2}\,(\%) = \frac{[CO_2]}{n \times [\text{VOC}]_{\text{in}} \times \text{conv}} \times 100 \tag{4.2}$$

where $[\text{VOC}]_{\text{out}}$ is the concentration of the VOC in the effluent gas, $[\text{VOC}]_{\text{in}}$ is the initial concentration of the VOC, $[CO_2]$ is the concentration of CO_2 in the effluent gas, and n is the number of carbon atoms in the VOC molecule.

The reaction products resulting from VOC oxidation in plasma as well as the product distribution depend on many factors such as the type of VOC, its initial concentration, the input energy, and so on. The most common reaction product that appears, besides CO_2, is carbon monoxide. Other gaseous carbon-containing compounds have been also reported by some authors as a result of VOC oxidation in nonthermal plasma. Solid reaction by-products have been observed frequently as well, especially in the case of aromatic VOCs.

It was generally observed that the concentrations of by-products resulting from VOC plasma oxidation are diminished in the presence of catalysts. The combination of nonthermal plasma and heterogeneous catalysis leads to a large improvement in the selectivity to CO_2, shifting the process toward total oxidation. A more detailed account of the reaction by-products detected from VOC oxidation in plasma and in plasma-catalytic systems is given in Section 4.3.8.

4.2.3
Types of Catalysts

Various types of catalysts have been used in combination with plasma for VOC decomposition, such as noble metals, transition and group IV and V metal oxides, and zeolites.

Noble metal catalysts, especially Pt and Pd, are known for their high efficiency for VOC abatement by catalytic oxidation [29, 30]. Plasma-catalytic systems using noble metals such as Pt, Pd, Au, Ag, and Rh were investigated in numerous works. The noble metals were deposited on various support materials, such as Al_2O_3 [31–38], TiO_2 [34, 35, 39, 40], or zeolites [41]. In most cases, the catalysts were placed directly in the plasma zone in ferroelectric packed-bed reactors. The effect of these catalysts placed downstream of the plasma reactor on VOC oxidation was investigated in [32] for Pd and in [33] for Au catalysts.

Transition metals such as Co, Cu, Mo, and Ni supported on Al_2O_3 have been investigated as well for the plasma-catalytic oxidation of benzene, and the results were similar to those obtained with noble metals under the same operation conditions [31].

Transition metal oxides are reported as promising for VOC-catalytic oxidation [42]. This review summarizes recent results obtained with CeO_2, mixed oxides containing Ce and Zr, as well as Mn oxides mixed with Zr, Fe, Co, and Cu oxides, showing very good activity for the catalytic combustion of various VOCs [42]. Transition metal oxides have also been investigated in combination with nonthermal plasma. Some of these oxides have the ability to decompose ozone, thus promoting VOC oxidation on the catalyst surface with ozone generated in the plasma [43–48]. Numerous authors studied the decomposition of various VOCs in plasma-catalytic hybrid systems containing MnO_2 placed either inside the plasma region [37, 46, 49–53] or downstream of the plasma reactor [41, 44, 48, 50, 54–57]. Mn oxides mixed with Cu oxides or Fe oxides have been investigated as well in combination with nonthermal plasma [32, 43, 56–58], and CuO and Fe_2O_3 have been used in [45] downstream of a DBD reactor. Other metal oxides have been studied as well in plasma-catalytic systems, such as Cr_2O_3 [59], $CuO-Cr_2O_3$ [43], $Ba-CuO-Cr_2O_3$ [60], CoO_x [51, 52], and V_2O_5 [61].

TiO_2 has been widely used as a photocatalyst for decomposing VOCs [62–64]. It is known that electrical discharges in air emit UV radiation in the wavelength range appropriate for TiO_2 activation; therefore, plasma can be used as a UV source for the photocatalyst. Plasma-catalytic oxidation of VOCs with TiO_2 pellets was studied in packed-bed reactors [34, 35, 39, 65] or corona discharges [66]. Van Durme *et al.* [58] used an extruded TiO_2 catalyst placed between the electrodes of a corona discharge. In other experimental works, TiO_2 was coated on the wall of the plasma reactor [46], on the inner electrode [67] or on glass beads [68, 69], silica gel pellets [46], Al_2O_3 pellets [68], and glass fibers [70, 71] in packed-bed, DBD or corona discharges.

Adsorbents have been used as well in combination with nonthermal plasma for the oxidation of VOCs. Francke *et al.* [43] investigated the synergy between plasma and adsorbent materials (activated carbon and Zeosorb 5A) located downstream of a DBD reactor in case of butyl acetate removal. Other zeolites such as Na–Y, H–Y, H–mordenite, and ferrierite, packed inside the plasma reactor have been used in [72] for toluene decomposition. Kim *et al.* [34] reported a cycled system based on adsorption of VOCs on ferrierite, MS-13X and H–Y, followed by treatment of the adsorbed VOCs by oxygen plasma. A similar cycled operation was investigated with HZSM-5 zeolite by Fan *et al.* [41] for the removal of benzene. The effect of several molecular sieves (MS-3A, MS-4A, MS-5A, and MS-13X) packed between the $BaTiO_3$ pellets of a ferroelectric packed-bed reactor on benzene oxidation was studied in [31].

4.2.4 Single-Stage Plasma-Catalytic Systems

Plasma-catalytic systems, where the catalyst is placed directly inside the plasma region, are generally called *single-stage systems* [27, 60]. Several other terms have been used to describe single-stage plasma-catalytic systems, such as in-plasma catalysis (IPC) [25, 28, 32], plasma-driven catalysis (PDC) [34, 39, 40], and one-stage plasma catalysis [37]. The most common way to introduce the catalyst in the plasma reactor is in the form of pellets, which can either completely fill the discharge region [36, 40, 68] or can occupy only a part of the plasma zone [60, 72]. Catalytic pellets can be also used in mixtures with noncatalytic material, as in ferroelectric packed-bed reactors [35, 37, 39]. Catalysts can be also introduced in the plasma region as coatings on the inner wall of the plasma reactor [46] or as coatings on the inner electrode [51–53, 67, 73]. Catalysts deposited on ceramic honeycomb monoliths [74, 75] or on foam [76] have also been used in these plasma-catalytic systems.

In single-stage systems, the presence of catalytic material can influence plasma properties, modifying the electron energy distribution or changing the discharge type. On the other hand, plasma can influence catalyst characteristics, can lead to thermal activation or UV light activation of catalysts, and can affect the adsorption process [25, 27]. Short-lived chemically active species generated in plasma, as well as species with longer lifetimes can react with VOCs on the catalyst surface and decompose them. The mechanism of plasma-catalytic processes is discussed in Chapter 5, so they are not detailed here.

4.2.5 Two-Stage Plasma-Catalytic Systems

Plasma-catalytic systems where the catalyst is located outside of the plasma region are generally called *two-stage systems* [27, 60]. Another term used by some authors is postplasma catalysis [25, 28, 32, 65]. Usually, the catalyst is placed downstream of the plasma reactor in order to remove undesired by-products formed in plasma, such as NO_x, O_3, and to shift the process toward total oxidation.

Unlike single-stage plasma-catalytic systems, in two-stage systems the short-lived chemically active species generated in the plasma disappear before reaching the catalyst. Therefore, only chemical species with sufficiently long lifetimes, such as ozone, can contribute to VOC oxidation on the catalyst surface. The plasma changes the composition of the gas entering the catalytic reactor; besides ozone formation, it also has the role to partially convert the VOCs, while the main role of the catalyst is to enhance process selectivity and efficiency. An advantage of these systems is that they can combine the optimum operation conditions of the plasma and catalyst [19].

4.3
VOC Decomposition in Plasma-Catalytic Systems

4.3.1
Results Obtained in Single-Stage Plasma-Catalytic Systems

Recent results obtained in single-stage plasma-catalytic systems are summarized in Table 4.1. The comparison with results obtained with plasma alone, in the absence of catalysts, is also included in the table, where available.

In the absence of catalysts, both VOC conversion and CO_2 selectivity increase with increasing SIE, since higher input energy leads to the generation of larger amounts of chemically active species responsible for VOC oxidation. Several other parameters influence VOC oxidation in plasma and plasma-catalytic systems, such as the chemical structure of the VOC, its initial concentration, the gas composition (O_2 partial pressure, presence of water vapor), and temperature, which are discussed in Sections 4.3.3 and 4.3.4.

Numerous attempts have been made in order to optimize the plasma reactor for enhancing the efficiency of VOC oxidation. It was found that the geometry of the ground electrode influences VOC decomposition efficiency [78]: insufficient contact between the ground electrode and the dielectric can lead to the formation of parasitic discharge outside of the plasma reactor and consequently to waste of energy, whereas reducing the air gap between them resulted in higher energy efficiency. It was also observed that the gap length in DBD reactors influences VOC conversion [53].

It was reported that packed-bed reactors outperform gas-phase reactors [26, 28, 69] even when the packing material is not catalytically active. Chang and Lin [69] reported a marked increase in the decomposition efficiency of toluene and acetone in a glass-packed-bed reactor as compared to the results achieved in the unpacked reactor. Other authors found similar results for VOC oxidation in the empty DBD reactor and the reactor packed with glass [26, 28]. However, they obtained an improvement in conversion and especially in the CO_2 selectivity when packing the reactor with ferroelectric material ($PbZrO_3-PbTiO_3$ or $BaTiO_3$) [26, 28]. The nature of the ferroelectric material, in particular its dielectric constant, was found also to affect VOC decomposition [31].

Table 4.1 Results obtained in single-stage plasma-catalytic systems and comparison with plasma alone.

VOC	Initial concentration (ppm)	Plasma reactor	Catalyst	SIE ($J\,l^{-1}$)	Plasma alone			Plasma + catalyst			References
					Conversion (%)	S_{CO_2} (%)	S_{CO} (%)	Conversion (%)	S_{CO_2} (%)	S_{CO} (%)	
Acetone	1100	DBD/packed bed (glass)	TiO_2/glass	60–1200	2–15/2–45	10–40	20–40	2–45	35–76	20–42	[69]
Formaldehyde	140	DBD/packed bed	α-Al_2O_3 (70 °C) γ-Al_2O_3 (70 °C)	47–108	54–76	–	–	56–84 63–92	–	–	[77]
Propane	100	$BaTiO_3$ packed bed	MnO_2/γ-Al_2O_3 (180–300 °C) Pt/γ-Al_2O_3 (180–260 °C) mixed with $BaTiO_3$ beads	60	54–70	40	–	43–90 53–90	60–100 60–100	–	[37]
Isopropanol	100–1000 250	DBD	CoO_x MnO_x coated on inner electrode	195 160–295	80 – 50 52 – 100	40 – 30 22 – 42	50 – 45 43 – 58	100 – 52 100 – 55 62 – 100 70 – 100	83 – 40 87 – 42 37 – 58 48 – 74	17 – 50 13 – 50 45 – 42 35 – 26	[73]

(*continued overleaf*)

Table 4.1 (*continued*)

VOC	Initial concentration (ppm)	Plasma reactor	Catalyst	SIE ($J\,l^{-1}$)	Plasma alone			Plasma + catalyst			References
					Conversion (%)	S_{CO_2} (%)	S_{CO} (%)	Conversion (%)	S_{CO_2} (%)	S_{CO} (%)	
Dichloromethane	200	Packed bed	Ba–CuO–Cr_2O_3/ Al_2O_3 (100 °C)	60–180	19–39	–	–	15–41	–	–	[60]
Trichloroethylene	200	DBD	MnO_2 coated on inner electrode	150–720	73–99	15–25	15–20	94–98	25–60	12–18	[53]
Trichloroethylene	250	DBD	TiO_2 MnO_2 MnO_2–TiO_2 coated on inner electrode	140–1090	100	6–22	50–78	100	15–40 43–65 40–80	41–60 19–35 25 – 20	[67]
Benzene	200	$BaTiO_3$ packed bed	M/γ-Al_2O_3, M = Ag, Co, Cu, Mo, Ni, Pd, Pt, Rh mixed with $BaTiO_3$ beads	1800	90	67	33	100	74–78	22–26	[31]
Benzene	200	$BaTiO_3$ packed bed	γ-Al_2O_3 (500 °C) Ag/γ-Al_2O_3 (500 °C) TiO_2 (500 °C) 2% Ag/TiO_2 (500 °C) mixed with $BaTiO_3$ beads	60	–	–	–	72 78 65 72	–	–	[35]
Benzene	203–210	Packed bed ($BaTiO_3$)	TiO_2 (100 °C) Ag/TiO_2 (100 °C) Pt/TiO_2 (100 °C)	125–408	24–65	53–56	40–41	51–82 56–90 48–80	54–63 67–72 65–79	29–37 29–32 27 – 22	[39]

Benzene	105	DBD	TiO_2 – coated on the tube inner wall TiO_2/silica gel pellets MnO_2 pellets	900 320 360	34	–	–	60 50 54	–	–	[46]
Benzene	100	Packed bed	TiO_2 – coated on γ-Al_2O_3 beads (30 °C)	140	30	20	4	48	55	4	[68]
Phenol	185	$BaTiO_3$ packed bed	$BaTiO_3 + LaCoO_3$ (two separate layers) $LaCoO_3$	460–920	85–95	20–30	15–20	98–100 97–100	37–65 32–39	20 – 15 12 – 11	[28]
Toluene	200	$BaTiO_3$ packed bed	γ-Al_2O_3 (400–500 °C) Ag/γ-Al_2O_3 (400–500 °C) Pd/γ-Al_2O_3 (200–300 °C) Pt/γ-Al_2O_3 (200–300 °C) mixed with $BaTiO_3$ beads	60	–	–	–	65–100 84–100 65–96 80–100	–	–	[35]
Toluene	1000	Packed bed	MnO_2/γ-Al_2O_3 TiO_2/γ-Al_2O_3 γ-Al_2O_3	100–700	–	–	–	50–97 31–87 29–75	–	–	[49]
Toluene	100	DBD	MnO_x CoO_x coated on inner electrode	160–295	67–100	22–51	40–49	78–100 86–100	52–80 52–71	23–29 30 to 20	[51]

(continued overleaf)

Table 4.1 *(continued)*

VOC	Initial concentration (ppm)	Plasma reactor	Catalyst	SIE ($J\,l^{-1}$)	Plasma alone			Plasma + catalyst			References
					Conversion (%)	S_{CO_2} (%)	S_{CO} (%)	Conversion (%)	S_{CO_2} (%)	S_{CO} (%)	
Toluene	0.5	Corona	TiO_2	17	27	–	–	83	–	–	[58]
Toluene	200	Packed bed	$Ba–CuO–Cr_2O_3/Al_2O_3$ (100 °C)	60–180	30–86	–	–	35–86	–	–	[60]
Toluene	80–100	Corona	TiO_2	180	45	–	–	76	–	–	[66]
Toluene	1100	DBD/packed bed (glass)	TiO_2/glass	60–1200	2–35/2–78	2–31	8–25	10–78	5–48	8–25	[69]
Toluene	200	Packed bed	Na–Y	600	16	57	43	78	60	40	[72]
			H–Y					82–87	38	62	
			H–Mordenite					42	52	48	
			Ferrierite					22	67	33	
Styrene	132	DBD/packed bed	Al_2O_3	900–4000	100	10–40	10–31	100	45–62	20–30	[36]
			$Pt–Pd/Al_2O_3$						67–96	0–2	

As seen in Table 4.1, the addition of catalysts generally leads to an enhancement of VOC conversion and/or CO_2 selectivity as compared to the results obtained with plasma in the absence of catalysts.

Owing to its photocatalytic properties, TiO_2 was investigated in combination with plasma for VOC oxidation by numerous authors. Photocatalytic activity of TiO_2 is related to its ability to create electron–hole pairs as a result of exposure to ultraviolet radiation with the wavelength shorter than that corresponding to the bandgap. The resulting free radicals are very efficient oxidizers of organic matter. Since plasma emits UV light, it can activate the photocatalyst, so it was thought their combination would be successful for VOC removal. Significant increase of VOC conversion as compared to plasma alone was found when introducing TiO_2 catalysts as pellets or extrudate in the plasma reactor [39, 58, 66]. The selectivity to CO_2 was improved by supporting Ag or Pt on TiO_2 [39]. Chang and Lin [69] reported that coating TiO_2 on glass pellets did not enhance the conversion of toluene and acetone as compared to the glass packing but improved the carbon balance and the CO_2 selectivity. The total oxidation of benzene was more efficient in a plasma reactor packed with TiO_2 coated on γ-Al_2O_3 beads than in the empty reactor [68] or in the reactor packed with γ-Al_2O_3 pellets. Subrahmanyam *et al.* [67] used an inner electrode made of stainless steel fibers coated with TiO_2 or TiO_2–MnO_2 for the oxidation of TCE. The selectivity toward CO and CO_2 is shown in Figure 4.4 for an ordinary DBD reactor with a Cu inner electrode and for the reactor with a modified inner electrode. The CO_2 selectivity was improved by 10–15% with the TiO_2-coated electrode, while with TiO_2–MnO_2 coating much better results were obtained.

Futamura *et al.* [46] introduced TiO_2 as a coating on the inner wall of the DBD reactor and achieved a benzene conversion approximately two times higher than with the original DBD reactor, as well as increased CO_2 selectivity. The authors found that TiO_2-catalytic activity remains relatively high over time in the plasma–TiO_2 combination, while in a classical photooxidation experiment the photocatalyst is rapidly deactivated.

Several authors suggested that the better performance achieved in the presence of TiO_2 is due to its activation by UV light emitted by plasma [67, 68]. However, experiments performed in Ar–O_2 mixtures showed higher VOC decomposition than in air, even if the radiation emitted by the plasma is not in the UV spectral range [40]. This proves that the radiation emitted by plasma has a negligible contribution to TiO_2 activation. Thevenet *et al.* [70] also concluded that plasma-emitted UV light is not able to activate the photocatalyst; however, additional external UV light induces a significant photocatalytic effect, improving VOC total oxidation.

Noble metal catalysts, such as mainly Pt and Pd and also Ag, have been investigated in combination with nonthermal plasma, since they are well known from VOC-catalytic oxidation. Generally, these metals were supported on alumina pellets and packed inside the plasma region, either alone [36, 39] or in combination with ferroelectric materials [31, 35, 37]. All authors reported good conversion and especially high values of CO_2 selectivity, significantly improved as compared to the results with plasma alone. The improvement was more marked for Pt and Pd than

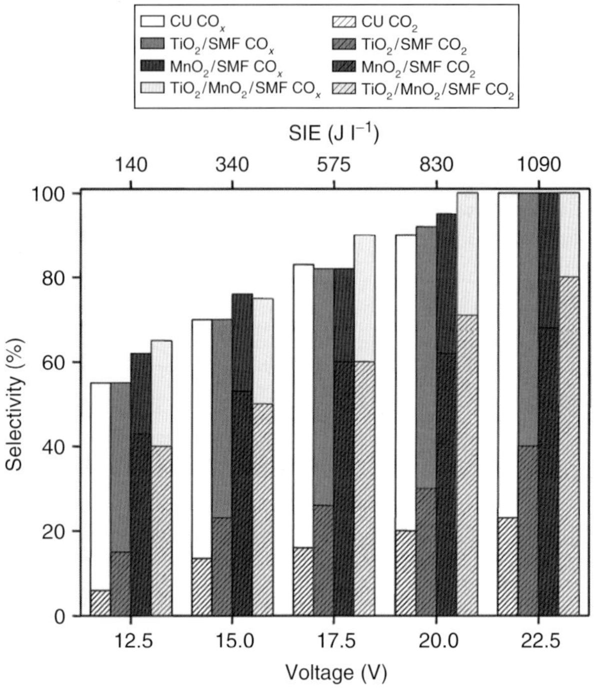

Figure 4.4 Selectivity to CO and CO_2 obtained from trichloroethylene oxidation in a DBD reactor with TiO_2 and TiO_2-MnO_2 coating on the stainless steel fibers (SMF) inner electrode and comparison with the uncoated electrode [67].

for Ag [35]. This effect is clearly due to the metal itself, since reactors packed only with the support material yield inferior results [35, 36, 39].

In numerous studies, transition metal oxides (especially manganese oxides and cobalt oxides) have been used in combination with plasma for VOC removal because of their ability to decompose ozone formed in the plasma-generating highly reactive atomic oxygen, which can oxidize VOCs on the catalyst surface. Mn oxides have been introduced in the plasma reactor as MnO_2 pellets [46], supported on γ-Al_2O_3 and either mixed with ferroelectric material or not [37, 49] in packed-bed reactors, or coated on the inner electrode of a DBD reactor [51–53, 67, 73]. Similar results as those achieved with noble metals have been obtained in [37]. All authors reported considerable enhancement of VOC total oxidation by the addition of Mn oxide catalyst. To illustrate the positive effect of these catalysts, the selectivity to CO and CO_2 are plotted in Figures 4.5 and 4.6 for the oxidation of toluene [51] and TCE [53], respectively. The addition of Co oxides also improved the results obtained in plasma, but to a lesser extent than Mn oxides [51, 52, 73].

Higher energy efficiency than in plasma oxidation of VOCs has been reported as well when introducing zeolites in the plasma reactor [31, 72]. Oh *et al.* [72]

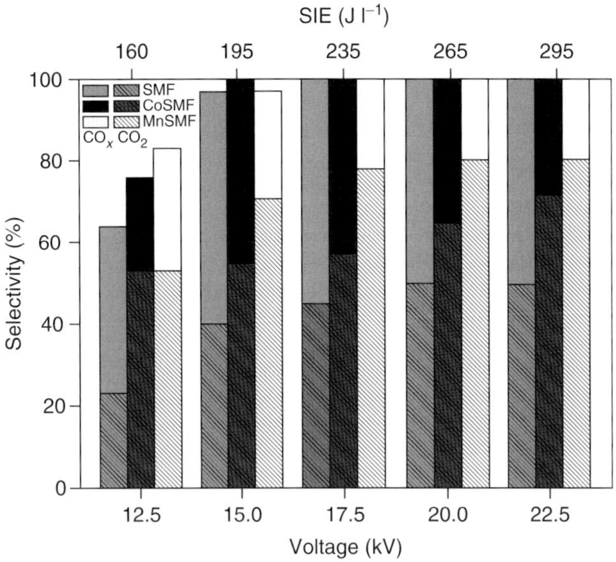

Figure 4.5 Selectivity to CO and CO_2 obtained from toluene oxidation in a DBD reactor with Mn oxide and Co oxide coating on the stainless steel fibers (SMF) inner electrode and comparison with the uncoated electrode [51].

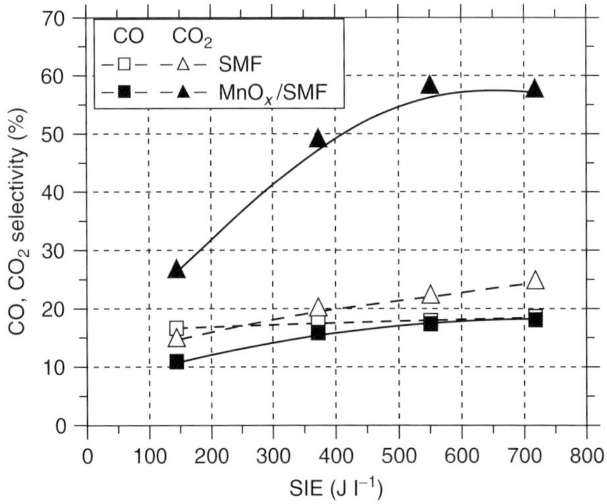

Figure 4.6 Selectivity to CO and CO_2 obtained from trichloroethylene oxidation in a DBD reactor with Mn oxide coating on the stainless steel fibers (SMF) inner electrode and comparison with the uncoated electrode [53].

reported a dramatic increase in toluene conversion in the presence of Na–Y and H–Y zeolites, while the use of H–Mordenite and Ferrierite gave less spectacular results. This suggests that adsorption plays an important role in plasma-catalytic reactions.

4.3.2 Results Obtained in Two-Stage Plasma-Catalytic Systems

Results obtained in two-stage plasma-catalytic systems are summarized in Table 4.2. The comparison with results obtained with plasma alone, in the absence of catalysts, is also included in the table, where available.

In two-stage plasma-catalytic systems, the catalyst is not subjected to the direct influence of electrical discharge and of the short-lived chemically active species generated in plasma, since these reactive species disappear before reaching the catalyst. The role of plasma in this case is to modify the composition of the gas that enters the catalytic reactor by converting the VOCs into chemical compounds that are more easily oxidized by the catalyst and by generating long-lived oxidizers (e.g., ozone), which further contribute to VOC decomposition on the catalyst surface [25, 27].

In order to separate these two effects, several authors introduced the VOCs after the plasma reactor and thus investigated only the role of ozone on catalytic oxidation of the VOCs [55, 56, 60, 82]. Catalysts that have the ability to decompose ozone have been used in these studies. Ogata *et al.* [60] investigated the oxidation of toluene and dichloromethane in a plasma-catalytic system with Ba–CuO–Cr_2O_3/Al_2O_3 catalyst and achieved higher conversion of these organic compounds when introducing these VOCs between the plasma reactor and the catalyst as compared with the results obtained in single-stage configuration. Several other VOCs (TCE, benzene, 1,2-dichloroethane) were successfully decomposed on MnO_2 catalysts with ozone generated in the plasma [82]. However, the concentration of by-products resulting from TCE oxidation, especially trichloroacetaldehyde, was considerably higher when TCE was introduced after the plasma reactor than in the direct plasma-catalytic process [82]. Similar results have been obtained by other authors [55, 56] in the presence of MnO_2 catalysts: the decomposition efficiency was very good regardless of the VOC introduction in plasma, but the carbon balance was appreciably improved when the pollutant was first oxidized in the plasma and then further treated on the catalyst. These results show that ozone has an essential contribution to VOC decomposition on these catalysts, but short-lived chemically active species that formed in the plasma are also essential for the total oxidation of VOCs.

In many of the reported studies, transition metal oxides, especially Mn oxides, have been used in two-stage plasma-catalytic systems because of their ozone decomposition properties. These oxide catalysts were found to be very effective and large improvement of VOC conversion in the plasma-catalytic system as compared to plasma alone has been reported even at room temperature, as seen in Table 4.2. The effect of Mn oxide addition on the product distribution is less investigated;

Table 4.2 Results obtained in two-stage plasma-catalytic systems and comparison with plasma alone.

VOC	Initial concentration (ppm)	Plasma reactor	Catalyst	SIE ($J\,l^{-1}$)	Plasma alone			Plasma + catalyst			References
					Conversion (%)	S_{CO_2} (%)	S_{CO} (%)	Conversion (%)	S_{CO_2} (%)	S_{CO} (%)	
Propane	200	Corona	MnO_2/γ-Al_2O_3	100–300	19–65	4–21	2–30	67–92	15–65	2–35	[55]
Butyl acetate	53	Corona	Zeosorb 5A (120 °C)	54	20	–	–	35	–	–	[43]
			Activated carbon (120 °C)					50			
			Cu/Cr oxide (120 °C)					55			
			Cu/Mn oxide (120 °C)					72			
			Fe/Mn oxide (120 °C)					77			
Butyl acetate	120	Corona	Pt/Al_2O_3 (210 °C)	30–660	2–25	–	–	37–66	99	1	[79]
Dichloromethane	200	Surface discharge	Ba–CuO–Cr_2O_3/Al_2O_3	60–360	19–39	–	–	43–70	–	–	[60]
Trichloroethylene	250	DBD	MnO_2	240	100	46		100	64		[45]
			Fe_2O_3						52		
			CuO						48		
Trichloroethylene	250	DBD	MnO_2	60–400	80–100	19–55		98–100	36–98		[54]
Cyclohexane	88	Packed bed (glass)	MnO_2/Al honeycomb monolith	600	24	13	9	100	19	7	[56]
Benzene	106	DBD	MnO_2	300–1500	5–30	–	–	36–99.5	32–46	12 – 19	[44]
Benzene	500	Packed bed ($BaTiO_3$)	TiO_2 (19–613 °C)	60	–	–	–	34–85	–	–	[65]
			Ag/TiO_2 (19–613 °C)					46–95			
			γ-Al_2O_3 (19–613 °C)					28–100			
			Ag/γ-Al_2O_3 (19–613 °C)					39–99			
Benzene	106	DBD	MnO_2	300–1700	5–30	–	–	30–95	32–46	12–19	[80]
Benzene toluene	1–1.5	Corona	$MnOx/\gamma$-Al_2O_3	10	21949	10	20	94	90	0	[41]
p-xylene								97	95		

(continued overleaf)

Table 4.2 *(continued)*

VOC	Initial concentration (ppm)	Plasma reactor	Catalyst	SIE ($J\,l^{-1}$)	Plasma alone			Plasma + catalyst			References
					Conversion (%)	S_{CO_2} (%)	S_{CO} (%)	Conversion (%)	S_{CO_2} (%)	S_{CO} (%)	
Toluene	0.5	Corona	Pd/Al_2O_3, $Cu-Mn/TiO_2$, $Fe_2O_3-MnO_2$, $CuO-MnO_2$	2.5	<10	–	–	>90	–	–	[32]
Toluene	70	Packed bed (glass)	MnO_2/Al honeycomb monolith	342	36	11	9	100	16	10	[56]
			MnO_2-CuO						15	1.6	
			MnO_2/Al honeycomb monolith + MnO_2-CuO						22	2	
Toluene	240	Packed bed (glass)	$Fe_2O_3-MnO_2$	172	36	17	22	76	31	22	[57]
			γ-Al_2O_3					74	15	11	
			MnO_2/γ-Al_2O_3					88	20	16	
			Activated carbon					98.5	29	22	
			MnO/activated carbon					99.7	30	25	
Toluene	0.5	Corona	$CuO-MnO_2/TiO_2$	2.5	2	–	–	78	–	–	[58]
Toluene	200	Surface discharge	$Ba-CuO-Cr_2O_3/Al_2O_3$	60–360	30–88	–	–	55–96	–	–	[60]
Toluene	500	Packed bed ($BaTiO_3$)	TiO_2 (19–613 °C)	60	13	–	–	47–95	–	–	[65]
			Ag/TiO_2 (19–613 °C)					39–96			
			γ-Al_2O_3 (19–613 °C)					24–100			
			Ag/γ-Al_2O_3 (19–613 °C)					28–100			
Toluene	50	DBD	Ag–Al	75–370	18–56	35–39	65 – 61	18–58	57 – 45	43–55	[81]
			Ag–Al (100 °C)	75–370	–	–	–	20–60	65 – 50	35–50	

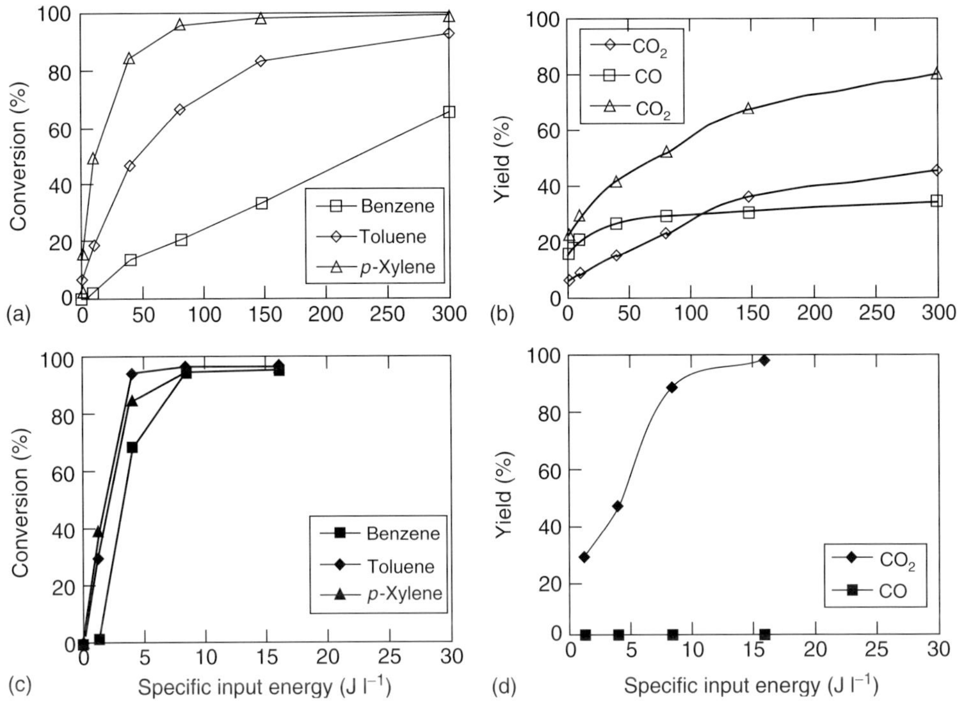

Figure 4.7 Conversion of low concentrations (1–1.5 ppm) of benzene, toluene, and xylene, and carbon oxides yields in a corona discharge (a,b) and in the presence of Mn oxide catalysts placed downstream of the plasma reactor (c,d) [41].

however, the process is obviously shifted toward total oxidation [41, 55]. Almost 100% selectivity to CO_2 has been reported using Pt catalyst downstream of the plasma, heated to 210 °C [79]. An example of the beneficial effect of Mn oxide catalyst on the oxidation of a mixture of benzene, toluene, and xylene is shown in Figure 4.7 [41].

Adsorbent materials have been studied as well, downstream of plasma for VOC removal. Francke *et al.* [43] did not observe a synergetic effect using activated carbon or Zeosorb 5A, while the results achieved with mixed transition metal oxides were much better. On the contrary, Delagrange *et al.* [57] obtained good conversion with activated carbon and only a slight improvement with MnO deposited on activated carbon. They concluded therefore that the activity of the catalyst depends mainly on the support used. Holzer *et al.* [83] found that in a two-stage plasma-catalytic system, the porosity of the catalyst is not a determining parameter, the only essential property being its ability to decompose ozone. However, in single-stage configuration, the porosity is important for the selectivity to CO_2.

4.3.3 Effect of VOC Chemical Structure

According to their chemical structure, hydrocarbons can be classified into two groups: aromatic hydrocarbons and aliphatic hydrocarbons, the latter being further divided into saturated and unsaturated hydrocarbons. It was found that the chemical structure of VOCs influences their decomposition efficiency in nonthermal plasma [19, 84–86]. Futamura *et al.* [84] investigated the decomposition in a ferroelectric packed-bed reactor of several aliphatic hydrocarbons (methane, ethane, butane, and ethylene) and several chlorinated VOCs (1,1,2-trichloroethane, TCE, and tetrachloroethylene). They observed that the conversion of alkanes increases with the increasing number of carbon atoms in the VOC molecule. The conversions of methane, ethane, and butane are much lower than that of ethylene, even if the C=C bond is stronger than the C–C bond and the dissociation energy of C–H bonds in ethylene and ethane are similar. It was also found that the chlorinated compounds are removed easier as compared to the nonchlorinated ones. Pekarek *et al.* [85] investigated the decomposition of a complex mixture of volatile hydrocarbons from gasoline, containing paraffins, iso-paraffins, naphthenes, aromatics, olefins, and methyl-*tert*-butyl-ether. Their results confirmed differences in decomposition efficiency of various hydrocarbon types. These authors also foundsignificantly higher decomposition of olefins as compared to the other VOCs, suggesting that the stability of molecules with double C=C bond is lower than that of saturated hydrocarbons or aromatics. It was also observed that for each type of hydrocarbon, the decomposition increases with the number of carbon atoms.

A relationship between the reactivity of various VOC molecules to O and OH radicals (illustrated by the rate constants) and the ionization potential of the respective VOCs (i.e., the energy required to remove an electron from the VOC molecule) is given in [19, 87]. The authors showed that VOCs with lower ionization potential have higher probability of radical attack. This inverse relationship is applicable to alkanes, alkenes, and aromatic compounds, and, according to the authors, is a consequence of the electrophilic character of radicals, which tend to attack chemical structures with large electron density [19]. It is known that within a group of similar molecules, the ionization potential decreases with increasing molecular size, since in larger molecules there are more electrons available for ionization without disrupting the molecule stability. The increase in VOC conversion obtained in [84–86] with increasing molecular size suggests that radical reactions play an important role in the decomposition process [19, 87].

On the contrary, no correlation between the ionization potential and the decomposition was found in a plasma-catalytic system, consisting of a $BaTiO_3$ packed-bed reactor with Ag/TiO_2 catalysts placed in the plasma region, among the ferroelectric pellets [87]. This shows that the decomposition mechanism is different in gas-phase plasma reactors as compared to plasma-catalytic systems, where catalytic reactions play an important role.

4.3.4
Effect of Experimental Conditions

4.3.4.1 Effect of VOC Initial Concentration

The influence of the initial concentration of VOCs on their total oxidation in plasma-catalytic systems has been addressed in many works, for various compounds such as benzene [40], toluene [48, 51, 57], styrene [36], formaldehyde [77], isopropanol [73], and so on. It was generally found that VOC conversion and/or CO_2 selectivity decrease significantly with the increase in the initial concentration. An illustrative example of this effect is presented in Figure 4.8 [51].

Delagrange *et al.* [57] noticed that in the presence of a catalyst, the decrease in VOC conversion toward higher concentration was less marked than with plasma alone. Kim *et al.* [40] observed that the amount of removed benzene in a plasma reactor packed with Ag/TiO_2 catalysts is determined only by the SIE, regardless of the VOC initial concentration. However, in other studies, it was usually found that at constant SIE, the amount of converted VOCs showed a significant increase with increasing initial VOC concentration [48, 51, 57, 73, 77]. Therefore, the energy efficiency for VOC removal is higher for higher VOC concentration. On the other hand, the formation of by-products, which also increases significantly with VOC input concentration, must be also considered when determining an optimum concentration range.

4.3.4.2 Effect of Humidity

It was observed that the presence of water vapor in the treated gas can also influence VOC removal in plasma-catalytic experiments [32, 44, 58, 71, 88]. Oh *et al.* [72] found that low water content (up to 2%) did not affect the toluene decomposition efficiency in plasma. Moderate humidity slightly improved toluene conversion in plasma in the absence of catalysts [58, 88]. However, in plasma-catalytic systems, the VOC conversion was lower in humid air than in dry air [32, 44, 58, 71]. Futamura *et al.* [44] noticed that the negative effect of water vapor was less important at very high input energy, and, in addition, the selectivity to CO_2 was improved as compared to results obtained in dry air. Humidity had a detrimental effect on the plasma-catalytic mechanism both in single-stage and two-stage systems for different catalysts [32, 58], as shown in Figure 4.9 [58].

The authors affirm that this effect is due to adsorption of water molecules on the catalyst surface, blocking catalytic active sites and therefore reducing catalyst activity [58]. Owing to the competition between VOC molecules and H_2O molecules for adsorption on free catalytic sites, it is clear that the VOC adsorption is essential for its efficient decomposition. A logarithmic correlation between the equilibrium sorption coefficient and the VOC removal efficiency was found for constant SIE [32]. On the basis of an analysis of the sorption behavior of three VOC molecules on different catalysts in the presence of water vapor, Van Durme *et al.* [32] concluded that the negative effect of humidity on plasma-catalytic decomposition of VOCs is mainly due to van der Waals interactions.

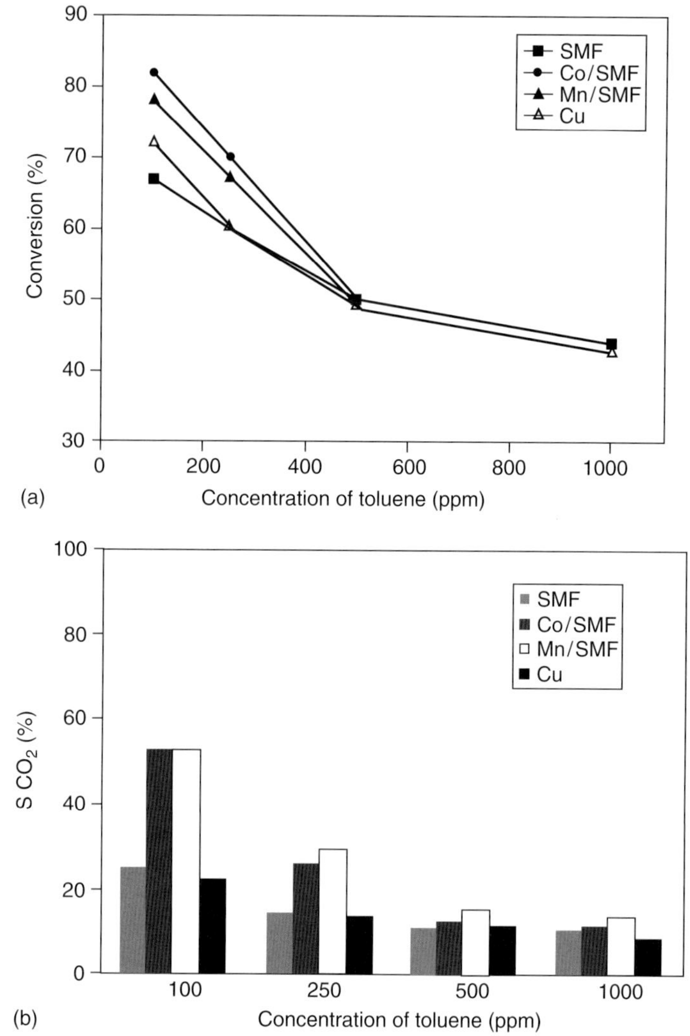

Figure 4.8 (a) Toluene conversion and (b) selectivity to CO_2 in a DBD and in plasma-catalytic experiments, as a function of toluene initial concentration [51].

4.3.4.3 Effect of Oxygen Partial Pressure

The influence of O_2 content on the oxidation of VOCs in plasma-catalytic systems has been investigated by several authors [34, 77, 89]. It was generally found that VOC decomposition in plasma shows a maximum at 2–5% O_2 and decrease for higher O_2 concentration [34, 90]. In order to explain this decrease, the authors suggested that the increase in oxygen partial pressure enhances the consumption of atomic oxygen or ozone formation, which does not contribute to VOC decomposition in gas phase. Kim *et al.* [34] observed that the O_2 partial pressure did not significantly

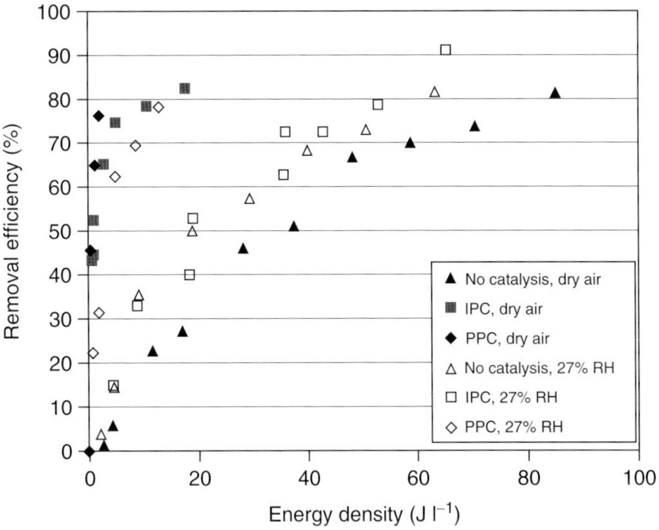

Figure 4.9 Effect of humidity on toluene conversion in a corona discharge and in plasma-catalytic experiments with TiO_2 catalysts placed in plasma and $CuO-MnO_2/TiO_2$ catalysts postplasma [58]. RH, relative humidity.

affect the selectivities for CO and CO_2 obtained from benzene and toluene oxidation in plasma.

In plasma-catalytic systems, different and sometimes contradictory results are obtained by various authors. Some researchers reported a similar trend as in plasma experiments for the removal of toluene [89] and formaldehyde [77], with a maximum toluene conversion achieved at 5% O_2, followed by a decrease at higher oxygen content. An enhancement of the selectivity to CO_2 was found when increasing O_2 content [77]. Kim *et al.* [34] investigated the effect of O_2 concentration on the oxidation of benzene and toluene in a plasma reactor packed with various catalysts, such as Al_2O_3, TiO_2, zeolites, and metals (Ag, Au, Pt, Pd, and Ni) supported on these materials. They found that the increase in O_2 partial pressure enhanced both the VOC conversion and, to a lesser extent, the CO_2 selectivity, regardless of the catalyst used. Some of their results are illustrated in Figure 4.10 [34].

The strong dependence of VOC decomposition on O_2 content was explained by increased formation of reactive oxygen species and their subsequent reactions with VOC molecules on the catalyst surface [34].

4.3.4.4 **Effect of Catalyst Loading**

The amount of catalyst and the concentration of the catalytically active material may influence VOC oxidation in plasma-catalytic systems. More active sites are available for higher catalyst loading, which is expected to have a positive effect on VOC decomposition.

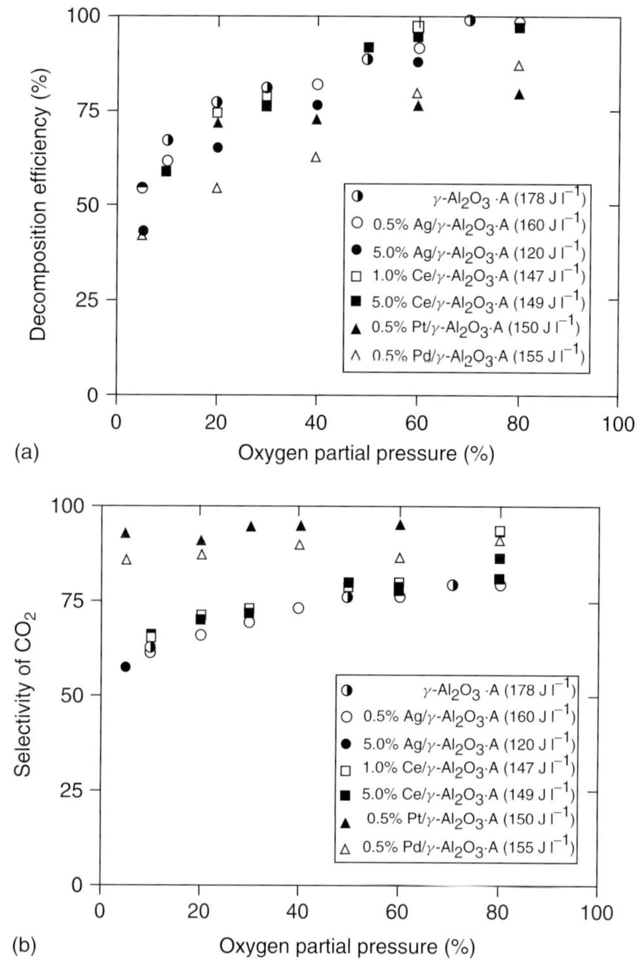

Figure 4.10 Effect of oxygen partial pressure on (a) benzene conversion and on (b) CO_2 selectivity obtained using a plasma reactor packed with various catalysts supported on γ-Al_2O_3 [34].

Ogata *et al.* [60] studied the influence of catalyst amount on toluene and dichloromethane removal in a plasma-catalytic system with Ba–CuO–Cr_2O_3/Al_2O_3 catalyst. They found that when the catalyst was placed in the plasma region, increasing the amount of catalyst had a detrimental effect on VOC decomposition. On the contrary, a higher catalyst amount in the two-stage configuration led to an important enhancement of VOC conversion. Futamura *et al.* [44] observed also an improvement in benzene conversion with the increase in the amount of MnO_2 catalyst up to 1.5 g, while a further increase did not have any noticeable effect.

Gold nanoparticles confined in the walls of mesoporous silica have been used in [33] for the removal of TCE. Catalysts containing different weight percentage

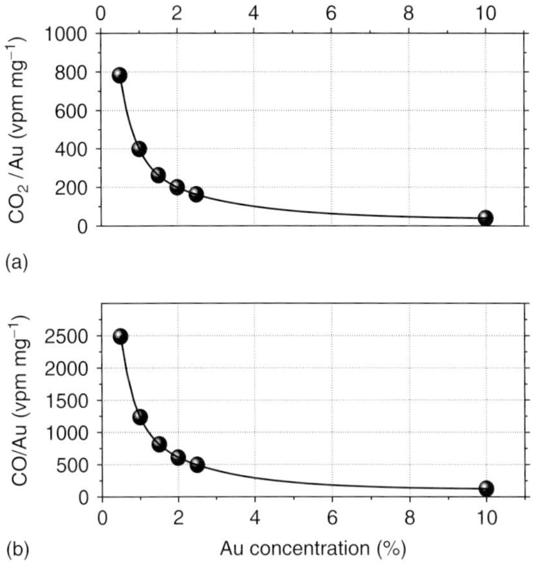

Figure 4.11 The concentrations of (a) evolved CO_2 and (b) CO divided by the amount of Au used, as a function of the Au concentration in the catalysts investigated [33].

of Au (0.5–10%) have been tested in a two-stage plasma-catalytic configuration. As shown in Figure 4.11, by correlating the resulting concentrations of CO_2 and CO with the amount of Au present in the catalysts, a decrease in the oxidation ability of the catalysts was observed with an increase in the Au content. Therefore, the catalyst containing the least amount of Au was the most effective for TCE oxidation.

The effect of the loading amount of Ag supported on TiO_2 and γ-Al_2O_3 on benzene oxidation in a packed-bed reactor has been investigated by Kim *et al.* [34]. Increasing the Ag content from 0 to 2% did not generate any major differences in the decomposition efficiency, and higher Ag concentration even reduced benzene conversion. The authors suggested that this effect might be due to the decrease in the BET surface area of catalysts with increasing Ag content [34].

4.3.5 Combination of Plasma Catalysis and Adsorption

Adsorption is one of the established methods for VOC removal. In addition, it was previously shown that total oxidation of VOCs in plasma-catalytic systems is also strongly influenced by their adsorption on the catalyst material. The total oxidation of hydrocarbons immobilized on porous materials using plasma has been proved very effective [83]. The authors attributed the remarkably higher selectivity to CO_2 observed in the presence of porous catalysts to longer residence time of intermediates in the discharge zone as compared to the case of nonporous materials or gas-phase discharge reactors [83].

A cycled system has been proposed by Kim *et al.* [34, 91], consisting of an adsorption step followed by subsequent decomposition of adsorbed VOCs using O_2 plasma catalysis. Various materials have been tested in this cycled system, such as metals supported on TiO_2 or on γ-Al_2O_3 and zeolites packed in the plasma reactor. The best results were obtained with Ag, Au, and Pt supported on TiO_2. This cycled system proved very promising for VOC oxidation, since it is able to oxidize completely the adsorbed VOCs to CO_2 with high energy efficiency. In addition, operating the plasma in oxygen prevents the formation of nitrogen oxides, which is otherwise unavoidable [34].

A similar system has been employed by Fan *et al.* [41] for the oxidation of low concentrations of benzene in a DBD reactor packed with Ag/HZSM-5. They found that the Ag loading of HZSM zeolite enhanced significantly its adsorption capacity even in humid air. Complete oxidation of the benzene to CO_2 has also been achieved and the catalyst was stable over several operation cycles [41].

4.3.6
Comparison between Catalysis and Plasma Catalysis

Francke *et al.* [43] showed a comparison between the conventional catalytic combustion and plasma catalysis for the removal of VOCs from the off-gas of an air-stripping groundwater-cleaning plant. Using a pulsed corona discharge in combination with a Fe–Mn oxide catalyst at 90 °C, they obtained over 85% removal of all contaminants in the gas stream (toluene, xylene, ethylbenzene, vinyl chloride, and 1,2-dichloroethene). In conventional catalysis, the catalyst was heated to 420 °C, while under plasma-catalytic conditions the energy deposited in the plasma corresponded to an equivalent temperature increase of only 35 °C [43].

Several authors compared plasma catalysis and conventional catalysis for the decomposition of benzene and toluene [34, 35, 65], propane and propene [37], and dichloromethane [38]. In these experiments, various catalysts (MnO_2/γ-Al_2O_3, Pt/γ-Al_2O_3, Ag/γ-Al_2O_3, Ag/γ-TiO_2, γ-Al_2O_3, and TiO_2) have been placed in a $BaTiO_3$ packed-bed reactor, among the $BaTiO_3$ pellets. The results obtained by Blackbeard *et al.* [37] for propane oxidation are shown in Figure 4.12.

It was generally found that plasma catalysis is more efficient than conventional catalysis over the entire temperature range investigated. The difference is more remarkable at low temperature, below the threshold for thermal activation of the catalyst, where plasma processes and plasma activation of the catalyst are dominant. Further increase in the temperature results in thermal activation of the catalyst and the difference between the results obtained with plasma catalysis and with conventional catalysis is less striking [37, 65]. Whitehead [38] reported that over 30% energy saving can be achieved using plasma catalysis, since lower electrical power supplied to the heater is required when using plasma to achieve the same conversion as in thermal catalysis.

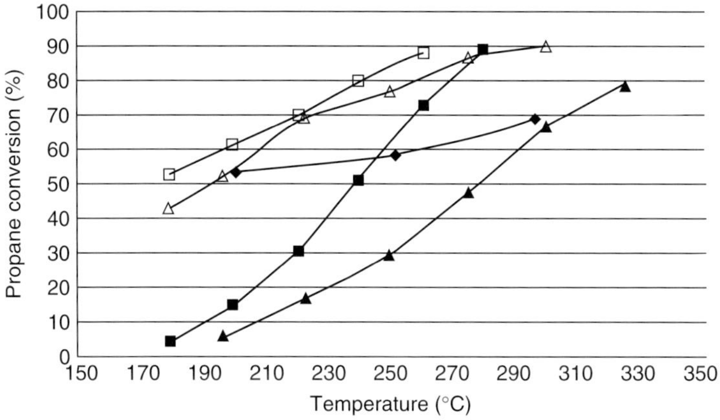

Figure 4.12 Conversion of propane by catalysis and plasma catalysis as a function of temperature using Pt and MnO_2 catalysts (♦ plasma alone; ■ Pt/γ-Al_2O_3 thermal; □ Pt/γ-Al_2O_3 plasma catalysis; ▲ MnO_2/γ-Al_2O_3 thermal; and Δ MnO_2/γ-Al_2O_3 plasma catalysis) [37].

4.3.7 Comparison between Single-Stage and Two-Stage Plasma Catalysis

Several studies reported comparisons between single-stage and two-stage plasma-catalytic systems from the point of view of the performance of VOC removal. This is of interest both for optimization purposes and for getting an insight into the mechanisms responsible for VOC decomposition in the two situations. The results differ mainly as a function of the catalyst used. Harling *et al.* [65] used Ag/TiO_2 and Ag/Al_2O_3 catalysts over a wide temperature range and obtained higher decomposition efficiency of benzene and toluene in single-stage configuration as compared to the two-stage system. A similar behavior has been observed by Van Durme *et al.* [58] in a corona discharge with TiO_2 catalysts. On the contrary, with a CuO–MnO_2/TiO_2 catalysts, the authors achieved significantly higher toluene removal in the two-stage configuration, even in humid air [58]. Mixed transition metal oxides (CuO–Cr_2O_3/Al_2O_3) have been also used in [60]. It was found that placing the catalyst in the discharge region did not improve toluene conversion as compared to the plasma results. In two-stage configuration, an enhancement of the removal efficiency could be achieved with a relatively high amount of catalyst [60].

Roland *et al.* [28, 92] proved that in two-stage configuration the ability of the catalyst to decompose ozone is essential for the effective oxidation of VOCs. Several nonvolatile hydrocarbons were successfully decomposed using γ-Al_2O_3 downstream of the plasma reactor, while in the presence of α-Al_2O_3, silica gel, or quartz no conversion was detected. However, in single-stage configuration, all the investigated materials performed equally well, so the authors concluded that in this case, short-lived species formed in the plasma generate other reaction pathways for the oxidation of hydrocarbons [28, 92].

4.3.8 Reaction By-Products

4.3.8.1 Organic By-Products

Various gaseous-carbon-containing compounds have been reported as a result of VOC oxidation in nonthermal plasma. Table 4.3 summarizes hydrocarbon by-products detected in experiments of plasma oxidation of several VOCs.

It was generally observed that the concentrations of gaseous hydrocarbons resulting from VOC plasma oxidation are diminished in the presence of catalysts [36, 68]. For example, Chang *et al.* [36] reported that the aromatic hydrocarbons obtained from styrene oxidation in a DBD can be completely oxidized when the plasma reactor was packed with Pt–Pd/Al_2O_3 pellets. Lee *et al.* [68] found that the formation of by-products from benzene treatment in a DBD was suppressed in the presence of TiO_2/Al_2O_3 catalysts. On the contrary, Jarrige and Vervisch [55] detected acetone as the main reaction product of isopropyl alcohol oxidation in a pulsed corona discharge, and they found that the addition of MnO_2/Al_2O_3 catalysts downstream of the plasma reactor led to an increase in the formation yield of acetone under their experimental conditions.

In case of TCE oxidation, Oda *et al.* [54] observed a significant decrease in the concentration of dichloroacetyl chloride (DCAC) in the presence of MnO_2 catalysts, while the concentration of trichloroacetaldehyde (TCAA) shows a marked increase at low input energy. At higher input energy, both these chlorinated by-products are

Table 4.3 Hydrocarbon by-products detected in experiments of plasma oxidation of VOCs.

VOC	Type of discharge	Reaction by-products	References
Propane	Pulsed corona	Formaldehyde, methyl nitrate	[55]
Isopropyl alcohol		Acetone, formaldehyde	
Acetylene	DBD	Formic acid	[70]
Trichloroethylene	DBD	Dichloro acetylchloride, trichloroacetaldehyde, phosgene	[45, 54]
Trichloroethylene	Pulsed corona	Dichloro acetylchloride	[93]
Trichloroethylene	DBD	Dichloroacetyl chloride, trichloroacetyl chloride, trichloroacetaldehyde, chloro-octane, chlorononane	[53]
Benzene	DBD	Phenol, benzenediol, benzaldehyde, benzoic acid, pyridine	[68]
Toluene	DC corona	Formic acid, benzaldehyde, benzyl alcohol, nitrophenol, methylnitrophenols, methyldinitrophenols, methylpropylfuran	[88]
Styrene	DBD packed bed (glass)	Benzene, toluene, benzene acetaldehyde	[36]

oxidized on the catalyst surface. Complete decomposition of chlorine-containing products of TCE plasma oxidation was reported in [53] for MnO_x catalysts, while on the contrary, on Au catalysts the authors reported a slight increase in chlorinated by-product concentrations [33].

The formation of solid reaction by-products have also been reported frequently, especially for the oxidation of aromatic VOCs [28, 36, 46, 57, 69]. For high VOC concentrations, these polymeric deposits can represent the main products of VOC treatment. Chang *et al.* [36] reported that solid-phase products account for 72–91% of the styrene decomposed in a DBD for initial styrene concentrations in the range 132–1008 ppm. Similar results were obtained for toluene oxidation in a DBD, where up to 80% of the converted toluene is transformed into solid-polymeric deposits [28]. Delagrange *et al.* [57] identified benzoic acid as the main component of the carbonaceous deposit formed on Al_2O_3 pellets during nonthermal plasma oxidation of toluene. Various carboxylic acids containing one, two, or four carbon atoms have been identified on TiO_2 coated glass fibers placed in a DBD reactor during plasma-catalytic oxidation of acetylene [70].

Roland *et al.* [28] found that the formation of polymeric deposits during the oxidation of phenol and toluene is diminished in a ferroelectric packed-bed reactor ($PbZrO_3$–$PbTiO_3$ packing) as compared to an empty discharge reactor. The addition of a $LaCoO_3$ catalyst led to a further reduction in the polymeric deposits [28]. Chang *et al.* [36] also reported smaller amounts of solid-phase products in a packed-bed reactor (glass or Al_2O_3 packing) than in a nonpacked DBD. They succeeded in reducing further the carbonaceous deposits by adding a Pt–Pd/Al_2O_3 catalyst.

4.3.8.2 **Inorganic By-Products**

The most common reaction product that results from VOC oxidation in plasma, besides CO_2, is carbon monoxide. Numerous authors have reported high concentrations of CO, comparable to or exceeding the CO_2 concentrations, as shown in Tables 4.1 and 4.2. The results summarized in these tables prove that the addition of catalysts increases significantly the CO_2 to CO ratio in most cases.

When the plasma is operated in air, the formation of ozone and nitrogen oxides is unavoidable. These are also harmful by-products, and therefore the need to remove them from the effluent gas of plasma-catalytic VOC treatment should not be neglected.

In most studies, it was found that the concentration of ozone generated in the plasma increases linearly with the SIE [26, 32, 46, 41, 58]. Values between several hundreds and several thousands parts per million of ozone have been reported. Holzer *et al.* [26] observed a different trend in a ferroelectric packed-bed reactor, where the O_3 concentration reached a maximum at intermediate SIE and then decreased with further increasing SIE. The authors suggested that this behavior is due to thermal decomposition of ozone in high temperature regions associated with hot spots [26].

Some of the catalysts used in combination with plasma for VOC oxidation have a beneficial effect on the O_3 removal as well. Van Durme *et al.* [32] concluded that mainly the active compound, rather than the support material, determines the rate of O_3 decomposition. One of the most effective materials from the point of view of ozone removal was MnO_2, which can successfully decompose ozone both when placed directly in the plasma region [49, 52, 67] and downstream of the plasma reactor [44, 46, 55, 56, 41]. Ozone decomposition was promoted by increasing the MnO_2 loading [44, 49]. Mixed transition metal oxides such as $CuO-Cr_2O_3$ [60], $CuO-MnO_2$ [32, 56, 58], $Fe_2O_3-MnO_2$ [32] and noble metals such as Pt and Pd [32, 36] were also reported to reduce O_3 concentration in plasma-catalytic experiments for VOC oxidation. On the contrary, it was found that TiO_2 catalysts did not produce a significant O_3 reduction [46, 58, 82].

The formation of nitrogen oxides in plasma-catalytic experiments for VOC oxidation, as well as the possibilities of their removal, has been investigated. It was found that the formation of nitrogen oxides strongly depends on the plasma operating conditions and gas composition. Nitric oxide (NO) was generally not detected due to efficient oxidation of NO in the presence of ozone and atomic oxygen [41, 58]. Nitrogen dioxide (NO_2) and nitrous oxide (N_2O) have been detected in many works [31, 34, 40, 44, 41, 58], and the reported concentrations reach several tens of parts per million [34, 40]. The formation of N_2O showed a linear relation to the SIE, while for NO_2 a quadratic increase was observed [40]. Kim *et al.* [34] investigated the formation of NO_2 and N_2O in a packed-bed reactor as a function of the O_2 partial pressure. They found that the formation of NO_2 changed drastically with O_2 content: a maximum was observed at 10–20% O_2 followed by a decreased with a further increase in O_2. N_2O concentration was affected by the O_2 partial pressure below 10%. Clearly, the increase in O_2 partial pressure can reduce the formation of nitrogen oxides, but for their complete elimination, O_2 plasma is necessary [34]. The presence of water vapor also reduces significantly the formation of NO_2 [58].

The presence of catalysts was found to be beneficial for the removal of nitrogen oxides. Ogata *et al.* [31] reported that in the presence of γ-Al_2O_3 pellets placed among the ferroelectric material in a packed-bed reactor the formation of N_2O was reduced. This effect was enhanced by using Ag, Cu, Mo, and Ni supported on Al_2O_3. Suppression of NO_x formation was obtained using molecular sieves [31]. NO_2 concentrations were also reduced by using ferrierite or Ag/γ-Al_2O_3 as packing material [34] or with the TiO_2 catalyst placed inside the discharge region [58]. Effective removal of NO_2 was obtained as well using Mn oxide [44, 41] and $CuO-MnO_2$ catalysts [58] placed downstream of the plasma reactor.

4.4 Concluding Remarks

The decomposition of VOCs is an intensely studied topic because of its relevance to environmental research. During the past years, an increased interest was directed to the oxidation of VOCs by nonthermal plasma or by the combination of nonthermal plasma and heterogeneous catalysis. Plasma treatment shares the advantage of thermal combustion: no condensate or saturated adsorber needs to be disposed of.

On the other hand, the application of plasmas may be limited by their consumption of electrical energy. In addition, owing to their high-energy nature, some difficult to control by-products such as carbon monoxide, nitrogen oxides, or ozone are formed, which must be controlled. A promising solution to these problems is the use of catalysts in combination with plasma. Enhanced decomposition and a shift of the products distribution toward total oxidation have been generally observed in plasma-catalytic hybrid systems, as compared with plasma alone. In comparison with conventional catalysis, plasma catalysis improves the results especially at low temperature.

References

1. Kesselmeier, J. and Staudt, M. (1999) Biogenic volatile organic compounds (VOC): an overview on emission, physiology and ecology. *J. Atmos. Chem.*, **33** (1), 23–88.
2. Karl, T., Guenther, A., Spirig, C., Hansel, A., and Fall, R. (2003) Seasonal variation of biogenic VOC emissions above a mixed hardwood forest in northern Michigan. *Geophys. Res. Lett.*, **30** (23), 2186–2189.
3. Kesselmeier, J., Kuhn, U., Wolf, A., Andreae, M.O., Ciccioli, P., Brancaleoni, E., Frattoni, M., Guenther, A., Greenberg, J., De Castro Vasconcellos, P., Telles de Oliva Tavares, T., Artaxo, P. (2000) Atmospheric volatile organic compounds (VOC) at a remote tropical forest site in central Amazonia. *Atmos. Environ.*, **34** (24), 4063–4072.
4. Theloke, J. and Friedrich, R. (2007) Compilation of a database on the composition of anthropogenic VOC emissions for atmospheric modeling in Europe. *Atmos. Environ.*, **41** (19), 4148–4160.
5. Jenkin, M.E. and Clemitshaw, K.C. (2000) Ozone and other secondary photochemical pollutants: chemical processes governing their formation in the planetary boundary layer. *Atmos. Environ.*, **34** (16), 2499–2527.
6. Bowman, F.M., Pilinis, C., and Seinfeld, J.H. (1995) Ozone and aerosol productivity of reactive organics. *Atmos. Environ.*, **29** (5), 579–589.
7. Bowman, F.M. and Seinfeld, J.H. (1994) Ozone productivity of atmospheric organics. *J. Geophys. Res.*, **99** (D3), 5309–5324.
8. Sillman, S. (1999) The relation between ozone, NO_x and hydrocarbons in urban and polluted rural environments. *Atmos. Environ.*, **33** (12), 1821–1845.
9. Ito, A., Sillman, S., and Penner, J.E. (2007) Effects of additional nonmethane volatile organic compounds, organic nitrates, and direct emissions of oxygenated organic species on global tropospheric chemistry. *J. Geophys. Res.*, **112** (D6), D06309. doi:
10. Lippmann, M. (1991) Health effects of tropospheric ozone. *Environ. Sci. Technol.*, **25** (12), 1954–1962.
11. Ostro, B.D. (1993) Examining acute health outcomes due to ozone exposure and their subsequent relationship to chronic disease outcomes. *Environ. Health Perspect.*, **101** (4), 213–216.
12. Wolkoff, P., Wilkins, C.K., Clausen, P.A., and Nielsen, G.D. (2006) Organic compounds in office environments – sensory irritation, odor, measurements and the role of reactive chemistry. *Indoor Air*, **16** (1), 7–19.
13. Ernstgård, L., Löf, A., Wieslander, G., Norbäck, D., and Johanson, G. (2007) Acute effects of some volatile organic compounds emitted from water-based paints. *J. Occup. Environ. Med.*, **49** (8), 880–889.
14. O'Toole, T.E., Conklin, D.J., and Bhatnagar, A. (2008) Environmental risk factors for heart disease. *Rev. Environ. Health*, **23** (3), 167–202.
15. Irigaray, P., Newby, J.A., Clapp, R., Hardell, L., Howard, V., Montagnier,

L., Epstein, S., and Belpomme, D. (2007) Lifestyle-related factors and environmental agents causing cancer: an overview. *Biomed. Pharmacother.*, **61** (10), 640–658.

16. Clapp, R.W., Howe, G.K., and Jacobs, M.M. (2007) Environmental and occupational causes of cancer: a call to act on what we know. *Biomed. Pharmacother.*, **61** (10), 631–639.
17. Rumchev, K., Brown, H., and Spickett, J. (2007) Volatile organic compounds: do they present a risk to our health? *Rev. Environ. Health*, **22** (1), 39–55.
18. Bernstein, J.A., Alexis, N., Bacchus, H., Bernstein, I.L., Fritz, P., Horner, E., Li, N., Mason, S., Nel, A., Oullette, J., Reijula, K., Reponen, T., Seltzer, J., Smith, A., and Tarlo, S.M. (2008) The health effects of nonindustrial indoor air pollution. *J. Allergy Clin. Immunol.*, **121** (3), 585–591.
19. Kim, H.-H., Ogata, A., and Futamura, S. (2006) in *Trends in Catalysis Research* (ed. L.P. Bevy), Nova Science Publishers, Inc., pp. 1–50.
20. Urashima, K. and Chang, J.-S. (2000) Removal of volatile organic compounds from air streams and industrial flue gases by non-thermal plasma technology. *IEEE Trans. Dielectrics Electr. Insul.*, **7** (5), 602–614.
21. Schnelle, K.B., Jr. and Brown, C.A. (2002) *Air Pollution Control Technology Handbook*, CRC Press LLC, ISBN-0-8493-9588-7.
22. Eliasson, B. and Kogelschatz, U. (1991) Nonequilibrium volume plasma chemical processing. *IEEE Trans. Plasma Sci.*, **19** (6), 1063–1077.
23. Rutgers, W.A. and van Veldhuizen, E.M. (2000) in *Electrical Discharges for Environmental Purposes: Fundamentals and Applications* (ed. E.M. van Veldhuizen), Nova Science Publishers, Inc., pp. 5–20.
24. Kogelschatz, U. and Salge, J. (2008) in *Low Temperature Plasmas: Fundamentals, Technologies and Techniques* (eds R. Hippler, H. Kersten, M. Schmidt, and K.H. Schoenbach), Wiley-VCH Verlag GmbH, pp. 439–462.
25. Van Durme, J., Dewulf, J., Leys, C., and Van Langenhove, H. (2007) Combining non-thermal plasma with heterogeneous catalysis in waste gas treatment: a review. *Appl. Catal. B: Environ.*, **78** (3–4), 324–333.
26. Holzer, F., Kopinke, F.D., and Roland, U. (2005) Influence of ferroelectric materials and catalysts on the performance of non-thermal plasma (NTP) for the removal of air pollutants. *Plasma Chem. Plasma Process.*, **25** (6), 595–611.
27. Chen, H.L., Lee, H.M., Chen, S.H., Chang, M.B., Yu, S.J., and Li, S.N. (2009) Removal of volatile organic compounds by single-stage and two-stage plasma catalysis systems: a review of the performance enhancement mechanisms, current status, and suitable applications. *Environ. Sci. Technol.*, **43** (7), 2216–2227.
28. Roland, U., Holzer, F., and Kopinke, F.-D. (2002) Improved oxidation of air pollutants in a non-thermal plasma. *Catal. Today*, **73** (3–4), 315–323.
29. Spivey, J.J. (1987) Complete catalytic oxidation of volatile organics. *Ind. Eng. Chem. Res.*, **26** (11), 2165–2180.
30. Everaert, K. and Baeyens, J. (2004) Catalytic combustion of volatile organic compounds. *J. Hazard. Mater.*, **109** (1–3), 113–139.
31. Ogata, A., Einaga, H., Kabashima, H., Futamura, S., Kushiyama, S., and Kim, H.-H. (2003) Effective combination of nonthermal plasma and catalysts for decomposition of benzene in air. *Appl. Catal. B: Environ.*, **46** (1), 87–95.
32. Van Durme, J., Dewulf, J., Demeestere, K., Leys, C., and Van Langenhove, H. (2009) Post-plasma catalytic technology for the removal of toluene from indoor air: Effect of humidity. *Appl. Catal. B: Environ.*, **87** (1–2), 78–83.
33. Magureanu, M., Mandache, N.B., Hu, J., Richards, R., Florea, M., and Parvulescu, V.I. (2007) Plasma-assisted catalysis total oxidation of trichloroethylene over gold nano-particles embedded in SBA-15 catalysts. *Appl. Catal. B: Environ.*, **76** (3–4), 275–281.
34. Kim, H.-H., Ogata, A., and Futamura, S. (2008) Oxygen partial pressure-dependent behavior of various catalysts for the total oxidation of VOCs using cycled system of adsorption and

oxygen plasma. *Appl. Catal. B: Environ.*, **79** (4), 356–367.

35. Harling, A.M., Kim, H.-H., Futamura, S., and Whitehead, J.C. (2007) Temperature dependence of plasma-catalysis using a nonthermal, atmospheric pressure packed bed; the destruction of benzene and toluene. *J. Phys. Chem. C*, **111** (13), 5090–5095.
36. Chang, C.-L., Bai, H., and Lu, S.-J. (2005) Destruction of styrene in an air stream by packed dielectric barrier discharge reactors. *Plasma Chem. Plasma Process.*, **25** (6), 641–656.
37. Blackbeard, T., Demidyuk, V., Hill, S.L., and Whitehead, J.C. (2009) The effect of temperature on the plasma-catalytic destruction of propane and propene: a comparison with thermal catalysis. *Plasma Chem. Plasma Process.*, **29** (6), 411–419.
38. Whitehead, J.C. (2010) Plasma catalysis: a solution for environmental problems. *Pure Appl. Chem.*, **82** (6), 1329–1336.
39. Kim, H.-H., Lee, Y.-H., Ogata, A., and Futamura, S. (2003) Plasma-driven catalyst processing packed with photocatalyst for gas-phase benzene decomposition. *Catal. Commun.*, **4** (7), 347–351.
40. Kim, H.-H., Oh, S.-M., Ogata, A., and Futamura, S. (2005) Decomposition of gas-phase benzene using plasma-driven catalyst (PDC) reactor packed with Ag/TiO_2 catalyst. *Appl. Catal. B: Environ.*, **56** (3), 213–220.
41. Fan, X., Zhu, T.L., Wang, M.Y., and Li, X.M. (2009) Removal of low-concentration BTX in air using a combined plasma catalysis system. *Chemosphere*, **75** (10), 1301–1306.
42. Li, W.B., Wang, J.X., and Gong, H. (2009) Catalytic combustion of VOCs on non-noble metal catalysts. *Catal. Today*, **148** (1–2), 81–87.
43. Francke, K.-P., Miessner, H., and Rudolph, R. (2000) Plasmacatalytic processes for environmental problems. *Catal. Today*, **59** (3–4), 411–416.
44. Futamura, S., Zhang, A., Einaga, H., and Kabashima, H. (2002) Involvement of catalyst materials in nonthermal plasma chemical processing of hazardous air pollutants. *Catal. Today*, **72** (3–4), 259–265.
45. Han, S.-B. and Oda, T. (2007) Decomposition mechanism of trichloroethylene based on by-product distribution in the hybrid barrier discharge plasma process. *Plasma Sources Sci. Technol.*, **16** (2), 413–421.
46. Futamura, S., Einaga, H., Kabashima, H., and Hwan, L.Y. (2004) Synergistic effect of silent discharge plasma and catalysts on benzene decomposition. *Catal. Today*, **89** (1–2), 89–95.
47. Magureanu, M., Mandache, N.B., Elloy, P., Gaigneaux, E.M., and Parvulescu, V.I. (2005) Plasma-assisted catalysis for volatile organic compounds abatement. *Appl. Catal. B: Environ.*, **61** (1–2), 12–20.
48. Magureanu, M., Mandache, N.B., Gaigneaux, E.M., Paun, C., and Parvulescu, V.I. (2006) Toluene oxidation in a plasma-catalytic system. *J. Appl. Phys.*, **99** (12), 301–308.
49. Zhu, T., Li, J., Liang, W., and Jin, Y. (2009) Synergistic effect of catalyst for oxidation removal of toluene. *J. Hazard. Mater.*, **165** (1–3), 1258–1260.
50. Chang, C.-L. and Lin, T.-S. (2005) Elimination of carbon monoxide in the gas streams by dielectric barrier discharge systems with Mn catalyst. *Plasma Chem. Plasma Process.*, **25** (4), 387–401.
51. Subrahmanyam, C., Magureanu, M., Renken, A., and Kiwi-Minsker, L. (2006) Catalytic abatement of volatile organic compounds assisted by non-thermal plasma. Part 1: a novel dielectric barrier discharge reactor containing catalytic electrode. *Appl. Catal. B: Environ.*, **65** (1–2), 150–156.
52. Subrahmanyam, Ch., Renken, A., and Kiwi-Minsker, L. (2006) Catalytic abatement of volatile organic compounds assisted by non-thermal plasma: Part II. Optimized catalytic electrode and operating conditions. *Appl. Catal. B: Environ.*, **65** (1–2), 157–162.
53. Magureanu, M., Mandache, N.B., Parvulescu, V.I., Subrahmanyam, Ch., Renken, A., and Kiwi-Minsker, L. (2007) Improved performance of non-thermal plasma reactor during decomposition of trichloroethylene: optimization of the reactor geometry and introduction

of catalytic electrode. *Appl. Catal. B: Environ.*, **74** (3–4), 270–277.

54. Han, S.-B., Oda, T., and Ono, R. (2005) Improvement of the energy efficiency in the decomposition of dilute trichloroethylene by the barrier discharge plasma process. *IEEE Trans. Ind. Appl.*, **41** (5), 1343–1349.
55. Jarrige, J. and Vervisch, P. (2009) Plasma-enhanced catalysis of propane and isopropyl alcohol at ambient temperature on a MnO_2-based catalyst. *Appl. Catal. B: Environ.*, **90** (1–2), 74–82.
56. Harling, A.M., Glover, D.J., Whitehead, J.C., and Zhang, K. (2009) The role of ozone in the plasma-catalytic destruction of environmental pollutants. *Appl. Catal. B: Environ.*, **90** (1–2), 157–161.
57. Delagrange, S., Pinard, L., and Tatibouet, J.-M. (2006) Combination of a non-thermal plasma and a catalyst for toluene removal from air: Manganese based oxide catalysts. *Appl. Catal. B: Environ.*, **68** (3–4), 92–98.
58. Van Durme, J., Dewulf, J., Sysmans, W., Leys, C., and Van Langenhove, H. (2007) Efficient toluene abatement in indoor air by a plasma catalytic hybrid system. *Appl. Catal. B: Environ.*, **74** (1–2), 161–169.
59. Mok, Y.S. (2006) Behaviour of trichloroethylene decomposition in a plasma-catalytic combined process. *Plasma Sci. Technol.*, **8** (6), 661–665.
60. Ogata, A., Saito, K., Kim, H.-H., Sugasawa, M., Aritani, H., and Einaga, H. (2010) Performance of an ozone decomposition catalyst in hybrid plasma reactors for volatile organic compound removal. *Plasma Chem. Plasma Process.*, **30** (1), 33–42.
61. Oda, T., Yamaji, K., and Takahashi, T. (2004) Decomposition of dilute trichloroethylene by nonthermal plasma processing – gas flow rate, catalyst, and ozone effect. *IEEE Trans. Ind. Appl.*, **40** (2), 430–436.
62. Zou, L., Luo, Y., Hooper, M., and Hu, E. (2006) Removal of VOCs by photocatalysis process using adsorption enhanced TiO_2-SiO_2 catalyst. *Chem. Eng. Process.*, **45** (11), 959–964.
63. Mo, J., Zhang, Y., Xu, Q., Joaquin Lamson, J., and Zhao, R. (2009) Photocatalytic purification of volatile organic compounds in indoor air: a literature review. *Atmos. Environ.*, **43** (14), 2229–2246.
64. Zhao, J. and Yang, X. (2003) Photocatalytic oxidation for indoor air purification: a literature review. *Build. Environ.*, **38** (5), 645–654.
65. Harling, A.M., Demidyuk, V., Fischer, S.J., and Whitehead, J.C. (2008) Plasma-catalysis destruction of aromatics for environmental clean-up: effect of temperature and configuration. *Appl. Catal. B: Environ.*, **82** (3–4), 180–189.
66. Li, D., Yakushiji, D., Kanazawa, S., Ohkubo, T., and Nomoto, Y. (2002) Decomposition of toluene by streamer corona discharge with catalyst. *J. Electrostat.*, **55** (3–4), 311–319.
67. Subrahmanyam, Ch., Magureanu, M., Laub, D., Renken, A., and Kiwi-Minsker, L. (2007) Nonthermal plasma abatement of trichloroethylene enhanced by photocatalysis. *J. Phys. Chem. C*, **111** (11), 4315–4318.
68. Lee, B.-Y., Park, S.-H., Lee, S.-C., Kang, M., and Choung, S.-J. (2004) Decomposition of benzene by using a discharge plasma--photocatalyst hybrid system. *Catal. Today*, **93–95**, 769–776.
69. Chang, C.-L. and Lin, T.-S. (2005) Decomposition of toluene and acetone in packed dielectric barrier discharge reactors. *Plasma Chem. Plasma Process.*, **25** (3), 227–243.
70. Thevenet, F., Guaitella, O., Puzenat, E., Herrmann, J.-M., Rousseau, A., and Guillard, C. (2007) Oxidation of acetylene by photocatalysis coupled with dielectric barrier discharge. *Catal. Today*, **122** (1–2), 186–194.
71. Thevenet, F., Guaitella, O., Puzenat, E., Guillard, C., and Rousseau, A. (2008) Influence of water vapour on plasma/photocatalytic oxidation efficiency of acetylene. *Appl. Catal. B: Environ.*, **84** (3–4), 813–820.
72. Oh, S.-M., Kim, H.-H., Ogata, A., Einaga, H., Futamura, S., and Park, D.-W. (2005) Effect of zeolite in surface discharge plasma on the decomposition of toluene. *Catal. Lett.*, **99** (1–2), 101–104.

73. Subrahmanyam, Ch., Renken, A., and Kiwi-Minsker, L. (2007) Novel catalytic dielectric barrier discharge reactor for gas-phase abatement of isopropanol. *Plasma Chem. Plasma Process.*, **27** (1), 13–22.
74. Jeon, S.G., Kim, K.-H., Shin, D.H., Nho, N.-S., and Lee, K.-H. (2007) Effective combination of non-thermal plasma and catalyst for removal of volatile organic compounds and NO_X. *Korean J. Chem. Eng.*, **24** (3), 522–526.
75. Ayrault, C., Barrault, J., Blin-Simiand, N., Jorand, F., Pasquiers, S., Rousseau, A., and Tatibouët, J.M. (2004) Oxidation of 2-heptanone in air by a DBD-type plasma generated within a honeycomb monolith supported Pt-based catalyst. *Catal. Today*, **89** (1–2), 75–81.
76. Guo, Y.-F., Ye, D.-Q., Chen, K.-F., and He, J.-C. (2007) Toluene removal by a DBD-type plasma combined with metal oxides catalysts supported by nickel foam. *Catal. Today*, **126** (3–4), 328–337.
77. Ding, H.-X., Zhu, A.-M., Yang, X.-F., Li, C.-H., and Xu, Y. (2005) Removal of formaldehyde from gas streams via packed-bed dielectric barrier discharge plasmas. *J. Phys. D Appl. Phys.*, **38** (23), 4160–4167.
78. Kim, H.-H., Oh, S.-M., Ogata, A., and Futamura, S. (2004) Decomposition of benzene using ag/tio2 packed plasma-driven catalyst reactor: influence of electrode configuration and Ag-loading amount. *Catal. Lett.*, **96** (3–4), 189–194.
79. Demidiouk, V., Moon, S.I., and Chae, J.O. (2003) Toluene and butyl acetate removal from air by plasma-catalytic system. *Catal. Commun.*, **4** (2), 51–56.
80. Francke, K.-P., Miessner, H., and Rudolph, R. (2000) Cleaning of air streams from organic pollutants by plasma – catalytic oxidation. *Plasma Chem. Plasma Process.*, **20** (3), 393–403.
81. Magureanu, M., Piroi, D., Mandache, N.B., Parvulescu, V.I., Parvulescu, V., Cojocaru, B., Cadigan, C., Richards, R., Daly, H., and Hardacre, C. (2011) In-situ study of ozone and hybrid plasma catalyst oxidation of toluene with Ag-Al catalysts: evidence of the nature of the active sites. *Appl. Catal. B: Environ.*, **104** (1–2), 84–90.
82. Oda, T., Takahashi, T., and Yamaji, K. (2004) TCE decomposition by the nonthermal plasma process concerning ozone effect. *IEEE Trans. Ind. Appl.*, **40** (5), 1249–1256.
83. Holzer, F., Roland, U., and Kopinke, F.-D. (2002) Combination of non-thermal plasma and heterogeneous catalysis for oxidation of volatile organic compounds Part 1. Accessibility of the intra-particle volume. *Appl. Catal. B: Environ.*, **38** (3), 163–181.
84. Futamura, S., Zhang, A.H., and Yamamoto, T. (1997) The dependence of nonthermal plasma behavior of VOCs on their chemical structures. *J. Electrostat.*, **42** (1–2), 51–62.
85. Pekarek, S., Kriha, V., Pospisil, M., and Viden, I. (2001) Multi hollow needle to plate plasmachemical reactor for pollutant decomposition. *J. Phys. D Appl. Phys.*, **34** (22), L117–L121.
86. Ogata, A., Ito, D., Mizuno, K., Kushiyama, S., Gal, A., and Yamamoto, T. (2002) Effect of coexisting components on aromatic decomposition in a packed-bed plasma reactor. *Appl. Catal. A: Gen.*, **236** (1–2), 9–15.
87. Kim, H.-H., Ogata, A., and Futamura, S. (2005) Atmospheric plasma-driven catalysis for the low temperature decomposition of dilute aromatic compounds. *J. Phys. D Appl. Phys.*, **38** (8), 1292–1300.
88. Van Durme, J., Dewulf, J., Sysmans, W., Leys, C., and Van Langenhove, H. (2007) Abatement and degradation pathways of toluene in indoor air by positive corona discharge. *Chemosphere*, **68** (10), 1821–1829.
89. Guo, Y.-F., Ye, D.-Q., Chen, K.-F., He, J.-C., and Chen, W.-L. (2006) Toluene decomposition using a wire-plate dielectric barrier discharge reactor with manganese oxide catalyst in situ. *J. Mol. Catal. A: Chem.*, **245** (1–2), 93–100.
90. Falkenstein, Z. (1999) Effects of the O_2 concentration on the removal efficiency of volatile organic compounds with dielectric barrier discharges in Ar and N_2. *J. Appl. Phys.*, **85** (1), 525–529.

91. Kim, H.H., Oh, S.M., Ogata, A., and Futamura, S. (2005) Decomposition of gas-phase benzene using plasma-driven catalyst reactor: complete oxidation of adsorbed benzene using oxygen plasma. *J. Adv. Oxidation Technol.*, **8** (2), 226–233.

92. Roland, U., Holzer, F., and Kopinke, F.-D. (2005) Combination of non-thermal plasma and heterogeneous catalysis for oxidation of volatile organic compounds Part 2. Ozone decomposition and deactivation of γ-Al_2O_3. *Appl. Catal. B: Environ.*, **58** (3–4), 217–226.

93. Kirkpatrick, M.J., Finney, W.C., and Locke, B.R. (2003) Chlorinated organic compound removal by gas phase pulsed streamer corona electrical discharge with reticulated vitreous carbon electrodes. *Plasmas Polym.*, **8** (3), 165–177.

5
VOC Removal from Air by Plasma-Assisted Catalysis: Mechanisms, Interactions between Plasma and Catalysts

Christophe Leys and Rino Morent

5.1
Introduction

In the previous chapter, different hybrid plasma-catalytic reactor schemes for volatile organic compound (VOC) abatement were discussed. Ample experimental evidence indicates that the combination of nonthermal plasma (NTP) with heterogeneous catalysts yields an approach that outperforms plasma-alone solutions on a number of aspects including energy cost, mineralization rate, and by-product formation. In some cases, synergetic effects are observed in the sense that, for example, the removal rates obtained with a plasma-catalytic setup are higher than those predicted by simply adding the effects of plasma and catalyst. In this chapter, an account is given of the present understanding of the mechanisms involved in plasma-catalytic processes.

The combination of NTP with heterogeneous catalysts can be divided into two categories depending on the location of the catalyst (Figure 5.1): in-plasma catalysis (IPC) and postplasma catalysis (PPC). The latter is a two-stage process where the catalyst is located downstream of the plasma reactor, while the former is a single-stage process with the catalyst being exposed to the active plasma.

As in classical heterogeneous catalysis, the catalyst material can be introduced in the hybrid system in different ways for both IPC and PPC: in the form of pellets (a so-called packed bed configuration) [1–5], foam [6, 7], or honeycomb monolith [8, 9] as a layer of catalyst material [10] or as a coating on the reactor wall [11, 12] or electrodes [13–16].

In an IPC reactor, the catalyst region can partially or entirely occupy the plasma zone. When pellets are used, the reactor can be packed with purely catalytic pellets or with a mixture of catalytic and noncatalytic pellets. The mechanisms involved in IPC are manifold, as bulk plasma processes and heterogeneous catalytic processes occur simultaneously.

For a PPC reactor, the plasma reactor could be located either downstream or upstream from the catalyst. The latter is by far the most occurring configuration for two major reasons [17]. At first, reaction products generated by a catalyst are typically more stable than those generated by the plasma. Therefore, when the

Plasma Chemistry and Catalysis in Gases and Liquids, First Edition.
Edited by Vasile I. Parvulescu, Monica Magureanu, and Petr Lukes.

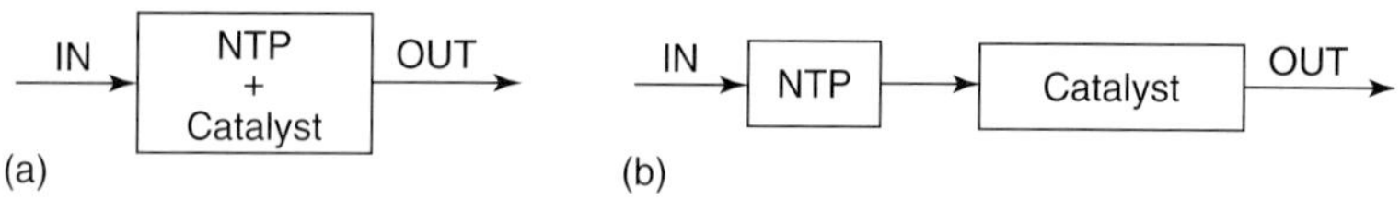

Figure 5.1 Plasma–catalyst combinations: (a) in-plasma catalysis (IPC) and (b) postplasma catalysis (PPC).

plasma reactor is located downstream, the catalytic reaction products will have a minor influence on the subsequent plasma-chemical processes. In contrast, the plasma produces a mixture of active species, some of them sufficiently long living to have an impact on the downstream catalysis. Secondly, unwanted by-products of plasma treatment (e.g., CO, NO_x, O_3, other VOCs, etc.) can be selectively removed by a catalyst placed after the plasma reactor.

The performance enhancement mechanisms in PPC are, in comparison with IPC, relatively simple, as the catalyst is not directly exposed to short-living active species produced in the plasma. The key role fulfilled by the plasma in this case is to alter the gas composition that is fed to the catalyst, by the addition of chemically reactive species or through the conversion of the pollutant into a compound that is easier to decompose catalytically. An advantage of PPC is that optimum working conditions (e.g., temperature) can be independently realized for both the NTP and catalyst.

Plasma-catalytic mechanisms obviously involve catalytic surface reactions, but adsorption processes and the influence of the catalyst on the discharge may play an equally important role in certain aspects of plasma-catalytic synergy [18–20]. In the following sections, we investigate how combining catalysts with NTP affects the physicochemical characteristics of both plasma and catalyst and how these interactions contribute to plasma-catalytic reaction schemes.

Although an attempt is made to present generic mechanisms in this chapter, it is to be noted that the discussion is often based on experimental findings that pertain to a specific combination of VOC, catalyst, and discharge type. On the other hand, it is known that, for instance, for a given catalyst, plasma–catalyst interactions may differ widely depending on the VOC involved (Kirkpatrick *et al.* [13]). Nevertheless, for the full set of experimental details, which are largely omitted in the following sections, the reader is referred to the source paper.

5.2
Influence of the Catalyst in the Plasma Processes

5.2.1
Physical Properties of the Discharge

When a dielectric surface is introduced in the gap of a streamer-type discharge, the discharge mode at least partially changes from bulk streamers to more intense streamers running along the surface (surface flashover) [20]. An example of how the presence of a catalyst in the discharge zone influences the discharge pattern is shown in Figure 5.2. A streamer discharge propagates from the tip of a needle electrode to a mesh electrode. In configuration (a), the mesh is covered with a bed of TiO_2 pellets, partially filling the discharge gap (IPC), while in configuration (b), the pellets are located under the mesh, downstream of the discharge zone (PPC). In the IPC configuration, bright spots appear on the surface of the catalytic bed, marking the existence of regions of enhanced ionization.

In the presented case, the pellets are arranged in a packed bed configuration. In dielectric barrier discharge (DBD) packed bed reactors, microdischarges are generated in the voids between pellets because of the high electric field near the contacting points. The macroscopic effect is that a chemically active discharge can be sustained at lower voltages [22] or, when keeping the voltage constant, the enhanced microdischarge activity leads to an increase in the energy density.

In DBDs, the average energy of the electrons can be influenced by varying the similarity parameter *pd*, the product of pressure and gap length [23]. By introducing, as described in [24], a ceramic foam with an average pore size of 420 μm in a 6 mm gap, the effective gap length is reduced by a factor of 15. The resulting increase in breakdown field leads to a shift in the electron energy distribution function toward higher energies.

Similar field effects can lead to higher average electron energies when the discharge zone is filled with ferroelectric pellets, leading to a more oxidative discharge [1]. An increase in electron temperature is also observed in the presence of zeolite catalysts, which have a very strong electrical field, up to 10 V nm^{-1}, within the zeolite structure [25].

Parameters that influence the effect of the packed bed on the discharge are the dielectric constant of the pellet material, and the size and shape of the pellets. The dielectric constant affects the electric field in the void between the pellets and thereby the mean electron energy. With increasing pellet size, the number of microdischarges decreases, but the amount of charge that is transferred per microdischarge increases. Chen *et al.* [26] concluded that there exists an optimum

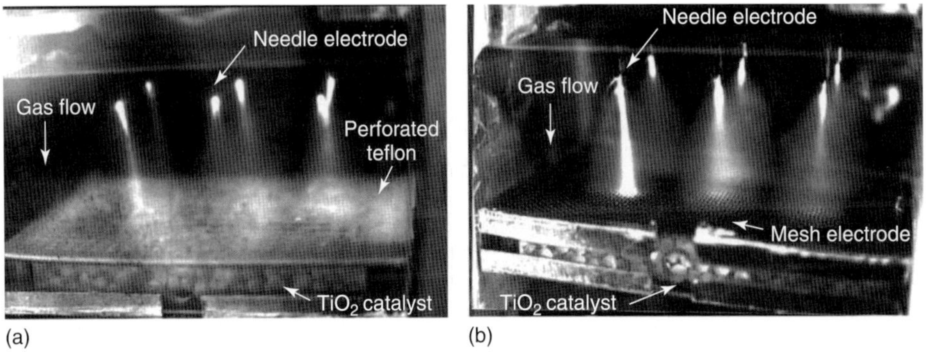

Figure 5.2 Photographs of streamer discharges in (a) IPC and (b) PPC reactor [21].

pellet size for a specific packing material and reactor geometry. They introduced the ratio of the gap length to the pellet size or void fraction as a useful parameter in optimizing packed bed reactors. Finally, the shape of the catalytic pellets matters, as sharp edges induce high local electrical fields and thereby electrons with high energies.

5.2.2 Reactive Species Production

Obviously, when introducing a heterogeneous catalyst changes the physical characteristics of the discharge, the chemical activity will be affected as well.

Roland *et al.* [19] studied the oxidation of various organic substances immobilized on porous and nonporous alumina and silica catalysts and concluded that short-living active species are formed in the pore volume of porous materials when exposed to NTP.

Next to merely enhancing the production of active species, the presence of a particular catalyst can favor the formation of specific species, and thereby influence the abatement mechanism. Chavadej *et al.* [27] showed that when using a TiO_2 catalyst with 1% Pt, the superoxide radical anion $O_2^{\bullet -}$ is preferentially produced, which resulted in an increased CO_2 selectivity.

On the other hand, introducing a catalyst can reduce the concentration of ionic species [28]. However, this effect did not impair the catalyst's role in reducing the emissions of ozone and carbon monoxide for this particular application (indoor air control).

5.3 Influence of the Plasma on the Catalytic Processes

5.3.1 Catalyst Properties

NTPs are used for different surface treatment applications [29–32], including catalyst preparation [33–38]. In the following text, we list a number of examples of plasma-induced modifications of the catalyst surface.

Plasma treatment of the catalyst enhances the dispersion of active catalytic components [6, 39] and influences the stability and catalytic activity of the exposed catalyst material [40].

The oxidation state of the catalyst can be altered by NTP. For instance, when a Mn_2O_3 catalyst is exposed to DBD plasma for 40 h with an energy density of 756 J/l, X-ray diffraction (XRD) spectra reveal the presence of Mn_3O_4, a lower valent manganese oxide with a larger oxidation capability. Owing to plasma–catalyst interactions, less parent Ti–O bonds are found on TiO_2 surfaces after several hours of discharge operation [41]. Even new types of active sites with unusual properties may be formed [18], such as stable Al–O–O* with a lifetime exceeding more than two weeks, as observed in the pores of Al_2O_3 in IPC experiments [18].

Demidyuk and Whitehead [42] have studied the activation mechanism of silver–alumina and manganese oxide catalysts. Plasma treatment decreases the activation energy of silver–alumina because of oxygen radicals generated at the surface [43]. In the case of manganese oxide, the activation energy is unaltered. The catalytic destruction efficiency is increased due to the formation of additional active centers when the oxidation state of the manganese ions is changed [42]. Magureanu *et al.* [44] observed only a dispersion of manganese on the surface of manganese oxide-coated electrodes, but no changes in the oxidation state.

Plasma exposure can result in an increase in the specific surface area or in a change of catalyst structure as evidenced by Guo *et al.* [6, 40] through scanning electron microscopic (SEM) and XRD measurements. Figure 5.3 shows SEM measurements of different catalysts before and after plasma treatment. The granularity of the grain on the catalyst surface becomes smaller and the distribution is more uniform after discharge exposure. This results in the formation of ultrafine particles with higher specific surface and a crystal lattice with a large number of vacancies. These physical changes induce a higher catalytic activity, partially explaining the synergetic effect for this particular plasma-catalytic system. Depending on the catalyst used, plasma treatment can also lead to a decrease in the surface area. For a zeolite catalyst (HZSM-5), the surface area is almost halved, while the reduction in the case of alumina and TiO_2 catalysts is not significant [45].

5.3.2 Adsorption

Adsorption processes play an important role in plasma-catalytic reaction mechanisms. If the catalyst has a significant adsorption capacity for pollutant molecules, it prolongs the pollutant retention time in the reactor. In the case of IPC, the pollutant concentration in the discharge zone is increased. The resulting higher collision probability between pollutant molecules and active species enhances the removal efficiency.

Adsorption of VOC and active species increases with the porosity of the catalyst [46]. Owing to the very large specific surface, zeolites with micropores may adsorb a high amount of VOC over a long period of time, and this can adversely affect the abatement efficiency [47]. However, positive use can be made of the large VOC adsorption on zeolites when the catalyst is placed at the downstream end of an IPC reactor or when operating in a PPC configuration [48, 49]. This is the case for a VOC that is decomposed by active species such as atomic oxygen or ozone, rather than by direct collisions with high-energy electrons.

Under conditions where plasma-generated ozone is not effective in itself to destroy pollutants, high decomposition rates are obtained because of the adsorption of ozone on the catalyst surface and the subsequent dissociation into atomic oxygen species [50]. As ozone is one of the major unwanted by-products in

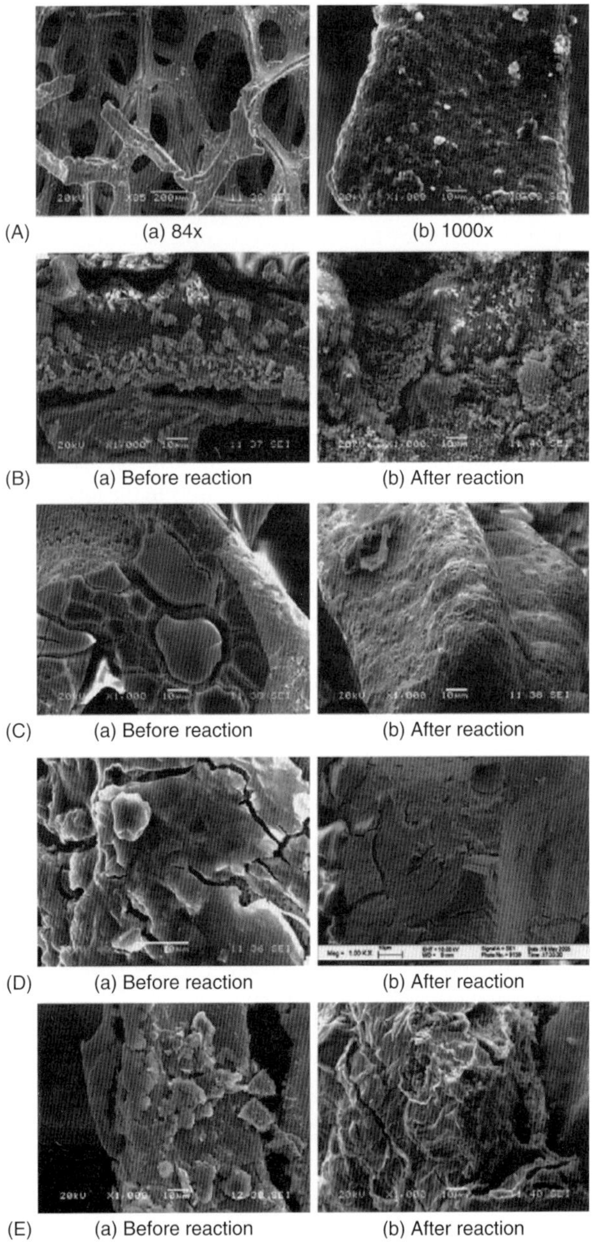

Figure 5.3 (A) SEM micrographs of nickel foam: (a) 85× and (b) 1000×. (B) SEM micrographs of CuO/Al_2O_3/NF catalyst (1000×): (a) before reaction and (b) after reaction. (C) SEM micrographs of Fe_2O_3/Al_2O_3/NF catalyst (1000×): (a) before reaction and (b) after reaction. (D) SEM micrographs of Co_3O_4/Al_2O_3/NF catalyst (1000×): (a) before reaction and (b) after reaction. (E) SEM micrographs of Mn_2O_3/Al_2O_3/NF catalyst (1000×): (a) before reaction and (b) after reaction [40].

NTP-driven gas cleaning, this scenario in which the ozone is decomposed on the catalyst surface, thereby creating active species to oxidize adsorbed VOC, is quite attractive for applications and partly accounts for the synergy observed in different plasma-catalytic systems. Given the long lifetime of ozone near room temperature, this mechanism is at stake both in IPC and PPC configurations. In fact, from experiments with a Cu–Cr catalyst, it was concluded that the ozone decomposition into atomic oxygen is more effective in the two-stage combination than in the single-stage one [51].

The catalyst characteristics that determine the ozone decomposition efficiency are specific surface area and the type of active compound [50]. For instance, ozone is effectively decomposed on a MnO_x/Al_2O_3 catalyst. The proposed mechanism is an electron transfer from Mn^{n+} to an ozone molecule [52]. The ozone is then decomposed into an oxygen molecule and an oxygen atom. Next, unstable $O_2{}^{2-}Mn^{(n+2)+}$ complexes decompose by reducing the oxidized manganese back to Mn^{2+} and desorption of an oxygen molecule.

Humidity is a critical parameter in plasma-catalytic processes. The adsorption of water on the catalyst surface results in a decrease in the reaction probability of the VOC with the surface and therefore reduces the catalyst activity [7]. Van Durme *et al.* [50] investigated the effect of humidity on both ozone and toluene removal for several catalysts located downstream of a DC corona discharge. While humidity had no sizeable effect on catalytic ozone decomposition, the toluene removal efficiency was adversely affected. Moreover, as different toluene removal efficiencies were determined for different catalysts, even when identical masses of ozone were converted, it was concluded that not the amount of decomposed ozone but VOC adsorption is the more critical parameter determining the efficiency of catalytic VOC oxidation. The measured equilibrium adsorption constants were found to differ significantly from catalyst to catalyst and to decrease with increasing humidity, explaining the observed differences at dry and humid conditions between the tested catalysts.

The equilibrium between adsorption and desorption is influenced by the presence of a discharge [8]. According to Lin *et al.* [53], the ionic wind may increase absorption. NTP can also promote desorption of reaction products, which is instrumental in maintaining the catalytic activity.

5.4
Thermal Activation

The prolonged operation of a discharge increases the gas temperature due to elastic and inelastic electron–molecule collisions and vibrational–translational energy relaxation processes. The temperature increase in a given reactor depends on the energy density, that is, the ratio of the input power to the flow rate. Measured temperature rises in plasma-catalytic reactors are of the order of 10–15 K at 10 J l^{-1} [54] and 70 K at 200 J l^{-1} [55].

Although gas heating will result in higher catalyst surface temperatures [56], the heating effect is, in general, too small to account for thermal activation of the catalyst. However, hot spots can be formed in packed bed reactors as a result of localized heating by intense microdischarges that run between sharp edges and corners of adjacent pellets. Evidence of these hot spots was found by Holzer *et al.* [1] when studying ozone degradation in a $BaTiO_3$ packed bed. With the discharge, running ozone degradation was observed at a packed bed temperature of 330 K, whereas without discharge, the packed bed had to be heated to 370 K for the ozone degradation to occur.

Increased catalyst temperatures can promote catalytic VOC removal. This effect is described by Kim *et al.* [55] for benzene decomposition on a Pt/Al_2O_3 catalyst.

5.5 Plasma-Mediated Activation of Photocatalysts

In photocatalysis, VOCs are adsorbed on the surface of a porous semiconductor material that is exposed to UV radiation. The UV photons generate electron–hole pairs, inducing the subsequent oxidation of the adsorbed VOC by valence band holes. In the final step, the oxidation products are desorbed. Among other photocatalysts (e.g., ZnO, ZnS, CdS, Fe_2O_3, and WO_3), TiO_2 is one of the most efficient for the decomposition of a wide range of VOCs. Moreover, the combination of TiO_2 with NTP results in higher oxidation efficiencies and better selectivity into CO_2. In the following paragraph, we focus on this poorly understood case of plasma-catalytic synergy.

As the anatase phase of TiO_2 has a band gap of 3.2 eV, it takes a photon with a wavelength shorter than 388 nm to create an electron–hole pair. Although there are excited nitrogen states that emit light in this wavelength range, there is experimental evidence that photocatalysis induced by UV light from the plasma cannot explain the observed synergy in several hybrid plasma/TiO_2 systems reported in the literature. In an effort to quantify the contribution of plasma-driven photoactivation of TiO_2 to the decomposition of acetaldehyde, Sano *et al.* [12] measured the UV intensity emitted by a surface discharge in N_2/O_2 mixtures. Owing to quenching of the excited N_2 levels, the intensity of UV emission decreased with increasing O_2 content. In atmospheric air, the photocatalytic decomposition rate was calculated to be less than 0.2% of the estimated decomposition rate by the plasma itself, as inferred from literature data on the plasma decomposition of other VOCs in a similar surface discharge reactor [57].

In some cases, TiO_2 shows plasma-induced catalytic activity under conditions where there is no UV or very little emitted by the plasma [27, 58]. Direct plasma activation has been observed when TiO_2 is exposed to an atmospheric pressure argon discharge at room temperature [59]. The question then arises how the plasma-exposed TiO_2 is activated, if not by UV photons. Different mechanisms to bridge the TiO_2 band gap by plasma-driven processes can be envisaged, but to date there is insufficient information to elaborate on the relative importance of electrons, ions, metastables, charging effects, surface recombination, and so on.

Even when the plasma is not effective in activating the photocatalyst, plasma species may still enhance the reactivity of TiO_2, activated by external UV lamps [60, 61]. The synergy between plasma and irradiated TiO_2, as observed in experiments on the plasma-photocatalytic decomposition of acetylene in dry air, is tentatively ascribed to the adsorption of atomic oxygen on TiO_2 nanoparticles supported on silica fibers. When water vapor is added to the gas stream, OH becomes the dominant oxidative species, and as OH radicals, chemisorbed on the catalyst surface, react two orders of magnitude slower with acetylene than with O^*, the synergetic effect disappears. This is another example of how water vapor, a frequent component of industrial effluents, may drastically alter the underpinning mechanisms that apply to synthetic dry air experiments. We conclude our account of the acetylene case study by noting that the prominent role of O^* adsorption on TiO_2 has recently been confirmed in a series of experiments using plasma-treated catalysts [62]. It is revealed that a TiO_2 catalyst pretreated in oxygen-containing plasma promotes C_2H_2 removal in the absence of UV activation, an effect that can be explained by the permanence after plasma exposure of weakly bonded oxygen atoms on the catalyst surface.

5.6 Plasma-Catalytic Mechanisms

In plasma-catalysis, synergetic effects are related to the activation of the catalyst by the plasma. Activation mechanisms include ozone, UV, local heating, changes in work function, activation of lattice oxygen, plasma-induced adsorption/desorption, creation of electron–hole pairs, and direct interaction of gas-phase radicals with adsorbed pollutants [63].

The plasma–catalyst interactions described in the previous sections contribute to one or more of these catalyst activation mechanisms. The presented experimental findings, applying to specific working conditions, may appear as scattered pieces of information. Indeed, further research is needed to connect the loose ends and unravel the detailed mechanisms. However, it is meaningful to try and extract some general pathways at this stage.

Kim *et al.* [63] propose the reaction scheme visualized in Figure 5.4. Two surface reaction models are believed to play a role in plasma-catalysis. In the Langmuir–Hinshelwood (LH) model, both reactants (e.g., a plasma-generated radical and an organic compound) are adsorbed on the catalyst and then migrate to an active site. In the Eley–Rideal (ER) model, reactions occur between an adsorbed reactant and a reactant in the gas phase [64, 65].

In plasma-catalytic systems, reactive surface species (e.g., O^-, O_2^-) can be generated from the lattice oxygen [66] or from plasma-generated O_3 [67] and O [68]. These adsorbed reactive species migrate and eventually recombine or react with adsorbed VOCs. According to Kim *et al.* [55, 63], this occurs mainly on the surface

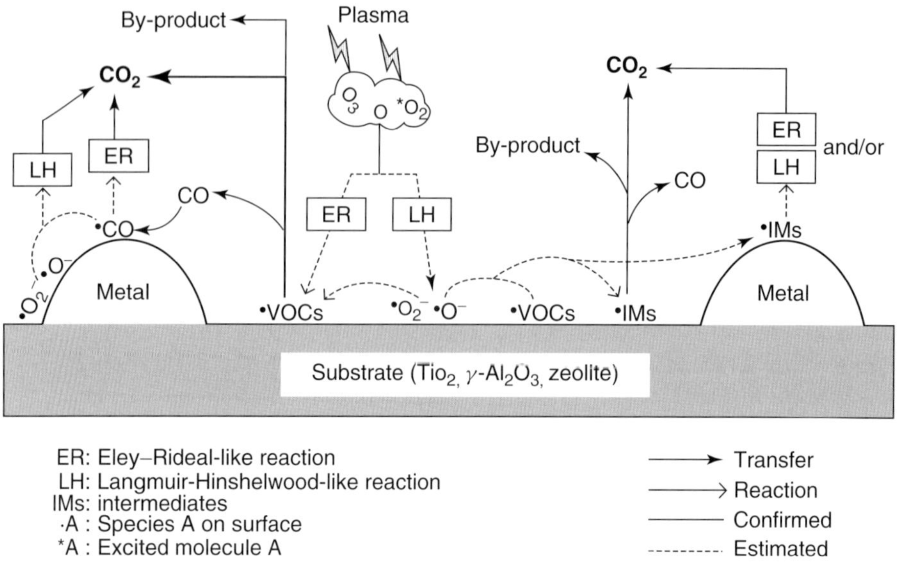

Figure 5.4 Reaction pathways in plasma-driven catalysis. Solid lines are based on experimental observations [63]. Dotted lines are expected pathways without experimental evidence.

of the bare catalysts (TiO_2, γ-Al_2O_3, and zeolite), rather than on the supported metal catalysts. The latter is effective in further oxidizing gas-phase CO. For CO oxidation on the supported metal catalyst, the ER mechanism is invoked, based on experiments with a gold nanoparticle catalyst supported on TiO_2 [69]. However, the LH pathway cannot be excluded. As for other dotted-line reaction pathways in Figure 5.4, much depends on the unresolved behavior of the reactive-oxygen species on the surface.

References

1. Holzer, F., Kopinke, F.D., and Roland, U. (2005) Influence of ferroelectric materials and catalysts on the performance of non-thermal plasma (NTP) for the removal of air pollutants. *Plasma Chem. Plasma Process.*, **25**, 595–611.
2. Morent, R., Dewulf, J., Steenhaut, N., Leys, C., and Van Langenhove, H. (2006) Hybrid plasma-catalyst system for the removal of trichloroethylene in air. *J. Adv. Oxidation Technol.*, **9**, 53–58.
3. Takaki, K., Urashima, K., and Chang, J.S. (2004) Ferro-electric pellet shape effect on C2F6 removal by a packed-bed-type nonthermal plasma reactor. *IEEE Trans. Plasma Sci.*, **32**, 2175–2183.
4. Urashima, K., Kostov, K.G., Chang, J.S., Okayasu, Y., Iwaizumi, T., Yoshimura, K., and Kato, T. (2001) Removal of C_2F_6 from a semiconductor process flue gas by a ferroelectric packed-bed barrier discharge reactor with an adsorber. *IEEE Trans. Ind. Appl.*, **37**, 1456–1463.
5. Morent, R., Leys, C., Dewulf, J., Neirynck, D., Van Durme, J., and Van Langenhove, H. (2007) DC-excited non-thermal plasmas for VOC

abatement. *J. Adv. Oxidation Technol.*, **10**, 127–136.

6. Guo, Y.F., Ye, D.Q., Chen, K.F., He, J.C., and Chen, W.L. (2006) Toluene decomposition using a wire-plate dielectric barrier discharge reactor with manganese oxide catalyst in situ. *J. Mol. Catal. A-Chem.*, **245**, 93–100.
7. Guo, Y.F., Ye, D.Q., Chen, K.F., and Tian, Y.F. (2006) Humidity effect on toluene decomposition in a wire-plate dielectric barrier discharge reactor. *Plasma Chem. Plasma Process.*, **26**, 237–249.
8. Blin-Simiand, N., Tardiveau, P., Risacher, A., Jorand, F., and Pasquiers, S. (2005) Removal of 2-heptanone by dielectric barrier discharges – The effect of a catalyst support. *Plasma Process. Polym.*, **2**, 256–262.
9. Krawczyk, K., Ulejczyk, B., Song, H.K., Lamenta, A., Paluch, B., and Schmidt-Szalowski, K. (2009) Plasma-catalytic reactor for decomposition of chlorinated hydrocarbons. *Plasma Chem. Plasma Process.*, **29**, 27–41.
10. Hensel, K., Katsura, S., and Mizuno, A. (2005) DC microdischarges inside porous ceramics. *IEEE Trans. Plasma Sci.*, **33**, 574–575.
11. Futamura, S., Einaga, H., Kabashima, H., and Hwan, L.Y. (2004) Synergistic effect of silent discharge plasma and catalysts on benzene decomposition. *Catal. Today*, **89**, 89–95.
12. Sano, T., Negishi, N., Sakai, E., and Matsuzawa, S. (2006) Contributions of photocatalytic/catalytic activities of TiO_2 and gamma-Al_2O_3 in nonthermal plasma on oxidation of acetaldehyde and CO. *J. Mol. Catal. A-Chem.*, **245**, 235–241.
13. Kirkpatrick, M.J., Finney, W.C., and Locke, B.R. (2004) Plasma-catalyst interactions in the treatment of volatile organic compounds and NOx with pulsed corona discharge and reticulated vitreous carbon Pt/Rh-coated electrodes. *Catal. Today*, **89**, 117–126.
14. Subrahmanyam, C., Magureanu, M., Laub, D., Renken, A., and Kiwi-Minsker, L. (2007) Nonthermal plasma abatement of trichloroethylene enhanced by photocatalysis. *J. Phys. Chem. C*, **111**, 4315–4318.
15. Subrahmanyam, C., Renken, A., and Kiwi-Minsker, L. (2007) Novel catalytic non-thermal plasma reactor for the abatement of VOCs. *Chem. Eng. J.*, **134**, 78–83.
16. Subrahmanyarn, C., Renken, A., and Kiwi-Minsker, L. (2007) Novel catalytic dielectric barrier discharge reactor for gas-phase abatement of isopropanol. *Plasma Chem. Plasma Process.*, **27**, 13–22.
17. Chen, H.L., Lee, H.M., Chen, S.H., Chang, M.B., Yu, S.J., and Li, S.N. (2009) Removal of volatile organic compounds by single-stage and two-stage plasma catalysis systems: a review of the performance enhancement mechanisms, current status, and suitable applications. *Environ. Sci. Technol.*, **43**, 2216–2227.
18. Pribytkov, A.S., Baeva, G.N., Telegina, N.S., Tarasov, A.L., Stakheev, A.Y., Tel'nov, A.V., and Golubeva, V.N. (2006) Effect of electron irradiation on the catalytic properties of supported Pd catalysts. *Kinet. Catal.*, **47**, 765–769.
19. Roland, U., Holzer, F., and Kopinke, E.D. (2005) Combination of non-thermal plasma and heterogeneous catalysis for oxidation of volatile organic compounds Part 2. Ozone decomposition and deactivation of gamma-Al_2O_3. *Appl. Catal. B-Environ.*, **58**, 217–226.
20. Malik, M.A., Minamitani, Y., and Schoenbach, K.H. (2005) Comparison of catalytic activity of aluminum oxide and silica gel for decomposition of volatile organic compounds (VOCs) in a plasmacatalytic reactor. *IEEE Trans. Plasma Sci.*, **33**, 50–56.
21. Li, D.A., Yakushiji, D., Kanazawa, S., Ohkubo, T., and Nomoto, Y. (2002) Decomposition of toluene by streamer corona discharge with catalyst. *J. Electrostat.*, **55**, 311–319.
22. Oda, T., Takahahshi, T., and Yamaji, K. (2002) Nonthermal plasma processing for dilute VOCs decomposition. *IEEE Trans. Ind. Appl.*, **38**, 873–878.
23. Kogelschatz, U., Eliasson, B., and Egli, W. (1997) Dielectric-barrier discharges. Principle and applications. *J. Phys. IV*, **7**, 47–66.

24. Kraus, M., Eliasson, B., Kogelschatz, U., and Wokaun, A. (2001) CO_2 reforming of methane by the combination of dielectric-barrier discharges and catalysis. *Phys. Chem. Chem. Phys.*, **3**, 294–300.

25. Liu, C.J., Wang, J.X., Yu, K.L., Eliasson, B., Xia, Q., Xue, B.Z., and Zhang, Y.H. (2002) Floating double probe characteristics of non-thermal plasmas in the presence of zeolite. *J. Electrostat.*, **54**, 149–158.

26. Chen, H.L., Lee, H.M., Chen, S.H., and Chang, M.B. (2008) Review of packed-bed plasma reactor for ozone generation and air pollution control. *Ind. Eng. Chem. Res.*, **47**, 2122–2130.

27. Chavadej, S., Saktrakool, K., Rangsunvigit, P., Lobban, L.L., and Sreethawong, T. (2007) Oxidation of ethylene by a multistage corona discharge system in the absence and presence of Pt/TiO_2. *Chem. Eng. J.*, **132**, 345–353.

28. Chae, J.O., Demidiouk, V., Yeulash, M., Choi, I.C., and Jung, T.G. (2004) Experimental study for indoor air control by plasma-catalyst hybrid system. *IEEE Trans. Plasma Sci.*, **32**, 493–497.

29. Morent, R., De Geyter, N., Verschuren, J., De Clerck, K., Kiekens, P., and Leys, C. (2008) Non-thermal plasma treatment of textiles. *Surf. Coat. Technol.*, **202**, 3427–3449.

30. Simor, M., Rahel, J., Vojtek, P., Cernak, M., and Brablec, A. (2002) Atmospheric-pressure diffuse coplanar surface discharge for surface treatments. *Appl. Phys. Lett.*, **81**, 2716–2718.

31. Wagner, H.E., Brandenburg, R., Kozlov, K.V., Sonnenfeld, A., Michel, P., and Behnke, J.F. (2003) The barrier discharge: basic properties and applications to surface treatment. *Vacuum*, **71**, 417–436.

32. Pochner, K., Neff, W., and Lebert, R. (1995) Atmospheric-pressure gas-discharges for surface-treatment. *Surf. Coat. Technol.*, **74–75**, 394–398.

33. Wang, J.G., Liu, C.J., Zhang, Y.P., Yu, K.L., Zhu, X.L., and He, F. (2004) Partial oxidation of methane to syngas over glow discharge plasma treated Ni-Fe/Al_2O_3 catalyst. *Catal. Today*, **89**, 183–191.

34. Li, Z.H., Tian, S.X., Wang, H.T., and Tian, H.B. (2004) Plasma treatment of Ni catalyst via a corona discharge. *J. Mol. Catal. A-Chem.*, **211**, 149–153.

35. Zhu, X.L., Huo, P.P., Zhang, Y.P., and Liu, C.J. (2006) Characterization of argon glow discharge plasma reduced Pt/Al_2O_3 catalyst. *Ind. Eng. Chem. Res.*, **45**, 8604–8609.

36. Liu, C.J., Zou, J.J., Yu, K.L., Cheng, D.G., Han, Y., Zhan, J., Ratanatawanate, C., and Jang, B.W.L. (2006) Plasma application for more environmentally friendly catalyst preparation. *Pure Appl. Chem.*, **78**, 1227–1238.

37. Ratanatawanate, C., Macias, M., and Jang, B.W.L. (2005) Promotion effect of the nonthermal RF plasma treatment on Ni/Al_2O_3 for benzene hydrogenation. *Ind. Eng. Chem. Res.*, **44**, 9868–9874.

38. Zhu, Y.R., Li, Z.H., Zhou, Y.B., Lv, J., and Wang, H.T. (2005) Plasma treatment of Ni and Pt catalysts for partial oxidation of methane. *React. Kinet. Catal. Lett.*, **87**, 33–41.

39. Zhang, Y.P., Ma, P.S., Zhu, X.L., Liu, C.J., and Shen, Y.T. (2004) A novel plasma-treated Pt/NaZSM-5 catalyst for NO reduction by methane. *Catal. Commun.*, **5**, 35–39.

40. Guo, Y.F., Ye, D.Q., Chen, K.F., and He, J.C. (2007) Toluene removal by a DBD-type plasma combined with metal oxides catalysts supported by nickel foam. *Catal. Today*, **126**, 328–337.

41. Wallis, A.E., Whitehead, J.C., and Zhang, K. (2007) Plasma-assisted catalysis for the destruction of CFC-12 in atmospheric pressure gas streams using TiO_2. *Catal. Lett.*, **113**, 29–33.

42. Demidyuk, V. and Whitehead, J.C. (2007) Influence of temperature on gas-phase toluene decomposition in plasma-catalytic system. *Plasma Chem. Plasma Process.*, **27**, 85–94.

43. Demidiouk, V. and Chae, J.O. (2005) Decomposition of volatile organic compounds in plasma-catalytic system. *IEEE Trans. Plasma Sci.*, **33**, 157–161.

44. Magureanu, M., Mandache, N.B., Parvulescu, V.I., Subrahmanyam, C., Renken, A., and Kiwi-Minsker, L. (2007)

Improved performance of non-thermal plasma reactor during decomposition of trichloroethylene: optimization of the reactor geometry and introduction of catalytic electrode. *Appl. Catal. B-Environ.*, **74**, 270–277.

45. Wallis, A.E., Whitehead, J.C., and Zhang, K. (2007) The removal of dichloromethane from atmospheric pressure nitrogen gas streams using plasma-assisted catalysis. *Appl. Catal. B-Environ.*, **74**, 111–116.
46. Rousseau, A., Guaitella, O., Ropcke, J., Gatilova, L.V., and Tolmachev, Y.A. (2004) Combination of a pulsed microwave plasma with a catalyst for acetylene oxidation. *Appl. Phys. Lett.*, **85**, 2199–2201.
47. Oda, T. and Yamaji, K. (2003) Dilute trichloroethylene decomposition in air by using non-thermal plasma – catalyst effect. *J. Adv. Oxidation Technol.*, **6**, 93–99.
48. Oh, S.M., Kim, H.H., Einaga, H., Ogata, A., Futamura, S., and Park, D.W. (2006) Zeolite-combined plasma reactor for decomposition of toluene. *Thin Solid Films*, **506**, 418–422.
49. Oh, S.M., Kim, H.H., Ogata, A., Einaga, H., Futamura, S., and Park, D.W. (2005) Effect of zeolite in surface discharge plasma on the decomposition of toluene. *Catal. Lett.*, **99**, 101–104.
50. Van Durme, J., Dewulf, J., Demeestere, K., Leys, C., and Van Langenhove, H. (2009) Post-plasma catalytic technology for the removal of toluene from indoor air: effect of humidity. *Appl. Catal. B-Environ.*, **87**, 78–83.
51. Ogata, A., Saito, K., Kim, H.H., Sugasawa, M., Aritani, H., and Einaga, H. (2010) Performance of an ozone decomposition catalyst in hybrid plasma reactors for volatile organic compound removal. *Plasma Chem. Plasma Process.*, **30**, 33–42.
52. Radhakrishnan, R., Oyama, S.T., Ohminami, Y., and Asakura, K. (2001) Structure of $MnOx/Al_2O_3$ catalyst: a study using EXAFS, in situ laser raman spectroscopy and ab initio calculations. *J. Phys. Chem. B*, **105**, 9067–9070.
53. Lin, H., Huang, Z., Shangguan, W.F., and Peng, X.S. (2007) Temperature-programmed oxidation of diesel particulate matter in a hybrid catalysis-plasma reactor. *Proc. Combust. Inst.*, **31**, 3335–3342.
54. Hammer, T., Kappes, T., and Baldauf, M. (2004) Plasma catalytic hybrid processes: gas discharge initiation and plasma activation of catalytic processes. *Catal. Today*, **89**, 5–14.
55. Kim, H.H., Ogata, A., and Futamura, S. (2006) Effect of different catalysts on the decomposition of VOCs using flow-type plasma-driven catalysis. *IEEE Trans. Plasma Sci.*, **34**, 984–995.
56. Lu, B., Zhang, X., Yu, X., Feng, T., and Yao, S. (2006) Catalytic oxidation of benzene using DBD corona discharges. *J. Hazard. Mater.*, **137**, 633–637.
57. Kim, H.H., Lee, Y.H., Ogata, A., and Futamura, S. (2003) Plasma-driven catalyst processing packed with photocatalyst for gas-phase benzene decomposition. *Catal. Commun.*, **4**, 347–351.
58. Kim, H.H., Oh, S.M., Ogata, A., and Futamura, S. (2005) Decomposition of gas-phase benzene using plasma-driven catalyst (PDC) reactor packed with Ag/TiO_2 catalyst. *Appl. Catal. B-Environ.*, **56**, 213–220.
59. Ogata, A., Kim, H.H., Oh, S.M., and Futamura, S. (2006) Evidence for direct activation of solid surface by plasma discharge on CFC decomposition. *Thin Solid Films*, **506**, 373–377.
60. Guaitella, O., Thevenet, F., Puzenat, E., Guillard, C., and Rousseau, A. (2008) C_2H_2 oxidation by plasma/TiO_2 combination: Influence of the porosity, and photocatalytic mechanisms under plasma exposure. *Appl. Catal. B-Environ.*, **80**, 296–305.
61. Thevenet, F., Guaitella, O., Puzenat, E., Guillard, C., and Rousseau, A. (2008) Influence of water vapour on plasma/photocatalytic oxidation efficiency of acetylene. *Appl. Catal. B-Environ.*, **84**, 813–820.
62. Guaitella, O., Lazzaroni, C., Marinov, D., and Rousseau, A. (2010) Evidence of atomic adsorption on TiO_2 under plasma exposure and related C_2H_2

surface reactivity. *Appl. Phys. Lett.*, **97**, 011502.

63. Kim, H.H., Ogata, A., and Futamura, S. (2008) Oxygen partial pressure-dependent behavior of various catalysts for the total oxidation of VOCs using cycled system of adsorption and oxygen plasma. *Appl. Catal. B-Environ.*, **79**, 356–367.

64. Miranda, B., Diaz, E., Ordonez, S., Vega, A., and Diez, F.V. (2007) Oxidation of trichloroethene over metal oxide catalysts: kinetic studies and correlation with adsorption properties. *Chemosphere*, **66**, 1706–1715.

65. Oguz, H., Koch, S., and Weisweiler, W. (2000) Comparison of mechanistic models for the catalytic oxidation of trichloroethylene over Cr/Al_2O_3 and Al-Cr/porous glass catalysts. *Chem. Eng. Technol.*, **23**, 395–400.

66. Ogata, A., Kim, H.H., Futamura, S., Kushiyama, S., and Mizuno, K. (2004) Effects of catalysts and additives on fluorocarbon removal with surface discharge plasma. *Appl. Catal. B-Environ.*, **53**, 175–180.

67. Konova, P., Stoyanova, M., Naydenov, A., Christoskova, S., and Mehandjiev, D. (2006) Catalytic oxidation of VOCs and CO by ozone over alumina supported cobalt oxide. *Appl. Catal. A-Gen.*, **298**, 109–114.

68. Roland, U., Holzer, F., and Kopinke, F.D. (2002) Improved oxidation of air pollutants in a non-thermal plasma. *Catal. Today*, **73**, 315–323.

69. Kim, H.H., Tsubota, S., Date, M., Ogata, A., and Futamura, S. (2007) Catalyst regeneration and activity enhancement of Au/TiO_2 by atmospheric pressure nonthermal plasma. *Appl. Catal. A-Gen.*, **329**, 93–98.

6
Elementary Chemical and Physical Phenomena in Electrical Discharge Plasma in Gas–Liquid Environments and in Liquids

Bruce R. Locke, Petr Lukes, and Jean-Louis Brisset

6.1
Introduction

Electrical discharges in gas–liquid environments and in liquids (primarily water, but in some cases also organic liquids) have been studied for a number of years for applications in electrical transmission, chemical destruction in pollution control, chemical synthesis, polymer surface treatment, biological inactivation, biomedical treatment, material and nanoparticle synthesis, and chemical analysis of liquid solutions [1–14]. Much of this work has focused on discharges in, over, and in contact with liquid water, but discharges in and over organic liquids (e.g., various hydrocarbons, alcohols, ionic liquids) in electrical transmission and material and chemical synthesis are also important [15, 16]. In addition to the type of liquid and the liquid composition, these types of discharges can be classified by the electrode configuration [1] and the phase distribution [7].

Figure 6.1 shows the wide variety of electrode configurations that have been studied in such systems ranging from point to plane, point to point, and wire cylinder geometry in gas, liquid, and gas–liquid systems. Some other configurations include diaphragm discharge directly in water, gas-phase gliding arc discharge in contact with liquid, and plasma jet formed inside the liquid with gas injection. Generally, the contact between gas and liquid phases can be envisioned in two ways as conceptually shown in Figure 6.2, whereby in the first mode, the plasma is formed in the continuous gas phase with the plasma contacting molecules, clusters, aerosols, droplets, and planar surfaces of the liquid and in the second mode, the liquid phase is a continuous domain with the plasma formed inside bubbles introduced into the liquid or the plasma forms channels directly inside the liquid. In this chapter, we discuss the basic aspects of both the chemical and physical factors involved in such discharges with an emphasis on liquid water and atmospheric pressure and temperature conditions.

Electrical discharge plasma formed in a *gas phase in contact* with a liquid (i.e., over a liquid surface, in a preexisting bubble, or interacting with water droplets and sprays) has significant differences from those formed *directly* in the liquid phase. The general means for plasma generation and gas phase breakdown over

Plasma Chemistry and Catalysis in Gases and Liquids, First Edition.
Edited by Vasile I. Parvulescu, Monica Magureanu, and Petr Lukes.

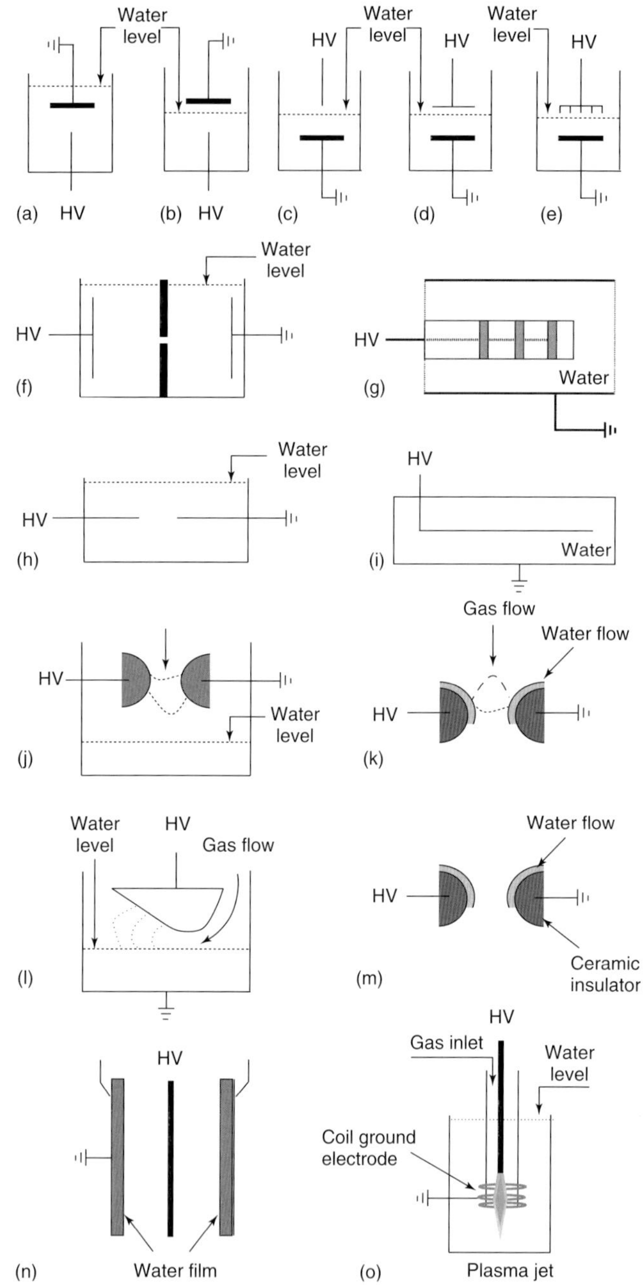

Figure 6.1 (a–o) Various electrode configurations. HV, high voltage electrode. (Source: Adapted from Ref. [1]. Adapted with permission from [1]. Copyright (2006) American Chemical Society; (m), Ref. [17]; (o), Ref. [18].)

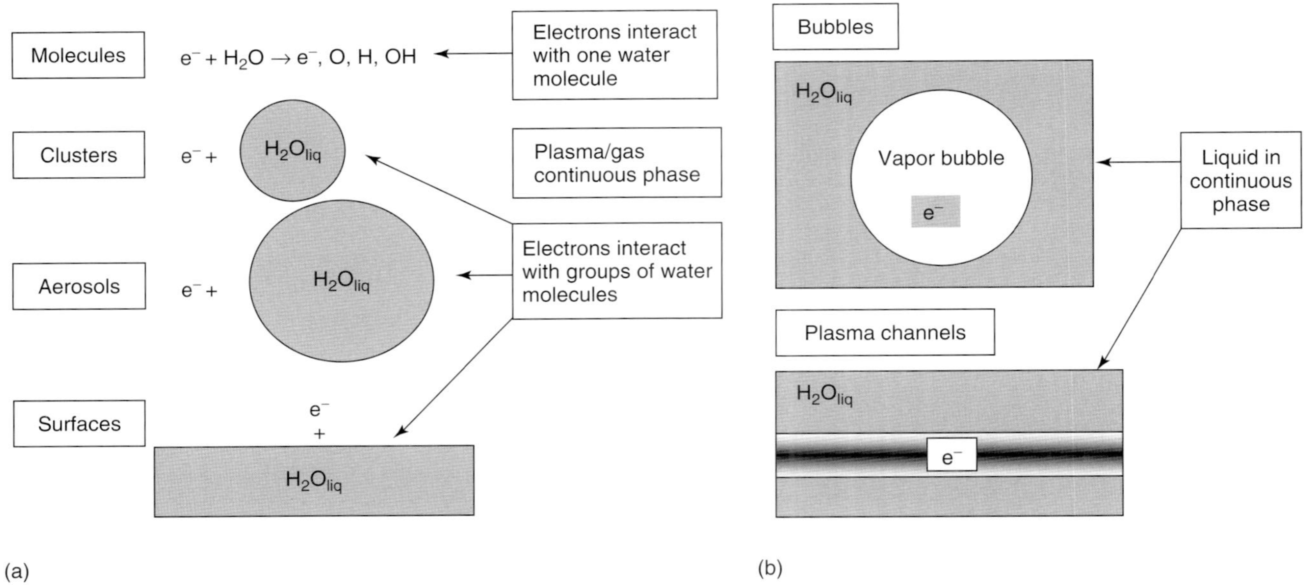

Figure 6.2 (a,b) General means for gas and liquid phase contact with plasma. (Source: Reprinted from [7]. Copyright (2011) with permission of IOP Publishing.)

or in contact with a liquid are very similar to those discussed in Chapter 1 dealing with gas-phase plasma because the breakdown strength in the atmospheric gas (tens of kilovolts per centimeter) is generally much lower than that in the liquid (e.g., for liquid water, it is $>1\,MV\,cm^{-1}$ [3, 19]). The major difference between gas-phase discharge where both electrodes are in the gas phase in the absence of liquid surfaces and a discharge in the gas phase over or in contact with liquid interfaces is the existence of the liquid surface functioning as an electrode; it is the interaction of the gas-phase-generated plasma with the liquid surface, that is, gas–liquid interface, that provides new and interesting physics and chemistry, as well as applications. Nevertheless, because of the similarities among all cases where the discharge is formed in the gas phase, as discussed in later sections of this chapter, mathematical models and computer simulations of plasma dynamics have been adapted from gas-phase work to describe the interactions of the gas-phase plasma with the condensed surfaces [20–26].

The mechanism for discharge generation directly in the liquid phase, however, is significantly different from that in the gas since the liquid density is 10^3 times larger than the gas and the liquid can have widely variable conductivity and other properties. Nevertheless, the breakdown in the liquid phase is intimately coupled to the existence and formation of microbubbles and regions of low liquid density [19, 27, 28] and liquid heating by the input energy. Further, in the cases of liquid water, all the above types of discharges have certain common chemical features, albeit with putatively different reaction mechanisms. For example, these systems generally lead to the formation of molecular and radical species such as hydrogen peroxide (H_2O_2), molecular hydrogen (H_2), molecular oxygen (O_2), and hydroxyl radicals ($OH^\bullet$). This chapter seeks to address these commonalities and differences and to provide fundamental background on the current state of knowledge of the chemistry and physics of these various types of discharges.

6.2
Physical Mechanisms of Generation of Plasma in Gas–Liquid Environments and Liquids

6.2.1
Plasma Generation in Gas Phase with Water Vapor

The effects of water vapor on gas plasma propagation have been studied in the context of atmospheric physics of lightning [29–32], electrostatic precipitators [33, 34], laboratory corona reactors [35, 36], ozone generation [37–39], and basic plasma physics [40, 41]. The general features of plasma generation and streamer propagation are similar to those of dry gases reported elsewhere [42–45], but some specific aspects of humidity on discharge initiation [41], ion mobility [46], and formation of ionic clusters can be mentioned here [39, 47–49]. Recently, for example, through modeling and experimental analysis of pressure and humidity on DC corona in humid air with (+) DC point-to-plane electrode configuration, it was shown that as humidity increases the corona inception voltage decreases because of lower ionization coefficients and the lower ion mobility at higher humidity also decreases the corona current [49]. In DC parallel plate discharge with a uniform field, the electric field for stable discharge streamer formation and the breakdown increases with increasing humidity [31, 50]. In other studies using model simulations, it was found that the suppression of streamer propagation by increasing humidity is due to increased electron ion recombination reactions rather than negative ion reaction rates [51]. Humidity was also found to lower the number of streamer pulses and affect the surface dielectric properties in point-to-plane silent discharge [52]. Breakdown in rod-to-plane gaps [53, 54], electric fields for streamer propagation [55, 56], and sparking conditions [57] are affected by humidity. In summary, humidity can affect the physical nature of gas-phase discharge, including initiation, propagation, and time lags, through a number of factors discussed above.

6.2.2 Plasma Generation in Gas–Liquid Systems

Electrical discharge formed in the gas phase whereby at least one of the electrodes contacts the liquid phase or is placed inside the liquid is very similar to discharges directly in only the gas phase. The discharge is formed in the gas phase where electrical breakdown occurs, and thereafter, the discharge impinges on the liquid. A variety of such systems (Figure 6.1) have been studied (see also reviews [1, 5]), including point or multipoint discharges over the liquid phase [58, 59], with the ground electrode placed in the liquid (Figure 6.1c, e), and the high-voltage electrode in the liquid phase with the ground electrode in the gas phase (Figure 6.1b) [60, 61]. Gas discharge over the liquid can also be formed in gliding arc (Figure 6.1j, k, l) and falling film (Figure 6.1n) configurations. In some cases, both electrodes are in the liquid phase with an intervening gas gap (Figure 6.1k,m). Gas injection from an electrode tip through a concentric band of ring barrier electrodes can form a plasma jet that extends into the liquid (Figure 6.1o) [18]. Various types of power inputs including AC, DC, pulsed, RF, and microwave discharge over water surfaces have been investigated in a wide range of electrode configurations [1]. The properties of the plasma can also vary significantly and may depend on the liquid properties (conductivity, pH, and temperature), gas composition, electrode materials, and field polarity [1, 5, 62].

6.2.2.1 Discharge over Water

Discharges formed directly over water surfaces are of interest for fundamental studies of electrohydrodynamics and applications of chemical reactions at gas–liquid interfaces [63–66]. Figure 6.3 [67] shows examples of discharges generated over a liquid water surface using a pulsed power supply (capacitor discharge with rotating spark gap) and a reticulated carbon (multipoint) electrode in the gas phase. The interactions of the discharge channels propagating along the surface of the water are clearly shown and suggest similarities with discharges over solid [25]

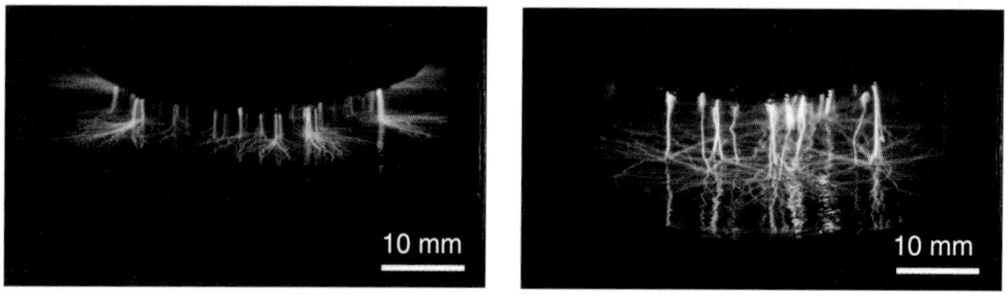

(a) (b)

Figure 6.3 Gas-phase pulsed corona discharge generated in air above liquid using planar electrode from reticulated vitreous carbon. Effect of discharge gap spacing of (a) 4 mm and (b) 10 mm. Water conductivity was 100 μS/cm. (Source: © 2011 IEEE. Reprinted, with permission, from [67].)

surfaces. Figure 6.3a shows discharge with a smaller gap distance between the gas phase electrode and the liquid surface, and Figure 6.3b shows the case with a larger gap distance. For the narrow gap case, very fine brushlike discharges are seen propagating along the surface of the water, and as the gap distance is increased, the discharge channels become both longer and wider and have less fine structure. The branching at the surface resembles Lichtenberg figures [63], and as the conductivity increases, with fixed gap discharge, the channels become thicker, brighter, and shorter. While both conductivity and gap distance affect the discharge characteristics, changing gap distance cannot compensate for the effect of conductivity on the discharge structure. The roles of the electrode placement and formation of the discharge over the surface on the generation of ozone was emphasized in these studies [67, 68]. Other work has shown that as the gas gap is reduced, the thickness of the water layer has a larger effect on the corona to spark transition [69]. The formation of Taylor cones [70, 71], that is, liquid instabilities due to electrohydrodynamics [72], the flow induced by the electric field [73, 74], at the gas–liquid interface for discharge over a planar water surface has also been observed, and this instability can affect the breakdown conditions in such discharges [75–78].

Babaeva and Kushner [25, 26] developed a multicomponent two-dimensional hydrodynamics model of the formation of the electric field for a gas discharge over a dielectric surface. The conservation balances for the gas components, including positive and negative ions and neutral and excited state species, coupled to the Poisson equation for the electric field propagation were solved accounting for electron energy conservation and determining the electron energy distribution from the Boltzmann equation. Photoionization by the streamer was incorporated, and for the gas-phase discharge in dry air, seven species (N_2, N_2^*, N_2^{**}, N_2^+, O_2, O_2^+, and electrons) were included. Electric fields $>200\,kV\,cm^{-1}$ were found at the heads of the dielectric barrier discharge (DBD) filaments propagating to the surface, and it was shown that the electric field increased to as high as $400–800\,kV\,cm^{-1}$ as the filament interacts with the surface. Thus, the dielectric surface can affect the physical and electrical properties of the plasma filaments, and, for dielectric surfaces of high dielectric constant (i.e., similar to that of liquid water), they found that the gaseous ions reach energies of 150 eV as they impinge onto the surface. Although these simulations do not account for the specific chemical aspects of water discharge, they demonstrate that the ionic chemistry can be very important and that the dielectric properties of the boundary have a strong effect on the electric field. Further work to include the other species, including water and the dissociation products of water in the plasma, is needed to address specific aspects of the liquid water surface and to make direct comparison with experimental measurements of reaction products and plasma properties for discharge over the water surface.

6.2.2.2 **Discharge in Bubbles**

Discharges directly inside bubbles have also presented an interesting case for fundamental studies and potential applications because of large surface areas and the presence of the gas phase for ease of discharge initiation [79–94]. Figure 6.4 shows an example discharge generated within a bubble formed inside a liquid. In this case, the solution temperature was controlled to near boiling so that large bubbles formed, which only contained water vapor and possible trace gases that were dissolved in and released from the liquid [90]. It can be noted that the discharge initiates in the bubble and propagates along the inside of the bubble at the gas–liquid interface; the plasma has a filamentary structure similar to those that occurs for the over-surface discharge shown in Figure 6.3b (both of these example

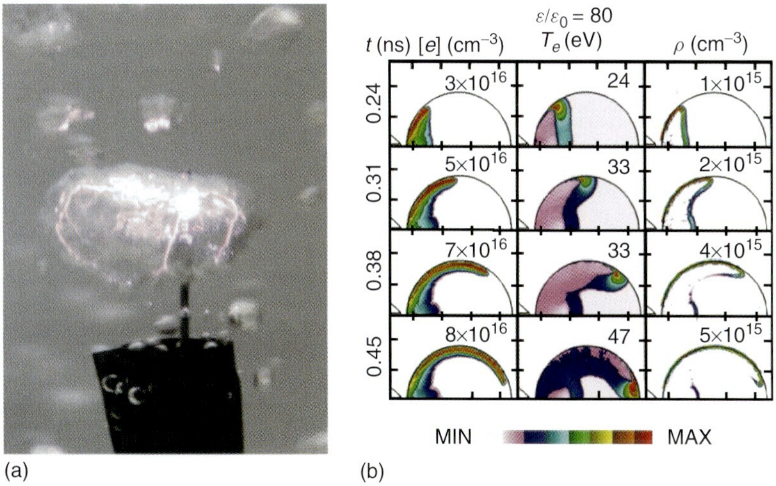

Figure 6.4 (a) Point-to-plane discharge in liquid water, with bubbles showing discharge along the inside. Large bubble size about 2 cm. The water is near boiling point, and the bubble contains only water vapor and reaction products from discharge [90]. (b) Simulations of electric field propagation inside bubbles. (Source: (b) Reprinted from [24]. Copyright (2009), with permission of IOP Publishing.)

cases use pulsed power supply of nominally 60–100 W with pulse energies of 1 J). Under near-boiling conditions (95°C), the discharge can easily form from a long needle electrode submerged in a moderately conductive solution, whereas under ambient conditions, such a discharge could not form because of the high surface area of the electrodes. The higher temperature leads to nearly double the production rates of H_2, O_2, and H_2O_2 compared to the same discharge at ambient temperature [90]. In the case of injected argon bubbles, at ambient temperature, the energy yields for the production of H_2, O_2, and H_2O_2 were independent of input power and were, in general, lower than those for direct discharge in water and over water surfaces for the same power [92].

Model simulations using the multicomponent hydrodynamics plasma model mentioned above but for discharges within humid air in bubbles [22–24] showed significant similarities to the experimental results [90, 92, 95, 96] whereby the discharge channels propagate along the interface on the gas side of the bubble (Figure 6.4). More components were included to account for water vapor and ozone, but not nitrogen oxides or hydrogen peroxide. The model included N_2, $N_2(v)$, N_2^*, N_2^{**}, N_2^+, N, N^*, N^+, N_4^+, O_2, O_2^*, O_2^+, O_2^-, O^-, O, O^*, O^+, O_3, H_2O, H_2O^+, H_2, $H^\bullet$, $OH^\bullet$, and electrons. Experimental results showed that the conductivity (150–3000 μS cm^{-1}) has a small effect on H_2 and H_2O_2 generation in bubbles and over surfaces with argon gas but strongly reduces the generation of these species directly in the liquid [92]. It was also found experimentally that generating discharges within bubbles, either those introduced by bubbling argon into the liquid or those formed by maintaining the solution temperature near boiling, are visually very similar, but the higher temperature case produces about double the H_2, O_2, and H_2O_2. This increase in molecular yield was partially attributed to the larger amount of water vapor contacting the plasma inside the bubble [90, 92]. Other experimental work demonstrated significant effects of the gas composition on the nature of the discharge inside bubbles; molecular gases (N_2 and O_2) lead to discharge propagation along the gas–liquid interface, whereas noble gases (He, Ne, Ar) produce more volumetric discharges within the gas bubble [95]. Further, high-speed imaging of discharges inside air bubbles demonstrated capillary oscillations of the gas–liquid surface within the bubble arising from the electric field of the propagating streamer [96]. As in the discharge over water, additional work to directly compare modeling efforts with experimental results on chemical species production and plasma properties is needed.

Many other studies of discharges over surfaces and in bubbles have been reported, and we discuss various aspects of the effect of water surface on the chemistry and physics of the discharge in the following sections of this chapter [5, 78, 85, 86, 88, 97–102].

6.2.2.3 Discharge with Droplets and Particles

The effects of small droplets and particles in gas-phase discharge are of importance to topics such as atmospheric chemistry (lightning) [103], environmental analysis [104], electrostatic precipitation [105], and material synthesis [106]. Electric fields can lead to droplet breakup by hydrodynamic instabilities [71], and some plasma conditions can lead to vaporization of water as well as the effects of water droplets on streamer propagation [20, 22]. An example of the generation of a gas-phase discharge in the presence of small aerosol droplets is shown in Figure 6.5 for a pulsed gliding arc discharge [107, 108]. It is not possible to see the very small aerosol droplets in this photograph (e.g., 20–50 μm mean diameter), but as discussed below, diagnostic tools such as emissions spectroscopy show some similarities and some significant differences from the other types of discharges with liquid water (see discussion related to Figure 6.11).

Computational analyses of gas-phase plasma containing solid particles and liquid droplets with the multiple-component hydrodynamics plasma model mentioned above have shown how the propagating discharge channel interacts with micron-scale particles suspended in the gas phase [21–24]. Simulations have shown that the smaller particles are incorporated within the plasma streamers and that the larger particles can initiate streamers. The permittivity and capacitance of the particles have strong effects on the streamer propagation, and multiple particles can lead to large increases of the streamer speed. Experimental results on water sprays in plasma have focused on the efficiencies of formation of hydrogen peroxide and hydrogen in gliding arc and pulsed gliding arc reactors with water aerosols in the range of 20–50 μm [107, 108]. Further work on the physical characterization of such systems and the coupling of the chemical models (and experiments) with the physical models are needed.

6.2.3
Plasma Generation Directly in Liquids

In general, the liquid phase can have properties spanning those of a highly conducting material (e.g., high salt containing liquid) to a more insulating surface

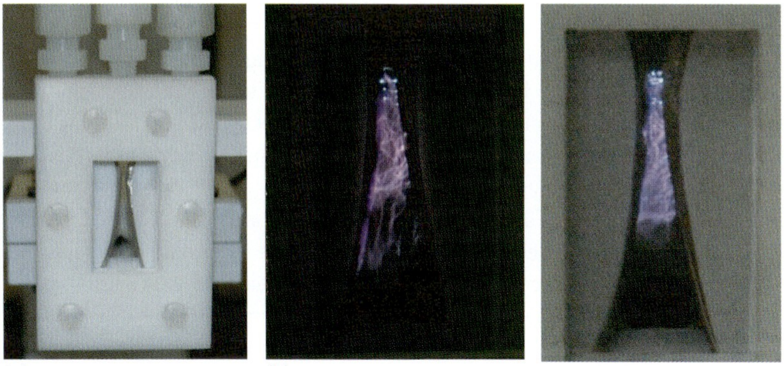

(a) (b) (c)

Figure 6.5 Gliding arc discharge (a) reactor, (b) in flowing argon, and (c) with water spray flowing into an argon carrier gas. The fine water droplets of 20 μm are not visible in (c) (Source: Picture taken at Florida State University by B. R. Locke.)

(e.g., organic liquid or highly purified water), and in some cases, the properties of the liquid can change between these limits during the time of the discharge (e.g., ionic reaction products increase conductivity and may affect the solution pH). The existence of the liquid can have significant effects on both the chemical reactions and the physical features of the discharge. Some of the key physical aspects for direct discharge in water are discussed in this section, and we seek to address the major factors on the chemical effects in Section 6.4.

The generation of an electrical discharge directly in the liquid depends on a number of factors, including the type of power supply (pulsed, AC, DC, radio frequency (RF), microwave (MW)), the nature and characteristics of the energy input (energy per pulse, overall power, rise time, pulse width, current and voltage waveforms), the electrode geometry and material properties, and the solution properties (conductivity, pressure, pH, and temperature). Since very large localized electric fields of the order of megavolts per centimeter are needed for electrical breakdown of water, the pulsed high voltage and electric field enhancing electrode systems (point-plate, wire-cylinder, hole-plate, rod-rod, ring-cylinder) are often used to generate electrical (electrohydraulic) discharge in water. In addition to the electrical parameters, bulk properties, such as hydrostatic pressure, temperature, solution conductivity, and deaeration (preexistence of microbubbles) of water, and electrode conditions, such as electrode material (i.e., work function), electrode conditions (surface asperities), interfacial properties, and field polarity, are important factors in the electrical breakdown of water [19, 28].

Several models for the mechanism of breakdown in water stressed to high voltages have been proposed, and these models, in general, assume formation of low-density localized regions at (or near) the site of the field enhancement, for example, at an electrode tip or a bubble interface. The interfaces could either be the metal–liquid boundary at the electrode surface or the liquid–vapor boundary at a microbubble site. Field-driven molecular dissociation, as well as localized heating and electron injection into microbubbles, can play a role. The large electric field, together with the electronic conditions at the interface boundaries, triggers streamer initiation. Phase instabilities, heating, local vaporization leading to formation of localized low-density regions, bubble creation, field-assisted molecular/dipolar dissociation, and electron transfer between metal and liquid are some of the phenomena that may play a role in the complex initiation physics [27].

More recently, the roles of surface chemistry at the ceramic–electrolyte interface and the formation of an electrical double layer at the electrode surface in the electrical breakdown of water using metallic rod electrodes coated by a thin layer of porous ceramic (composite electrodes) were also demonstrated [109]. Such design of electrodes has been developed recently to generate a relatively large volume of plasma in water [110–112]. The role of the ceramic layer is to redistribute the electric field on the electrode during the predischarge phase. Owing to the differences in conductivity and permittivity between the water ($\epsilon_{rw} = 81$) and the ceramic layer ($\epsilon_{rc} = 8\text{–}12$, $\sigma_c = 2\,\mu S/cm$), the electric field strength at the surface of a composite electrode is many times enhanced in comparison with a metallic electrode of the same dimensions. Thus, a large number of discharge channels, distributed almost homogeneously along the entire surface of the ceramic layer, can be generated from the pores at relatively moderate levels of applied voltage (20–30 kV).

Figure 6.6 shows examples of a pulsed discharge with point-to-plane electrode configuration, and Figure 6.7 shows a multipoint discharge utilizing a porous ceramic coated on a metal electrode [112]. With the wide range of variables to consider, it is not currently possible to propose a general theoretical analysis that explains all the phenomena [19]. Most experimental studies have focused on point-to-plane geometry [113] with pulsed capacitive discharges since the small

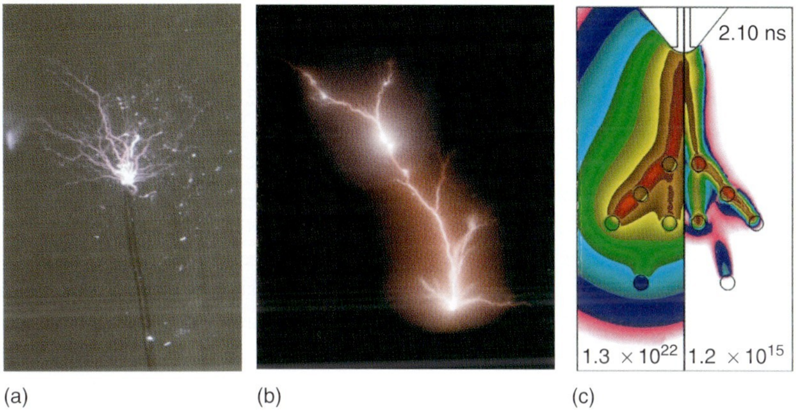

Figure 6.6 Point-to-plane discharge in liquid water showing filamentous structure. (a) Discharge-time-averaged photo in low conductivity water, (b) single-shot photo of discharge directly in liquid and (c) model simulations of branching patterns due to preexisting bubbles. (Source: a) Picture was taken at Florida State University by Kai-Yuan Shih. b) Picture was taken at the Technical University of Eindhoven by Dr. E. van Veldhuizen with Dr. B. R. Locke. (c) © 2008 IEEE. Reprinted, with permission, from [21].)

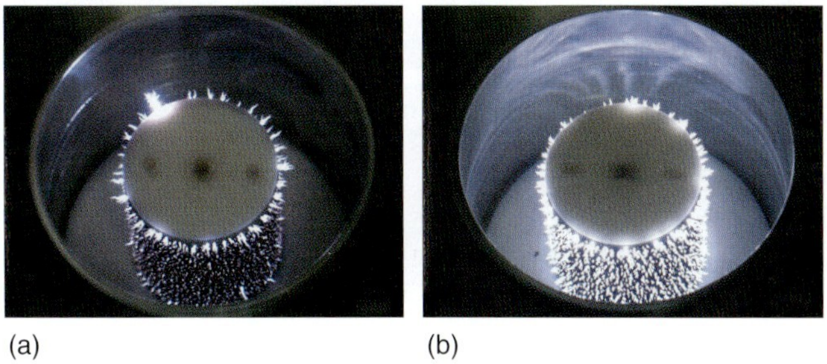

Figure 6.7 Multichannel pulsed discharge in water using porous ceramic-coated metal electrodes and pulsed capacitive discharge in water. Effect of water conductivity on discharge characteristics: (a) 1.5 $mS\,cm^{-1}$ and (b) 6 $mS\,cm^{-1}$. (Source: © 2008 IEEE. Reprinted, with permission, from [112].)

needle-point electrodes have very high curvature and can concentrate the electric field to very high values (Figures 6.6 and 6.8). Another commonly used electrode configuration utilizes the pinhole (diaphragm) or capillary discharge, whereby two planar electrodes are separated by an insulating sheet (diaphragm) with small holes or with capillary, respectively (Figure 6.1f) [88, 114–122]. In this case, large electric fields are generated in connecting holes in the diaphragm layer or in the capillary and electrical breakdown of water (i.e., discharge formation) proceeds by thermal process when sufficiently high predischarge current in the liquid causes strong heating in the small hole or the narrow capillary, inducing formation of vapor bubble inside the diaphragm (capillary) and its subsequent breakdown. Pulsed, DC, or AC power supplies can be used in this case. Reviews of many of these configurations have been reported [1–5], and the fundamental issues revolve around how electrons reach sufficient energy to cause ionization, leading to breakdown and plasma streamerlike propagation [19, 27].

Traditional explanations of electrical breakdown in liquids have considered two competing theories. The first is akin to the classical gas-phase breakdown mechanism, and the second involves formation of small bubbles, primarily through Joule heating. Detailed discussion of both these theories have been reported [19], but in this chapter, we focus on the more recent experimental data and conceptual framework [27]. Owing to the high density of liquids, with corresponding low mean free paths, high rates of electron solubilization by water, and isotropic scattering of electrons and generated ions, there is a very low probability that the energy required for electron or ion impact ionization can be reached in liquids [27]. It has therefore been suggested that regions of low-density fluid must occur where the electric field enhancement is high in order for the discharge to initiate. These local regions of low density can be formed by vaporization, chemical decomposition, or mechanical

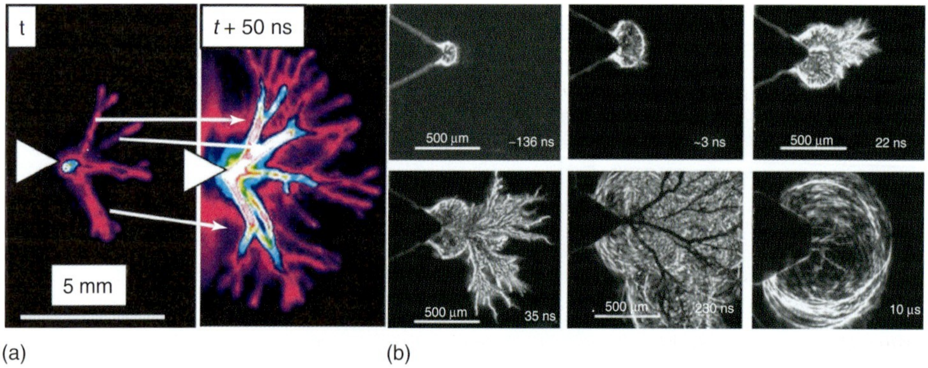

Figure 6.8 Time-resolved images of discharge growth in water using emission spectroscopy and Schlieren photography. (a) Discharge branching patterns development and an increase in the emission intensity in the parent discharge filament within 50 ns. (b) Schlieren images of primary and secondary streamer development at different times. (Source: (a) Reprinted from [123]. Copyright (2010), with permission of IOP Publishing. (b) Reprinted with permission from [124]. Copyright (2007) American Institute of Physics.)

motion [19]. In cases where microbubbles are already present, electrons can impact with these preexisting microbubbles [125]. Figure 6.6c shows simulations using the gas-phase multicomponent hydrodynamics model, discussed above [21], of branching patterns that arise through electric field propagation in a gas phase with preexisting low-density inhomogeneities. It was argued that the similarities between observed branching patterns formed in discharge propagation in liquids and gases point to a common mechanism, and therefore, the gas-phase simulations provide insight into the discharge directly in the liquid. Such simulations do not account for the liquid phase, and, in general, the issue of liquid to vapor (and plasma) phase transition is key to analyzing direct discharge in the liquid phase. This transition can be strongly affected by the solution properties, particularly conductivity, the existence of microbubbles, and perhaps pressure, as well as by the nature of the electric field application (pulse rise time, energy input).

With the advent of high-speed photography and imaging techniques, significant advances have been made in experimentally analyzing the formation and propagation of direct discharges in liquid water (Figure 6.8) [123, 124]. Time-resolved imaging has shown that the propagation velocity ($30\,\mathrm{km\,s^{-1}}$) does not depend on solution conductivity, but the discharge length decreases strongly with conductivity between 10 and $1000\,\mu\mathrm{S\,cm^{-1}}$ [123]. The time delay in the discharge initiation was found to be consistent with the time for microbubbles to form by Joule heating [123]. Schlieren photography (Figure 6.8b), used to determine time and spatially resolved pressure fields, supports the bubble mechanism for primary streamer formation but not for secondary streamer formation and the fact that the local pressure in the liquid can be very high [124].

Recent experiments using extremely short pulses (400 ps pulse width and 150 ps rise time) in low-conductivity pure water ($5\,\mu\mathrm{S\,cm^{-1}}$) have shown that discharges in water can be formed in the absence of gas bubble formation and with minimal thermal effects [126]. Discharge channels with 0.5–0.8 mm length and up to $100\,\mu\mathrm{m}$ diameter propagate at $5000\,\mathrm{km\,s^{-1}}$ with radial expansion of $250\,\mathrm{km\,s^{-1}}$. These velocities are significantly larger than the nanosecond pulses in water of $10–30\,\mathrm{km\,s^{-1}}$ [123, 127, 128]. For a 220 kV pulse, the electric field at the estimated $10\,\mu\mathrm{m}$ tip of the plasma channel was $220\,\mathrm{MV\,cm^{-1}}$, and since this is much higher than the required $30\,\mathrm{MV\,cm^{-1}}$ needed for ionization in the condensed phase, it was concluded that plasma propagation by direct ionization, as in gas-phase nonequilibrium plasma [45, 129, 130], can occur in the liquid. Although longer width pulses can lead to thermal effects, heating of the liquid, and bubble formation, clearly for very fast pulses, there is evidence for plasma propagation in the liquid by electron impact ionization as occurs in the gas phase.

A model to account for the liquid to gas phase transition in pulsed discharge (longer pulse widths) for high salt solutions in electrosurgical applications [131–135] based on thermal energy balances and Maxwell's equations applied to the region adjacent to the stressed electrode was developed [136]. The model included Gauss's law to determine the electric field (Eq. (6.1)), Ohm's law to couple the current density to the electric field (Eq. (6.2)), and the thermal energy balance with

Joule heating source term (Eq. (6.4)), and this system of equations was solved in a 2D axisymmetric cylindrical geometry using a finite element code (COMSOL, Multiphysics). The model included the following equations:

$$\nabla \cdot \left(\epsilon(T)\overline{E}\right) = 0 \tag{6.1}$$

$$\overline{J} = \sigma(T)\overline{E} \tag{6.2}$$

$$\overline{E} = -\nabla V \tag{6.3}$$

$$\rho(T)C_p(T)\frac{\partial T}{\partial t} - \nabla \cdot \left(k(T)\nabla T\right) = \overline{J} \cdot \overline{E} \tag{6.4}$$

where J is the current density vector, E is the electric field vector, $\epsilon(T)$ is the dielectric constant, $\rho(T)$ is the electrical conductivity, V is the electric potential, T is the temperature, $k(T)$ is the thermal conductivity, and $C_p(T)$ is the heat capacity. Simulations considered the transient time scale up to 1 ms for a 10 ms duration pulse in 0.9% NaCl solution. As the temperature increases to the boiling point, Heaviside step functions are used for the physical properties (conductivity, permittivity, density, specific heat, thermal conductivity) to switch from the liquid state to the vapor state. Model data comparison of the current suggested that the superheating in the liquid might occur because of the very high rates of temperature increase (10^6 K s^{-1}).

In order to account for pressure and volume expansion as well as the overall electrical circuit and other factors such as radiation transport and water evaporation and condensation, alternative model formulations based on overall energy, mass, charge, and momentum conservation balances coupled to equations of state for the plasma have been applied to a plasma channel [137–142]. No single formulation can describe all types of discharges in water, and it is important to note that as the energy deposition into the liquid varies from <1 J $pulse^{-1}$ to 1 kJ $pulse^{-1}$ or greater, significant changes in temperature, radiation emissions [143], pressure shockwaves [144], and the electrical circuit [69, 145] must be considered in the model formulation.

The thermal energy balance (Eq. (6.4)) is the basis for another model wherein the temporal function of the input energy from a capacitor coupled to an assumed cylindrical geometry and size of a plasma channel are used to estimate the temperature rise in the plasma channel and to couple this temperature to the chemical kinetics of water plasma reactions [146, 147]. Physical observations were used to estimate channel size, and experimental measurements of current and voltage waveforms were used to estimate the transient temperature rise in the plasma channel. Estimates of temperatures in the plasma channels of over 5000 K were found based on this simple energy balance. In Section 6.4, we discuss the chemical aspects in more detail; however, the chemical model requires knowledge of the thermal physical properties of H_2O and its decomposition products (e.g., H_2, O_2, $H^\bullet$, $O^\bullet$, $OH^\bullet$) over very wide range of temperatures (ambient to >10 000 K) and pressures (ambient to >10 MPa). Data from thermal arc [148, 149] and other literature [150] can be used to estimate physical properties, including thermal conductivity and heat capacity.

Figure 6.9 Hybrid reactors. (a) Series reactor showing discharge over water surface and in liquid. Pulsed discharge using 1 J pulse^{-1} and 60 Hz. Photo is a long-time exposure showing multiple discharge changes in the liquid and the gas. The gas-phase electrode is made of reticulated vitreous carbon. (b) Parallel reactor. (c) Series–parallel reactor. ((a) Source: Photo taken at Florida State University by M. Sahni.)

The thermal energy analysis has been applied for the cases of high-power pulsed discharge in water, by including radiation losses from the plasma channel, and the dynamic nature of the plasma channel expansion [137, 138] (see also [1] for a review of earlier literature on this subject). For a high-frequency (13.6 MHz) discharge, relatively large bubbles are formed, and a model coupling the bubble dynamics [151] and species balances with chemical reactions and thermal energy balance has been developed [91].

It is also useful to note that hybrid reactors combining the gas and liquid phase discharge have been investigated (Figure 6.9) [60, 61]. Figure 6.9a shows a series reactor wherein the high-voltage needle electrode is placed in the liquid and the ground multipoint electrode is in the gas above the liquid. In Figure 6.1b, the parallel reactor, high-voltage electrodes are placed in both the gas and liquids and the ground is placed at the gas–liquid interface. A combined case is shown in Figure 6.9c. Such reactors may have significant effects on chemical reactions in the gas and liquid and also present interesting systems for fundamental and practical study.

6.3 Formation of Primary Chemical Species by Discharge Plasma in Contact with Water

6.3.1 Formation of Chemical Species in Gas Phase with Water Vapor

The earlier chapters dealt with specific aspects of plasma chemistry generated directly in the gas phase, and reviews are available on this subject [152]. Table 6.1 shows some examples of the electron collision products of water [153]. A more comprehensive list of 577 reactions with 46 species for water vapor in a He plasma

Table 6.1 Selected water-electron collision reactions [153].

No.	Reaction	No.	Reaction
1	$H_2O + e^- \longrightarrow OH^\bullet + H^\bullet + e^-$	9	$H_2O + e^- \longrightarrow H^+ + O^\bullet + H^\bullet + 2\,e^-$
2	$H_2O + e^- \longrightarrow O^\bullet + 2\,H^\bullet + e^-$	10	$H_2O + e^- \longrightarrow H_2^+ + O^\bullet + 2\,e^-$
3	$H_2O + e^- \longrightarrow O^\bullet + H_2 + e^-$	11	$H_2O + e^- \longrightarrow OH^- + H^+ + e^-$
4	$H_2O + e^- \longrightarrow H_2O^+ + 2\,e^-$	12	$H_2O + e^- \longrightarrow O^- + H^+ + H^\bullet + e^-$
5	$H_2O + e^- \longrightarrow OH^+ + H^\bullet + 2\,e^-$	13	$H_2O + e^- \longrightarrow O^- + H_2^+ + e^-$
6	$H_2O + e^- \longrightarrow O^+ + 2\,H^\bullet + 2\,e^-$	14	$H_2O + e^- \longrightarrow H^- + OH^+ + e^-$
7	$H_2O + e^- \longrightarrow O^+ + H_2 + 2\,e^-$	15	$H_2O + e^- \longrightarrow H^- + O^+ + H^\bullet + e^-$
8	$H_2O + e^- \longrightarrow H^+ + OH^\bullet + 2\,e^-$	16	$H_2O + e^- \longrightarrow H^- + O^\bullet + H^+ + e^-$

is given in [154], including electron impact excitation, attachment, dissociative attachment, dissociate recombination, and ion reactions. Of particular note, it was found that simplified models with as few as (i) 21 species and 28 reactions for 1–30 ppm water vapor and (ii) 29 species and 58 reactions for 30–3000 ppm were able to predict the main chemical species found in the full model. Of significant importance to water vapor plasma chemistry is the existence of water clusters and ions [36, 154, 155].

Generally, analysis of the collision of plasma-generated electrons with molecular species requires knowledge of the collision cross section $\sigma(e)$ and the electron energy distribution function (EEDF) $f(e)$ to determine the reaction rate constant k by [156]

$$k = \int_0^\infty \sigma(e) f(e)\, de \tag{6.5}$$

Collisional cross-sectional data, either simulated or experimentally measured, is available for many species including water, oxygen, and nitrogen [153, 157–163]. Figure 6.10 shows the cross sections as functions of the electron energy for some reactions of electrons with water in the range of 0.1–100 eV [160]. As shown, vibrational excitation is important between 0.2 eV and over 10 eV, while electron attachment is significant between 4 eV and over 10 eV, with ionization playing an important role above 10 eV. The electron energy distribution is often assumed to be Maxwellian but can also be determined from solution of the Boltzmann equation [164]. For a gas-phase system with Maxwellian distributions of electrons, semiempirical rate constants dependent on input power, W, have been successfully utilized [165]

$$k[e] = \beta \sqrt{\frac{1}{\alpha P}}\, W^{0.75} \exp\left(- \frac{\alpha P}{W} \right) \tag{6.6}$$

where P is pressure, β is a constant dependent on reactor and electrode properties, and α is a model constant parameter.

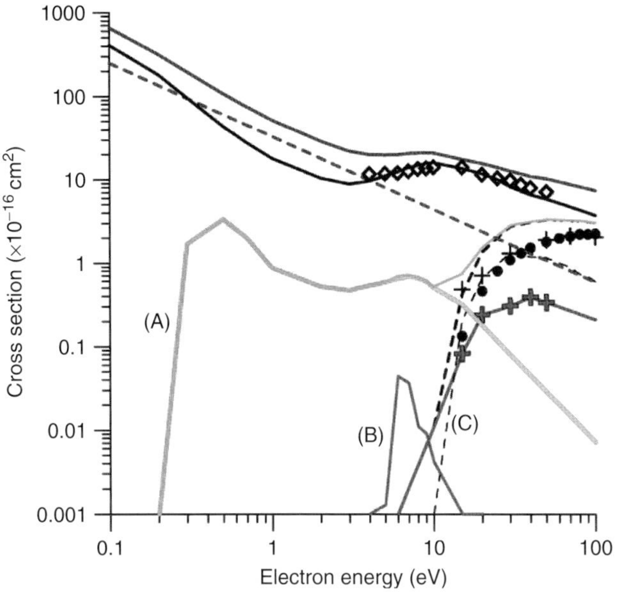

Figure 6.10 Scattering cross-sectional data for water-electron collisions: (A) vibrational excitation, (B) electron attachment, and (C) ionization (dashed line with solid circles). For explanation of other lines see original source [161]. (Source: Reprinted from [161]. Copyright (2008), with permission from Elsevier.)

6.3.1.1 Gas-Phase Chemistry with Water Molecules

The addition of water molecules into a plasma leads to water dissociation reactions (and other reactions shown in Table 6.1, see also [166]) with electrons such as the following:

$$H_2O + e^- \longrightarrow H^\bullet + OH^\bullet + e^- \tag{6.7}$$

$$H_2O + e^- \longrightarrow H^\bullet + O^\bullet + H^\bullet + e^- \tag{6.8}$$

The rates of these reactions clearly depend on the water content as well as the electron energy and collision cross sections for water molecules with electrons as mentioned above. There is abundant evidence that gas-phase discharge with water vapor leads to the formation of $OH^\bullet$ radicals, and an example emission spectra showing the characteristic OH band at 308 nm is shown in Figure 6.11a [167–169]. Other discharges with liquid water (e.g., bubbles, directly in water, and droplets) are shown in Figure 6.11b,c,d. In the case of liquids, such as water droplets, water surfaces, and even water clusters [170, 171], the cross-sectional data is not available fully and determination of reaction rates requires future research to obtain such cross-sectional data and to assess reaction processes.

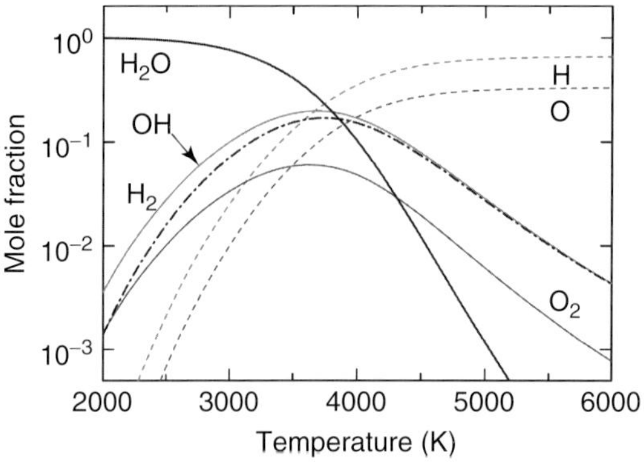

Figure 6.11 Thermal equilibrium concentration of water and its decomposition products. (Source: Reprinted from [175]. Copyright (2011), with permission of IOP Publishing.)

Thermal dissociation may play a role in very high temperature plasmas (over 3000 K [166, 172])

$$H_2O + M \longrightarrow H^\bullet + OH^\bullet + M \tag{6.9}$$

Figure 6.12 shows the equilibrium distribution of the dissociation of water as a function of temperature, indicating that above 2–3×10³ K, water and molecular hydrogen and oxygen are dissociated into H and O and that maximum H_2 and O_2 occur just above 3500 K. For many practical applications, the recombination reactions that lead to stable molecular products may control or limit the process efficiency and effectiveness of reactions that require hydroxyl radicals

$$H^\bullet + H^\bullet \longrightarrow H_2 \tag{6.10}$$

$$OH^\bullet + OH^\bullet \longrightarrow H_2O_2 \tag{6.11}$$

$$H^\bullet + OH^\bullet \longrightarrow H_2O \tag{6.12}$$

The formation of molecular oxygen is also an important aspect to consider since it can occur directly from a variety of reactions including

$$OH^\bullet + O^\bullet \longrightarrow O_2 + H^\bullet \tag{6.13}$$

$$H_2O_2 \longrightarrow H_2O + \tfrac{1}{2}\,O_2 \tag{6.14}$$

Measurements of key stable molecular species such as O_2, H_2, and H_2O_2 can provide insight into the differences and common features between various types of plasma [91, 173] and some insight into the reaction mechanisms [146, 147]. To date, these molecular products have not been linked to the basic plasma properties, including electron density and temperature, but emissions from $OH^\bullet$, $H^\bullet$, and $O^\bullet$ have been correlated to these molecular products [174]. In general, very few studies have provided both molecular product measurements and plasma properties for the same experimental system.

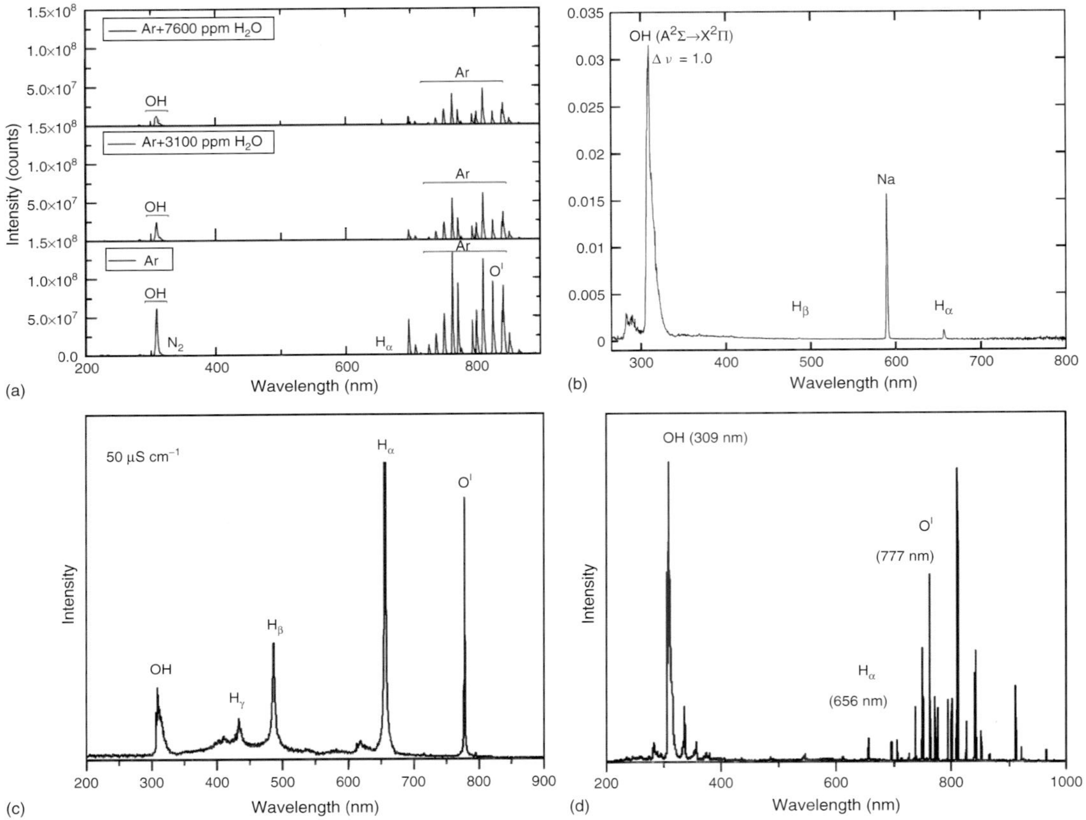

Figure 6.12 Sample emissions spectra for water plasma cases. (a) Atmospheric pressure plasma jet in mixtures of Ar with water vapor. (b) DC discharge in water vapor bubble in capillary filled with NaCl solution (300 µS cm^{-1}). (c) Pulsed discharge directly in water (NaH_2PO_4 solution, 50 µS cm^{-1}). (d) Discharge in argon carrier gas with water droplet spray. (Source: (a) Reprinted from [169]. Copyright (2011), with permission of IOP Publishing. (b) Reprinted from [88]. Copyright (2008), with permission of IOP Publishing. (c) IPP Prague. (d) FSU Tallahassee.)

Analysis of many gas-phase plasma chemical reactions, beyond the electron collision reactions mentioned above, utilizes the extensive literature on the chemical reactions and their reaction rate coefficients from atmospheric chemistry. Typically, in the atmosphere, the concentrations of radicals may be lower than in some local regions of a plasma discharge. For example, because of the differences in the radical concentrations between typical plasma reactors and the atmosphere, the major pathway for hydrogen peroxide generation in the atmosphere is by $HO_2^\bullet$ radical recombination (Eq. (6.15)) [176], while that in plasma is by $OH^\bullet$ radical recombination (Eq. (6.11))

$$HO_2^\bullet + HO_2^\bullet \longrightarrow H_2O_2 + O_2 \tag{6.15}$$

$$OH^\bullet + OH^\bullet \longrightarrow H_2O_2 \tag{6.11}$$

Analysis of lightning discharge, on the other hand, in the atmosphere may have many commonalities including the need to account for the plasma properties [177, 178]. General atmospheric chemistry models have been developed that include the effects of water vapor on many reactions with nitrogen oxides and other components of the atmosphere with, for example, simulations including up to 486 species [177, 179–182]. These models can be useful in developing models of plasma reactors; however, further work is needed to perform parametric sensitivity analysis [183], to determine the effects of different feed compositions, and to evaluate the electron energy distributions and concentration input/distribution specific for given applications as well as to include aerosols and gas–liquid reactions.

Analysis of helium plasma with molecules of water [154] has recently shown the importance of water clusters. Recent emissions spectroscopy of an Ar–water vapor plasma 12.7 W capillary discharge showed significant cooling of the plasma with increasing water content to 4 500 ppm and decreases in the $OH^\bullet$ and other radical intensities in the gas phase with water content [168]. Clearly, the water content in the plasma, even as a vapor, can have significant effects on the plasma chemical properties.

The hydroxyl radical is a key species for many chemical reactions generated in plasma with water vapor and Figure 6.11a shows an example emissions spectrograph showing spectral lines for $H^\bullet$, $OH^\bullet$, and $O^\bullet$, as well as those for argon [169]. For vapor-phase plasma, the reported range of $OH^\bullet$ radical densities are in the range of 10^{16} to 10^{21} m^{-3} [166]. Various methods have been used to estimate the hydroxyl radical concentration in gas-phase plasma with water molecules. For example, laser-induced florescence experimental techniques were used by Ono and Oda [184] to show that the concentration of $OH^\bullet$ radicals in a gas-phase humid air pulsed corona plasma is 7×10^{20} m^{-3}, and this magnitude is in good agreement with experimental studies using CO reactions [185] of 9×10^{20} m^{-3} of pulsed discharge in argon with water vapor and computational studies [154] in a humid helium plasma of 10^{20} m^{-3}. Although the value of hydroxyl radical concentration is expected to be highly dependent on the many chemical reactions that occur, particularly if nitrogen oxides or organic compounds are present, the relatively close agreement between these various experiments and models are significant.

Bruggeman and Schram discuss $OH^{\bullet}$ production in plasma containing water, and they provide key insights on the needed plasma diagnostics to determine electron density and temperature and gas-phase temperature [166]. Key reactions for the formation of $OH^{\bullet}$ radical from water are given and include major pathways through thermal dissociation, electron dissociation and attachment, as well as ion and metastable pathways. Simulations suggest that only above 3000 K is thermal dissociation (Eq. (6.9)) comparable to ionic pathways. Key loss pathways for $OH^{\bullet}$ radicals are also discussed and are strongly dependent on the chemical nature of the gas (i.e., particularly the presence of nitrogen oxides). Hydroxyl radical densities in the range of $10^{21}-10^{24}$ m^{-3} can occur depending on the degree of ionization. The $OH^{\bullet}$ radical densities fall within the range of $10^{19}-10^{22}$ m^{-3}, while electron densities are fairly narrowly distributed between 10^{22} to 10^{23} m^{-3} for gliding arc, pulsed arc, pulse streamer, dielectric barrier (AC and pulsed), and microwave plasma [166]. Electron temperatures are typically about 1 eV, but can vary from 1 to 2 eV, 1 to 10 eV, and 1 to 5 eV for pulsed streamers, AC DBD, and pulsed DBD, respectively.

Figure 6.11a shows a sample emission spectra for water vapor in an argon carrier for different amounts of water [169]. The argon lines are clearly visible, while at the high water content (0.76% H_2O), the OH bands and H_α lines become apparent. In comparison, discharge inside a vapor bubble shows the very small H_β line as well as the H_α line and a strong emission of the OH(A-X) band (Figure 6.11b) [88]. At high conductivity, adjusted with NaCl, the Na line (sodium ion) is seen. Other spectra for discharge in bubbles with different gases (N_2, O_2, He, Ne, and Ar) show different intensities for the OH, O, and H lines [81, 82, 95, 186]. Pulsed capacitative discharges directly in water with energy input of 1 J $pulse^{-1}$ lead to relatively higher H_α lines in comparison to the OH bands than discharges in gas bubbles and humid gas (Figure 6.11c) (see [113, 174, 187, 188] for other example spectra of discharge directly in water). Figure 6.11d shows the emissions for a discharge in gas (argon) with water droplets, and while the relative magnitudes of the H_α line to OH bands are similar to those in gas-phase discharge (with water vapor or in bubbles), the relative magnitude of the O line (777 nm) compared to the OH band is more like the direct discharge in liquid water.

Optical emissions spectroscopy is a key diagnostic tool to determine plasma properties such as electron temperatures and densities [166, 168, 169, 189–191]. The Stark broadening of the H_α line [124, 188, 192] and H_β line [191, 193] have been used to estimate the electron density, n_e, from the full width at half maximum (FWHM) for discharges directly in water and in bubbles. As shown in Figure 6.11, the H_α line is generally more intense and thus may be easier to use. However, for example, H_β was used with Eq. (6.16) to determine electron densities [191, 193].

$$\text{FWHM} = 4.800\ \text{nm}(n_e/10^{23})^{0.68116} \tag{6.16}$$

It was noted that information from *both* H_α and H_β required a two-component fit to the experimental data and suggested that there are two electron densities implying different populations of electrons in the plasma [191]. Sunka *et al.* [188] found using the H_α line electron densities for direct discharge in water to vary from 10^{23} to 10^{25} m^{-3} depending on solution conductivity. In bubbles of He and Ar/O_2, Bruggeman *et al.* [191] found electron densities of 10^{21} and 10^{22} to 10^{23} m^{-3}, respectively.

6.3.1.2 Gas-Phase Chemistry with Water Molecules, Ozone, and Nitrogen Species

In gas-phase plasma with water vapor (and indeed with liquid surfaces), the composition of the gas is critical to the chemistry. For example, in the case of air or nitrogen carrier gases, nitrogen oxides can be formed in significant amounts, and these species can strongly affect the formation and destruction of components such as hydroxyl radicals and hydrogen peroxide [7]. Ozone generation in plasma has been well studied and includes the key reactions [194–200]

$$O_2 + e^- \longrightarrow 2\,O + e^- \quad (6.17)$$

$$O + O_2 + M \longrightarrow O_3 + M \quad (6.18)$$

Of particular note for gas-phase ozone generation is the effect of water vapor on suppressing ozone generation and increasing H_2O_2 through $OH^\bullet$ radical recombination (Eq. (6.11)) [201, 202].

Using a computational model with 43 reactions and 18 species (and the Boltzmann equation for electron energy distribution and the coupled charge carrier continuity equations with Maxwell's equation for the electron density) of a DC corona discharge in 100% relative humidity air, it was found that ozone suppression was due to the direct reaction with hydroxyl radicals [39]

$$O_3 + OH^\bullet \longrightarrow HO_2 + O_2 \quad (6.19)$$

and that the reaction of NO with ozone (Eq. (6.29)) (see below) was insignificant at high humidity but predominated in dry gas.

While most work has dealt with the application of plasma with water vapor or liquids to remove nitrogen oxides [181, 203–211], the generation of nitrogen oxides in gas-phase plasma in the presence of liquid water is also of interest. For example, from a practical perspective, the Birkeland–Eyde process utilized a thermal plasma arc for the generation of nitrate for use in fertilizers at the beginning of the twentieth century. This process operated commercially for about 20 years before being supplanted by the high-temperature Haber catalytic process [152, 212]. The plasma arc first generated nitric oxide (NO) by the overall and limiting endothermic reaction

$$\tfrac{1}{2}\,N_2 + \tfrac{1}{2}\,O_2 \longrightarrow NO \qquad \Delta H = 96\ \text{kJ mol}^{-1}(1\ \text{eV per molecule}) \quad (6.20)$$

Downstream of the plasma reactor in the original process, the above reaction was followed by the two exothermic, nonplasma reactions involving first, high-temperature oxidation with oxygen and then, reaction with water

$$NO + \tfrac{1}{2}\,O_2 \longrightarrow NO_2 \qquad \Delta H = -56\ \text{kJ mol}^{-1} \quad (6.21)$$

$$3\,NO_2 + H_2O \longrightarrow 2\,HNO_3 + NO \qquad \Delta H = -138\ \text{kJ mol}^{-1} \quad (6.22)$$

The generation of NO from N_2 and O_2 in a plasma reactor has been extensively reviewed [152] (primarily for the dry gas synthesis of NO), and the energy cost for the production of NO by the Birkeland–Eyde process was 25 eV per molecule. This result corresponds to a relatively low 4% energy yield based on the enthalpy of reaction. Thermal plasma systems at very high pressure (2–3 MPa) and temperature (3000–3500 K) have subsequently been found to lead to an energy efficiency of 11% (9 eV per molecule), and very fast quenching rates (large temporal gradients, $10^8\,K\,s^{-1}$, in the plasma temperature to the ambient) are needed to achieve this efficiency (for comparison, the commercial production of ozone from oxygen, one of the most successful applications of nonthermal plasma in industrial processes, also has an energy yield of about 10% of thermodynamic limits, but experimental systems can reach over 40% of the thermodynamic limit [213]). In the dry air plasma, oxygen is dissociated through either dissociative attachment (Eq. (6.23)) or through direct electron dissociation (Eq. (6.17)) (in addition to other electron collision reactions with molecular oxygen) [158]

$$O_2 + e^- \longrightarrow O + O^- \tag{6.23}$$

The Zeldovich mechanism through vibrationally excited nitrogen (generically known as N_2^*) has two steps, the first, a limiting endothermic reaction (Eq. (6.24)) (with $\Delta H = 289\,kJ\,mol^{-1}$ or 3 eV per molecule) and the second, exothermic reaction

$$O^\bullet + N_2^* \longrightarrow NO + N^\bullet \tag{6.24}$$

$$N^\bullet + O_2 \longrightarrow NO + O^\bullet \tag{6.25}$$

Key reactions that suppress the formation of NO and thus limit the achievable energy efficiencies are the in-plasma and outside-of-plasma reverse reactions

$$N^\bullet + NO \longrightarrow N_2 + O \tag{6.26}$$

$$NO + NO \longrightarrow N_2 + O_2 \tag{6.27}$$

The latter reaction, (Eq. (6.27)), occurs strongly at temperatures above 1500 K, and the former reaction, (Eq. (6.26)), coupled with the branching reaction

$$NO^* + O_2 \longrightarrow NO_2 + O^\bullet \tag{6.28}$$

leads to high atomic oxygen in the plasma resulting in high ozone concentrations [152], although typically, ozone does not coexist with NO [214, 215] because of the rapid reaction (Eq. (6.29)). For the generation of nitrate, the formation of NO_2 in the plasma would be desirable, and ozone and atomic oxygen can further oxidize both NO and NO_2 by reactions (Eqs. 6.29–6.32)

$$NO + O_3 \longrightarrow O_2 + NO_2 \tag{6.29}$$

$$NO_2 + O_3 \longrightarrow O_2 + NO_3 \tag{6.30}$$

$$NO + O^\bullet \longrightarrow NO_2 \tag{6.31}$$

$$NO_2 + O^\bullet \longrightarrow NO_3 \tag{6.32}$$

Some experimental analysis and kinetic simulations of nitrogen oxide reactions [214] in the context of nitrogen oxide pollution abatement support the key role of reaction (Eq. (6.24)) in NO formation and that oxygen concentrations above 2.5% (in nitrogen carrier) favor the formation of nitrogen oxides in dry gas mixtures. Combustion models with superimposed plasma also support reaction (Eq. (6.24)) [216]. However, in dry N_2/O_2 mixtures [165, 214, 217–220] it was found that excited states of nitrogen have negligible contributions to NO_x formation and destruction and that reaction (Eq. (6.25)) (following from reaction (Eq. (6.33)) is of more importance in NO generation. Some atmospheric chemistry models also support the key role of reaction (Eq. (6.25)) [182, 221]. Thus, the importance of reaction (Eq. (6.24)) compared to the direct electron collisional dissociation of nitrogen or excited nitrogen [152, 214] by reaction (Eq. (6.33)) followed by reaction (Eq. (6.25)) needs to be further assessed to resolve the differences in the literature.

$$N_2 + e^- \longrightarrow 2\,N^\bullet + e^- \tag{6.33}$$

Itikawa and Mason [222] show that vibrational excitation cross sections are higher than direct electron dissociation by reaction (Eq. (6.33)) in the electron energy range of 1–4 eV and that higher energies can lead to direct dissociation into neutrals and ions. For example, in a 2.4 GHz HF plasma with O_2/N_2, Kuhn *et al.* [223] showed a broad electron velocity distribution and only used reaction (Eq. (6.33)) to simulate NO and O_3 formation. In a more extensive simulation of nitrogen–oxygen plasma reactions (in the dry case), Kossyi *et al.* [224] developed a model with over 400 reactions, including neutrals and ions. They demonstrated the importance of the reaction of excited state N_2 with oxygen atoms by reaction (Eq. (6.24)) to double NO production.

The addition of water vapor leads to the formation of hydroxyl radicals through water dissociation and other [157] reactions with electrons (i.e., (Eqs. (6.7)–(6.9)) and Table 6.1). A wide range of reactions (Eqs. (6.34)–(6.37)) with nitrogen oxides and hydroxyl and hydroperoxyl radicals lead to the formation of nitrite (NO_2^-) and nitrate (NO_3^-)

$$NO + OH^\bullet \longrightarrow HNO_2 \tag{6.34}$$

$$HNO_2 + OH^\bullet \longrightarrow NO_2 + H_2O \tag{6.35}$$

$$NO_2 + OH^\bullet \longrightarrow HNO_3 \tag{6.36}$$

$$NO + HO_2^\bullet \longrightarrow NO_2 + OH^\bullet \tag{6.37}$$

Comparison of the pathways to nitrite and nitrate through ozone with those through $OH^\bullet$ radicals can assist in the analysis of pathways in a given water–plasma environment. Using the rate constants for reaction of NO with ozone k_{29} (Eq. (6.29)) and $OH^\bullet$ radical k_{34} (Eq. (6.34)) [225], the relative rates are given by

$$k_{29}/k_{34} = 5 \times 10^{-4}[O_3]/[OH^\bullet] \tag{6.38}$$

Radical measurements and model estimates, discussed above, for $OH^\bullet$ radical concentrations can be used to determine the relative rate (Eq. (6.38)) for comparing the $OH^\bullet$ and ozone pathways. For a 1% ozone concentration, as found in the output of a typical ozone generator, the concentration of $OH^\bullet$ radicals needs to be about 10^{21} m^{-3} for reaction (Eq. (6.34)) to be comparable to that of reaction (Eq. (6.29)), and these conditions are typically met. It can also be concluded that when water vapor is present, the ozone concentration will be lower than used in the above estimate and the $OH^\bullet$ radical pathway would be favored.

In humid oxygen/nitrogen mixtures reactions (Eqs. (6.31) and (6.32)) through dissociated oxygen formed by reactions (Eqs. (6.17) and (6.23)) are perhaps more likely than direct ozone reactions; however, the roles of these and the other reactions must be more fully assessed through the combination of experimental and modeling approaches. Recent emissions spectroscopy of an Ar–water vapor plasma 12.7 W capillary discharge showed significant cooling of the plasma with increasing water content to 4500 ppm and decreases in the $OH^\bullet$ and other radical intensities with water content [168].

Gas-phase plasma reactors also form UV light emissions [226], although the amounts and intensities vary with the type of reactor, gas composition, and other conditions. The addition of water vapor leads to characteristic emissions of $OH^\bullet$ (at 284 and 309 nm) [227, 228] (see also Section 6.3.2 and Figure 6.11). It is known from atmospheric chemistry that hydrogen peroxide, for example, can be dissociated by UV at 190–350 nm [229]

$$H_2O_2 + h\nu \longrightarrow OH^\bullet + OH^\bullet \tag{6.39}$$

In addition, water and oxygen can be dissociated by UV with high quantum yields below 190 nm [230] by

$$O_2 + h\nu \longrightarrow O^\bullet + O^\bullet \tag{6.40}$$

$$H_2O + h\nu \longrightarrow H^\bullet + OH^\bullet \tag{6.41}$$

The cross section for reaction (Eq. (6.40)) spans 10^{-17}–10^{-18} cm^2 [231], with a quantum yield ~1 at 120–170 nm [232], and the cross section for reaction (Eq. (6.41)) is 10^{-17}–10^{-18} cm^2 [233], with a quantum yield ~0.4–1 at 120–170 nm [234]. Reactions that might cause back reactions are given below [229]

$$NO_2 + h\nu \longrightarrow NO + O^\bullet \qquad 202 - 422\ \text{nm} \tag{6.42}$$

$$NO_3 + h\nu \longrightarrow NO + O_2 \qquad 600 - 670\ \text{nm} \tag{6.43}$$

For combustion gas treatment containing NO in the feed, Mok *et al.* [209] and Hu *et al.* [220] found decreases in NO conversion with increasing water vapor. Mok *et al.* found that increasing H_2O decreases the production of ozone by

$$OH^\bullet + O_3 \longrightarrow HO_2 + O_2 \tag{6.19}$$

and that in this reactor, ozone generally causes NO oxidation by reaction (Eq. (6.28)). They also found that the energy delivered by the pulsed corona decreases with increasing water content, and this result is consistent with that of Hu *et al.* [220] (both groups used pulsed corona wire-cylinder type discharges). Yin *et al.* [235] found in DBD that there was no effect of H_2O on discharge power but there was a reduced NO conversion in the $N_2/NO/H_2O$ plasma with up to 2% water vapor; however, the simultaneous effects of oxygen were not reported. This reduced conversion of NO with the addition of water is also consistent with the reported *increased* NO production in an inductively coupled radio frequency (ICRF) plasma by reaction (Eq. (6.44)) [236]

$$N^\bullet + OH^\bullet \longrightarrow NO + H^\bullet \tag{6.44}$$

Such increases in NO are not desirable in many of the applications for NO_x removal by plasma in pollution control; however, they are important in forming NO and converting the formed NO into nitrates. Reaction (Eq. (6.44)) could provide significant enhancement of NO formation by the $N^\bullet$ formed in Eqs. (6.33) and (6.24).

A recent model of DBD in air at 300–500 K and 1.5 W cm^{-3} utilized 68 chemical reactions of neutral species and the Boltzmann equation to determine the electron energy distribution function [237]. Increasing the relative humidity from 20% to 80% led to approximately three times higher peak concentrations of hydroxyl radicals (1.6×10^{19} m^{-3}) and hydrogen peroxide (3.5×10^{19} m^{-3}), while most of the other species were not strongly affected by this change in humidity. The model showed relatively good comparison with experimental measurements on O_3, HNO_3, HNO_2, N_2O_5, and NO_3 for both levels of humidity, suggesting that most of the key reactions were included in the model and that the plasma conditions were accurately simulated.

Clearly, the chemistry of gas-phase plasma with water vapor is quite complex even in the absence of a liquid. While key reactions in ozone, nitrogen oxide, and hydrogen peroxide chemistry have been used to analyze and interpret experimental results in water vapor plasma, further work is needed to develop fully predictive models coupling the physics and chemistry for a wide range of reactors. In the presence of a liquid phase, as shown in Figure 6.2, much of the basic gas-phase chemistry is similar to that discussed above; however, the interactions of the molecules, radicals, and other active species of the gas with the liquid, the transport of liquid-phase reactants to the gas phase, and the subsequent reactions in both the gas and the liquid become important and are discussed in the following sections.

6.3.2
Plasma-Chemical Reactions at Gas–Liquid Interface

When a liquid phase is placed in contact with a plasma region, it is possible for the plasma-generated electrons and ions to interact with the liquid water and for species from the liquid phase to, in turn, transfer into the plasma and affect its properties. In addition, gas- and plasma-phase chemical species, for example, ozone and hydrogen peroxide [238–240], can strongly affect liquid-phase reactions [68, 241]. The direct collision of high-energy electrons with liquid surfaces has some similarities to radiation chemistry; however, the electron energies in most gas-phase plasma (e.g., 1–10 eV) may not, in general, be as high as those in liquid-phase radiation chemistry (kiloelectronvolts to megaelectronvolts) [242, 243].

Nevertheless, much of the basic chemistry of plasma interactions with liquid water and water vapor stems from the study of radiation chemistry of water [242, 244] (for additional review with more historical references, see [7]). As mentioned above, one aspect that has not received sufficient attention is the analysis of the direct collisions of electrons with the condensed water phase and with clusters of water molecules in the electron energies between about 0.1 and 10 eV, and perhaps future work using molecular dynamics can lead to progress on this aspect [170, 171, 245–248]. Ionization cross sections, for example, have been determined for electron impact with ice and liquid water for 10 eV–1000 eV and show higher, significantly larger cross sections for ice than the condensed phase [247, 248].

Emissions spectroscopy and laser-induced fluorescence (LIF) spectroscopy have been conducted for electrical discharges over water surfaces, and, as in water vapor discharges, radicals such as $H^\bullet$, $O^\bullet$, and $OH^\bullet$ have been identified and are similar to those shown in Figure 6.11 [190, 249–252]. Using LIF combined with a chemical probe to assess liquid-phase $OH^\bullet$ concentration, it was suggested that for a plasma propagating over a water surface the $OH^\bullet$ is formed in the gas phase, after which it diffuses into the liquid [253]. Of course, the gas-phase atmosphere (e.g., air, nitrogen, oxygen, argon, etc.) will have a large effect on the formation of chemical species, and of key importance is the formation of ozone as mentioned above [68]. In analysis of the gas- and liquid-phase chemical reactions, it is necessary to carefully account for the formation of ozone and hydrogen peroxide and to consider solution properties such as pH, interfacial surface area, and mass transport [61, 241, 254].

The interaction of the plasma-formed species with the liquid phase is of importance in order to determine if the chemistry at the gas–liquid interface is governed by the gas-phase discharge above the water surface and to determine the role of mass transfer from the gas into the liquid. For example, the penetration of the plasma-formed radicals into the liquid through the interface and the effects of the radicals of the plasma on the bulk-phase reactions in the liquid are not well known. These are the key factors for the kinetic laws observed in the solution (overall zero and first order). The gas-phase reactants impinging on the liquid–gas interface may also affect temperature gradients and quenching reactions. Recent work in atmospheric chemistry on the interactions of hydroxyl radicals with water surfaces [255] has shown that hydroxyl radicals can transfer into the liquid phase (or into clusters of water molecules) and form local hydrogen bonds with water molecules near the surface [255–257]. Such molecular simulations are only guides since they do not explicitly account for the plasma environment above the liquid surface.

Glow discharge electrolysis involves the formation of a glow discharge over a liquid solution, and the analysis of the mechanism of chemical processes at the interface between a gas and liquid relies extensively on concepts from radiation chemistry. In this work, it is stated that 10–100 eV positive ions formed in the gas phase impinge on the liquid surface leading to water excitation and ionization and, ultimately, secondary electron emissions from the water cathode [12]. This *ion* energy is higher than the average *electron* energies reported for gas-phase pulsed corona, dielectric barrier, and gliding arc discharges, which are all reported in the range of 1–5 eV [166, 258]. Owing to the significant role of acidity, it is expected that ionization is the major pathway for the formation of the final products in the liquid phase of H^+, H_2O_2, and H_2. In contrast to direct discharge in water described elsewhere where only 20% of the molecular oxygen is formed from H_2O_2, in the glow discharge electrolysis, all molecular oxygen is considered to be formed from H_2O_2. Another important difference with direct discharge in water and the glow discharge electrolysis is that no pH change (or very small) is observed in direct discharge in water in the absence of any added gases (for more about acidic effects of plasma in water, see Chapter 7). The general pathway for the glow discharge in water [12] is given by

$$\begin{aligned} &P^+ + xH_2O \longrightarrow xH_2O^* + H_2O^+ \longrightarrow (n+1)H_2O^+ + ne^-_{aq} \longrightarrow \\ &(n+1)OH^\bullet + (n+1)H^+ + ne^-_{aq} \longrightarrow (n+1)/2H_2O_2 + nH^\bullet + H^+ \end{aligned} \tag{6.45}$$

where P^+ is the positive ion, x is the total number of water molecules that are impacted by the positive ions, the coefficient n accounts for reverse reactions, and e^-_{aq} are aqueous electrons. The n moles of hydrogen ($H^\bullet$) transport to the gas phase where secondary electrons are formed by

$$nH^\bullet \longrightarrow (n-\gamma)H^\bullet + \gamma e^-_g + \gamma H^+ \tag{6.46}$$

where γ is the secondary electron emission coefficient. In the glow discharge above the water surface, the positive ions impinge on the gas–liquid interface, but the plasma does not penetrate into the liquid phase, and therefore the reactions in the liquid phase may follow lower-temperature ambient conditions in contrast to the high-temperature reactions in the corona discharge model in liquid described above. As mentioned in Section 6.2.2, recent simulations of DBD gas discharges over dielectric surfaces by Babaeva and Kushner [25] support the possibility of high-energy ion formation over the surface and impingement onto the surface. These results are quite supportive of the general aspects of high-energy ions impinging on the surface, although details of reaction chemistry, specifically for water vapor and water surfaces, were not considered in these simulations.

Thagard *et al.* [62] showed that hydrogen peroxide formation and emissions spectroscopy for different polarity discharges over water surfaces are strong functions of the solution pH and conductivity. These results were interpreted within the context of changing secondary emissions coefficients with pH, and the highest production of hydrogen peroxide occurred at the lower pH because of the high production of gas-phase ions. In negative polarity discharges over the liquid–water surface, it was suggested that electrons generated in the gas phase propagate into the liquid phase, where they react; positive polarity discharges over the liquid showed no similar effects.

The reactions of ozone (generated by the gas-phase plasma) in the liquid phase and liquid-phase-induced electrical discharge reactions to form H_2O_2 and $OH^\bullet$ for the oxidation of phenol were included in a model of plasma discharge in hybrid (Figure 6.1b) gas–liquid reactors and in reactions with oxygen flowing through the high-voltage needle electrodes [60, 259]. Relative reactions of ozone from the gas phase and $OH^•$ in the liquid are quite important in assessing reaction pathways of organic compounds in the liquid.

6.3.3
Plasma Chemistry Induced by Discharge Plasmas in Bubbles and Foams

The basic chemistry of discharges in bubbles and foams is generally expected to be similar to that in the gas phase with water vapor (Section 6.3.1) and in gas-phase discharge over a liquid surface (Section 6.3.2). Electrical discharges in single bubbles and in foams and bubbly liquids may lead to increased mass transfer rates of gaseous phase species into the liquid and reduced energy requirements to generate the plasma compared to directly in the liquid. Analysis of fundamental aspects of the gas–liquid interfacial plasma may also be facilitated through utilization of these contacting patterns. It can be noted that experiments with bubbles formed in the liquid by the electrical discharge, with preexisting bubbles that may arise in solutions near the boiling point, can be distinguished from those where the bubbles are introduced [90, 92].

Discharges in single bubbles suspended in the liquid or in a capillary tube have been analyzed with regard to emissions spectroscopy, hydroxyl radical formation, and hydrogen peroxide formation [79–82, 85, 86, 88, 89, 94, 95, 101] (Figure 6.11b). Emissions spectra clearly show the existence of $OH^•$ radicals and have been used to estimate electron density and temperature [85, 101]. Discharges in foams and bubbly liquids have been developed and characterized with regard to chemical species formation including hydrogen peroxide and ozone [80–82, 84, 87, 93, 260–264].

In general, two recent reviews [7, 265] have suggested that plasma reactors with high surface area of gas–liquid contact, but with the discharge formed in the gas, are generally more efficient for both chemical species formation from liquid water, for example, hydrogen peroxide, and for initial dye decolorization. Figure 6.13 shows the range of energy efficiency, given by energy yield, for hydrogen peroxide formation by a wide variety of electrical discharge reactors in contact with liquids. Clearly, the most efficient is that with gas-phase plasma in contact with water aerosols flowing through the plasma. It is suggested that plasma-quenching steps limit efficiency [7] in the peroxide formation. In contrast, for high degrees of mineralization, more intense, and higher energy density and temperature, plasma is required. Further, the energy yields for the formation of hydrogen peroxide, hydrogen, and oxygen were found to be independent of input power over the range of conditions whereby discharges were formed in the bubbles [90, 92].

A recent model was developed that includes gas-phase plasma chemistry inside a bubble coupled with liquid-phase reactions and the mass transfer between the gas and the liquid [266]. Water (Eq. (6.7)), oxygen (Eq. (6.17)), and ozone reactions with electrons were included in the gas phase along with 37 gas-phase reactions and 16 liquid-phase reactions of species, including some ions in the liquid and all neutrals in the gas phase. In order to account for an increased amount of $OH^•$ radicals, it was postulated that additional $OH^•$ radicals are produced at the

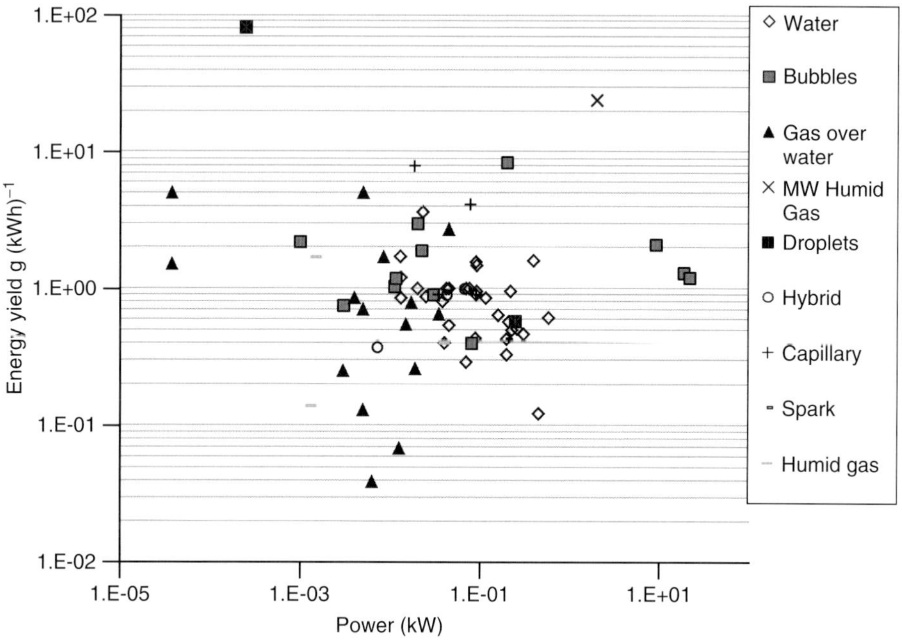

Figure 6.13 Energy yield for hydrogen peroxide generation by a wide range of gas–liquid plasma systems. (Source: Reprinted from [7]. Copyright (2011), with permission of IOP Publishing.)

gas–liquid interface by the reaction of plasma-generated O

$$H_2O(l) + O^\bullet(g) \longrightarrow 2OH^\bullet(l) \tag{6.47}$$

It was necessary to adjust the rate constant for this reaction to fit the experimental data on H_2O_2 formation and acetic acid degradation. Ozone generation in the gas phase and mass transfer into the liquid with subsequent liquid-phase reactions were also found to be important.

Oxygen bubbled through high-voltage electrodes immersed in water has been shown to lead to ozone formation and dissolution into the liquid [60, 80, 113, 259, 267] and bubble and gas–liquid discharge systems can lead to the cogeneration of ozone and hydrogen peroxide [259, 268, 269]. In some discharges in water with bubbles where the temperature can be high, radical and molecular decomposition reactions can be supplemented by thermal decomposition. For example, thermal pyrolysis in plasma generated in bubbles and $OH^\bullet$ radical reactions in the liquid near the interface were assessed for phenol degradation [270]. It was found that for phenol concentrations below 1 mmol l^{-1}, 75%–80% of the degradation occurs in the liquid phase by radical reactions, while at 10 mmol l^{-1}, there is about 50% degradation in the liquid by radical reactions and the rest by thermal pyrolysis, and as concentration increases polymerization of phenol occurs and liquid-phase reactions dominate again.

6.3.4
Plasma Chemistry Induced by Discharge Plasmas in Water Spray and Aerosols

The injection of liquid or solid particles into plasma and/or the corresponding formation of liquid or solid products from reactions in the plasma have many interesting applications. For example, plasma generated in the presence of particulate matter [20], for example, dusty plasma, is of interest in processes for material synthesis [106] such as the generation of airborne nanoparticles and air pollution control [271]. Liquid water injection into thermal plasma arcs have been analyzed and utilized for waste treatment and gasification [272–278]. In the case of organic liquid droplets introduced into the plasma, there is increasing interest in the field of plasma-assisted combustion as well as gasification [275, 279–282]. The introduction of water droplets (aerosols) into the plasma has potential applications in water cleaning [108, 283, 284], disinfection [285], air cleaning [286], and direct chemical synthesis of useful compounds such as hydrogen peroxide, hydrogen, and nitrates [107]. It is important to note that the conditions for destruction of chemical or biological materials and for the synthesis of chemical products can be dramatically different [7]. To synthesize a given species, high product selectivity is required. To destroy and mineralize materials, selectivity is less important.

High-temperature thermal plasma conditions (>10 000–50 000 K) generally have sufficient energy to vaporize most solids and liquids, and the plasma quenching or transition zone (either spatial or temporal) may play a significant role in the final product composition and form [152]. Thermal plasma arc water properties up to very high temperature have been estimated [148, 149] and may be useful in some arclike plasma discharges with liquid water where temperatures can be very high. For example, the temperatures of gliding arc reactors (in the absence of liquid water) have been estimated to exceed 5 000 K [287–289].

In high-power plasma arcs, for example, gliding arc discharges (400 to 500 W), with the addition of liquid water the energy introduced into the plasma is sufficient to vaporize completely the water droplets and can lead to effective pollutant mineralization [290, 291]. In such plasma, whether thermal or nonthermal or with regions of both thermal and nonthermal conditions, the reactor conditions can be analyzed using more conventional gas-phase plasma techniques for thermal plasma [292–294] and for nonthermal plasma (see Section 6.3.1). Experimental studies with 400–500 W gliding arc discharges have shown the formation of H_2O_2, H_2 from water droplets with noble gases and oxygen [295]. Modeling efforts of chemical kinetics have shown the importance of quenching on thermal reactions and formation of stable molecular species [172].

On the other hand, if the input energy of the plasma is insufficient to vaporize the injected water droplets, the multiphase nature of the plasma–aerosol interactions must be considered [7]. A water droplet flowing through the plasma with insufficient energy to vaporize the droplet can interact with the plasma in several ways. In the dilute solution limit where the water droplets have little direct effect on the plasma properties, some water may evaporate into the gas-phase plasma where it participates in direct reactions with electrons (and other species) generated in the plasma. In such a case, the water droplets serve as a source of reactants (here water) and can deliver significantly more water molecules into the plasma than occur with humidified gases.

Further, these water droplets can function as a product collector. For example, in the case of hydrogen peroxide generated in water-spray plasma, the high solubility of hydrogen peroxide in the liquid droplet and the lower concentration of hydroxyl radicals in the liquid phase allows for high-efficiency production of hydrogen peroxide. The hydrogen peroxide formed in the gas phase or at the interface between the gas and liquid phase transports into the liquid droplet where it is protected from the plasma radicals that tend to degrade the product [7]. The plasma is effectively quenched by the water droplets. There is also evidence that H_2O_2 found in rainwater is produced by thunderstorms [296].

Other applications of this concept are the removal of nitrogen oxides by discharges in contact with liquid surfaces and droplets [210, 211, 286, 297] and the production of nitrate from water droplets sprayed into air. Following the discussion of Section 6.3.1 on nitrate chemistry in vapor phase plasma, the presence of water droplets will strongly influence the overall reaction processes. For example, Yan *et al.* [298] found that gas-phase water concentrations over 10% enhance NO removal in $N_2/O_2/H_2O/NO$ mixtures and suggested that reactions (Eqs. (6.34) and (6.36)) with $OH^\bullet$ radicals and further, reactions (Eqs. (6.48) and (6.49)) by the N_2O_5 pathway

$$NO_2 + NO_3 \longrightarrow N_2O_5(g) \tag{6.48}$$

$$N_2O_5(g) \rightleftarrows N_2O_5(l) \tag{6.49}$$

$$N_2O_5(l) + H_2O(l) \longrightarrow 2HNO_3(l) \tag{6.50}$$

can be enhanced in the presence of liquid water, which absorbs the reaction products. It is possible that the production of N_2O from

$$N^\bullet + NO_2 \longrightarrow N_2O + O^\bullet \tag{6.51}$$

may also be suppressed in some reactors because of the rapid conversion of $NO_2^\bullet$ by $OH^\bullet$ radicals. Certainly, from an environmental perspective, N_2O formation would be undesirable since it is a greenhouse gas. Since nitrate is highly soluble in the liquid phase, any nitrate, like hydrogen peroxide, formed in the gas-phase plasma will rapidly absorb into the liquid. Hydrogen peroxide and nitrate have very high Henry's Law constants (0.75 and 2.1 mol $(l\ Pa)^{-1}$) in comparison to NO, NO_2, N_2O, $OH^\bullet$, and HNO_2 (1.9×10^{-8}, 1.0×10^{-7}, 2.5×10^{-7}, 2.5×10^{-4}, and 4.9×10^{-4} mol $(l\ Pa)^{-1}$, respectively) [229], and thus under atmospheric conditions, the former two species are rapidly absorbed into the liquid phase. Although some H_2O_2 and no HNO_3 will exist in the gas phase, all the other species remain in the gas.

It is useful to point out that one other example is known where the existence of a water droplet enhances the plasma chemical conversion. In the oxidation of SO_2 in a plasma reactor with a superimposed electron beam (to generate electrons and radicals in the liquid phase droplets formed from the sulfuric acid), [299] shows that the cyclic chain mechanism with oxygen enhances efficiency and leads to 0.8 eV per molecule. This work suggests that the presence of water droplets (*formed* in the sulfuric acid case or *introduced* in other cases) enhances gas–plasma production of condensable products.

A number of important issues remain in the study of discharges with water droplets. The effects of droplet size and size distribution on the chemical species formation and energy yields should be determined. Analysis and identification of the plasma reaction zone near the particle–gas interface should be experimentally studied. Determination of the mass transport of water from the droplet to the plasma and of plasma products into the liquid droplet is necessary. As in the discharge over a water surface, the importance of the penetration of radicals, ions, and electrons into the liquid droplet from the gas-phase plasma still needs both experimental and computational verification.

6.4 Chemical Processes Induced by Discharge Plasma Directly in Water

6.4.1 Reaction Mechanisms of Water Dissociation by Discharge Plasma in Water

Figure 6.11c shows a sample emission spectrum for a pulsed discharge directly in the liquid water. Emission spectroscopy has demonstrated radiation from the pulsed corona discharge in a wide range of wavelengths (200–1000 nm), which is dominated by the spectral lines of hydrogen (peaks at 434, 486, 656 nm) and oxygen atom (777 nm) and by emission from $OH^{\bullet}$ radical (309 nm). Electron density above $10^{24}\,m^{-3}$ in the streamer discharge has been determined from the H_{α} spectral line profile, and this density is a strong function of solution conductivity. From the molecular OH band spectra, a rotational temperature 2000–5000 K of the $OH^{\bullet}$ radicals has been estimated. With increasing water conductivity, stronger radiation and higher electron density in the discharge have been determined (above $10^{25}\,m^{-3}$) [3, 124, 187, 188, 192]. In contrast to the water vapor emission spectra shown in Figure 6.11a, the H_{α} peak is much higher than the OH peak in the direct discharge in water. Analysis of the $O^{\bullet}$, $H^{\bullet}$, and $OH^{\bullet}$ emission intensities relative to an argon probe was correlated with the direct measurement of the chemical species (H_2, O_2, and H_2O_2) to show various reaction pathway changes as the solution conductivity increased [174].

Early efforts to model plasma chemistry directly in the liquid phase (for low-frequency pulsed discharge) utilized concepts and reaction rates and mechanisms from the radiation chemistry literature coupled with semiempirical overall rates of formation of H_2O_2, $OH^{\bullet}$, and e^-_{aq} [300]. These early modeling efforts could describe overall reaction chemistry based on H_2O_2 and $OH^{\bullet}$ radical [60, 259, 301] but did not account properly for subsequent measurements of molecular gas products (O_2 and H_2) [173]. When more accurate measurements of $OH^{\bullet}$ radical formation and reducing species were determined by chemical means [302–304], it became necessary to develop a more structured model that dealt with physical and thermal aspects of the discharge plasma formation [146, 147]. It can be noted that a wide range of $OH^•$ radical scavengers including alcohol (methanol, propanol, ethanol, acetone) [305, 306] and dimethyl sulfoxide (DMSO) and terephthalic acid (NaTA) [302] support the mechanism for H_2O_2 production by $OH^•$ radical recombination (Eq. (6.11)).

Mededovic and Locke [146, 147] proposed a model that assumes that small regions with temperatures of up to 5000 K can be formed in liquid water with 1 J per pulse electrical discharges with microsecond pulse widths. The localized nature and small volume of the discharge channel (with widths in the tens to hundreds of micrometers scale) strongly supports these high temperatures in the plasma channels. At such temperatures, thermal dissociation of water, as well as possible direct electron collisions, is expected and hydroxyl radicals and atomic hydrogen and oxygen are formed through reactions (Eqs. (6.7)–(6.9)). Through the thermal energy balance (Eq. (6.4)) and kinetic modeling with the thermal dissociation reaction (Eq. (6.7)) in the high-temperature range of 2000–5000 K, other important reactions were found to be Eqs. (6.10), (6.11), and (6.13). In the high-temperature zone, reaction (Eq. (6.10)) leads to 80% of the total molecular hydrogen formed and reactions (Eqs. (6.13) and (6.14)) lead to 50% of the total molecular oxygen formed. It was postulated that the short-lived high-temperature zone rapidly decays in time and space to a somewhat cooler recombination zone (ambient to 2000 K) where hydrogen peroxide is formed primarily by hydroxyl radical recombination, Eq. (6.11), and in this zone, 20% of the total molecular hydrogen is formed by reaction (Eq. (6.52))

$$H^• + OH^• \longrightarrow O^• + H_2 \tag{6.52}$$

In the low-temperature zone, 20% of the molecular oxygen is formed by reactions (Eq. (6.13)) and

$$OH^• + O^• \longrightarrow O_2 + H^• \tag{6.53}$$

but additional oxygen (30%) is formed from hydrogen peroxide by

$$H_2O_2 + OH^• \longrightarrow HO_2 + H_2O \tag{6.54}$$

$$HO_2 + OH^• \longrightarrow H_2O + O_2 \tag{6.55}$$

These modeling efforts are directly supported by experimental measurements of the formation of hydrogen peroxide, molecular oxygen and hydrogen, hydroxyl radicals, and oxygen radical ion (see below) [173, 302–304]. These experimental methods utilized chemical probes to quantify the concentrations of these radicals and direct measurements by chromatography and colorimetric assays to determine the molecular species. The experimentally measured stoichiometry of $H_2 : H_2O_2 : O_2$ was 4 : 2 : 1, with current efficiencies of 26 : 13 : 5 moles per mole of electron, much higher than expected by Faraday's law. The only adjustable parameter in the model, other than the assumed temperature profiles and size of the plasma region, is the pressure. The pressure required to match the model results with the experimentally observed quantities of hydrogen peroxide, hydrogen, and oxygen, was 1.4 MPa, and while smaller than the measured localized transient pressures [124], it is consistent with the significantly higher ambient pressure in the localized region of the plasma. Once the pressure is set, the model is able to predict the amounts of hydroxyl radical and oxygen radical ion formed as an independent check of the model. Although further modeling refinement is necessary to account for higher pressures observed in some experiments and to account for possible electron collisions with water, the model provides an excellent base to build more detailed descriptions.

Figure 6.14 shows example data for the production of $OH^{\bullet}$ radicals, H_2O_2, and reductive species for direct discharge in liquid water with point-to-plane geometry

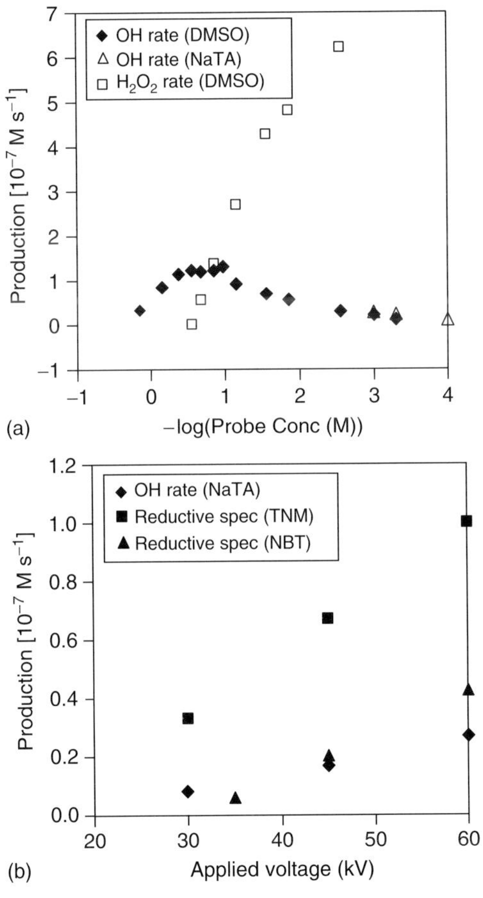

Figure 6.14 Molecular and radical species (from chemical probes) formation in direct discharge in water at 1 J per pulse in point-to-plane geometry. (a) OH and H_2O_2 [302] and (b) reductive species [303]. (Source: (a) Adapted with permission from [302]. Copyright (2006) American Chemical Society.)

and 1 J per pulse. The effect of probe concentration (DMSO and NaTA) on OH$^•$ and H_2O_2 production is shown in Figure 6.14a [302], and as the OH$^•$ radical probe concentration increases, the H_2O_2 generation decreases, supporting the hypothesis that H_2O_2 is generated primarily by OH$^•$ radical recombination. Evidence for reductive species (Figure 6.14b), including H$^•$ and $O_2^{•-}$ radicals, was demonstrated using two different probes (tetranitromethane (TNM) and nitroblue tetrazolium chloride (NBT)) [303]. Figure 6.15 shows the effects of conductivity on the formation of molecular species (O_2, H_2, and H_2O_2) and radical species (O$^•$, H$^•$, and OH$^•$ –from emissions spectroscopy) [174]. At the highest conductivity, all molecular and radical species generation rates are reduced, while in the intermediate range (150 μS cm^{-1}), H and O intensities were the largest, due to lower power input and incomplete capacitor discharge below this conductivity.

For a high-frequency discharge in water with a single bubble formation at the tip of the electrode, another model was developed to account for chemical species

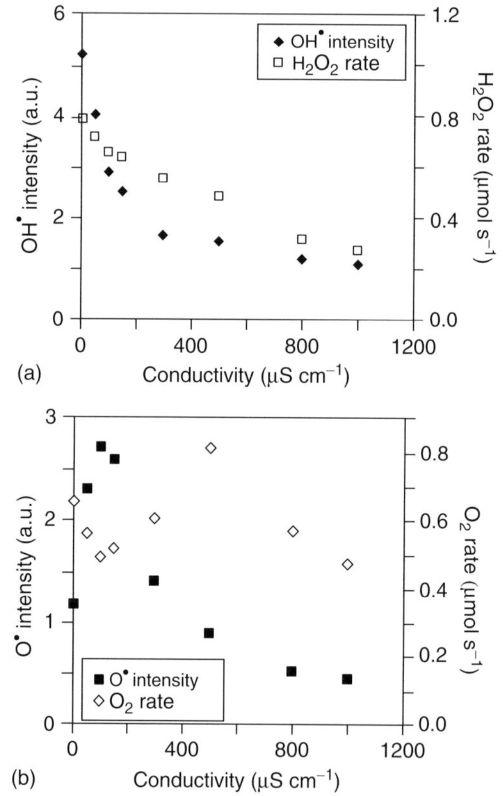

Figure 6.15 (a,b) The effects of conductivity on molecular and radical (from emissions spectroscopy) species formation in direct discharge in water at 1 J per pulse in point-to-plane geometry. (Source: © 2011 IEEE. Reprinted, with permission, from [174].)

formation in the plasma in the growing bubble [91]. The model predicted large temperature gradients inside the bubble, with peak temperatures of near 4500 K at 300 W input power. In addition, the model predicted $H_2 : O_2$ ratios of 1 : 0.3, while the experimental results gave the ratio 1 : 0.04. The H_2O_2 concentration in the model was 10^4 times lower than measured, but the OH values were comparable between the model and experiment. It is interesting to note that this 13.6 MHz discharge in water gives H_2O_2 energy yields of about 1 g kWh^{-1}, which is similar to a wide range of other discharges directly in water [7], including pulsed coronalike discharge [173], but the stoichiometries for H_2, O_2, and H_2O_2 are significantly different.

Another issue in discharges in water is the role of the aqueous or hydrated electrons. In the radiation chemistry of water [242, 244], liquid water molecules subject to ionizing radiation lead to

$$H_2O + \text{radiation} \longrightarrow H_2O^+ + e^- \tag{6.56}$$

The H_2O^+ has lifetimes below 10^{-14} s, and over $<10^{-12}$ s, the electron becomes solvated to form the aqueous or hydrated electron [242]

$$e^- \longrightarrow e^-_{aq} \tag{6.57}$$

Joshi *et al.* [300] proposed in analogy with the radiation chemistry of water that aqueous electrons might be formed in direct discharge in water. However, experimental measurements with the addition of N_2O into the liquid to generate additional hydroxyl radicals [242] by

$$e^-_{aq} + N_2O \longrightarrow N_2 + O^- \tag{6.58}$$

$$O^- + H_2O \longrightarrow OH^\bullet + OH^- \tag{6.59}$$

failed to show additional hydroxyl radicals. There are many other reactions of the aqueous electron that occur in the radiation chemistry of liquid [242], but one recent study has suggested by experimentally measured reduction of silver ions in solution that aqueous electrons can be formed in negative polarity discharges over water surfaces [62]; positive polarity discharge in the same system did not result in silver ion reduction.

It is possible that aqueous electrons can react with dissolved oxygen in water by

$$O_2 + e^-_{aq} \longrightarrow O_2^{\bullet-} \tag{6.60}$$

However, it should be noted that the oxygen radical anion can be formed from atomic hydrogen [303] and molecular oxygen by

$$H^\bullet + O_2 \longrightarrow HO_2^\bullet \tag{6.61}$$

$$HO_2^\bullet \rightleftarrows H^+ + O_2^{\bullet-} \tag{6.62}$$

In other considerations, while direct discharge in water using high-voltage pulses does not have significant electrolysis reactions [3], glow discharge electrolysis and similar processes may have features of both electrolytic reactions and radiation reaction [307–310].

6.4.2
Effect of Solution Properties and Plasma Characteristics on Plasma Chemical Processes in Water

Solution conductivity is one of the most important solution-phase properties that affect direct discharge in water. In earlier work, Jones and Kunhardt [311, 312] found that solution conductivity did not affect the time lag for breakdown, but that increasing pressure increased the time lag for breakdown initiation. Using the Hα line broadening to estimate electron density, Sunka *et al.* [3] showed that the electron density increased from $2\times10^{23}\,m^{-3}$ to over $10^{25}\,m^{-3}$ as conductivity increased from 0.1 to $1\,mS\,cm^{-1}$. Conductivity also has a strong effect on the formation of molecular species, including H_2O_2, H_2, and O_2 [174] (Figure 6.15). Direct measurements of these chemical species was compared to optical emissions spectroscopy of radical species ($H^•$, $O^•$, and $OH^•$) in the conductivity range from $5\,\mu S\,cm^{-1}$ to $1\,mS\,cm^{-1}$. As conductivity increased, both H_2O_2 and $OH^•$ monotonically decreased, supporting the rather close association of these two species. However, the changes in $O^•$ intensity with conductivity were the inverse of those for O_2 generation in the range up to $500\,\mu S\,cm^{-1}$, and above this conductivity, both species decreased. The patterns for $H^•$ and H_2 were similar to those for $O^•$ and O_2. Overall, these patterns of molecular and radical species formation suggested a change in the discharge at about $150\,\mu S\,cm^{-1}$ where an optimal amount of radicals ($H^•$ and $O^•$) were produced at the maximum discharge power. This work also demonstrated the value of using argon emissions to calibrate the measurements [174].

The radiant energy and photon flux of UV radiation for pulsed discharge directly in the liquid phase was measured by chemical actinometry with potassium ferrioxalate [313]. It was found that in the range of 190–280 nm, the photon flux of UV radiation, J in quanta per pulse, scaled with the discharge mean power, P_p, by

$$J = 44.33\ P_p^{2.11} \tag{6.63}$$

The generation of hydrogen peroxide decreased with increasing solution conductivity in the range of $100–500\,\mu S\,cm^{-1}$, and this decrease was partly attributed to the decomposition of H_2O_2 by UV photolytic reactions such as Eq. (6.39) [313].

While electrical breakdown in aqueous salt solutions under high pressure has been used to analyze the mechanism of discharge formation [311], high pressure has not led to large changes in aqueous solution chemistry [314]. Electrical discharge plasma can be formed more easily in supercritical CO_2 than in water because of the lower critical point [315–317]. Nanoparticles [318] can be synthesized and phenol polymerized [319] in supercritical CO_2 with plasma under these conditions. In gas-phase electrical discharge for a large range of pressure, as the pressure increases the gap distance between electrodes necessary for generating electrical discharge decreases [130]. Alternatively, the electric field required for breakdown increases. Much more complicated behavior was found near the supercritical region of CO_2 [320], where the breakdown voltage exhibits a sharp minimum near the critical point. It was suggested that this minimum was due to clustering of CO_2 molecules causing more liquidlike behavior. The breakdown behavior of electrical discharges in water as a function of pressure up to 40 MPa has been shown to be independent of liquid conductivity [312]. In water having pressure up to 1.4 MPa, the initiation voltage from a pulsed capacitor discharge (1 J per pulse at 60 Hz) was linearly related to the pressure, but once a stable discharge was formed, the current and voltage characteristics as well as the production of H_2O_2 were not affected by the externally applied pressure [314]. In a 27 MHz discharge directly in the liquid phase with the formation of small microbubbles, pressure between 0.1 and 0.4 MPa caused the $OH^\bullet$ temperature to increase from 3500 to 5000 K and the electron density to increase over 10 times from 3.5×10^{20} to $5.8 \times 10^{21}\ m^{-3}$ while the electron temperature dropped from 3700 to 3200 K [175]. Clearly, the type of discharge has some effect on these plasma properties, but the relative effects of the discharge type on the reaction processes under high pressure in liquids are not known.

The material properties of the electrodes can have a significant effect on the chemical reactions in electrical discharge in water [19, 111, 112, 321–324]. The first evidence of this effect was in the use of steel electrodes and the demonstration that the hydrogen peroxide formed by the discharge reacts with ferrous iron released by the electrode [301, 325] by

$$Fe^{2+} + H_2O_2 \longrightarrow Fe^{3+} + OH^- + OH^\bullet \tag{6.64}$$

It was subsequently found that platinum particles could sputter from solid platinum electrode surfaces [323] and thereafter cause heterogeneous catalytic reactions to reduce ferrous iron to ferric iron by reaction with the plasma-generated H_2 [326–328]

$$H_2 + 2\,Pt \longrightarrow 2\,Pt - H \tag{6.65}$$

$$Fe^{3+} + Pt - H \longrightarrow Fe^{2+} + Pt + H^+ \tag{6.66}$$

A catalytic cycle is set between reactions (Eqs. (6.64)–(6.66)) whereby the Fenton reaction (Eq. (6.64)) is sustained to continuously produce hydroxyl radicals. Reactions of H_2O_2 with tungsten (tungstate ions) has also been demonstrated for direct discharge in water to increase the oxidation of the organic compound DMSO [324] by

$$(CH_3)_2SO + H_2O_2 + WO_4^{2-} \longrightarrow (CH_3)_2SO_2 + H_2O + WO_4^{2-} \tag{6.67}$$

The initial rates of H_2O_2 formation with Ti and W electrodes were the same, but the tungsten electrodes showed a decrease in H_2O_2 after a certain period of operating time [324]; similar results were found in a gas–liquid discharge (electrode series configuration shown in Figure 6.1B) where Ti, Ni–Cr, Cu, stainless steel, tungsten–copper, and tungsten carbide had the same initial rates of formation, but the tungsten-containing electrodes showed a drop in H_2O_2 at longer times [321]. Other heterogeneous catalysts, including activated carbon, zeolites, and TiO_2 photocatalysts have shown varying degrees of activity in electrical discharges in water [259, 323, 328–333]. Further details on catalytic reactions are discussed in Chapter 7.

A fabric electrode was prepared, whereby conducting metal wires were interwoven with insulating fibers [334]. On application of a DC field, microbubbles of H_2 were formed on the electrode surface and a microplasma was generated after imposition of a 1.3 kV 10 kHz AC field. The resulting hydrogen plasma in water was suggested to lead to interesting chemical reduction chemistry based on the atomic hydrogen and other species. Very limited analysis of the effect of liquid temperature on the chemical reactions in direct discharge in water has been performed. The main result of varying temperature to near the boiling point was to facilitate the formation of bubbles in the liquid and ease the generation of plasma [90].

6.5
Concluding Remarks

Electrical discharges in liquids and in gas–liquid environments are of interest for many potential applications and are currently under intensive investigations by a number of laboratories throughout the world. There is a growing interest in this field, and the complex nature of the subject presents many challenges for both experimentalists and modelers. While a complete understanding of this rather general and broad field is not available, many specific aspects have been well studied, and recent experiments and modeling efforts have shed great insight into both the chemical and the physical aspects of such systems. Discharges in gases in contact with liquid water have much in common with those directly in the gas, but the presence of the condensed phase can affect both the physical nature of the discharge formation and the chemical reactions induced by the discharge. While discharge propagation along gas–liquid interfaces and the subsequent reactions at the interface require significantly more experimental and modeling studies, gas-phase plasma interactions with water molecules and the subsequent reaction chemistry are reasonably well understood. Gas-phase discharges over water can be strongly influenced by both the gas composition and the liquid properties. Discharges directly in the liquid phase may present significantly different physics from those in the gas phase. Such direct discharges in water are strongly influenced by the nature of the power input (e.g., pulse duration, energy level), type and configuration of electrodes, and other solution-phase properties. Much of the knowledge of the chemistry of all such systems in water and in gas–liquid liquid environments is based on radiation chemistry, radical chemistry, photochemistry, atmospheric chemistry, and gas-phase plasma chemistry.

Acknowledgments

Locke would like to gratefully acknowledge financial support from the US National Science Foundation under grant number CBET-0932481. Lukes would like to acknowledge financial support from the Grant Agency of Academy of Sciences of the Czech Republic (project No. IAAX00430802).

References

1. Locke, B.R., Sunka, P., Sato, M., Hoffmann, M., and Chang, J.S. (2006) Electrohydraulic discharge and non thermal plasma for water treatment. *Ind. Eng. Chem. Res.*, **45**, 882–905.
2. Malik, M.A., Ghaffar, A., and Malik, S.A. (2001) Water purification by electrical discharges. *Plasma Sources Sci. Technol.*, **10**, 82–91.
3. Sunka, P. (2001) Pulse electrical discharges in water and their applications. *Phys. Plasmas*, **8**, 2587–2594.
4. Akiyama, H. (2000) Streamer discharges in liquids and their applications. *IEEE Trans. Dielectr. Electr. Insul.*, **7**, 646–653.
5. Bruggeman, P., and Leys, C. (2009) Non-thermal plasmas in and in contact with liquids. *J. Phys. D: Appl. Phys.*, **42**, 1–28.
6. Brisset, J.L., Moussa, D., Doubla, A., Hnatiuc, E., Hnatiuc, B., Youbi, G.K., Herry, J.M., Naitali, M., and Bellon-Fontaine, M.N. (2008) Chemical reactivity of discharges and temporal post-discharges in plasma treatment of aqueous media: Examples of gliding discharge treated solutions. *Ind. Eng. Chem. Res.*, **47**(16), 5761–5781.
7. Locke, B.R. and Shih, K.Y. (2011) Review of the methods to form hydrogen peroxide in electrical discharge plasma with liquid water. *Plasma Sources Sci. Technol.*, **20**(3), 034006.
8. Sato, M. (2008) Environmental and biotechnological applications of high-voltage pulsed discharges in water. *Plasma Sources Sci. Technol.*, **17**(2), 024021.
9. Laroussi, M. (2002) Nonthermal decontamination of biological media by atmospheric-pressure plasma: Review, analysis and prospects. *IEEE Trans. Plasma Sci.*, **30**(4), 1409–1415.
10. Fridman, G., Friedman, G., Gutsol, A., Shekhter, A.B., Vasilets, V.N., and Fridman, A. (2008) Applied plasma medicine. *Plasma Process. Polym.*, **5**(6), 503–533.
11. Graham, W.G.G.W.G. and Stalder, K.R. (2011) Plasmas in liquids and some of their applications in nanoscience. *J. Phys. D: Appl. Phys.*, **44**(17), 174037.
12. Mezei, P., and Cserfalvi, T. (2007) Electrolyte cathode atmospheric glow discharges for direct solution analysis. *Appl. Spectrosc. Rev.*, **42**(6), 573–604.
13. Mottaleb, M.A., Yang, J.S., and Kim, H.J. (2002) Electrolyte-as-cathode glow discharge (ELCAD)/glow discharge electrolysis at the gas-solution interface. *Appl. Spectrosc. Rev.*, **37**(3), 247–273.
14. Brablec, A., Slavicek, P., Stahel, P., Cizmar, T., Trunec, D., Simor, M., and Cernak, M. (2002) Underwater pulse electrical diaphragm discharges for surface treatment of fibrous polymeric materials. *Czech. J. Phys.*, **52** (Suppl. D), 491–500.
15. Hatakeyama, R., Kaneko, T., Oohara, W., Li, Y.F., Kato, T., Baba, K., and Shishido, J. (2008) Novel-structured carbon nanotubes creation by nanoscopic plasma control. *Plasma Sources Sci. Technol.*, **17**(2), 024009.
16. Thagard, S.M., Takashima, K., and Mizuno, A. (2009) Electrical discharges in polar organic liquids. *Plasma Process. Polym.*, **6**(11), 741–750.
17. Barinov, Y.A. and Shkol'nik, S.M. (2002) Probe measurements in a discharge with liquid nonmetallic electrodes in air at atmospheric pressure. *Tech. Phys.*, **47**, 313–319.
18. Foster, J., Sommers, B., Weatherford, B., Yee, B., and Gupta, M. (2011) Characterization of the evolution of underwater DBD plasma jet. *Plasma Sources Sci. Technol.*, **20**(3), 034018.
19. Kolb, J.F., Joshi, R.P., Xiao, S., and Schoenbach, K.H. (2008) Streamers in water and other dielectric liquids. *J. Phys. D: Appl. Phys.*, **41**(23), 234007.
20. Babaeva, N.Y., Bhoj, A.N., and Kushner, M.J. (2006) Streamer dynamics in gases containing dust particles. *Plasma Sources Sci. Technol.*, **15**(4), 591–602.
21. Babaeva, N.Y. and Kushner, M.J. (2008) Streamer branching: the role of inhomogeneities and bubbles. *IEEE Trans. Plasma Sci.*, **36**(4), 892–893.

22. Babaeva, N.Y. and Kushner, M.J. (2009) Effect of inhomogeneities on streamer propagation: I. Intersection with isolated bubbles and particles. *Plasma Sources Sci. Technol.*, **18**(3), 035009.
23. Babaeva, N.Y. and Kushner, M.J. (2009) Effect of inhomogeneities on streamer propagation: II. Streamer dynamics in high pressure humid air with bubbles. *Plasma Sources Sci. Technol.*, **18**(3), 035010.
24. Babaeva, N.Y. and Kushner, M.J. (2009) Structure of positive streamers inside gaseous bubbles immersed in liquids. *J. Phys. D: Appl. Phys.*, **42**(13), 132003.
25. Babaeva, N.Y. and Kushner, M.J. (2011) Ion energy and angular distributions onto polymer surfaces delivered by dielectric barrier discharge filaments in air: I. Flat surfaces. *Plasma Sources Sci. Technol.*, **20**(3), 035017.
26. Babaeva, N.Y. and Kushner, M.J. (2011) Ion energy and angular distributions onto polymer surfaces delivered by dielectric barrier discharge filaments in air: II. Particles. *Plasma Sources Sci. Technol.*, **20**(3), 035018.
27. Joshi, R.P., Kolb, J.F., Xiao, S., and Schoenbach, K.H. (2009) Aspects of plasma in water: Streamer physics and applications. *Plasma Process. Polym.*, **6**(11), 763–777.
28. Lewis, T.J. (2003) Breakdown initiating mechanisms at electrode interfaces in liquids. *IEEE Trans. Dielectr. Electr. Insul.*, **10**(6), 948–955.
29. Griffiths, R.F. and Phelps, C.T. (1976) Effects of air-pressure and water-vapor content on propagation of positive corona streamers, and their implications to lightning initiation. *Q. J. R. Meteorol. Soc.*, **102**(432), 419–426.
30. Phelps, C.T. and Griffiths, R.F. (1976) Dependence of positive corona streamer propagation on air-pressure and water-vapor content. *J. Appl. Phys.*, **47**(7), 2929–2934.
31. Mikropoulos, P.N., Stassinopoulos, C.A., and Sarigiannidou, B.C. (2008) Positive streamer propagation and breakdown in air: the influence of humidity. *IEEE Trans. Dielectr. Electr. Insul.*, **15**(2), 416–425.
32. Abdelsalam, M. (1985) Positive wire-to-plane coronas as influenced by atmospheric humidity. *IEEE Trans. Ind. Appl.*, **21**(1), 35–40.
33. Nouri, H., Zebboudj, Y., Zouzou, N., Moreau, E., and Dascalescu, L. (2010) Effect of relative humidity on the collection efficiency of a wire-to-plane electrostatic precipitator, in *Proceedings 2010 IEEE Industrial Applications Society Annual Meeting (IAS)*, IEEE, IEEE Industry Applications Society Annual Meeting, pp. 1–6.
34. Abdelsalam, M. (1992) Influence of humidity on charge-density and electric-field in electrostatic precipitators. *J. Phys. D: Appl. Phys.*, **25**(9), 1318–1322.
35. Bian, X.M., Meng, X.B., Wang, L.M., MacAlpine, J.M.K., Guan, Z.C., and Hui, J.F. (2011) Negative corona inception voltages in rod-plane gaps at various air pressures and humidities. *IEEE Trans. Dielectr. Electr. Insul.*, **18**(2), 613–619.
36. Pavlik, M. and Skalny, J.D. (1997) Generation of $[H_3O]^+.(H_2O)_n$ clusters by positive corona discharge in air. *Rapid Commun. Mass Spectrom.*, **11**, 1757–1766.
37. Viner, A.S., Lawless, P.A., Ensor, D.S., and Sparks, L.E. (1992) Ozone generation in DC-energized electrostatic precipitators. *IEEE Trans. Ind. Appl.*, **28**(3), 504–512.
38. Held, B. and Peyrous, R. (1999) A systematic parameters study from analytic calculations to optimize ozone concentration in an oxygen-fed wire-to-cylinder ozonizer. *Eur. Phys. J. Appl. Phys*, **7**(2), 151–166.
39. Chen, J.H. and Wang, P.X. (2005) Effect of relative humidity on electron distribution and ozone production by DC coronas in air. *IEEE Trans. Plasma Sci.*, **33**(2), 808–812.
40. Ryzko, H. (1965) Drift velocity of electrons and ions in dry and humid air and in water vapour. *Proc. Phys. Soc.*, **85**, 1283.
41. Hartmann, G. (1984) Theoretical evaluation of Peek's law. *IEEE Trans. Ind. Appl.*, **20**(6), 1647–1651.

42. Smirnov, B.M. (2009) Modeling of gas discharge plasma. *Phys. Usp.*, **52**(6), 559.
43. Lowke, J.J. and D'Alessandro, F. (2003) Onset corona fields and electrical breakdown criteria. *J. Phys. D: Appl. Phys.*, **36**(21), 2673–2682.
44. Gallimberti, I. (1972) A computer model for streamer propagation. *J. Phys. D: Appl. Phys.*, **5**, 2179–2189.
45. Gallimberti, I. (1988) Impulse corona simulation for flue gas treatment. *Pure Appl. Chem.*, **60**, 663–674.
46. Jones, J.E., Dupuy, J., Schreiber, G.O.S., and Waters, R.T. (1988) Boundary-conditions for the positive direct-current corona in a coaxial system. *J. Phys. D: Appl. Phys.*, **21**(2), 322–333.
47. MacAlpine, J.M.K. and Zhang, C.H. (2003) The effect of humidity on the charge/phase-angle patterns of AC corona pulses in air. *IEEE Trans. Dielectr. Electr. Insul.*, **10**(3), 506–513.
48. Takata, N. (1994) The effects of humidity on volume recombination in ionization chambers. *Phys. Med. Biol.*, **39**(6), 1047.
49. Bian, X.M., Wang, L.M., MacAlpine, J.M.K., Guan, Z.C., Hui, J.F., and Chen, Y. (2010) Positive corona inception voltages and corona currents for air at various pressures and humidities. *IEEE Trans. Dielectr. Electr. Insul.*, **17**(1), 63–70.
50. Mikropoulos, P.N., Stassinopoulos, C.A., Stapountzi, M.E., and Sarigiannidou, B.C. (2006) Streamer propagation and flashover along insulator surface in a uniform field in air: Influence of humidity. *Proceedings 41st International Universities Power Engineering Conference 2006 (UPEC'06)*, pp. 916–920.
51. Aleksandrov, N.L., Bazelyan, E.M., and Novitskii, D.A. (1998) Influence of moisture on the properties of long streamers in air. *Tech. Phys. Lett.*, **24**(5), 367–368.
52. Messaoudi, R., Younsi, A., Massines, F., Despax, B., and Mayoux, C. (1996) Influence of humidity on current waveform and light emission of a low-frequency discharge controlled by a dielectric barrier. *IEEE Trans. Dielectr. Electr. Insul.*, **3**(4), 537–543.
53. Allen, N.L. and Dring, D. (1984) Effect of humidity on the properties of corona in a rod plane gap under positive impulse voltages. *Proc. R. Soc. London, Ser. A: Math. Phys. Eng. Sci.*, **396**(1811), 281–295.
54. Allen, N.L. (1986) Corona, breakdown and humidity in the rod-plane gap. *IEE Proc. Sci. Meas. Technol.*, **133**(8), 562–568.
55. Allen, N.L. and Boutlendj, M. (1991) Study of the electric-fields required for streamer propagation in humid air. *IEE Proc. Sci. Meas. Technol.*, **138**(1), 37–43.
56. Hui, J.F., Guan, Z.C., Wang, L.M., and Li, Q.W. (2008) Variation of the dynamics of positive streamer with pressure and humidity in air. *IEEE Trans. Dielectr. Electr. Insul.*, **15**(2), 382–389.
57. Boutlendj, M., Allen, N.L., Lightfoot, H.A., and Neville, R.B. (1991) Positive DC corona and sparkover in short and long rod-plane gaps under variable humidity conditions. *IEE Proc. Sci. Meas. Technol.*, **138**(1), 31–36.
58. Goheen, S.C., Durham, D.E., McCulloch, M., and Heath, W.O. (1994) The degradation of organic dyes by corona discharge, in *Proceedings of the 2nd International Symposium on Chemical Oxidation: Technologies for the Nineties*, in (eds W.W. Eckenfelder, A.R. Bowers, and J.A. Roth), Techomoic AG, pp. 356–367.
59. Hoeben, W.F.L.M., van Veldhuizen, E.M., Rutgers, W.R., and Kroesen, G.M.W. (1999) Gas phase corona discharges for oxidation of phenol in an aqueous solution. *J. Phys. D: Appl. Phys.*, **32**, L133–L137.
60. Grymonpre, D.R., Finney, W.C., Clark, R.J., and Locke, B.R. (2004) Hybrid gas–liquid electrical discharge reactors for organic compound degradation. *Ind. Eng. Chem. Res.*, **43**(9), 1975–1989.
61. Lukes, P., Appleton, A., and Locke, B.R. (2004) Hydrogen peroxide and ozone formation in hybrid gas–liquid

electrical discharge reactors. *IEEE Trans. Ind. Appl.*, **40**(1), 60–67.
62. Thagard, S.M., Takashima, K., and Mizuno, A. (2009) Chemistry of the positive and negative electrical discharges formed in liquid water and above a gas–liquid surface. *Plasma Chem. Plasma Process.*, **29**(6), 455–473.
63. Belosheev, V.P. (1998) Leader discharge over a water surface in a Lichtenberg figure geometry. *Tech. Phys.*, **43**, 1329–1332.
64. Belosheev, V.P. (1998) Study of the leader of a spark discharge over a water surface. *Tech. Phys.*, **43**, 783–789.
65. Moon, J.D., Kim, J.G., and Lee, D.H. (1998) Electrophysicochemical characteristics of a waterpen point corona discharge. *IEEE Trans. Ind. Appl.*, **34**, 1212–1217.
66. Janca, J., Kuzmin, S., Maximov, A., Titova, J., and Czernichowski, A. (1999) Investigation of the chemical action of the gliding and "point" arcs between the metallic electrode and aqueous solution. *Plasma Chem. Plasma Process.*, **19**(1), 53–67.
67. Lukes, P., Clupek, M., and Babicky, V. (2011) Discharge filamentary patterns produced by pulsed corona discharge at the interface between a water surface and air. *IEEE Trans. Plasma Sci.*, **39**(11), 2644–2645.
68. Lukes, P., Clupek, M., Babicky, V., Janda, V., and Sunka, P. (2005) Generation of ozone by pulsed corona discharge over water surface in hybrid gas–liquid electrical discharge reactor. *J. Phys. D: Appl. Phys.*, **38**, 409–416.
69. Bozhko, I.V., Kondratenko, I.P., and Serdyuk, Y.V. (2011) Corona discharge to water surface and its transition to a spark. *IEEE Trans. Plasma Sci.*, **39**(5), 1228–1233.
70. Taylor, G. (1969) Electrically driven jets. *Proc. R. Soc. London, Ser. A: Math. Phys. Sci.*, **313**(1515), 453.
71. Taylor, G. (1964) Disintegration of water drops in electric field. *Proc. R. Soc. London, Ser. A: Math. Phys. Sci.*, **280**(1380), 383.
72. Castellanos, A. and Gonzalez, A. (1998) Nonlinear electrohydrodynamics of free surfaces. *IEEE Trans. Dielectr. Electr. Insul.*, **5**(3), 334–343.
73. Melcher, J.R. and Taylor, G.I. (1969) Electrohydrodynamics: A review of the role of interfacial shear stresses. *Annu. Rev. Fluid Mech.*, **1**, 111–146.
74. Melcher, J.R. and Smith, C.V. (1969) Electrohydrodynamic charge relaxation and interfacial perpendicular-field instability. *Phys. Fluids*, **12**, 778–790.
75. Robinson, J.A., Bergougnou, M.A., Cairns, W.L., Castle, G.S.P., and Inculet, I.I. (2000) Breakdown of air over a water surface stressed by a perpendicular alternating electric field, in the presence of a dielectric barrier. *IEEE Trans. Ind. Appl.*, **36**(1), 68–75.
76. Robinson, J.A., Bergougnou, M.A., Cairns, W.L., Castle, G.S.P., and Inculet, I.I. (1998) A new type of ozone generator using Taylor cones on water surfaces. *IEEE Trans. Ind. Appl.*, **34**, 1218–1223.
77. Bruggeman, P., Graham, L., Degroote, J., Vierendeels, J., and Leys, C. (2007) Water surface deformation in strong electrical fields and its influence on electrical breakdown in a metal pin-water electrode system. *J. Phys. D: Appl. Phys.*, **40**(16), 4779–4786.
78. Bruggeman, P., Van Slycken, J., Degroote, J., Vierendeels, J., Verleysen, P., and Leys, C. (2008) DC electrical breakdown in a metal pin water electrode system. *IEEE Trans. Plasma Sci.*, **36**(4), 1138–1139.
79. Kurahashi, M., Katsura, S., and Mizuno, A. (1997) Radical formation due to discharge inside a bubble in liquid. *J. Electrostatics*, **42**, 93–105.
80. Ihara, S., Miichi, T., Satoh, S., Yamabe, C., and Sakai, E. (1999) Ozone generation by a discharge in bubbled water. *Jpn. J. Appl. Phys.*, **38**, 4601–4604.
81. Miichi, T., Ihara, S., Satoh, S., and Yamabe, C. (2000) Spectroscopic measurements of discharges inside bubbles in water. *Vacuum*, **59**(1), 236–243.
82. Miichi, T., Hayashi, N., Ihara, S., Satoh, S., and Yamabe, C. (2002) Generation of radicals using discharge inside bubbles in water for water treatment. *Ozone Sci. Eng.*, **24**, 471–477.

83. de Baerdemaeker, F., Monte, M., and Leys, C. (2004) Capillary underwater discharges in repetitive pulse regime. *Czech. J. Phys.*, **54**, C1062–C1067.
84. Yamabe, C., Takeshita, F., Miichi, T., Hayashi, N., and Ihara, S. (2005) Water treatment using discharge on the surface of a bubble in water. *Plasma Process. Polym.*, **2**(3), 246–251.
85. Bruggeman, P.J., Leys, C.A., and Vierendeels, J.A. (2006) Electrical breakdown of a bubble in a water-filled capillary. *J. Appl. Phys.*, **99**(11), 116101.
86. Bruggeman, P., Leys, C., and Vierendeels, J. (2007) Experimental investigation of dc electrical breakdown of long vapour bubbles in capillaries. *J. Phys. D: Appl. Phys.*, **40**(7), 1937–1943.
87. Gershman, S., Mozgina, O., Belkind, A., Becker, K., and Kunhardt, E. (2007) Pulsed electrical discharge in bubbled water. *Contrib. Plasma Phys.*, **47**(1-2), 19–25.
88. Bruggeman, P., Degroote, J., Vierendeels, J., and Leys, C. (2008) DC-excited discharges in vapour bubbles in capillaries. *Plasma Sources Sci. Technol.*, **17**(2), 025008.
89. Sato, K. and Yasuoka, K. (2008) Pulsed discharge development in oxygen, argon, and helium bubbles in water. *IEEE Trans. Plasma Sci.*, **36**(4), 1144–1145.
90. Shih, K.Y. and Locke, B.R. (2009) Effects of electrode protrusion length, pre-existing bubbles, solution conductivity and temperature, on liquid phase pulsed electrical discharge. *Plasma Process. Polym.*, **6**(11), 729–740.
91. Mukasa, S., Maehara, T., Nomura, S., Toyota, H., Kawashima, A., Hattori, Y., Hashimoto, Y., and Yamashita, H. (2010) Growth of bubbles containing plasma in water by high-frequency irradiation. *Int. J. Heat Mass Transf.*, **53**(15-16), 3067–3074.
92. Shih, K.Y. and Locke, B.R. (2010) Chemical and physical characteristics of pulsed electrical discharge within gas bubbles in aqueous solutions. *Plasma Chem. Plasma Process.*, **30**(1), 1–20.
93. Yamabe, C. and Ihara, S. (2010) Electrical discharge characteristics using a bubble in water and their applications to the water treatment. *J. Adv. Oxid. Technol.*, **13**(1), 65–70.
94. Sato, K., Yasuoka, K., and Ishii, S. (2010) Water treatment with pulsed discharges generated inside bubbles. *Electr. Eng. Jpn.*, **170**(1), 1–7.
95. Tachibana, K., Takekata, Y., Mizumoto, Y., Motomura, H., and Jinno, M. (2011) Analysis of a pulsed discharge within single bubbles in water under synchronized conditions. *Plasma Sources Sci. Technol.*, **20**(3), 034005.
96. Sommers, B.S., Foster, J.E., Babaeva, N.Y., and Kushner, M.J. (2011) Observations of electric discharge streamer propagation and capillary oscillations on the surface of air bubbles in water. *J. Phys. D: Appl. Phys.*, **44**(8), 082001.
97. Shin, W.T., Yiacoumi, S., and Tsouris, C. (1997) Experiments on electrostatic dispersion of air in water. *Ind. Eng. Chem. Res.*, **36**, 3647–3655.
98. Jouve, G., Goldman, A., Goldman, M., and Haut, C. (2001) Surface chemistry induced by air corona discharges in a negative glow regime. *J. Phys. D: Appl. Phys.*, **34**, 218–221.
99. Verreycken, T., Schram, D.C., Leys, C., and Bruggeman, P. (2010) Spectroscopic study of an atmospheric pressure dc glow discharge with a water electrode in atomic and molecular gases. *Plasma Sources Sci. Technol.*, **19**(4), 045004.
100. Bruggeman, P., Ribezl, E., Degroote, J., Vierendeels, J., and Leys, C. (2008) Plasma characteristics and electrical breakdown between metal and water electrodes. *J. Optoelectron. Adv. Mater.*, **10**(8), 1964–1967.
101. Bruggeman, P., Degroote, J., Leys, C., and Vierendeels, J. (2008) Electrical discharges in the vapour phase in liquid-filled capillaries. *J. Phys. D: Appl. Phys.*, **41**(19), 194007.
102. Bruggeman, P., Ribezl, E., Maslani, A., Degroote, J., Malesevic, A., Rego, R., Vierendeels, J., and Leys, C. (2008) Characteristics of atmospheric pressure air discharges with a liquid cathode and a metal anode. *Plasma Sources Sci. Technol.*, **17**(2), 025012.

103. Vukovic, Z.R. and Curic, M. (2005) The influence of the acoustic-electric coalescence on phase and mass transfer of supercooled drops. *Pure Appl. Geophys.*, **162**(12), 2453–2477.

104. Tenberken, B. and Bachmann, K. (1998) Sampling and analysis of single cloud and fog drops. *Atmos. Environ.*, **32**(10), 1757–1763.

105. Mizuno, A. (2000) Electrostatic precipitation. *IEEE Trans. Dielectr. Electr. Insul.*, **7**(5), 615–624.

106. Borra, J.P. (2006) Nucleation and aerosol processing in atmospheric pressure electrical discharges: powders production, coatings and filtration. *J. Phys. D: Appl. Phys.*, **39**(2), R19–R54.

107. Burlica, R., Shih, K.Y., and Locke, B.R. (2010) Formation of H_2 and H_2O_2 in a water-spray gliding arc nonthermal plasma reactor. *Ind. Eng. Chem. Res.*, **49**(14), 6342–6349.

108. Burlica, R. and Locke, B.R. (2008) Pulsed plasma gliding arc discharges with water spray. *IEEE Trans. Ind. Appl.*, **44**(2), 482–489.

109. Lukes, P., Clupek, M., Babicky, V., and Sunka, P. (2009) The role of surface chemistry at ceramic/electrolyte interfaces in the generation of pulsed corona discharges in water using porous ceramic-coated rod electrodes. *Plasma Process. Polym.*, **6**(11), 719–728.

110. Sunka, P., Babicky, V., Clupek, M., Lukes, P., Siumek, M., and Brablec, A. (1999) New approach to generation of corona like discharge in water, in *Proceedings of the 14th International Symposium on Plasma Chemistry*, vol. **II** (eds M. Hrabovsky, M. Konrad, and V. Kopecky), pp. 1057.

111. Lukes, P., Clupek, M., Sunka, P., and Babicky, V. (2002) Effect of ceramic composition on pulse discharge induced processes in water using ceramic-coated wire to cylinder electrode system. *Czech. J. Phys.*, **52** (Suppl. D), D800–D806.

112. Lukes, P., Clupek, M., Babicky, V., and Sunka, P. (2008) Pulsed electrical discharge in water generated using porous-ceramic-coated electrodes. *IEEE Trans. Plasma Sci.*, **36**(4), 1146–1147.

113. Clements, J.S., Sato, M., and Davis, R.H. (1987) Preliminary investigation of prebreakdown phenomena and chemical reactions using a pulsed high-voltage discharge in water. *IEEE Trans. Ind. Appl.*, **IA-23**, 224–235.

114. Monte, M., De Baerdemaeker, F., Leys, C., and Maximov, A.I. (2002) Experimental study of a diaphragm discharge in water. *Czech. J. Phys.*, **52** (Suppl. D), 724–730.

115. Stara, Z. and Krcma, F. (2004) The study of H_2O_2 generation by DC diaphragm discharge in liquids. *Czech. J. Phys.*, **54**, C1050–C1055.

116. Prochazkova, J., Stara, Z., and Krcma, F. (2006) Optical emission spectroscopy of diaphragm discharge in water solutions. *Czech. J. Phys.*, **56**, B1314–B1319.

117. Makarova, E.M., Khlyustova, A.V., and Maksimov, A.I. (2009) Diaphragm discharge influence on physical and chemical properties of electrolyte solutions. *Surf. Eng. Appl. Electrochem.*, **45**(2), 133–135.

118. Sunka, P., Babicky, V., Clupek, M., Lukes, P., and Balcarova, J. (2003) Modified pinhole discharge for water treatment, in *PPC-2003: 14th IEEE International Pulsed Power Conference, Digest of Technical Papers*, Vols 1 and 2 (eds M. Giesselmann and A. Neuber) IEEE, New York, pp. 229–231.

119. Krcma, F., Stara, Z., and Prochazkova, J. (2010) Diaphragm discharge in liquids: Fundamentals and applications. *J. Phys.: Conf. Ser.*, **207**, 012010.

120. Joshi, R., Schulze, R.D., Meyer-Plath, A., and Friedrich, J.F. (2008) Selective surface modification of poly(propylene) with OH and COOH groups using liquid-plasma systems. *Plasma Process. Polym.*, **5**(7), 695–707.

121. Nikiforov, A.Y. and Leys, C. (2007) Influence of capillary geometry and applied voltage on hydrogen peroxide and OH radical formation in ac underwater electrical discharges. *Plasma Sources Sci. Technol.*, **16**(2), 273–280.

122. De Baerdemaeker, F., Simek, M., and Leys, C. (2007) Efficiency of hydrogen peroxide production by ac capillary discharge in water solution. *J. Phys. D: Appl. Phys.*, **40**(9), 2801–2809.

123. Ceccato, P.H., Guaitella, O., Le Gloahec, M.R., and Rousseau, A. (2010) Time-resolved nanosecond imaging of the propagation of a corona-like plasma discharge in water at positive applied voltage polarity. *J. Phys. D: Appl. Phys.*, **43**(17), 175202.
124. An, W., Baumung, K., and Bluhm, H. (2007) Underwater streamer propagation analyzed from detailed measurements of pressure release. *J. Appl. Phys.*, **101**(5), 053302.
125. Joshi, R.P., Qian, J., Zhao, G., Kolb, J., Schoenbach, K.H., Schamiloglu, E., and Gaudet, J. (2004) Are microbubbles necessary for the breakdown of liquid water subjected to a submicrosecond pulse? *J. Appl. Phys.*, **96**(9), 5129–5139.
126. Starikovskiy, A., Yang, Y., Cho, Y., and Fridman, A. (2011) Non-equilibrium plasma in liquid water: dynamics of generation and quenching. *Plasma Sources Sci. Technol.*, **20**(2), 024003.
127. Lisitsyn, I.V., Nomiyama, H., Katsuki, S., and Akiyama, H. (1999) Thermal processes in a streamer discharge in water. *IEEE Trans. Dielectr. Electr. Insul.*, **6**, 351–356.
128. Klimkin, V.F. (1990) Mechanisms of electric breakdown of water from pointed anode in the nanosecond range. *Sov. Tech. Phys. Lett.*, **16**, 146–148.
129. Raizer, Y.P. (1997) *Gas Discharge Physics*, Springer-Verlag, Berlin.
130. Nasser, E. (1971) *Fundamentals of Gaseous Ionization and Plasma Electronics*, Wiley-Interscience, New York.
131. Stalder, K.R., Woloszko, J., Brown, I.G., and Smith, C.D. (2001) Repetitive plasma discharges in saline solutions. *Appl. Phys. Lett.*, **79**, 4503–4505.
132. Woloszko, J., Stalder, K.R., and Brown, I.G. (2002) Plasma characteristics of repetitively-pulsed electrical discharges in saline solutions used for surgical procedures. *IEEE Trans. Plasma Sci.*, **30**, 1376–1383.
133. Stalder, K.R., McMillen, D.F., and Woloszko, J. (2005) Electrosurgical plasmas. *J. Phys. D: Appl. Phys.*, **38**(11), 1728–1738.
134. Stalder, K.R., Nersisyan, G., and Graham, W.G. (2006) Spatial and temporal variation of repetitive plasma discharges in saline solutions. *J. Phys. D: Appl. Phys.*, **39**(16), 3457–3460.
135. Stalder, K.R. and Woloszko, J. (2007) Some physics and chemistry of electrosurgical plasma discharges. *Contrib. Plasma Phys.*, **47**(1-2), 64–71.
136. Schaper, L., Graham, W.G., and Stalder, K.R. (2011) Vapour layer formation by electrical discharges through electrically conducting liquids-modelling and experiment. *Plasma Sources Sci. Technol.*, **20**(3), 034003.
137. Kratel, A.W.H. (1996) *Pulsed power discharge in water*. PhD Dissertation. California Institute of Technology, Pasadena, CA USA
138. Lu, X., Pan, Y., Liu, M., and Zhang, H. (2002) Spark model of pulsed discharge in water. *J. Appl. Phys.*, **91**, 24–31.
139. Lu, X.P. (2007) One-dimensional bubble model of pulsed discharge in water. *J. Appl. Phys.*, **102**(6), 063302.
140. Gidalevich, E., Boxman, R.L., and Goldsmith, S. (2004) Hydrodynamic effects in liquids subjected to pulsed low current arc discharges. *J. Phys. D: Appl. Phys.*, **37**(10), 1509–1514.
141. Ioffe, A.I. (1966) Theory of the initial state of an electrical discharge in water. *J. Appl. Mech. Tech. Phys.*, **7**(6), 69–27.
142. Lan, S., Yang, J.X., Samee, A., Jiang, J.L., and Zhou, Z.Q. (2009) Numerical simulation of properties of charged particles initiated by underwater pulsed discharge. *Plasma Sci. Technol.*, **11**(4), 481–486.
143. Zeldovich, Y.B. and Raizer, Y.P. (2001) *Physics of Shock Waves and High-Temperature Hydrodynamic Phenomena*, Dover Publications, Inc., Mineola, New York.
144. Kolacek, K., Babicky, V., Preinhaelter, J., Sunka, P., and Benes, J. (1988) Pressure distribution measurements at the shock wave focus in water by schlieren photography. *J. Phys. D: Appl. Phys.*, **21**, 463–469.
145. Rodriguez-Mendez, B., Lopez-Callejas, R., Pena-Eguiluz, R., Mercado-Cabrera, A., Alvarado, R.V., Barocio, S.R.,

de la Piedad-Beneitez, A., Benitez-Read, J.S., and Pacheco-Sotelo, J.O. (2008) Instrumentation for pulsed corona discharge generation applied to water. *IEEE Trans. Plasma Sci.*, **36**(1), 185–191.

146. Mededovic, S. and Locke, B.R. (2007) Primary chemical reactions in pulsed electrical discharge channels in water. *J. Phys. D: Appl. Phys.*, **40**(24), 7734–7746.

147. Mededovic, S. and Locke, B.R. (2009) Primary chemical reactions in pulsed electrical discharge channels in water (vol 40, pg 7734, 2007). *J. Phys. D: Appl. Phys.*, **42**(4), 049801.

148. Krenek, P. (2008) Thermophysical properties of H_2O-Ar plasmas at temperatures 400-50,000 K and pressure 0.1 MPa. *Plasma Chem. Plasma Process.*, **28**(1), 107–122.

149. Krenek, P. and Hrabovsky, M. (2010) Influence of non-equilibrium effects on plasma property functions in hybrid water-argon plasma torch. *High Temp. Mater. Proc.*, **14**(1-2), 95–100.

150. Matsunaga, N. and Nagashima, A. (1983) Prediction of the transport-properties of gaseous H_2O and its isotopes at high-temperatures. *J. Phys. Chem.*, **87**(25), 5268–5279.

151. Leighton, T.G. (1994) *The Acoustic Bubble*, Academic Press, San Diego, CA.

152. Fridman, A. (2008) *Plasma Chemistry*, Cambridge University Press, Cambridge.

153. Dolan, T.J. (1993) Electron and ion collisions with water-vapor. *J. Phys. D: Appl. Phys.*, **26**(1), 4–8.

154. Liu, D.X., Bruggeman, P., Iza, F., Rong, M.Z., and Kong, M.G. (2010) Global model of low-temperature atmospheric-pressure He + H_2O plasmas. *Plasma Sources Sci. Technol.*, **19**(2), 025018.

155. Sakata, S. and Okada, T. (1994) Effect of humidity on hydrated cluster-ion formation in a clean room corona discharge neutralizer. *J. Aerosol. Sci.*, **25**(5), 879.

156. Lieberman, M.A. and Lichtenberg, A.J. (1994) *Principles of Plasma Discharges and Materials Processing*, John Wiley & Sons, Inc., New York.

157. Itikawa, Y. (2009) Cross sections for electron collisions with oxygen molecules. *J. Phys. Chem. Ref. Data*, **38**(1), 1–20.

158. Itikawa, Y. (2006) Cross sections for electron collisions with nitrogen molecules. *J. Phys. Chem. Ref. Data*, **35**(1), 31–53.

159. Itikawa, Y. and Mason, N. (2005) Cross sections for electron collisions with water molecules. *J. Phys. Chem. Ref. Data*, **34**(1), 1–22.

160. Munoz, A., Oller, J.C., Blanco, F., Gorfinkiel, J.D., Limao-Vieira, P., and Garcia, G. (2007) Electron-scattering cross sections and stopping powers in H_2O. *Phys. Rev. A*, **76**(5), 052707.

161. Munoz, A., Blanco, F., Garcia, G., Thorn, P.A., Brunger, M.J., Sullivan, J.P., and Buckman, S.J. (2008) Single electron tracks in water vapour for energies below 100 eV. *Int. J. Mass Spectrom.*, **277**(1-3), 175–179.

162. Morgan, L.A. (1998) Electron impact excitation of water. *J. Phys. B: Atom. Mol. Opt. Phys.*, **31**(22), 5003–5011.

163. Thorn, P.A., Brunger, M.J., Kato, H., Hoshino, M., and Tanaka, H. (2007) Cross sections for the electron impact excitation of the $\tilde{a}^3B_1$, $\tilde{b}^3A_1$ and $\tilde{B}^1A_1$ dissociative electronic states of water. *J. Phys. B: Atom. Mol. Opt. Phys.*, **40**(4), 697–708.

164. Hagelaar, G.J.M. and Pitchford, L.C. (2005) Solving the Boltzmann equation to obtain electron transport coefficients and rate coefficients for fluid models. *Plasma Sources Sci. Technol.*, **14**(4), 722–733.

165. Zhao, G.B., Hu, X.D., Yeung, M.C., Plumb, O.A., and Radosz, M. (2004) Nonthermal plasma reactions of dilute nitrogen oxide mixtures: NO_x in nitrogen. *Ind. Eng. Chem. Res.*, **43**(10), 2315–2323.

166. Bruggeman, P. and Schram, D.C. (2010) On OH production in water containing atmospheric pressure plasmas. *Plasma Sources Sci. Technol.*, **19**(4), 045025.

167. Hibert, C., Gaurand, I., Motret, O., and Pouvesle, J.M. (1999) [OH(X)] measurements by resonant absorption spectroscopy in a pulsed dielectric barrier discharge. *J. Appl. Phys.*, **85**(10), 7070–7075.

168. Nikiforov, A.Y., Sarani, A., and Leys, C. (2011) The influence of water vapor content on electrical and spectral properties of an atmospheric pressure plasma jet. *Plasma Sources Sci. Technol.*, **20**, 015014.

169. Sarani, A., Nikiforov, A.Y., and Leys, C. (2010) Atmospheric pressure plasma jet in Ar and Ar/H_2O mixtures: Optical emission spectroscopy and temperature measurements. *Phys. Plasmas*, **17**(6), 063504.

170. Caprasecca, S., Gorfinkiel, J.D., Bouchiha, D., and Caron, L.G. (2009) Multiple scattering approach to elastic electron collisions with molecular clusters. *J. Phys. B: Atom. Mol. Opt. Phys.*, **42**(9), 095205.

171. Bouchiha, D., Caron, L.G., Gorfinkiel, J.D., and Sanche, L. (2008) Multiple scattering approach to elastic low-energy electron collisions with the water dimer. *J. Phys. B: Atom. Mol. Opt. Phys.*, **41**(4), 045204.

172. Locke, B.R. and Mededovic-Thagard, S. (2009) Analysis of chemical reactions in gliding arc reactors with water spray. *IEEE Trans. Plasma Sci.*, **37**(4), 494–501.

173. Kirkpatrick, M.J. and Locke, B.R. (2005) Hydrogen, oxygen, and hydrogen peroxide formation in aqueous phase pulsed corona electrical discharge. *Ind. Eng. Chem. Res.*, **44**(12), 4243–4248.

174. Shih, K.Y. and Locke, B.R. (2011) Optical and electrical diagnostics of the effects of conductivity on liquid phase electrical discharge. *IEEE Trans. Plasma Sci.*, **36**(3), 883–892.

175. Nomura, S., Mukasa, S., Toyota, H., Miyake, H., Yamashita, H., Maehara, T., Kawashima, A., and Abe, F. (2011) Characteristics of in-liquid plasma in water under higher pressure than atmospheric pressure. *Plasma Sources Sci. Technol.*, **20**(3), 034012.

176. Moller, D. (2009) Atmospheric hydrogen peroxide: evidence for aqueous-phase formation from a historic perspective and a one-year measurement campaign. *Atmos. Environ.*, **43**(37), 5923–5936.

177. Ebert, U., Montijn, C., Briels, T.M.P., Hundsdorfer, W., Meulenbroek, B., Rocco, A., and van Veldhuizen, E.M. (2006) The multiscale nature of streamers. *Plasma Sources Sci. Technol.*, **15**(2), S118–S129.

178. Raizer, Y.P., Milikh, G.M., and Shneider, M.N. (2010) Streamer- and leader-like processes in the upper atmosphere: Models of red sprites and blue jets. *J. Geophys. Res. Space Phys.*, **115**, A00E42.

179. Gordillo-Vazquez, F.J. (2008) Air plasma kinetics under the influence of sprites. *J. Phys. D: Appl. Phys.*, **41**(23), 234016.

180. Nijdam, S., van Veldhuizen, E.M., and Ebert, U. (2010) Comment on "NO_x production in laboratory discharges simulating blue jets and red sprites" by H. Peterson et al. *J. Geophys. Res. Space Phys.*, **115**, A12305.

181. Tas, M.A., van Hardeveld, R., and van Veldhuizen, E.M. (1997) Reactions of NO in a positive streamer corona plasma. *Plasma Chem. Plasma Process.*, **17**, 371–391.

182. Sentman, D.D., Stenbaek-Nielsen, H.C., McHarg, M.G., and Morrill, J.S. (2008) Plasma chemistry of sprite streamers. *J. Geophys. Res. Atmos.*, **113**(D11), D11112.

183. Varma, A., Morbidelli, M., and Wu, H. (1999) *Parametric Sensitivity in Chemical Systems*, Cambridge University Press, Cambridge.

184. Ono, R. and Oda, T. (2002) Dynamics and density estimation of hydroxyl radicals in a pulsed corona discharge. *J. Phys. D: Appl. Phys.*, **35**(17), 2133–2138.

185. Su, Z.Z., Ito, K., Takashima, K., Katsura, S., Onda, K., and Mizuno, A. (2002) OH radical generation by atmospheric pressure pulsed discharge plasma and its quantitative analysis by monitoring CO oxidation. *J. Phys. D: Appl. Phys.*, **35**, 3192–3198.

186. Bruggeman, P., Schram, D., Gonzalez, M.A., Rego, R., Kong, M.G., and Leys, C. (2009) Characterization of a direct dc-excited discharge in water by optical

emission spectroscopy. *Plasma Sources Sci. Technol.*, **18**(2), 025017.

187. Sun, B., Sato, M., and Clements, J.D. (1997) Optical study of active species produced by a pulsed streamer corona discharge in water. *J. Electrostatics*, **39**, 189–202.

188. Sunka, P., Babicky, V., Clupek, M., Lukes, P., Simek, M., Schmidt, J., and Cernak, M. (1999) Generation of chemically active species by electrical discharges in water. *Plasma Sources Sci. Technol.*, **8**, 258–265.

189. Bruggeman, P., Schram, D.C., Kong, M.G., and Leys, C. (2009) Is the rotational temperature of OH(A-X) for discharges in and in contact with liquids a good diagnostic for determining the gas temperature? *Plasma Process. Polym.*, **6**(11), 751–762.

190. Bruggeman, P., Walsh, J.L., Schram, D.C., Leys, C., and Kong, M.G. (2009) Time dependent optical emission spectroscopy of sub-microsecond pulsed plasmas in air with water cathode. *Plasma Sources Sci. Technol.*, **18**(4), 045023.

191. Bruggeman, P., Verreycken, T., Gonzalez, M.A., Walsh, J.L., Kong, M.G., Leys, C., and Schram, D.C. (2010) Optical emission spectroscopy as a diagnostic for plasmas in liquids: opportunities and pitfalls. *J. Phys. D: Appl. Phys.*, **43**(12), 124005.

192. Namihira, T., Sakai, S., Yamaguchi, T., Yamamoto, K., Yamada, C., Kiyan, T., Sakugawa, T., Katsuki, S., and Akiyama, H. (2007) Electron temperature and electron density of underwater pulsed discharge plasma produced by solid-state pulsed-power generator. *IEEE Trans. Plasma Sci.*, **35**(3), 614–618.

193. Gigosos, M.A., Gonzalez, M.A., and Cardenoso, V. (2003) Computer simulated Balmer-alpha, -beta and -gamma Stark line profiles for non-equilibrium plasmas diagnostics. *Spectrochim. Acta, Part B*, **58**(8), 1489–1504.

194. Egli, W. and Eliasson, B. (1989) Numerical calculation of electrical breakdown in oxygen in a dielectric-barrier discharge. *Helv. Phys. Acta*, **62**, 302–305.

195. Eliasson, B., Hirth, M., and Kogelschatz, U. (1987) Ozone synthesis from oxygen in dielectric barrier discharges. *J. Phys. D: Appl. Phys.*, **20**, 1421–1437.

196. Eliasson, B. and Kogelschatz, U. (1986) Electron impact dissociation in oxygen. *J. Phys. B: Atom. Mol. Opt. Phys.*, **19**(8), 1241–1247.

197. Eliasson, B. and Kogelschatz, U. (1986) *Basic Data for Modelling of Electrical Discharges in Gases: Oxygen, Research Report KLR 86-11 C*, Asea Brown Boveri Corporate Research.

198. Eliasson, B. and Kogelschatz, U. (1991) Modeling and applications of silent discharge plasmas. *IEEE Trans. Plasma Sci.*, **19**, 309–323.

199. Eliasson, B. and Kogelschatz, U. (1991) Ozone generation with narrow-band UV-radiation. *Ozone Sci. Eng.*, **13**(3), 365–373.

200. Kogelschatz, U. and Eliasson, B. (1995) Ozone generation and applications, in *Handbook of Electrostatic Processes* (eds J.S. Chang, A.J. Kelly, and J.M. Crowley), Marcel Dekker, Inc., New York, pp. 581–605. Chapter 26.

201. Peyrous, R. (1990) The effect of relative humidity on ozone production by corona discharge in oxygen or air - A numerical simulation, Part II: Air. *Ozone Sci. Eng.*, **12**, 40–64.

202. Peyrous, R., Pignolet, P., and Held, B. (1989) Kinetic simulation of gaseous species created by an electrical discharge in dry or humid oxygen. *J. Phys. D: Appl. Phys.*, **22**, 1658–1667.

203. Gentile, A.C. and Kushner, M.J. (1995) Reaction chemistry and optimization of plasma remediation of N_xO_y from gas streams. *J. Appl. Phys.*, **78**(3), 2074–2085.

204. Lowke, J.J. and Morrow, R. (1995) Theoretical analysis of removal of oxides of sulphur and nitrogen in pulsed operation of electrostatic precipitators. *IEEE Trans. Plasma Sci.*, **23**, 661–671.

205. Shimizu, K., Hirano, T., and Oda, T. (2001) Effect of water vapor and hydrocarbons in removing NO_x by using nonthermal plasma and catalyst. *IEEE Trans. Ind. Appl.*, **37**, 464–471.

206. van Veldhuizen, E.M. (2000) *Electrical Discharges for Environmental Purposes, Fundamentals and Applications*, Nova Science Publishers, Inc., Huntington, New York.
207. van Veldhuizen, E.M., Rutgers, W.R., and Bityurin, V.A. (1996) Energy efficiency of NO removal by pulsed corona discharges. *Plasma Chem. Plasma Process.*, **16**(2), 227–247.
208. Gasparik, R., Ihara, S., Yamabe, C., and Satoh, S. (2000) Effect of CO_2 and water vapors on NO_x removal efficiency under conditions of DC corona discharge in cylindrical discharge reactor. *Jpn. J. Appl. Phys.*, **39**, 306–309.
209. Mok, Y.S., Kim, J.H., Nam, I.S., and Ham, S.W. (2000) Removal of NO and formation of byproducts in a positive-pulsed corona discharge reactor. *Ind. Eng. Chem. Res.*, **39**(10), 3938–3944.
210. Fujii, T. and Rea, M. (2000) Treatment of NO_x in exhaust gas by corona plasma over water surface. *Vacuum*, **59**, 228–235.
211. Fujii, T., Aoki, Y., Yoshioka, N., and Rea, M. (2001) Removal of NO_x by DC corona reactor with water. *J. Electrostatics*, **51-52**, 8–14.
212. Bakken, J.A. (1994) High-temperature processing and numerical modeling of thermal plasmas in Norway. *Pure Appl. Chem.*, **66**(6), 1239–1246.
213. Wang, D.Y., Matsumoto, T., Namihira, T., and Akiyama, H. (2010) Development of higher yield ozonizer based on nano-seconds pulsed discharge. *J. Adv. Oxid. Technol.*, **13**(1), 71–78.
214. Zhao, G.B., Garikipati, S.V.B., Hu, X.D., Argyle, M.D., and Radosz, M. (2005) Effect of oxygen on nonthermal plasma reactions of nitrogen oxides in nitrogen. *AIChE J.*, **51**(6), 1800–1812.
215. Sathiamoorthy, G., Kalyana, S., Finney, W.C., Clark, R.J., and Locke, B.R. (1999) Chemical reaction kinetics and reactor modeling of NO_x removal in a pulsed streamer corona discharge reactor. *Ind. Eng. Chem. Res.*, **38**, 1844–1855.
216. Rao, X., Matveev, I.B., and Lee, T. (2009) Nitric oxide formation in a premixed flame with high-level plasma energy coupling. *IEEE Trans. Plasma Sci.*, **37**(12), 2303–2313.
217. Zhao, G.B., Hu, X.D., Plumb, O.A., and Radosz, M. (2004) Energy consumption and optimal reactor configuration for nonthermal plasma conversion of N_2O in nitrogen and N_2O in argon. *Energ. Fuel.*, **18**(5), 1522–1530.
218. Zhao, G.B., Hu, X.D., Argyle, M.D., and Radosz, M. (2004) N atom radicals and $N_2(A^3\Sigma_u^+)$ found to be responsible for nitrogen oxides conversion in nonthermal nitrogen plasma. *Ind. Eng. Chem. Res.*, **43**(17), 5077–5088.
219. Zhao, G.B., Argyle, M.D., and Radosz, M. (2007) Optical emission study of nonthermal plasma confirms reaction mechanisms involving neutral rather than charged species. *J. Appl. Phys.*, **101**(3), 033303.
220. Hu, X.D., Zhao, G.B., Legowski, S.F., and Radosz, M. (2005) Moisture effect on NO_x conversion in a nonthermal plasma reactor. *Environ. Eng. Sci.*, **22**(6), 854–869.
221. Sentman, D.D., Stenbaek-Nielsen, H.C., McHarg, M.G., and Morrill, J.S. (2008) Correction to "plasma chemistry of sprite streamers" (vol 113, art d11112, 2008). *J. Geophys. Res. Atmos.*, **113**(D14), D14399.
222. Itikawa, Y. and Mason, N. (2005) Rotational excitation of molecules by electron collisions. *Phys. Rep.*, **414**(1), 1–41.
223. Kuhn, S., Bibinov, N., Gesche, R., and Awakowicz, P. (2010) Non-thermal atmospheric pressure HF plasma source: generation of nitric oxide and ozone for bio-medical applications. *Plasma Sources Sci. Technol.*, **19**(1), 015013.
224. Kossyi, I.A., Kostinsky, A.Y., Matveyev, A.A., and Silakov, V.P. (1992) Kinetic scheme of the non-equilibrium discharge in nitrogen-oxygen mixtures. *Plasma Sources Sci. Technol.*, **1**(3), 207–220.
225. NIST Chemical Kinetics Database, (2011) NIST Standard Reference Database 17, Version 7.0 (Web Version), Release 1.6.3, Data Version 2011.06, *http://kinetics.nist.gov/kinetics*.

226. Lu, X.P. and Laroussi, M. (2005) Optimization of ultraviolet emission and chemical species generation from a pulsed dielectric barrier discharge at atmospheric pressure. *J. Appl. Phys.*, **98**(2), 023301.

227. Shuaibov, A.K., Minya, A.J., Gomoki, Z.T., Shevera, I.V., and Gritsak, R.V. (2011) Vacuum-UV emitter using low-pressure discharge in helium-water vapor mixture. *Tech. Phys. Lett.*, **37**(2), 126–127.

228. Shuaibov, A.K., Shimon, L.L., Dashchenko, A.I., and Shevera, I.V. (2001) A water-vapor electric-discharge vacuum ultraviolet source. *Tech. Phys. Lett.*, **27**(8), 642–643.

229. Seinfeld, J.H. and Pandis, S.N. (1998) *Atmospheric Chemistry and Physics*, John Wiley & Sons, Inc., New York.

230. Zvereva, G.N. (2010) Investigation of water decomposition by vacuum ultraviolet radiation. *Opt. Spectrosc.*, **108**(6), 915–922.

231. Watanabe, K., Inn, E.C.Y., and Zelikoff, M. (1953) Absorption coefficients of oxygen in the vacuum ultraviolet. *J. Chem. Phys.*, **21**(6), 1026–1030.

232. Atkinson, R., Baulch, D.L., Cox, R.A., Crowley, J.N., Hampson, R.F., Hynes, R.G., Jenkin, M.E., Rossi, M.J., and Troe, J. (2004) Evaluated kinetic and photochemical data for atmospheric chemistry: Volume I - gas phase reactions of O_x, HO_x, NO_x and SO_x species. *Atmos. Chem. Phys.*, **4**, 1461–1738.

233. Watanabe, K. and Zelikoff, M. (1953) Absorption coefficients of water vapor in the vacuum ultraviolet. *J. Opt. Soc. Am.*, **43**(9), 753–755.

234. Heit, G., Neuner, A., Saugy, P.Y., and Braun, A.M. (1998) Vacuum-UV (172 nm) actinometry. The quantum yield of the photolysis of water. *J. Phys. Chem. A*, **102**(28), 5551–5561.

235. Yin, S.E., Sun, B.M., Gao, X.D., and Xiao, H.P. (2009) The effect of oxygen and water vapor on nitric oxide conversion with a dielectric barrier discharge reactor. *Plasma Chem. Plasma Process.*, **29**(6), 421–431.

236. Morgan, M.M., Cuddy, M.F., and Fisher, E.R. (2010) Gas-phase chemistry in inductively coupled plasmas for NO removal from mixed gas systems. *J. Phys. Chem. A*, **114**(4), 1722–1733.

237. Soloshenko, I.A., Tsiolko, V.V., Pogulay, S.S., Kalyuzhnaya, A.G., Bazhenov, V.Y., and Shchedrin, A.I. (2009) Effect of water adding on kinetics of barrier discharge in air. *Plasma Sources Sci. Technol.*, **18**(4), 045019.

238. Hoigne, J. (1988) The chemistry of ozone in water, in *Process Technologies for Water Treatment* (ed. S. Stucki), Plenum Press, New York, pp. 121–143.

239. Hoigne, J. and Bader, H. (1976) Role of hydroxyl radical reactions in ozonation processes in aqueous solutions. *Water Res.*, **10**(5), 377–386.

240. Staehelin, J. and Hoigne, J. (1982) Decomposition of ozone in water: rate of initiation by hydroxide ions and hydrogen peroxide. *Environ. Sci. Technol.*, **16**, 676–681.

241. Lukes, P. and Locke, B.R. (2005) Plasmachemical oxidation processes in a hybrid gas–liquid electrical discharge reactor. *J. Phys. D: Appl. Phys.*, **38**(22), 4074–4081.

242. Buxton, G.V. (1987) Radiation chemistry of the liquid state. (1) Water and homogeneous aqueous solutions, in *Radiation Chemistry, Principles and Applications* (eds N. Farhataziz and M.A.J. Rodgers), VCH, Weinheim, Germany, pp. 321–376.

243. Mozumder, A. (1999) *Fundamentals of Radiation Chemistry*, Academic Press, San Diego, CA.

244. Jonah, C.D., Bartels, D.M., and Chernovitz, A.C. (1989) Primary processes in the radiation chemistry of water. *Radiat. Phys. Chem.*, **34**, 145–156.

245. Trout, B.L. and Parrinello, M. (1999) Analysis of the dissociation of H_2O in water using first-principles molecular dynamics. *J. Phys. Chem. B*, **103**(34), 7340–7345.

246. Trout, B.L. and Parrinello, M. (1998) The dissociation mechanism of H_2O in water studied by first-principles

molecular dynamics. *Chem. Phys. Lett.*, **288**(2-4), 343–347.
247. Joshipura, K.N. and Vinodkumar, M. (1998) Ionizing collisions of electrons with H_2O molecules in ice and in water. *Int. J. Mass Spectrom.*, **177**(2-3), 137–141.
248. Vinodkumar, M., Joshipura, K.N., Limbachiya, C.G., and Antony, B.K. (2003) Electron impact ionization of H_2O molecule in crystalline ice. *Nucl. Instrum. Methods Phys. Res., Sect. B*, **212**, 63–66.
249. Hayashi, D., Hoeben, W.F.L.M., Dooms, G., van Veldhuizen, E.M., Rutgers, W.R., and Kroesen, G.M.W. (2000) LIF diagnostic for pulsed-corona-induced degradation of phenol in aqueous solution. *J. Phys. D: Appl. Phys.*, **33**, 1484–1486.
250. Hayashi, D., Hoeben, W.F.L.M., Dooms, G., van Veldhuizen, E.M., Rutgers, W.R., and Kroesen, G.M.W. (2000) Influence of gaseous atmosphere on corona-induced degradation of aqueous phenol. *J. Phys. D: Appl. Phys.*, **33**, 2769–2774.
251. Hoeben, W.F.L.M. (2000) *Pulsed corona-induced degradation of organic materials in water*. PhD Dissertation. Technische Universiteit Eindhoven, Eindhoven, NL.
252. Hayashi, D., Hoeben, W., Dooms, G., van Veldhuizen, E., Rutgers, W., and Kroesen, G. (2001) Laser-induced fluorescence spectroscopy for phenol and intermediate products in aqueous solutions degraded by pulsed corona discharges above water. *Appl. Optics*, **40**, 986–989.
253. Kanazawa, S., Kawano, H., Watanabe, S., Furuki, T., Akamine, S., Ichiki, R., Ohkubo, T., Kocik, M., and Mizeraczyk, J. (2011) Observation of OH radicals produced by pulsed discharges on the surface of a liquid. *Plasma Sources Sci. Technol.*, **20**(3), 034010.
254. Lukes, P. and Locke, B.R. (2005) Degradation of substituted phenols in a hybrid gas–liquid electrical discharge reactor. *Ind. Eng. Chem. Res.*, **44**(9), 2921–2930.
255. Vacha, R., Slavicek, P., Mucha, M., Finlayson-Pitts, B.J., and Jungwirth, P. (2004) Adsorption of atmospherically relevant gases at the air/water interface: Free energy profiles of aqueous solvation of N_2, O_2, O_3, OH, H_2O, HO_2, and H_2O_2. *J. Phys. Chem. A*, **108**(52), 11573–11579.
256. Du, S. and Francisco, J.S. (2008) Interaction between OH radical and the water interface. *J. Phys. Chem. A*, **112**(21), 4826–4835.
257. Du, S.Y., Francisco, J.S., and Kais, S. (2009) Study of electronic structure and dynamics of interacting free radicals influenced by water. *J. Chem. Phys.*, **130**(12), 124312.
258. Fridman, A. and Kennedy, L.A. (2004) *Plasma Physics and Engineering*, Taylor and Francis, New York.
259. Grymonpre, D.R., Finney, W.C., Clark, R.J., and Locke, B.R. (2003) Suspended activated carbon particles and ozone formation in aqueous phase pulsed corona discharge reactors. *Ind. Eng. Chem. Res.*, **42**(21), 5117–5134.
260. Pawlat, J., Hayashi, N., and Yamabe, C. (2001) Studies on electrical discharge in a foaming environment. *Jpn. J. Appl. Phys.*, **40**, 7061–7066.
261. Pawlat, J., Hayashi, N., Yamabe, C., and Pollo, I. (2002) Generation of oxidants with a foaming system and its electrical properties. *Ozone Sci. Eng.*, **24**, 181–191.
262. Pawlat, J., Hayashi, N., Ihara, S., Yamabe, C., and Pollo, I. (2003) Studies on oxidants' generation in a foaming column with a needle to dielectric covered plate electrode. *Plasma Chem. Plasma Process.*, **23**(3), 569–583.
263. Yamabe, C., Hirohata, T., Katanami, H., and Ihara, S. (2010) Electrical discharge with bubbles in water and their application - Water treatment by cavitation electrical discharge, in *International Workshop on Plasmas with Liquids* (ed K. Tachibana), Ehime University, Ehime, Japan, pp. 17–18.
264. Yamabe, C., Nakazaki, S., and Ihara, S. (2009) Water treatment including surfactant using a pulsed discharge. *Prz. Elektrotechniczny*, **85**(5), 105–108.
265. Malik, M.A. (2010) Water purification by plasmas: Which reactors are most

energy efficient. *Plasma Chem. Plasma Process.*, **30**(1), 21–31.

266. Matsui, Y., Takeuchi, N., Sasaki, K., Hayashi, R., and Yasuoka, K. (2011) Experimental and theoretical study of acetic-acid decomposition by a pulsed dielectric-barrier plasma in a gas–liquid two-phase flow. *Plasma Sources Sci. Technol.*, **20**(3), 034015.
267. Piskarev, I.M., Rylova, A.E., and Sevast'yanov, A.I. (1996) Formation of ozone and hydrogen peroxide during an electrical discharge in the solution-gas system. *Russ. J. Electrochem.*, **32**, 827–829.
268. Velikonja, J., Bergougnou, M.A., Castle, G.S.P., Cairns, W.L., and Inculet, I. (2001) Co-generation of ozone and hydrogen peroxide by dielectric barrier AC discharge in humid oxygen. *Ozone Sci. Eng.*, **23**(6), 467–478.
269. Anpilov, A.M., Barkhudarov, E.M., Bark, Y., Zadiraka, Y., Christofi, M., Kozlov, Y., Kossyi, I.A., Kop'ev, V.A., Silakov, V.P., Taktakishvili, M.I., and Temchin, S.M. (2001) Electric discharge in water as a source of UV radiation, ozone and hydrogen peroxide. *J. Phys. D: Appl. Phys.*, **34**, 933–999.
270. Polyakov, O.V., Badalyan, A.M., and Bakhturova, L.F. (2004) Relative contributions of plasma pyrolysis and liquid-phase reactions in the anode microspark treatment of aqueous phenol solutions. *High Energ. Chem.*, **38**(2), 131–133.
271. Dorai, R., Hassouni, K., and Kushner, M.J. (2000) Interaction between soot particles and NO_x during dielectric barrier discharge plasma remediation of simulated diesel exhaust. *J. Appl. Phys.*, **88**, 6060–6071.
272. Kezelis, R., Mecius, V., Valinciute, V., and Valincius, V. (2004) Waste and biomass treatment employing plasma technology. *High Temp. Mater. Proc.*, **8**(2), 273–282.
273. Nishikawa, H., Ibe, M., Tanaka, M., Ushio, M., Takemoto, T., Tanaka, K., Tanahashi, N., and Ito, T. (2004) A treatment of carbonaceous wastes using thermal plasma with steam. *Vacuum*, **73**(3-4), 589–593.
274. Hrabovsky, M., Konrad, M., Kopecky, V., Hlina, M., Kavka, T., van Oost, G., Beeckman, E., and Defoort, B. (2006) Gasification of biomass in water/gas-stabilized plasma for syngas production. *Czech. J. Phys.*, **56** (Suppl. B), B1199–B1206.
275. Serbin, S.I. (2006) Features of liquid-fuel plasma-chemical gasification for diesel engines. *IEEE Trans. Plasma Sci.*, **34**(6), 2488–2496.
276. Huang, J.J., Guo, W.K., and Xu, P. (2005) Thermodynamic study of water-steam plasma pyrolysis of medical waste for recovery of CO and H_2. *Plasma Sci. Technol.*, **7**(6), 3148–3150.
277. Kim, S.W., Park, H.S., and Kim, H.J. (2003) 100 kW steam plasma process for treatment of PCBs (polychlorinated biphenyls) waste. *Vacuum*, **70**(1), 59–66.
278. Mountouris, A., Voutsas, E., and Tassios, D. (2008) Plasma gasification of sewage sludge: Process development and energy optimization. *Energ. Convers. Manag.*, **49**(8), 2264–2271.
279. Starikovskaia, S.M. (2006) Plasma assisted ignition and combustion. *J. Phys. D: Appl. Phys.*, **39**(16), R265–R299.
280. Cathey, C.D., Tang, T., Shiraishi, T., Urushihara, T., Kuthi, A., and Gundersen, M.A. (2007) Nanosecond plasma ignition for improved performance of an internal combustion engine. *IEEE Trans. Plasma Sci.*, **35**(6), 1664–1668.
281. Chernyak, V.Y., Olszewski, S.V., Yukhymenko, V., Solomenko, E.V., Prysiazhnevych, I.V., Naumov, V.V., Levko, D.S., Shchedrin, A.I., Ryabtsev, A.V., Demchina, V.P., Kudryavtsev, V.S., Martysh, E.V., and Verovchuck, M.A. (2008) Plasma-assisted reforming of ethanol in dynamic plasma-liquid system: Experiments and modeling. *IEEE Trans. Plasma Sci.*, **36**(6), 2933–2939.
282. Fridman, A., Gutsol, A., Gangoli, S., Ju, Y.G., and Ombrellol, T. (2008) Characteristics of gliding arc and its application in combustion enhancement. *J. Propul. Power*, **24**(6), 1216–1228.

283. Burlica, R., Kirkpatrick, M.J., Finney, W.C., Clark, R.J., and Locke, B.R. (2004) Organic dye removal from aqueous solution by glidarc discharges. *J. Electrostatics*, **62**(4), 309–321.

284. Burlica, R., Kirkpatrick, M.J., and Locke, B.R. (2006) Formation of reactive species in gliding arc discharges with liquid water. *J. Electrostatics*, **64**(1), 35–43.

285. Burlica, R., Grim, R.G., Shih, K.Y., Balkwill, D., and Locke, B.R. (2010) Bacteria inactivation using low power pulsed gliding arc discharges with water spray. *Plasma Process. Polym.*, **7**(8), 640–649.

286. Ohneda, H., Harano, A., Sadakata, M., and Takarada, T. (2002) Improvement of NO_x removal efficiency using atomization of fine droplets into corona discharge. *J. Electrostatics*, **55**, 321–332.

287. Mutaf-Yardimci, O., Saveliev, A.V., Fridman, A.A., and Kennedy, L.A. (2000) Thermal and nonthermal regimes of gliding arc discharge in air flow. *J. Appl. Phys.*, **87**(4), 1632–1641.

288. Czernichowski, A., Nassar, H., Ranaivosoloarimanana, A., Fridman, A.A., Simek, M., Musiol, K., Pawelec, E., and Dittrichova, L. (1996) Spectral and electrical diagnostics of gliding arc. *Acta Phys. Pol. B*, **89**(5-6), 595–603.

289. Fridman, A.A., Petrousov, A., Chapelle, J., Cormier, J.M., Czernichowski, A., Lesueur, H., and Stevefelt, J. (1994) Model of the gliding arc. *J. Phys.*, **4**(8), 1449–1465.

290. Ghezzar, M.R., Abdelmalek, F., Belhadj, M., Benderdouche, N., and Addou, A. (2007) Gliding arc plasma assisted photocatalytic degradation of anthraquinonic acid green 25 in solution with TiO_2. *Appl. Catal. B: Environ.*, **72**(3-4), 304–313.

291. Poplin, M. (2006) *Removal of aqueous pollutants with gliding arc discharge*. BS Honors Thesis. Florida State University, Tallahassee, FL, USA.

292. Lowke, J.J., Kovitya, P., and Schmidt, H.P. (1992) Theory of free-burning arc columns including the influence of the cathode. *J. Phys. D: Appl. Phys.*, **25**(11), 1600–1606.

293. Lowke, J.J., Sansonnens, L., and Haidar, J. (1999) Arc modeling, in *Heat and Mass Transfer under Plasma Conditions: Part I. Plasma Torches*, Annals of the New York Academy of Sciences, **vol. 891**, The New York Academy of Sciences, pp. 1–13.

294. Wu, H.M., Carey, G.F., and Oakes, M.E. (1994) Numerical-simulation of AC plasma-arc thermodynamics. *J. Comput. Phys.*, **112**(1), 24–30.

295. Porter, D., Poplin, M., Holzer, F., Finney, W.C., and Locke, B.R. (2009) Formation of hydrogen peroxide, hydrogen, and oxygen in gliding arc electrical discharge reactors with water spray. *IEEE Trans. Ind. Appl.*, **45**(2), 623–629.

296. Zuo, Y.G. and Deng, Y.W. (1999) Evidence for the production of hydrogen peroxide in rainwater by lightning during thunderstorms. *Geochim. Cosmochim. Acta*, **63**(19-20), 3451–3455.

297. Daito, S., Tochikubo, F., and Watanabe, T. (2000) Improvement of NO_x removal efficiency assisted by aqueous-phase reaction in corona discharge. *Jpn. J. Appl. Phys., Part 1*, **39**(8), 4914–4919.

298. Yan, K., Kanazawa, S., Ohkubo, T., and Nomoto, Y. (1999) Oxidation and reduction processes during NO_x removal with corona-induced nonthermal plasma. *Plasma Chem. Plasma Process.*, **19**(3), 421–443.

299. Baranchicov, E.I., Belenky, G.S., Deminsky, M.A., Dorovsky, V.P., Erastov, E.M., Kochetov, V.A., Maslenicov, D.D., Potapkin, B.V., Rusanov, V.D., Severny, V.V., and Fridman, A.A. (1992) Plasma-catalysis SO_2 oxidation in an air stream by a relativistic electron beam and corona discharge. *Radiat. Phys. Chem.*, **40**, 287–294.

300. Joshi, A.A., Locke, B.R., Arce, P., and Finney, W.C. (1995) Formation of hydroxyl radicals, hydrogen peroxide and aqueous electrons by pulsed streamer corona discharge in aqueous solution. *J. Hazard. Mater.*, **41**, 3–30.

301. Grymonpre, D.R., Sharma, A.K., Finney, W.C., and Locke, B.R. (2001) The role of Fenton's reaction in aqueous phase pulsed streamer corona reactors. *Chem. Eng. J.*, **82**(1-3), 189–207.

302. Sahni, M. and Locke, B.R. (2006) Quantification of hydroxyl radicals produced in aqueous phase pulsed electrical discharge reactors. *Ind. Eng. Chem. Res.*, **45**(17), 5819–5825.

303. Sahni, M. and Locke, B.R. (2006) Quantification of reductive species produced by high voltage electrical discharges in water. *Plasma Process. Polym.*, **3**(4-5), 342–354.

304. Sahni, M. and Locke, B.R. (2006) The effects of reaction conditions on liquid-phase hydroxyl radical production in gas–liquid pulsed-electrical-discharge reactors. *Plasma Process. Polym.*, **3**(9), 668–681.

305. Polyakov, O.V., Badalyan, A.M., and Bakhturova, L.F. (2002) The water degradation yield and spatial distribution of primary radicals in the near-discharge volume of an electrolytic cathode. *High Energ. Chem.*, **36**(4), 280–284.

306. Sato, M., Ohgiyama, T., and Clements, J.S. (1996) Formation of chemical species and their effects on microorganisms using a pulsed high-voltage discharge in water. *IEEE Trans. Ind. Appl.*, **32**, 106–112.

307. Hickling, A. and Ingram, M.D. (1964) Contact glow-discharge electrolysis. *Trans. Faraday Soc.*, **60**, 783–793.

308. Hickling, A. (1971) Electrochemical processes in glow discharge at the gas-solution interface, in *Modern Aspects of Electrochemistry*, vol. **6** (eds J.O. Bockris and B.E. Conway), Plenum Press, New York, pp. 329–373.

309. Kanzaki, Y., Nishimura, N., and Matsumoto, O. (1984) On the yields of glow discharge electrolysis in various atmospheres. *J. Electroanal. Chem.*, **167**, 297–300.

310. Sengupta, S.K., and Singh, O.P. (1994) Contact glow discharge electrolysis: a study of its chemical yields in aqueous inert-type electrolytes. *J. Electroanal. Chem.*, **369**, 113–120.

311. Jones, H.M. and Kunhardt, E.E. (1995) Pulsed dielectric breakdown of pressurized water and salt solutions. *J. Appl. Phys.*, **77**, 795–805.

312. Jones, H.M. and Kunhardt, E.E. (1994) The influence of pressure and conductivity on the pulsed breakdown of water. *IEEE Trans. Dielectr. Electr. Insul.*, **1**(6), 1016–1025.

313. Lukes, P., Clupek, M., Babicky, V., and Sunka, P. (2008) Ultraviolet radiation from the pulsed corona discharge in water. *Plasma Sources Sci. Technol.*, **17**(2), 024012.

314. Shih, K.Y., Burlica, R., Finney, W.C., and Locke, B.R. (2009) Effect of pressure on discharge initiation and chemical reaction in a liquid-phase electrical discharge reactor. *IEEE Trans. Ind. Appl.*, **45**(2), 630–637.

315. Akiyama, H., Goto, M., Namihira, T., Sasaki, M., Heeren, T., Roy, B., Kyan, T., and Sunao, K. (2004) Plasma production in supercritical fluid, in *2nd International Workshop on Microplasmas*, Stevens Institute of Technology, p. 44.

316. Ito, T. and Terashima, K. (2002) Generation of micrometer-scale discharge in a supercritical fluid environment. *Appl. Phys. Lett.*, **80**(16), 2854–2856.

317. Lock, E.H., Saveliev, A., and Kennedy, L.A. (2005) Initiation of pulsed corona discharge under supercritical conditions. *IEEE Trans. Plasma Sci.*, **33**(2), 850–853.

318. Ito, T., Katahira, K., Shimizu, Y., Sasaki, T., Koshizaki, N., and Terashima, K. (2004) Carbon and copper nanostructured materials synthesis by plasma discharge in a supercritical fluid environment. *Chem. Mater.*, **14**(10), 1513–1515.

319. Kiyan, T., Sasaki, M., Ihara, T., Namihira, T., Hara, M., Goto, M., and Akiyama, H. (2009) Pulsed breakdown and plasma-aided phenol polymerization in supercritical carbon dioxide and sub-critical water. *Plasma Process. Polym.*, **6**(11), 778–785.

320. Ito, T., Fujiwara, H., and Terashima, K. (2003) Decrease of breakdown voltages for micrometer-scale gap electrodes for carbon dioxide near the critical point: Temperature and pressure dependences. *J. Appl. Phys.*, **94**(8), 5411–5413.

321. Holzer, F. and Locke, B.R. (2008) Influence of high voltage needle electrode

material on hydrogen peroxide formation and electrode erosion in a hybrid gas–liquid series electrical discharge reactor. *Plasma Chem. Plasma Process.*, **28**(1), 1–13.

322. Lukes, P., Clupek, M., Babicky, V., Sunka, P., Skalny, J., Stefecka, M., Novak, J., and Malkova, Z. (2006) Erosion of needle electrodes in pulsed corona discharge in water. *Czech. J. Phys.*, **56** (Suppl. B), B916–B924.

323. Kirkpatrick, M.J. and Locke, B.R. (2006) Effects of platinum electrode on hydrogen, oxygen, and hydrogen peroxide formation in aqueous phase pulsed corona electrical discharge. *Ind. Eng. Chem. Res.*, **45**(6), 2138–2142.

324. Lukes, P., Clupek, M., Babicky, V., Sisrova, I., and Janda, V. (2011) The catalytic role of tungsten electrode material in the plasmachemical activity of a pulsed corona discharge in water. *Plasma Sources Sci. Technol.*, **20**(3), 034011.

325. Sharma, A.K., Locke, B.R., Arce, P., and Finney, W.C. (1993) A preliminary study of pulsed streamer corona discharge for the degradation of phenol in aqueous solutions. *Hazard Waste Hazard Mater.*, **10**, 209–219.

326. Mededovic, S. and Locke, B.R. (2006) Platinum catalysed decomposition of hydrogen peroxide in aqueous-phase pulsed corona electrical discharge. *Appl. Catal. B: Environ.*, **67**, 149–159.

327. Mededovic, S. and Locke, B.R. (2007) The role of platinum as the high voltage electrode in the enhancement of fenton's reaction in liquid phase electrical discharge. *Appl. Catal. B: Environ.*, **72**, 342–350.

328. Sahni, M., Finney, W.C., and Locke, B.R. (2005) Degradation of aqueous phase polychlorinated biphenyls (PCB) using pulsed corona discharges. *J. Adv. Oxid. Technol.*, **8**(1), 105–111.

329. Grymonpre, D.R., Finney, W.C., and Locke, B.R. (1999) Aqueous-phase pulsed streamer corona reactor using suspended activated carbon particles for phenol oxidation: model-data comparison. *Chem. Eng. Sci.*, **54**(15-16), 3095–3105.

330. Lukes, P., Clupek, M., Sunka, P., Peterka, F., Sano, T., Negishi, N., Matsuzawa, S., and Takeuchi, K. (2005) Degradation of phenol by underwater pulsed corona discharge in combination with TiO_2 photocatalysis. *Res. Chem. Intermed.*, **31**(4-6), 285–294.

331. Kusic, H., Koprivanac, N., and Locke, B.R. (2005) Decomposition of phenol by hybrid gas/liquid electrical discharge reactors with zeolite catalysts. *J. Hazard. Mater.*, **125**(1-3), 190–200.

332. Kusic, H., Koprivanac, N., Peternel, I., and Locke, B.R. (2005) Hybrid gas/liquid electrical discharge reactors with zeolites for colored wastewater degradation. *J. Adv. Oxid. Technol.*, **8**(2), 172–181.

333. Peternel, I., Kusic, H., Koprivanac, N., and Locke, B.R. (2006) The roles of ozone and zeolite on reactive dye degradation in electrical discharge reactors. *Environ. Tech. Lett.*, **27**, 545–557.

334. Sakai, O., Kimura, M., Shirafuji, T., and Tachibana, K. (2008) Underwater microdischarge in arranged microbubbles produced by electrolysis in electrolyte solution using fabric-type electrode. *Appl. Phys. Lett.*, **93**(23), 231501.

7
Aqueous-Phase Chemistry of Electrical Discharge Plasma in Water and in Gas–Liquid Environments

Petr Lukes, Bruce R. Locke, and Jean-Louis Brisset

7.1
Introduction

Nonthermal electrical discharge plasma in liquids and at gas–liquid interfaces generate a number of chemical and physical processes (Chapter 6) that can affect both the chemical and biological components in water. Many organic compounds and microorganisms have been degraded and inactivated in water using various types of electrical discharges. The main mechanism of plasmachemical decomposition of organic compounds in water involves oxidation processes initiated by $OH^{\bullet}$ radicals, ozone, and hydrogen peroxide (especially in the presence of suitable catalysts such as iron, platinum, tungsten). In gas-phase discharges with gas–liquid interfaces in atmospheric air, nitrogen-based reactive species, especially peroxynitrites, also contribute to the plasmachemical oxidation processes in water. Reductive pathways can take place as well. In the case of microbial inactivation, the mechanism is more complex and physical processes such as ultraviolet (UV) photolysis, large electric fields, and shock waves may also contribute particularly for direct discharge in water –these mechanisms are elaborated in Chapter 8. In this chapter, we provide an overview of the chemistry and reaction kinetics of the main primary and secondary chemical species ($OH^{\bullet}$ radicals, ozone, hydrogen peroxide, peroxynitrite) generated by plasma discharge in water and gas–liquid interfaces. We also address the mechanisms of plasma interaction with the chemical compounds in water and the plasma-catalytic processes in liquid and gas–liquid environments.

7.2
Aqueous-Phase Plasmachemical Reactions

Depending on the type of discharge, its energy, and the chemical composition of the surrounding environment, various types of plasmachemical reactions can be initiated and a number of primary and secondary species can be formed by electrical discharges in liquids and in gas–liquid environments (Chapter 6). Chemical reactions induced by electrical discharges in water are dependent on several

Plasma Chemistry and Catalysis in Gases and Liquids, First Edition.
Edited by Vasile I. Parvulescu, Monica Magureanu, and Petr Lukes.

Table 7.1 The Main Species Produced by Discharges in Water and Gas–Liquid Environments Discussed in This Chapter.

Parent species	Primary species	Secondary species
H_2O, O_2, N_2	$OH^\bullet$, $H^\bullet$, $O^\bullet$, $N^\bullet$, $NO^\bullet$ and ions	H_2O_2, NO_x, O_3, HNO_2, HNO_3, ONOOH, and ions

factors. For example, the aqueous solution composition can affect the reactions through the presence of electrolytes and radical quenchers. Conductivity affects the electrical discharge in water, leading to lower rates of formation of some active species but higher rates of formation of UV light. The nature of the liquid-phase electrode can also affect the reactions in the liquid. Although direct electrochemical reactions may not be important, ions and particles released into solution from the electrode can affect solution chemistry. Furthermore, in gas–liquid discharge environments, the nature of the gas phase (typically air, oxygen, nitrogen, or argon) will affect the formation of gas-phase species, which, in turn, will transfer into the liquid. The liquid may also evaporate and affect gas-phase reactions. It is important to note that electrical discharge reactions in liquid and gas–liquid systems can lead to postdischarge reactions that may be due to the longer-lived radicals and reactive species formed in the discharge. Therefore, the yield of a total plasmachemical process is due to synergistic contributions of numerous different elementary reactions taking place simultaneously in a discharge system. In this chapter, we focus mostly on the aqueous-phase plasmachemical reactions induced by discharges generated in water and in gas–liquid environments. Table 7.1 lists some primary and secondary species generated under these conditions.

The most abundant primary species are $OH^\bullet$, $H^\bullet$, $O^\bullet$, and $NO^\bullet$ radicals. These species react with other primary species or with the surrounding gas or liquid molecules to produce oxygen- and nitrogen-based secondary species (H_2O_2, NO_x, O_3, HNO_2, HNO_3, ONOOH), which may react with the target molecules. Thus, the chemical effects induced by the particular type of discharge depend on the chemical properties of the active species directly or indirectly formed in the discharge, that is, the primary and secondary species. In the following sections, the plasma chemical reactions induced in aqueous solutions by discharges directly in water and/or in the gas phase in contact with the liquid are classified into four main categories.

1) Acid–base reactions due to acidic effects caused in water by the chemical products formed by the plasma (e.g., nitrous and nitric acids produced in water by air discharges) (Section 7.2.1)
2) Oxidation reactions caused by oxidative effects of reactive oxygen species (ROS) and reactive nitrogen species (RNS) produced by the plasma (e.g., $OH^\bullet$ radical, ozone, hydrogen peroxide, and peroxynitrite) (Section 7.2.2)
3) Reduction reactions caused by reductive species produced by the plasma (e.g., $H^\bullet$, $HO_2\bullet$ radicals) (Section 7.2.3)

4) Photochemical reactions initiated by UV radiation from the plasma (e.g., photolysis of H_2O_2 and ozone, photocatalysis) (Section 7.2.4).

7.2.1 Acid–Base Reactions

A large number of organic reactions, including oxidation–reduction reactions, involve hydrogen ions in the reaction medium. Since the reduction potential of organic species is often pH dependent, the reaction rates may be controlled by the acidity of the aqueous solution. For example, the basic form of phenol (i.e., phenolate ion) is more reactive than the associated acidic form because the electron liability of the π bonds in the phenol aromatic ring is enhanced [1, 2]. The reactivity and color stability of many organic dyes is strongly pH dependent, and, for example, $OH^{\bullet}$ radical reactions with some dyes are much faster at low pH. Solution acidity is therefore a very important parameter in the plasmachemical degradation of organic compounds in water and can be important for the plasma-induced inactivation of microorganisms in water.

The various ROS and RNS produced by discharges in water and in gas–liquid environments possess acid–base properties since they are able to release hydrogen ions into aqueous solutions. Table 7.2 shows acid–base reactions and the corresponding dissociation equilibrium constants, pK_a, for the most relevant chemical species produced by discharge plasmas discussed in this chapter [3].

The formation of acids and their effects on various reactions have been reported in many studies where aqueous solutions were exposed to pulsed corona, dielectric barrier discharge (DBD), or gliding arc discharges above water surfaces, and some reports have demonstrated pH effects when the discharge was generated directly in water [3–20]. Most of this work reported pH changes and their effects on chemical reactions; rarely were pH indicators used. For discharges generated in gas–liquid environments, the change in acidity of an aqueous solution was typically found for electric discharges formed in air. In this case the pH decrease was largely attributed to the formation of nitrous and nitric acids in the plasma-treated solutions. A decrease in pH was, however, also reported in solutions treated by the discharge generated in a nitrogen-free atmosphere [15–19] and in the case when the discharge was generated directly in water [20]. Even in the cases where the nitrous and nitric

Table 7.2 Acid–Base Dissociation Constants pK_a for Selected Chemical Species Produced by Discharges in Water and Gas–Liquid Environments.

$RH \leftrightarrow R^- + H^+$	pK_a	$RH \leftrightarrow R^- + H^+$	pK_a
$OH^{\bullet} \leftrightarrow O^{-\bullet} + H^+$	11.9	$HOONO \leftrightarrow ONO_2^- + H+$	6.8
$H_2O_2 \leftrightarrow HO_2^{-\bullet} + H^+$	11.75	$HONO \leftrightarrow ONO^- + H^+$	3.3
$HO_2\bullet \leftrightarrow O_2^{-\bullet} + H^+$	4.8	$HNO \leftrightarrow NO^- + H^+$	4.7

acids were formed, a complete analysis has not been conducted to establish if all of the pH drop is only due to these acids. Therefore, the gas-phase composition should be carefully assessed, and the presence of any impurities (e.g., nitrogen or other species) that might cause change of pH should be either measured or eliminated.

Among the first studies giving evidence of acidic effects induced by air plasma in water was that reported by Brisset *et al.* [4, 5] who used DC point-to-plane corona discharge in air above the surface of an aqueous solution to study air plasma acid–base properties (with primary interest in corrosion induced by these plasmas). Preliminary experiments were performed on drops of aqueous solutions containing acid–base indicators as targets treated by the corona discharge. Evidence for the presence of solvated protons in the solution under exposure to the discharge was found and used to explain the increase in acidity. The pH changes were found to depend on the polarity of the applied voltage, whereby a negative discharge induced a larger decrease of the solution pH than a positive discharge, thus indicating a significant role of ions. By varying the atmosphere by using air or oxygen it was concluded that in addition to the excited nitrogen species and their products (NO_2^-, NO_3^-), singlet oxygen contributed to the acidification of the solution. This was proved by experiments performed with the addition of NaN_3 used as quencher of singlet oxygen. Goldman *et al.* [6] utilized a system with a grid electrode placed between the high-voltage needle and the water surface (the ground) and a suction tube placed adjacent to the needle electrode. Using the suction to remove neutral species or the grid to remove charged species, they found that the combination of ions and neutrals formed in the gas discharge was responsible for pH changes in the liquid solution. Thus, activated neutral species had an important role in the pH decrease.

The formation of nitrate and nitrite ions and the increasing acidity in the liquid phase was reported by other researchers using corona discharge or DBD generated above a water surface under an air atmosphere [7–9]. The acidic effect and the formation of NO_2^-/NO_3^- ions in water accompanied by an increase in solution conductivity was also demonstrated for gliding arc discharges burning over water surface [3, 10–13] and for the case when water was sprayed directly through the gliding arc plasma zone [14–18]. For example, a drop in pH of more than 8 units was observed in a NaOH solution that was exposed to a gliding arc discharge for several minutes [3].

A general mechanism to describe the changes of pH and conductivity caused by air plasma in water is proposed in the following discussion. In addition to some of the gas-phase reaction pathways shown in Chapter 6, liquid-phase reactions can also be important. The NO that is formed in an air plasma by the gas-phase reactions of dissociated nitrogen and oxygen rapidly reacts with oxygen (Eq. (7.1)) or ozone (Eq. (7.2)) to yield NO_2.

$$2NO + O_2 \longrightarrow 2NO_2 \quad (7.1)$$

$$NO + O_3 \longrightarrow NO_2 + O_2 \quad (7.2)$$

The nitrogen dioxide can subsequently dissolve in water, leading to nitrite NO_2^- and nitrate NO_3^- via an electron capture by nitrogen dioxide NO_2 (Eq. (7.3)) or oxidation by nitric oxide NO (Eq. (7.4)) [21].

$$NO_2(aq) + NO_2(aq) + H_2O \longrightarrow NO_2^- + NO_3^- + 2H^+ \tag{7.3}$$

$$NO(aq) + NO_2(aq) + H_2O \longrightarrow 2NO_2^- + 2H^+ \tag{7.4}$$

The pH of the solution is then reduced, which, along with an increase of the solution conductivity, favors the disproportionation of nitrites into nitrates and nitric oxide

$$3\,NO_2^- + 3\,H^+ \longrightarrow 2\,NO + NO_3^- + H_3O^+ \tag{7.5}$$

The formation of nitrate may also proceed via the liquid-phase reaction of NO_2 with $OH^\bullet$ radicals to form peroxynitrous acid or its conjugate base peroxynitrite in a neutral or basic medium (Eq. (7.6)) (Section 7.2.2.4). This unstable moiety then isomerizes to the nitrate anion (Eq. (7.7)).

$$OH + NO_2(aq) \longrightarrow [O{=}N{-}OOH] \longrightarrow O{=}N{-}OO^- + H^+ \tag{7.6}$$

$$O{=}N{-}OO^- \longrightarrow NO_3^- \tag{7.7}$$

In the presence of hydrogen peroxide, peroxynitrite can also form by the reaction of nitrite anion with H_2O_2 (Section 7.2.2.4). On the other hand, in the presence of ozone, nitrites are rapidly oxidized to nitrates and oxygen, thereby eliminating their content in water (Eq. (7.8)).

$$NO_2^- + O_3 \longrightarrow NO_3^- + O_2 \tag{7.8}$$

In addition to the measurements of nitrate and nitrites in the solutions exposed to various types of air plasma, attempts were made to correlate changes of pH with conductivity observed in plasma-treated solutions [15–17]. Although it was demonstrated that a large part of the conductivity increase can be attributed to the pH change, the role of nitrates and nitrites was, however, found to be small and other unidentified ions were proposed to contribute to the conductivity changes. It can be noted that conductivity and pH are closely connected and that the contributions of H^+ and OH^- to conductivity Λ can be given by

$$\Lambda(\mu S\ cm^{-1}) = (349.82[H^+] + 198.6[OH^-]) \times 1000 + \Lambda_0 \tag{7.9}$$

where $[H^+]$ and $[OH^-]$ are the concentrations of hydrogen and hydroxyl ions, respectively, multiplied by the value of specific conductance of these ions in equivalents per liter, and Λ_0 is the initial solution conductivity. The effects of other ions (e.g., nitrate, nitrite, etc.) can be added to the above equation accounting for their concentrations and specific conductance. For example, nitrate can be accounted for by adding the term $71.4\,[NO_3^-] \times 1000$ [17].

Figure 7.1 shows the results of the above model for two cases, with initial solution conductivity $20\,\mu S\ cm^{-1}$ (at initial pH 5) and $150\,\mu S\ cm^{-1}$ (at initial pH 6), and comparison with experimental data, which were obtained for different types of electrical discharges generated either directly in water or in the combined gas–liquid environment in the laboratory at Florida State University, Tallahassee, USA (FSU). This figure shows that as the pH drops, the conductivity increases because of the higher mobility of H^+ relative to OH^-. This model relationship is largely sufficient to describe experimental changes in pH and conductivity determined in the case of direct discharge in water, where no or only small changes in pH or conductivity were observed (white circles in Figure 7.1), and in the case of hybrid series reactor (open and filled squares and triangles in Figure 7.1), in which discharges are simultaneously generated in the gas and the liquid (Chapter 6). Figure 7.1 shows the results for pH and conductivity in hybrid series reactor for Ar, O_2, N_2, and Ar/O_2 carrier gases. For all these hybrid reactor cases, the initial conditions were 150 μS cm^{-1} (at pH 6) and the pH dropped with corresponding increases in conductivity following, in general, the relationship (Eq. (7.9)). For gliding arc reactors with various carrier gases and water droplet spray, the data fall either above or below the model predictions depending on the input power (open diamands, solid circles, and crosses in Figure 7.1). For relatively low power (<1 W), the data fall below the model results, and in the case with a high power (>300 W), the data fall above. Some of the effects of the data above the model line are due to

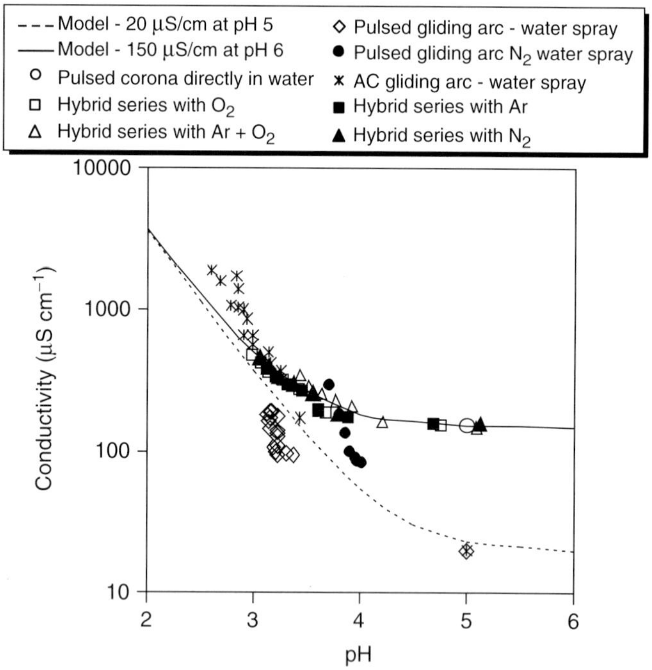

Figure 7.1 Correlation between conductivity and pH in the solutions treated by different types of electrical discharges, showing model results (lines) in comparison with experimental data (Source: Data from Refs. [15–17] and provided by Shih, K., FSU Tallahassee, USA.)

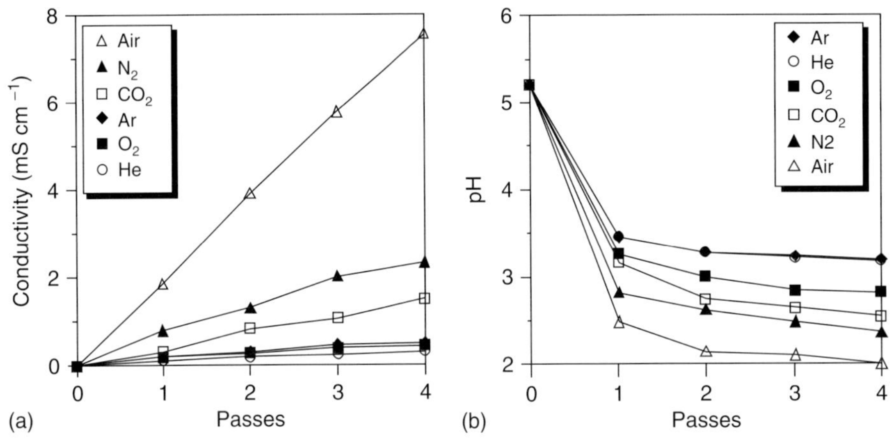

Figure 7.2 (a,b) Conductivity and pH changes per pass in the gliding arc discharge reactor with water spray for various carrier gases. (Source: © 2009 IEEE. Reprinted, with permission, from [17].)

the increase in conductivity due to nitrates; their inclusion into the model improved comparison in some cases [17] but not all cases [15].

Recently, Brisset *et al.* [3] suggested several reasons for the discrepancy between pH and conductivity measurements observed in the case of humid air gliding arc plasma burning over aqueous solution, which they attributed mainly to the peroxynitrite chemistry occurring in water; however, these are preliminary results (for more about peroxynitrite chemistry, see Section 7.2.2.4). Moreover, a decrease in pH was also reported in the solutions treated by the discharge generated in a nitrogen-free atmosphere [15–17]. O_2, Ar, He, or CO_2 when used as the working gas in gliding arc discharge with water spray were shown to lead to lower pH and higher conductivity in solutions exposed to plasma (Figure 7.2). The authors attributed these changes to H_3O^+ ions formed from the water because of the electronic and ionic bombardment since nitrogen oxides were not found in these cases.

A decrease of pH was also observed in the solutions treated by the discharge generated directly in water (i.e., without any gases added to the water other than those dissolved from the atmosphere) [20]. However, the change of pH in this case was much lower than in gas-phase discharges over water. Table 7.3 shows examples of pH changes in aqueous solutions of various electrolytes with different initial pH, which were treated by underwater pulsed corona discharge. Typically, the pH in the solutions of ionic salts of strong acids and bases (NaCl, Na_2SO_4, $NaClO_4$, $NaNO_3$) decreased after treatment by the discharge by about 0.8 units. Smaller changes of pH were observed in solutions of NaH_2PO_4 and Na_2CO_3. No change of pH was observed in the case of HCl solution, and only a small change was observed in the case of NaOH solution. The corresponding change of conductivity was about

Table 7.3 Change of pH of the Solution of Different Chemical Composition After 50 min Treatment with Underwater Corona Discharge (Solution Conductivity 100 μS cm^{-1}, Volume 1150 ml, Power Input 100 W, Applied Voltage 20 kV, Total Supplied Energy 300 kJ) [20].

solution	c [mmol/l]	pH_0	pH_{50}
HCl	0.35	3.2	3.2
NaCl	1.0	5.6	4.8
$NaClO_4$	1.0	5.6	4.7
Na_2SO_4	0.5	5.5	4.6
NaH_2PO_4	1.4	5.2	4.9
$NaNO_3$	1.0	5.6	4.6
Na_2CO_3	0.5	9.9	9.8

20-30 μS cm^{-1}. Some other work with similar pulsed corona discharges directly in the aqueous solution (Figure 7.1), indicates no change in pH or conductivity in underwater discharge (NaCl solution adjusted to 150 μS cm^{-1} initial conductivity), suggesting that further work is needed on this subject to identify discrepancies and differences.

The reasons for these pH changes are not fully understood. A decrease in pH might be partly accompanied by accumulation of hydrogen peroxide in water from the discharge. H_2O_2 is a weak acid, and depending on the concentration, its pH can be as low as 4.5. The final concentration of H_2O_2 in direct discharges in water was 2 mmol l^{-1} and this might be too low to cause an observed change in pH. The second possibility might be in the formation of hydrogen ions as a product of the reactions of primary and secondary species from underwater plasma. For example, perhydroxyl radical HO_2• dissociates at pH above 4 to superoxide ion $O_2^{\bullet -}$ (Eqs. 7.10–7.12).

$$OH^{\bullet} + H_2O_2 \longrightarrow HO_2^{\bullet} + H_2O \quad (7.10)$$

$$H^{\bullet} + O_2 \longrightarrow HO_2^{\bullet} \quad (7.11)$$

$$HO_2^{\bullet} \rightleftharpoons H^+ + O_2^{\bullet -} \quad (pK_a = 4.8) \quad (7.12)$$

Acid can also be formed by plasma-induced ionization of a water molecule (Eqs. (7.13) and (7.14)).

$$H_2O + e^- \longrightarrow H_2O^+ + 2e^- \quad (7.13)$$

$$H_2O^+ + H_2O \longrightarrow H_3O^+ + OH \quad (7.14)$$

The lack of pH change in the case of hydrochloric acid solution and only a small change in the case of NaOH solution might then be caused by a lower acidic strength of the produced hydrogen peroxide. Smaller changes of pH in the solutions of NaH_2PO_4 and Na_2CO_3 may be attributed to the buffer capacity of the $H_2PO_4^-/HPO_4^{2-}$ system (pK_a=7.21) and HCO_3^-/CO_3^{2-} system (pK_a=10.33), respectively.

7.2.2 Oxidation Reactions

The oxidizing properties of the chemically active species produced by discharges in water and in gas–liquid environments are most important for their applicability to the removal of organic compounds from water and for the inactivation of microorganisms. Table 7.4 shows the standard oxidation potentials E^0_{ox} for the most relevant chemical species produced by discharge plasmas concerned in this chapter, including ROS ($OH^\bullet$, $O^\bullet$ radicals, ozone, hydrogen peroxide and oxygen) and RNS (peroxynitrite, nitrite and nitrate ions, and nitrogen dioxide radical).

The oxidation potentials of these species are generally higher than the corresponding mean standard potentials of organic compounds, which are usually small ($E^0 < 0.5$ V per NHE). Therefore, the organic compounds are susceptible to strong oxidizers since the change in the standard state Gibbs free energy ΔG^0 of their reactions with oxidizing species is strongly negative, that is,

$$\Delta G^0 = -nF(E^0_{ox} - 0.5) \tag{7.15}$$

$$\Delta G = \Delta G^0 + RT \ln \prod (x_i)^{\upsilon i} \tag{7.16}$$

where n is the number of electrons transferred, F is the Faraday constant, x_i is the mole fraction of species i, and ν is the stoichiometric coefficient for the reaction. Thus, provided the concentrations of reactants and products lead to a negative Gibbs free energy ΔG by reaction (Eq. (7.16)), the oxidation of organic compounds by the discharge is thermodynamically favorable.

Among oxidation species discussed in this chapter, $OH^\bullet$ radicals and ozone are the key oxidants involved in most degradation processes induced by electrical discharges in water and gas–liquid environments. These species possess high

Table 7.4 Standard Oxidation Potentials E^0_{ox} for Selected Chemical Species Produced by Discharges in Water and Gas–Liquid Environments.

Ox + ne^- = Red	E^0_{ox} [V]	Ox + ne^- = Red	E^0_{ox} [V]
$OH^\bullet + H^+ + e^- = H_2O$	2.85	$O_3 + 6H^+ + 6e^- = 3H_2O$	1.51
$O^\bullet + 2H^+ + 2e^- = H_2O$	2.42	$HO_2 + H^+ + e^- = H_2O_2$	1.44
$ONOO^- + 2H^+ + e^- = NO_2 + H_2O$	2.41	$O_2 + 4H^+ + 4e^- = 2H_2O$	1.23
$O_3 + 2H^+ + 2e^- = O_2 + H_2O$	2.07	$HONO + H^+ + e^- = NO + H_2O$	1.00
$ONOOH + H^+ + e^- = NO_2 + H_2O$	2.02	$NO_3^- + 4H^+ + 3e^- = NO + 2H_2O$	0.96
$HO_2 + 3H^+ + 3e^- = 2H_2O$	1.70	$NO_3^- + 3H^+ + 2e^- = HONO + H_2O$	0.94
$H_2O_2 + 2H^+ + 2e^- = 2H_2O$	1.68	$NO_2\bullet + H^+ + e^- = HNO_2$	0.9

Table 7.5 Rate Constants for Reactions of $H^•$, $OH^•$ Radicals, and Ozone with Selected Inorganic and Organic Compounds in Water [l (mol s)$^{-1}$] [2, 22–25].

Compound	$H^•$	$OH^•$	O_3
H_2O_2	9×10^7	2.7×10^7	$<10^{-2}$
HO_2^-	1.2×10^9	7.5×10^9	5.5×10^6
O_2	2.1×10^{10}	-	-
O_3	3.8×10^{10}	1.1×10^8	-
H_2	-	4.2×10^7	-
OH^-	2.2×10^7	1.2×10^{10}	70
Fe^{2+}	7.5×10^6	4.3×10^8	$>5 \times 10^5$
Fe^{3+}	$<2 \times 10^6$	-	-
NO_2^-	7.1×10^8	9.1×10^9	3.7×10^5
NO_3^-	1.4×10^6	-	$<10^{-4}$
phenol	1.7×10^9	6.6×10^9	1.3×10^3
phenolate ion	-	9.6×10^9	1.4×10^9
4-chlorophenol	-	9.3×10^9	6×10^2
4-chlorophenolate ion	-	NR[a]	6×10^8
4-nitrophenol	-	3.8×10^9	<50
4-nitrophenolate ion	-	7.6×10^9	1.6×10^7
methanol	2.8×10^6	8.3×10^8	2×10^{-2}
dimethylsulfoxide	9.7×10^6	6.6×10^9	5.7
tetranitromethane	5.5×10^8	-	-
trichloroethylene	-	1.3×10^9	17
methylene blue	1.1×10^{10}	2.1×10^{10}	1.34×10^5[b]

[a] NR: rate constant not reported in the literature.
[b] Rate constant in $l^{0.5}/(mol^{0.5}\ s)$; at pH 7.

reactivity with a large number of organic compounds. Table 7.5 shows examples of the rate constants for $OH^•$ radical and ozone through direct reactions with selected organic and inorganic compounds in water.

7.2.2.1 **Hydroxyl Radical**

The hydroxyl radical is the major ROS found in electric discharge plasmas in water and gas–liquid environments. The $OH^•$ radical has a high oxidizing power (E^0= 2.85 V), and it is the strongest oxidant that can exist in an aqueous environment. It reacts unselectively with most organic and many inorganic compounds with rates that approach diffusion-controlled limits (Table 7.5). Its high reactivity causes a very short life time of close to 200 μs in the gas phase and less in aqueous solution, so that it may directly react with a target species only in its immediate surrounding, which is of special concern for discharges in water.

For the reactions of $OH^•$ radical with organic species, there are three common reaction pathways: (i) hydrogen abstraction (typically from alkyl or hydroxyl groups) (Eq. (7.17)), (ii) hydroxyl radical electrophilic addition to unsaturated systems (e.g., double bonds) (Eq. (7.18)), and (iii) direct electron transfer (Eq. (7.19)) [26].

$$OH^{\bullet} + R - H \longrightarrow R^{\bullet} + H_2O \quad (7.17)$$

$$OH^{\bullet} + R_2C{=}CR_2 \longrightarrow R_2^{\bullet}(OH)C{-}CR_2 \quad (7.18)$$

$$OH^{\bullet} + R - X \longrightarrow R - X^{+\bullet} + OH^{-} \quad (7.19)$$

Hydroxyl radicals react primarily by hydrogen abstraction with saturated aliphatic hydrocarbons and alcohols to yield H_2O and an organic radical $R^{\bullet}$ (Eq. (7.17)). On the other hand, in the case of olefins or aromatic hydrocarbons, $OH^{\bullet}$ radicals add to the C=C bonds of organic compounds to form a C-centered radical with a hydroxyl group at the –C atom (Eq. (7.18)). Reduction of hydroxyl radicals to hydroxide anions by an organic substrate (Eq. (7.19)) is of particular interest in the case where hydrogen abstraction or electrophilic addition reactions may be unfavorable due to multiple halogen substitutions or steric hindrances.

Organic radicals formed by such reactions may quickly add an oxygen molecule (if present in water) to form reactive organoperoxyl radicals ($ROO^{\bullet}$) that may become transformed into oxy radicals. Further reactions then often lead to an abstraction of an H atom by dissolved oxygen and the formation of hydroperoxyl radicals, hydrogen peroxide, and organic peroxides, aldehydes and acids. During the process, different types of radical species are simultaneously present. Typical reaction products include oxygenated and ring-opened species. In some cases, complete mineralization yields carbon dioxide as a final product (for more details see Section 7.3.1).

7.2.2.2 **Ozone**

Ozone is a powerful oxidant, and it has the highest standard redox potential among conventional oxidants ($E^0 = 2.07$ V) such as chlorine, chlorine dioxide, permaganate, and hydrogen peroxide. Although ozone is formed by the gas-phase discharge, when a discharge is generated in close proximity to the water surface, ozone can transfer from the gas phase into the liquid and subsequently oxidize organic compounds in the water (in addition to the case when ozone is formed directly in the liquid, e.g., by the discharge generated in oxygen bubbles). Ozone can oxidize organic matter in water either directly or through the hydroxyl radicals produced during the decomposition of ozone (Eq. 7.20–7.22). The direct oxidation with molecular ozone is of primary importance under acidic conditions; however, this pathway is relatively slow compared to the $OH^{\bullet}$ radical oxidation since ozone reacts directly very selectively and is pH dependent (Table 7.5) [27]. Ozone reacts through direct reactions with organic compounds with specific functional groups in their molecules. Examples are unsaturated and aromatic hydrocarbons with substituents such as hydroxyl, methyl, or amine groups. Ozone can react through 1,3-dipolar cycloaddition, electrophilic substitution and, rarely, nucleophilic reactions [26]. In water, only the first two reactions have been identified for most organics, and of these two, 1,3-dipolar cycloaddition is the most important (e.g., during plasma-induced degradation of aromatic hydrocarbons in water, Section 7.3.1.1).

Another group of ozone direct reactions are those with inorganic species such as Fe^{2+}, NO_2^-, OH^-, and HO_2^- (see Table 7.5) [23].

In neutral and basic solutions, ozone is unstable and decomposes via a series of chain reactions to produce hydroxyl radicals (Eq. (7.20)) [28].

$$O_3 + H_2O \xrightarrow{OH^-} {}^{\bullet}OH + O_2 + HO_2\bullet \tag{7.20}$$

The addition of hydrogen peroxide accelerates the decomposition of ozone and increases the hydroxyl radical concentration. The chemistry involved in the generation of $OH^{\bullet}$ radicals by the O_3/H_2O_2 process is assumed to be mainly the decomposition of ozone by the conjugate base of hydrogen peroxide (HO_2^-) to produce ozonate radical ($O_3^{\bullet-}$), which gives the hydroxyl radical by rapid reaction with water (Eq. (7.21) and (7.22)) [29].

$$O_3 + HO_2^- \longrightarrow HO_2^{\bullet} + O_3^{\bullet-} \tag{7.21}$$

$$O_3^{\bullet-} + H_2O \longrightarrow OH^{\bullet} + OH^- + O_2 \tag{7.22}$$

This process, also referred to as the *Peroxone process*, occurs very slowly at low pH, but at pH values above 5, it is greatly accelerated. The process is further enhanced by the photochemical generation of hydroxyl radicals in the $O_3/H_2O_2/UV$ process (Section 7.2.4). The presence of activated carbon (AC), silica gel, or zeolites can also stimulate the production of $OH^{\bullet}$ radicals from ozone (Sections 7.4.5–7.4.7).

7.2.2.3 **Hydrogen Peroxide**

Hydrogen peroxide is a very important species, which increases the collective oxidizing power of the plasma, especially in the case of underwater plasmas, in which hydrogen peroxide is the most abundant long-lived plasmachemical product. H_2O_2 reacts in the gas phase, at the liquid surface, and in the bulk target solution with solutes since it is highly soluble in water. Similar to ozone, hydrogen peroxide can react with organic matter through direct and indirect pathways. In direct mechanisms, hydrogen peroxide participates in redox reactions where it can behave as an oxidant ($E^0 = 1.77$ V) or as a reductant ($E^0 = -0.7$ V).

Although H_2O_2 has a relatively high oxidation potential, the direct oxidation effect of H_2O_2 in plasma-treated solutions is low since hydrogen peroxide does not significantly react with most organic compounds, at least at appreciable rates for water treatment. Nevertheless, H_2O_2 affects plasmachemistry in water very significantly and indirectly in a number of processes including (i) catalytic reactions with iron in a Fenton-type process via formation of $OH^{\bullet}$ radicals (Section 7.4.1) and with tungsten via formation of peroxotungstates (Section 7.4.3), (ii) with ozone in Peroxone process via formation of $OH^{\bullet}$ radicals (Eqs. (7.21) and (7.22)), (iii) with nitrites via formation of peroxynitrites (Section 7.2.2.4), or (iv) when it is photolyzed with UV radiation from plasma via formation of $OH^{\bullet}$ and $HO_2\bullet$ radicals (Section 7.2.4). These processes are described separately in the following sections of this chapter.

7.2.2.4 Peroxynitrite

It was described in Section 7.2.1 that electrical discharges generated in gas–liquid environments (especially those generated under an air atmosphere) generate highly acidic conditions in water, which are largely attributed to the formation of nitrous and nitric acids. In addition, the oxidation properties of nitrogen-containing reactive species, especially peroxynitrite, must be considered in aqueous solutions that are exposed to the electric discharge in air.

Peroxynitrite is produced by (i) the reaction of nitric oxide and superoxide anion radicals (Eq. (7.23)), (ii) the reaction of the nitrite anion with hydrogen peroxide (Eq. (7.24)), or (iii) the reaction of $NO_2\bullet$ with an $OH^\bullet$ radical (Eq. (7.25)) [30–33].

$$O_2^{-\bullet} + NO^\bullet \longrightarrow O{=}N{-}OO^- \quad (7.23)$$

$$NO_2^- + H_2O_2 \longrightarrow O{=}N{-}OO^- + H_2O \quad (7.24)$$

$$OH^\bullet + NO_2^\bullet \longrightarrow O{=}N{-}OO^- + H^+ \quad (7.25)$$

Among these three pathways, the second, utilizing hydrogen peroxide (Eq. (7.24)) is of particular importance in gas–liquid plasmas since a significant amount of H_2O_2 is produced by discharges in contact with water or in humid atmospheres.

The oxidant reactivity of peroxynitrite (E^0= 2.05 V) is highly pH dependent and both anionic ($O{=}N{-}OO^-$, peroxynitrite anion) and protonated forms ($O{=}N{-}OOH$, peroxynitrous acid) (pK_a=6.8) can participate in oxidation reactions. Peroxynitrite can react with organic compounds either directly via one- and two-electron oxidation reactions or indirectly through reactions of secondary radicals (nitrogen dioxide ($NO_2\bullet$), hydroxyl ($OH^\bullet$), or carbonate anion radical ($CO_3^{\bullet-}$), which are formed by H^+ or CO_2-catalyzed decomposition of peroxynitrite [30–33].

Direct reactions of peroxynitrite are more likely to occur just under alkaline conditions (pH>6.8). In acidic solutions, the protonated form of peroxynitrite predominates (Eq. (7.26)) and it decays into $OH^\bullet$ and $NO_2^\bullet$ radicals (Eq. (7.27)), which subsequently initiate indirect reactions.

$$O{=}N{-}OO^- + H^+ \longrightarrow O{=}N{-}OOH \quad (7.26)$$

$$O{=}N{-}OOH \longrightarrow OH^\bullet + NO_2^\bullet \quad (7.27)$$

$$OH^\bullet + NO_2^\bullet \longrightarrow NO_3^- + H^+ \quad (7.28)$$

$OH^\bullet$ and $NO_2^\bullet$ radicals can recombine to form nitric acid (Eq. (7.28)), however, the rate constant for this reaction in solution (>3×10^4 l (mol s)$^{-1}$; [30]) is much lower than that of most reactions involving the $OH^\bullet$ radical (Table 7.5). Thus, under acidic conditions, the reactivity of peroxynitrite is mainly determined by $OH^\bullet$ radicals formed by its decomposition via reaction (Eq. (7.27)) (which is the most likely scenario in the case of aqueous solutions exposed to air discharge plasmas, i.e. considering their acidic effects). The nitrogen dioxide radical is, however, also a highly reactive and potent oxidant (E^0= 0.9 V), capable of initiating, for example, fatty acid oxidation or the nitrosation of aromatic compounds (Section 7.3.1.1).

Therefore, peroxynitrite chemistry may have a significant role in the oxidation–degradation processes of pollutant abatement in water induced by air discharge plasma and even more importantly, in plasma interactions with living matter in water exposed to these discharges. For example, peroxynitrite reacts as a peroxidizing agent with the lipids present in the membranes of living cells and, thus, it can cause oxidative stress, for example, of bacteria in water. These effects are described in detail in Chapter 8.

7.2.3
Reduction Reactions

The presence of reductive species in a highly oxidizing environment offers the possibility that in a mixture of aqueous contaminants some pollutants degrade by reductive mechanisms, thereby increasing the degradation efficiency of the process. For example, the reductive species play an important role in the degradation of compounds with high electron affinity, primarily chlorinated compounds, and compounds containing nitrogroups [34, 35]. The reductive species were suggested to participate in degradation of aqueous phase trichloroethylene and polychlorinated biphenyls using pulsed corona discharge in water [36–38].

The likely reductive species produced from the discharge in water are hydrogen and superoxide $O_2^{\bullet -}$ radicals and molecular hydrogen gas. The production of the $H^{\bullet}$ radical in discharge plasma in water has been demonstrated using emission spectroscopy [39–42]. The production of hydrogen gas H_2 has also been quantified [43]. Production of superoxide radical by the discharge in water has been demonstrated using the chemical probes tetranitromethane (TNM) and nitroblue tetrazolium chloride by Sahni *et al.* [44]. They proposed the superoxide radical as the main reductive species in the underwater discharge plasma considering the fact that, in the oxygenated environment of the discharges in water, $H^{\bullet}$ radical reacts with oxygen to form peroxyl radical $HO_2^{\bullet}$, which is converted to $O_2^{\bullet -}$ (Eqs. (7.9) and (7.10)). Perhydroxyl radicals can be also produced photochemically during the photolysis of hydrogen peroxide by UV radiation from plasma in water [45] (Eqs. (7.38), and (7.10)). Another issue in discharges in water is the role of the aqueous or hydrated electrons. Joshi *et al.* [46] proposed in analogy with the radiation chemistry of water that aqueous electrons might be formed in direct discharge in water. However, experimental measurements with the addition of N_2O into the liquid to generate additional hydroxyl radicals failed to prove this hypothesis. One recent study has suggested by experimentally measured reduction of silver ions in solution that aqueous electrons can be formed in negative polarity discharges over water surfaces [47]; positive polarity discharge in the same system did not result in silver ion reduction (Chapter 6).

7.2.3.1 Hydrogen Radical

The $H^{\bullet}$ radical is a powerful reducing agent ($E^0 = -2.3$ V). The $H^{\bullet}$ radical undergoes two general types of reactions with organic compounds: (i) hydrogen addition and (ii) hydrogen abstraction. With saturated organic compounds, hydrogen atoms usually abstract hydrogen to give molecular hydrogen and an organic radical (Eq. (7.29)), while addition reactions occur with unsaturated and aromatic compounds

(Eq. (7.30)) [22, 48]

$$H^{\bullet} + R - H \longrightarrow R^{\bullet} + H_2 \quad (7.29)$$

$$H^{\bullet} + R_2C{=}CR_2 \longrightarrow R_2^{\bullet}(H)C - CR_2 \quad (7.30)$$

In this respect, the $H^{\bullet}$ radical resembles the hydroxyl radical (Eqs. (7.17) and (7.18)), although the hydroxyl radical is more reactive and less selective in abstraction reactions.

7.2.3.2 **Perhydroxyl/Superoxide Radical**

The perhydroxyl radical ($HO_2^{\bullet}$) and its conjugate base, the superoxide radical anion ($O_2^{\bullet-}$), act as oxidizing or reducing agents depending on the solute. The two forms rapidly equilibrate by the $HO_2^{\bullet}/O_2^{\bullet-}$ reaction (Eq. (7.10)). The pK_a of $HO_2^{\bullet}$ (4.8) is such that $HO_2^{\bullet}$ is the predominant form below about pH 4.5, while $O_2^{\bullet-}$ predominates above about pH 5. Both forms of the radical are inert toward organic compounds, unless they contain a relatively weakly bonded hydrogen atom so that hydrogen abstraction can occur (e.g., with hydroquinone). Electron transfer reactions can take place, for example, with TNM (Eq. (7.31)) [26, 48, 49].

$$C(NO_2)_4 + O_2^{\bullet-} \longrightarrow C(NO_2)_3^- + NO_2^{\bullet} + O_2 \quad (7.31)$$

$HO_2^{\bullet}$ can also react with organic peroxyradicals ($RO_2^{\bullet}$) giving organic peroxides and hydroperoxides

$$RO_2^{\bullet} + HO_2^{\bullet} \longrightarrow RO_2H + O_2 \quad (7.32)$$

7.2.4 Photochemical Reactions

A nonnegligible part of the energy in electrical discharge plasmas in water and gas–liquid environments is converted to UV light, which may induce photochemical reactions. Similar to ozonation and hydrogen peroxide oxidation, UV radiation may act on the organic compounds present in water in two different ways: direct photolysis or indirect photolysis (e.g., oxidation by $OH^{\bullet}$ radical formed from photolytic decomposition of ozone, H_2O_2 or TiO_2 photocatalysis).

Direct photooxidation of organic compounds with UV light on electronic excitation of the organic substrate (Eq. (7.33)) implies in most cases an electron transfer from the excited state (C^*) to the ground state molecular oxygen (Eq. (7.34)), with subsequent recombination of the radical ions or the hydrolysis of the radical cation or the homolysis of a carbon–halogen bond (Eq. (7.35)) to form radicals, which then react with oxygen [50].

$$R + h\nu \longrightarrow R^* \quad (7.33)$$

$$R^* + O_2 \longrightarrow R^{\bullet+} + O_2^{\bullet-} \quad (7.34)$$

$$R - X + h\nu \longrightarrow R^{\bullet} + X^{\bullet} \quad (7.35)$$

The use of UV radiation for the direct photooxidation of organic compounds in aqueous solutions is very limited since target organics must efficiently absorb the light required for photodissociation in competition with other absorbers, especially with water that absorbs significantly in the vacuum UV region. However, UV photolysis of pollutants may be important in cases where hydroxyl radical reactions are known to be slow, for example, highly fluorinated or chlorinated saturated aliphatic compounds may be efficiently eliminated on primary homolysis of a carbon–halogen bond. The corresponding domains of excitation are <190 nm for C–F and 210–230 nm for C–Cl bonds [50, 51].

The photolysis of ozone in the 200–280 nm region involves the light-induced homolysis of ozone and the subsequent production of $OH^\bullet$ radicals by the reaction of excited oxygen $O(^1D)$ with water (Eqs. (7.36) and (7.37)) [52].

$$O_3 + h\nu \longrightarrow O_2 + O(^1D) \tag{7.36}$$

$$O(^1D) + H_2O \longrightarrow 2\,OH^\bullet \tag{7.37}$$

The primary process of H_2O_2 photolysis in the 200–300 nm region is the dissociation of H_2O_2 to hydroxyl radicals with a quantum yield of two $OH^\bullet$ radicals formed per quantum of radiation absorbed (Eq. (7.38)) [26, 53, 54].

$$H_2O_2 + h\nu \longrightarrow 2\,OH^\bullet \tag{7.38}$$

The $OH^\bullet$ radicals thus formed enter a radical chain mechanism, in which the propagation cycle gives a high quantum yield of the photolysis of H_2O_2 (Eqs. (7.10) and (7.39)).

$$OH^\bullet + H_2O_2 \longrightarrow H_2O + HO_2^\bullet \tag{7.10}$$

$$HO_2^\bullet + H_2O_2 \longrightarrow O_2 + H_2O + OH^\bullet \tag{7.39}$$

This process has been demonstrated to occur in underwater plasma, especially in solutions of more conductive electrolytes [45].

Nitrites and nitrates can also act as indirect photosensitizers to produce secondary oxidants such as the oxide atom and superoxide and hydroxyl radicals via the following mechanism [26]:

$$NO_2^- + h\nu \longrightarrow NO^\bullet + O^\bullet - \tag{7.40}$$

$$NO_3^- + h\nu \longrightarrow NO_2^\bullet + O^{\bullet -} \tag{7.41}$$

$$NO_3^- + h\nu \longrightarrow NO_2^- + O \tag{7.42}$$

$$O + H_2O \longrightarrow 2\,OH^\bullet \tag{7.43}$$

$$O^{\bullet -} + H_2O \longrightarrow OH^\bullet + OH^- \tag{7.44}$$

$$2\,NO_2^\bullet + H_2O_2 \longrightarrow NO_2^- + NO_3^- + 2H^+ \tag{7.45}$$

Another possible photochemical reaction is the heterogeneous process with photocatalysts, for example, titanium dioxide (Section 7.4.4).

7.3
Plasmachemical Decontamination of Water

Electrical discharges in water and gas–liquid environments were successfully employed to eliminate a variety of organic compounds from water. For example, organic dyes are often chosen to demonstrate the water decontamination activity of newly developed or modified plasma reactors because these dyes are particularly vulnerable to attack by reactive species, there is negligible competition by the dye intermediates, and the primary dye concentration is easily evaluated by UV absorption spectroscopy. Phenols are also often used, especially in kinetic studies, to evaluate the oxidative effects of $OH^{\bullet}$ radicals and ozone caused by electrical discharges, because the radical oxidation chemistry is well known and the reaction products are relatively easily determined. The degradation of a number of other compounds, including highly persistent and toxic chemicals, has been performed in order to evaluate their degradability by the various types of discharges under different conditions. It is not the objective of this chapter to list all compounds treated by discharges in water and gas–liquid environments or to evaluate each of these methods in terms of the energy efficiency for removal in different types of discharge reactors. For such analysis, we refer the reader to two recent reviews [55, 56], which used organic dyes or hydrogen peroxide to assess the energy efficiency of different types of electrical discharge reactors used for water treatment. Several other papers compared the efficiency of different discharge reactors using phenol as a probe organic compound [57–60].

In this chapter, we are mainly concerned with the mechanisms of plasmachemical degradation processes occurring in aqueous solutions and the role of primary and secondary chemical species produced by different types of discharges in these processes. In general, both physical and chemical factors may be important in promoting desirable decontamination effects. Organic compounds either can be directly degraded by the discharge (e.g., pyrolysis reactions, photolysis reactions, direct electron impact collisions) or can be degraded indirectly through reactions with one or more of the primary and secondary molecular, ionic, or radical species produced by the discharges. The relative importance of these direct and indirect mechanisms is strongly dependent on the intensity of the energy input to the system as well as on the composition and concentrations of the reacting environment and the chemical structure of the treated compound. In the following sections, three categories of organic compounds have been chosen, aromatic hydrocarbons (especially various types of phenols, Section 7.3.1), organic dyes (Section 7.3.2), and several aliphatic hydrocarbons (Section 7.3.3), to present typical plasmachemical degradation mechanisms induced in aqueous solutions by discharges directly in water and in the gas phase in contact with the liquid.

7.3.1
Aromatic Hydrocarbons

Aromatic hydrocarbons include a large body of organic compounds that have been treated by various types of electrical discharge plasmas in water and gas–liquid environments. The discussion that follows examines in detail the plasmachemical degradation of phenol and several substituted phenols as models of more complicated aromatic compounds (Sections 7.3.1.1 and 7.3.1.2). The degradation of heterocyclic and polycyclic aromatic hydrocarbons is discussed in Section 7.3.1.3, and organic dyes are described separately in Section 7.3.2.

7.3.1.1 Phenol

Phenol degradation has been extensively studied in many plasma systems utilizing discharges generated directly in water or in gas–liquid environments [60–83], in the presence of various additives in the water (iron, titanium oxide, AC, zeolites, see Section 7.4) [81–103], and in the presence of various gases either bubbled through water or used as the working atmosphere of the discharge in the gas/liquid environment (oxygen, argon, air) [60–72]. In most of these studies, the oxidation of phenol was achieved through reaction with hydroxyl radicals, although phenol can also react at a slower rate by direct reactions with ozone (both pathways give catechol, hydroquinone, and 1, 4-benzoquinone as the primary hydroxylated aromatic products of phenol) or with peroxynitrite, giving, in addition to hydroxylated products, nitrated products. The degradation of phenol obeyed first-order kinetics (typical for electrophillic oxidations).

The mechanism of phenol **1** oxidation by $OH^{\bullet}$ radical is schematically shown in Figure 7.3. The $OH^{\bullet}$ radicals electrophillically attack phenol on the *ortho*- and *para*-positions relative to the phenolic OH group to yield a dihydroxycyclohexadienyl radical **2** ($DCHD^{\bullet}$). Depending on the pH and availability of the suitable oxidants or reductants, which both strongly influence further oxidation pathways, $DCHD^{\bullet}$ may further react by formation of a dioxygen radical adduct **3** (Figure 7.3, path A) or by oxidation to a cyclohexadienyl cation **4** (Figure 7.3, path D, OX=for example, quinone intermediates of phenol) [104–106].

The decay of such transients then leads to the formation of hydroxylated products of phenol such as catechol **6**, hydroquinone **7**, and 1, 4-benzoquinone **8** (Figure 7.3, path C, E). These products might be further oxidized to trihydroxybenzenes such as pyrogallol, hydroxyhydroquinone, and phloroglucinol [104]. At the same time, α, α′-endoperoxyalkyl radicals can be formed from $DCHD^{\bullet}$ radical. These radicals may scavenge another oxygen molecule to form endoperoxyalkylperoxyl radicals **5**, which through dimerization and subsequent decomposition into two endoperoxides (Figure 7.3, path B), result in the ring cleavage products of saturated and unsaturated aliphatic C1–C6 hydrocarbons with carboxyl-, aldehyde, ketone, or alkanol functional groups. Among these products, organic acids such as maleic, oxalic, and formic acids were typically reported in phenol reaction mixtures on treatment by the discharge generated directly in water or by gas-liquid phase discharge generated in an oxygen-free atmosphere.

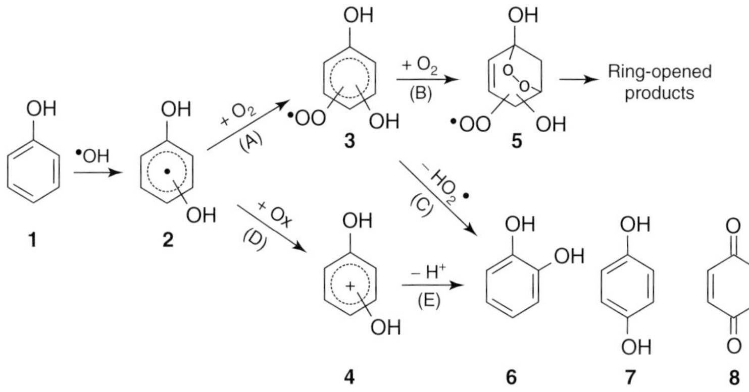

Figure 7.3 Mechanism of $OH^{\bullet}$ radical attack on phenol ring.

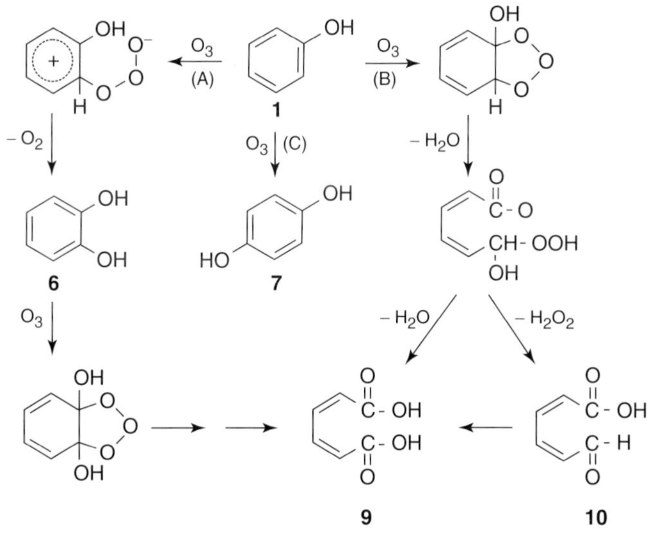

Figure 7.4 Mechanism of ozone radical attack on phenol ring.

Electrophilic attack by an ozone molecule on the aromatic ring of phenol proceeds preferentially on the *ortho-* and *para-* positions relative to the phenolic OH group to yield the same hydroxylated products as in the case of $OH^{\bullet}$ radical attack (Figure 7.4, path A, C). However, ozone may also react through a 1,3-dipolar cycloaddition mechanism, which causes direct ring cleavage of the aromatic ring (Figure 7.4, path B). Such a mechanism leads, in the first stage, to the formation of *cis,cis*-muconic acid **9** and/or muconaldehyde **10**, which then undergo further ozonation [107–109].

The effect of ozone (either bubbled through water or formed in an oxygen atmosphere of a gas-liquid-phase discharge) on the efficiency of plasmachemical degradation of phenol in water has been reported in a number of studies [60–70].

For example, in gas-phase pulsed corona discharge over water, it was determined that phenol removal was most rapid when the gas-phase atmosphere above the water surface was oxygen rather than argon and that hydroxyl radical reactions as well as ozone reactions lead to the degradation processes. In addition, different phenol degradation products were reported depending on the chemical composition of the atmosphere (argon or oxygen). Formation of *cis,cis*-muconic acid (along with its more stable cis, trans isomer) was reported by Lukes and Locke [67], who used combined gas-liquid-phase discharge generated simultaneously in water and above the water surface under an oxygen atmosphere. Since these products cannot be formed by $OH^{\bullet}$ radical attack, formation of *cis,cis*-muconic acid was clear evidence of the oxidation mechanism by ozone through the direct ring cleavage of the aromatic ring of phenol. The authors further proved the mechanism of phenol oxidation by ozone by the pH-dependent degradation of phenol observed under an oxygen atmosphere. The rate of phenol degradation increased with increasing pH and rose more than 3.5 times in the alkaline solution compared with the acidic case. On the other hand, only a negligible effect of pH on phenol removal was observed for the argon atmosphere. The reactivity of ozone with dissociating organic compounds considerably increases with the degree of their dissociation. For phenol ($pK_a = 9.89$), the rate constant of phenolate ion with ozone is as much as six orders of magnitude higher than that with phenol (Table 7.5) [2]. Therefore, the larger removal of phenol with increasing solution pH corresponds to a higher degree of phenol deprotonation and more efficient consumption of ozone by phenol in its phenolate form. With increasing pH, however, ozone may also react with phenol through indirect reactions of $OH^{\bullet}$ radicals resulting from the decomposition of ozone by hydroxide ions and hydrogen peroxide (via Peroxone process through reactions (Eqs. (7.21) and (7.22), see Section 7.2.2).

In humid air plasmas, such as a gliding arc discharge, 4-nitrosophenol **11** and 4-nitrophenol **13** were detected as by-products of phenol reaction, in addition to the hydroxylated products (catechol, hydroquinone, and 1,4-benzoquinone), on spraying of phenol solution through the plasma zone of the discharge [72, 88]. In this case, the acidic and oxidizing effects of the plasma (see Section 7.2) simultaneously controlled the phenol degradation. Regarding hydroxylated products of phenol, $OH^{\bullet}$ radicals were the major oxidizing agent formed either directly by the discharge or also through peroxynitrous acid decomposition (Eq. (7.27)) (Section 7.2.2.4). The occurrence of ozone in water and its contribution to the oxidation of phenol by air–liquid gliding arc discharge was negligible. The reason might be ozone elimination in the plasma zone of the discharge by NO and NO_2 (Chapter 6) or post-discharge reaction of ozone with nitrites in the liquid (Eq. (7.8)).

Nitrated products of phenol might result from (i) attack of $NO_2^{\bullet}$ or $NO^{\bullet}$ radicals formed in the plasma zone of the gliding discharge (Figure 7.5, path A), (ii) the postdischarge nitration occurring in the acidic solution by nitrous and nitric acid (Figure 7.5, path B), or (iii) attack by $NO_2^{\bullet}$ radical formed in the acidic solution by the decomposition of peroxynitrous acid in reaction (Eq. (7.27)) (Section 7.2.2.4) (Figure 7.5, path A). Nitration of phenol with these species occurs at *ortho-* and *para-* positions with respect to the OH group, giving 2- and 4-nitrophenol **13**,

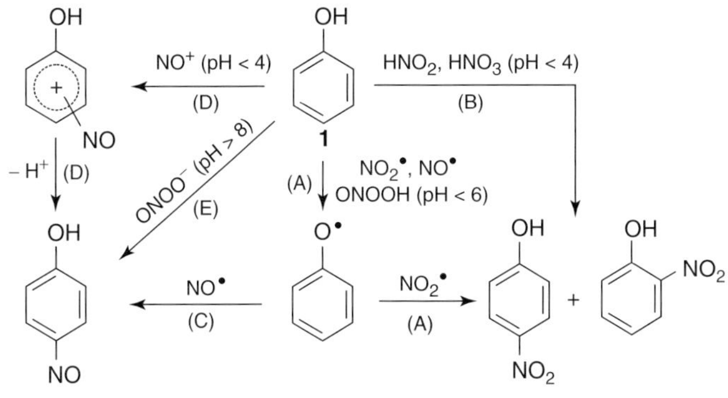

Figure 7.5 Mechanism of nitration and nitrosation of phenol.

14 (through the formation of a phenoxy radical intermediate **12** in the case of $NO_2^•$ and $NO^•$) [110–113]. Nitrosation of phenol takes place at the same positions giving preferentially 4-nitrosophenol **11**. Possible nitrosating agents include $NO^•$ radical (Figure 7.5, path C) and, under acidic conditions, also nitrosonium NO^+ ion formed by the protonation of nitrous acid (Figure 7.5, path D) [112, 113]. Under alkaline conditions, nitrosophenols might also be formed through the direct attack of phenol by peroxynitrite anion (Figure 7.5, path E) [113–115]. Uppu *et al.* [115] reported 4-nitrosophenol as the dominant product of the reaction of phenol with peroxynitrite at $pH > 8$, suggesting that in this pH range, peroxynitrite promotes nitrosation of phenolate ion through nucleophilic reaction.

In addition to the chemical mechanism of phenol degradation, in some discharges in water with bubbles where the temperature can be high, oxidative decomposition by radicals can be supplemented by thermal decomposition. For example, thermal pyrolysis in plasma generated in bubbles and $OH^•$ radical reactions in the liquid near the interface were assessed for phenol degradation [80]. It was found that for phenol concentrations below 1 mmol l^{-1}, 75%–80% of the degradation occurs in the liquid phase by radical reactions, while at 10 mmol l^{-1}, there is about 50% degradation in the liquid by radical reactions and the rest by thermal pyrolysis, and as concentration increases polymerization of phenol occurs and liquid-phase reactions dominate again.

7.3.1.2 **Substituted Aromatic Hydrocarbons**

Amongst various substituted aromatic hydrocarbons, the degradation of chlorinated and nitrated phenols was studied in many plasma systems utilizing discharges generated directly in water or in gas–liquid environments [116–131]. Most studies were conducted with 4-chlorophenol **17** [118–128]. In addition to phenols, plasmachemical degradation of aniline, benzoic acid, terephtalic acid, cresol, xylene, or acetophenone were examined [132–139]. The degradation of more substituted aromatic hydrocarbons including pentachlorophenol [140, 141] and 2,4,6-trinitrotoluene [119] were reported using corona or glow discharge above water and pulsed arc discharge in water, respectively. The general mechanisms for the degradation of these compounds were similar to that of phenol (Section 7.3.1.1), that is, initiated by $OH^{\bullet}$ radicals and/or by ozone attack, giving hydroxylated products. Several studies also reported the formation of nitrated products using combined gas–liquid discharge above water with air or nitrogen bubbling [120–122].

The degree of reactivity of the substituted aromatic compounds depends on the type of substituents present on the aromatic ring, which alter the electron density of the aromatic ring because of their resonance and inductive effects (such as hydroxyl, amine, methyl, chlorine, or nitro groups). For example, chlorine, as well as the nitro, functional groups are electron-withdrawing. They extract electron density from the aromatic ring and thus primarily reduce the reactivity of the aromatic ring. On the other hand, the hydroxyl, amine, or alkyl groups are electron-donating, which increase the aromatic electron density and renders the aromatic ring generally more susceptible toward oxidation. The number and position of the substituents on the aromatic ring also play an important role in their reactivity. In monosubstituted aromatic rings, the –OH, –Cl, $–NH_2$, or $–CH_3$ substituents are ortho/para directors of the electrophilic substitution, whereas $–NO_2$ or –C=O substituents are meta directors. Further, $OH^{\bullet}$ attack on bisubstituted aromatic rings is directed to the *ortho/para*- positions usually with respect to the –OH and $–NH_2$ groups due to the stronger directory effect of these groups than that of the other groups.

As a result, there can be a number of reaction pathways during degradation of substituted aromatic compounds, which yield a wide range of products depending on the type of the discharge and the reaction conditions. In the following paragraphs we discuss only several examples.

Detailed studies of plasmachemical degradation of different substituted aromatic compounds in aqueous solution were performed by Tezuka and Iwasaki [131–133] using contact glow discharge electrolysis (e.g., monochlorophenols, aniline, benzoic acid). The authors determined that the $OH^{\bullet}$ radical attack is the main mechanism for degradation of these compounds based on the identified hydroxylated products. In the case of monochlorophenols, the products were chlorohydroquinone, 3- and 4-chlorocatechol, and chlorine-free products including phenol, hydroquinone, and catechol. Further, hydroxylated products were hydroxyhydroquinone, pyrogallol, and phloroglucinol. Degradation of aniline yielded ortho, meta, and para isomers of aminophenol. Degradation of benzoic acid gave 4-hydroxybenzoic acid and 3-hydroxybenzoic acid as the primary products. Further oxidation of these products yielded 3,4-dihydroxybenzoic and 2,3-dihydroxybenzoic acids. Oxalic, formic, and malonic acids were determined as ring-opened products of all studied substituted aromatics.

Lukes and Locke [117] studied nonthermal plasma-induced decomposition of ortho, meta, and para isomers of hydroxy-, chloro-, and nitrophenols in water using the hybrid series gas–liquid electrical discharge reactor that generates gas-phase discharge above the water surface simultaneously with the electrical discharge directly in the liquid. Degradation of substituted phenols was evaluated for two gas-phase compositions, pure argon and pure oxygen, above the aqueous solution. Removal of all phenols was found to follow first-order kinetics. The rates of

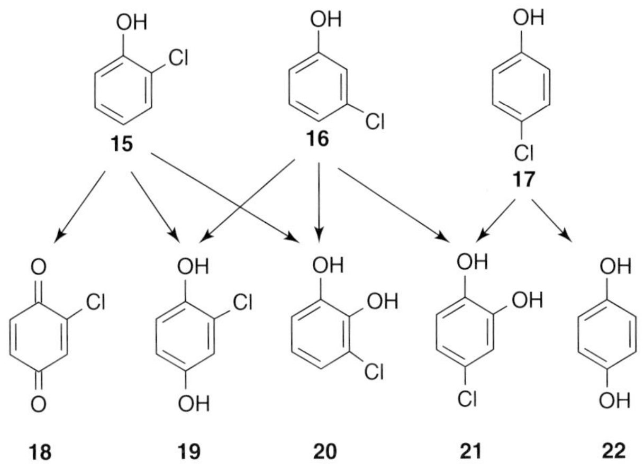

Figure 7.6 Primary products of plasmachemical degradation of 2-,3-,4-chlorophenol in combined gas–liquid phase discharge.

oxidation in the oxygen atmosphere of the hydroxylated phenols were greater than those obtained for the chloro- and the nitro- substituted phenols. The order of reactivity by substituent was –OH > –Cl > $-NO_2$. On the other hand, with use of an argon atmosphere, the oxidation of hydroxylated phenols was the slowest and the order of reactivity was –Cl > $-NO_2$ > –OH. Electrophilic attack by hydroxyl radicals and ozone were determined to be the main oxidation pathways for degradation of phenols in the hybrid series reactor under argon and oxygen atmospheres, respectively. Hydroxylated aromatic by-products were identified during degradation of all substituted phenols under both gas-phase compositions.

2-hydroxy-1,4-benzoquinone was detected during degradation of hydroxylated phenols. For chlorinated phenols **15-17**, the products were 2-chloro-1,4-benzoquinone **18**, chlorohydroquinone **19**, 3-chlorocatechol **20**, and 4-chlorocatechol **21** (Figure 7.6). For nitrophenols, they were nitrohydroquinone, 3-nitrocatechol, and 4-nitrocatechol. Dechlorinated and denitrated aromatic products such as hydroquinone **22** for 4-chlorophenol **17**; 2-hydroxy-1,4-benzoquinone and 1,4-benzoquinone for 3- and 4-nitrophenol were also detected. In addition, *cis,cis*-muconic acid was detected during degradation of catechol under an oxygen atmosphere. An electrophilic substitution reaction mechanism was also proved by the significant correlation between the relative rates of oxidation of substituted phenols obtained in the hybrid series reactor and the Hammett substituent constants. Dechlorination and denitration of chlorophenols and nitrophenols, respectively, corresponded typically to about 50% of total conversion of parent compounds.

7.3.1.3 **Polycyclic and Heterocyclic Aromatic Hydrocarbons**

Polycyclic and heterocyclic aromatic hydrocarbons are examples of highly persistent and toxic chemicals. A limited number of studies focused on plasmachemical degradation of these compounds in aqueous solution. Sahni *et al.* [37] conducted degradation experiments with the model polychlorinated biphenyl (PCB) compound, 2,2′,4,4′-tetrachlorobiphenyl (TCB) using direct liquid and combined gas–liquid discharges. The degradation mechanism in both reactor configurations was similar and was dominated by hydroxyl radical attack. In addition to hydroxyl radical attack, the authors suggested possible contribution of reductive species in TCB degradation. The rate of PCB degradation increased with lower number of chlorine substituents as determined by the comparison of TCB with the degradation of 2-chlorobiphenyl.

Mededovic and Locke [142] performed a detailed study of the degradation of atrazine by the discharge in water. The authors identified the main products of atrazine degradation and determined that the degradation followed an oxidative pathway by $OH^{\bullet}$ radical attack where processes such as dealkylation and dehalogenation of the *s*-triazine ring take place. In the first step, atrazine amide and simazine amide were formed by the alkyl chain oxidation. Further dealkylation yielded atrazine desethyl and atrazine deisopropyl. In the next step, these two compounds formed the final degradation by-product ammeline. The mechanism of atrazine degradation by $OH^{\bullet}$ radical in the discharge in water was confirmed by an experiment with *t*-butanol addition, showing no atrazine degradation. The same authors also studied degradation of *s*-triazine [143]. The Fenton reaction alone gave fairly poor mineralization of *s*-triazine, but mineralization was increased by a factor of 2 when the reaction was combined with electrical discharge. The authors also used sodium persulfate, whose addition led to increased *s*-triazine degradation by the discharge but not to decreases in total organic carbon (TOC). They suggested that thermal effects from plasma can locally induce decomposition of sodium persulfate, which produces sulfate radical anion and hydroxyl radical (Eqs. (7.46) and (7.47)).

$$S_2O_8^{2-} \xrightarrow{\Delta T} 2SO_4^{\bullet -} \tag{7.46}$$

$$SO_4^{\bullet -} + H_2O \longrightarrow OH^{\bullet} + HSO_4^- \tag{7.47}$$

Wang *et al.* [144] investigated the degradation of bisphenol A induced by glow discharge plasma in contact with an aqueous solution with or without the presence of ferrous or ferric ions. Four major aromatic intermediate products were identified: 2,2-bis(4-hydroxyphenyl) propanol, 2,2-bis(4-hydroxyphenyl) propanal, 2,2-bis(4-hydroxyphenyl) propanoic acid, and 2-(4hydroxylphenyl)-2-(3, 4-dihydroxyphenyl) propane. In addition, aliphatic organic acids such as oxalic, formic, malonic, and acetic acids were identified. A possible degradation mechanism of bisphenol was proposed involving primary $OH^{\bullet}$ radical attack on the aromatic ring and/or on the methyl group at the central carbon atom between the aromatic rings. Abdelmalek *et al.* [145] studied the degradation of the same compound by gas–liquid gliding arc discharge under different working gases (air, argon, oxygen–argon mixtures) and with and without the addition of ferrous ions. Under an air atmosphere, nitrated products 3,3-dinitrobisphenol A and 3-nitrobisphenol A were identified. Oxygen/argon (20/80) as the working gas in the presence of Fe(II) showed the best removal of bisphenol A.

The degradation of pharmaceutical compounds using nonthermal plasma has been recently studied by several authors [146–150]. Krause *et al.* [147, 148] used a corona discharge over water in order to remove several endocrine disrupting chemicals including carbamazepine, clofibric acid, and iopromide. Gerrity *et al.* [146] evaluated the degradation of meprobamate, dilantin, primidone, carbamazepine, atenolol, trimethoprim, and atrazine using pulsed corona discharge above water. Carbamazepine and trimethoprim were the compounds most susceptible to the degradation, while primidone, meprobamate, and atrazine were the most recalcitrant compounds. Magureanu *et al.* [149, 150] reported the degradation of pentoxifylline and three β-lactam antibiotics (amoxicillin, oxacillin, and ampicillin) in water using a pulsed DBD. The authors identified the degradation products resulting from the decomposition of the antibiotics and proposed the degradation mechanism by $OH^{\bullet}$ radicals and ozone.

7.3.2 Organic Dyes

Different types of electrical discharges have been successfully applied for the removal of a variety of organic dyes from model aqueous solutions as well as from from real wastewaters of dye effluents, for example, indigo carmine, methylene blue, acid orange 7, methyl orange, rhodamine B, reactive blue 137, methyl yellow, phenol red, and so forth [151–208]. The dye degradation was usually evaluated by the degree of decolorization of the dye solution as measured by absorption spectrophotometry at the wavelength of maximum absorption of the particular dye. The absorption spectrum of a dye is determined by the chemical structure of the chromophoric group(s) in the dye molecule (i.e., azo (–N=N–), carbonyl (–C=O), methine (–CH=), or nitro ($-NO_2$) groups). The change in absorbance and color of the dye is thus connected with the change/destruction of the dye chromophoric group(s). In some studies the degradation of the dye was evaluated using more advanced analytical techniques allowing identification of some degradation by-products. On the basis of the dye intermediates, several reaction mechanisms for the degradation of organic dyes in the discharge have been proposed depending on the chemical structure of the dye and the type of discharge. In general, the degradation of dyes by the discharge was mainly attributed to the oxidative attack by $OH^{\bullet}$ radical, ozone, or peroxynitrite, for example, [13, 151, 180]. Solution acidity (Section 7.2.1) was also found to contribute significantly to the degradation of dyes by the discharge [3, 11]. In less-colored solutions, dyes could be degraded photochemically by UV radiation from the discharge plasma. In the following sections, an overview of the chemical mechanisms assigned to the three basic categories of organic dyes mainly used in the plasmachemical studies, that is, azo (Section 7.3.2.1), carbonyl (Section 7.3.2.2), and aryl carbonium ion dyes (Section 7.3.2.3), is presented.

7.3.2.1 Azo Dyes

In general, several groups of organic dyes, distinguished on the basis of their molecular structure, are azo dyes, carbonyl dyes, phthalocyanines, polymethine dyes, arylcarbonium ion dyes, dioxazines, sulfur dyes, and nitro dyes [209]. Among these, azo dyes are the most important colorants, and they constitute a major part of all commercial dyes employed in a wide range of industrial processes. Azo dyes are characterized by the presence of the azo group (–N=N–) attached to two substituents, mainly benzene or naphthalene derivatives, containing electron-withdrawing and/or electron-donating groups (see e.g., molecular structures of azo dyes methyl orange **23** and acid orange 7 **24**; Figure 7.7).

The degradation of azo dyes was reported to proceed mainly by $OH^•$ radical attack on the carbon atom bearing the azo group (C–N=N–) [210]. The reactivity of the azo group with ozone is significantly lower compared to that of the $OH^•$ radical. Ozone attack is considered to proceed mainly on the aromatic rings of the azo dye, which are more reactive toward ozone than the azo group [211]. Oxidative attack of an azo dye leads to the breaking of the C–N= and –N=N– bonds and the corresponding fading of the dye. Primary intermediates from the degradation of azo dyes are derived from aromatic substituents attached to the azo group (i.e., benzene and naphtalene derivatives). They are either aromatic amines, naphthoquinones, or phenolic compounds. By subsequent oxidation, various aliphatic and carboxylic acids are formed.

Yan *et al.* [151] studied the degradation of acid orange 7 (AO7) using gliding arc discharge in air with water spray. The main intermediates of AO7 degradation were aromatic products including phenol, naphthalene, 2-naphthol **25**, phtalic acid **30**, 1,3-isobenzofurandione **31**, and 2-hydroxymethylbenzoic acid. The products of ring cleavage were oxalic acid, acetic acid, and malonic acid. Du *et al.* [152–154] reported for the same dye using gliding arc discharge in air burning above a water surface the aromatic products benzenesulfonic acid **26**, hydroquinone **7**, 1,4-benzoquinone **8**, and 1,2-naphthoquinone **28** and the ring cleavage products oxalic acid, acetic acid, and malonic acid. The degradation mechanism of the dye driven by $OH^•$ radical attack was proposed in both studies and is schematically shown in Figure 7.8 [210].

Mok *et al.* [155] reported degradation of AO7 using gas phase DBD reactor submerged in water combined with TiO_2 photocatalyst and ozone bubbling. Li *et al.* [156, 157] and Jin *et al.* [158] used for decolorization of the same dye pulsed corona discharge in water with TiO_2 photocatalyst and contact glow discharge electrolysis, respectively.

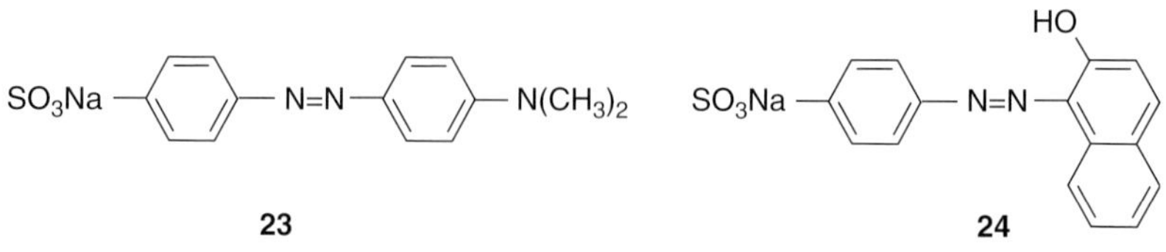

23 **24**

Figure 7.7 Methyl orange **23** and acid orange 7 **24** azo dyes.

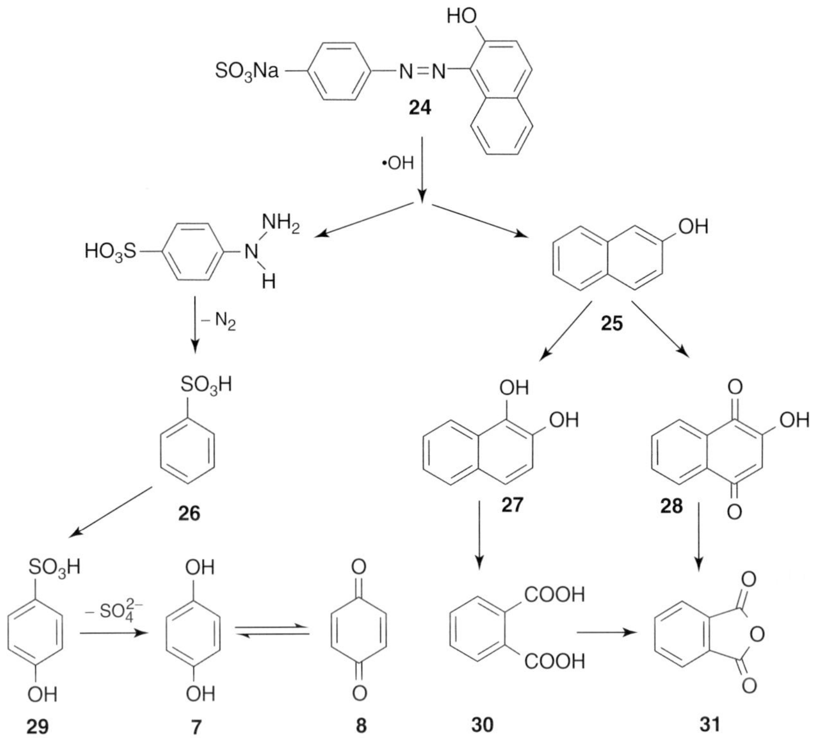

Figure 7.8 Scheme of major degradation pathways of acid orange 7. (Source: Modified from [210]. Copyright (2004), with permission from Elsevier.)

N,N-dimethyl-4-nitroaniline was identified as the primary reaction product of another azo dye methyl orange (MO), which was treated by the gliding discharge in air [13]. Several aliphatic oxidation products were reported for the same dye (2-*tert*-butoxyethanol, 3-methoxy-3-methyl-1-butanol, 2-ethoxy-2methylpropane), when degraded by the pulsed corona discharge in water bubbled with oxygen [159]. Degradation of MO was also performed by other types of discharges, for example, by pulsed discharge in water, by contact glow discharge electrolysis, or by corona discharge above a water surface with air and oxygen atmospheres [160–166].

The solution acidity was determined to have a significant effect on MO dye degradation by the discharge in water. The decolorization rate of MO increased with a decreasing initial pH of the dye solution [160, 162, 212]. This effect was mainly attributed to the pH-dependent reactivity of MO, which is determined by acid–base properties of the azo group in this dye molecule ($pK_a = 3.76$).

The same behavior was also observed for other dyes, which were treated by the discharge in water [161, 168, 212]. The effect of solution acidity on the decolorization rate of MO was also demonstrated for the gliding discharge in air above a water surface [13]. In this case, other processes in addition to the pH-dependent reactivity of the dye have been demonstrated to contribute to the dye degradation. Specifically, peroxynitrite and the nitrous acid were suggested to be the main oxidizing agents of the dye, most likely through indirect reactions of $OH^\bullet$ and $NO_2^\bullet$ radicals formed on pH-dependent homolysis of peroxynitrous acid (Section 7.2.2.4). Ozone was determined to have a negligible effect on the dye degradation in this case. Similar results were demonstrated for other dyes exposed to the gliding discharge in air (e.g., alizarin S, bromothymol blue) [3, 167].

Kozakova *et al.* [169] studied the decomposition of disazo dye direct red 79 by underwater diaphragm discharge and identified eight aromatic degradation products of this dye using LC–MS method. For the degradation of the disazo dye molecule, they proposed a mechanism involving the primary $OH^\bullet$ attack at one of the azo groups followed by detachment of the naphthalene substituent from the azo group and formation of amine and hydroxylated aromatic derivates. By subsequent oxidation, the second azo group breakdown occurred and lower-molecular-weight products were formed from defragmented naphthalene and benzene derivatives. Other azo dyes that were treated in aqueous solutions by electrical discharges include methyl yellow, methyl red, chicago sky blue, eriochrome black T, acid yellow 23, acid red 87, and many reactive azo dyes [14, 15, 159–161, 168–178].

7.3.2.2 **Carbonyl Dyes**

Carbonyl dyes are characterized by the presence of a carbonyl group (–C=O) as the essential chromophoric unit. These dyes possess significantly higher reactivity with ozone than azo dyes. Carbonyl dyes include mainly anthraquinones, indigoids, benzodifuranones, coumarins, naphthalimides, quinacridones, perylenes, perinones, and diketopyrrolopyrroles [209]. Indigo trisulfonate **32** and anthraquinonic acid blue 40 were one of the first dyes that were decolorized by the discharge in water [179]. Specifically, pulsed electrical discharge was operated directly in water with oxygen flowing through the nozzle needle electrode. The main effect of the discharge on the dye decolorization was attributed to the ozone produced by the discharge. With N_2 bubbling the discharge produced no decolorization. The indigo molecule contains only one C=C double bond, which reacts with ozone with a very high rate [213, 214]. Ozone attack on the C=C bond proceeds by a 1,3-dipolar cycloaddition and produces sulfonated isatin **33a**, **33b** as the primary product and eliminates the blue color of the dye (Figure 7.9). A high reactivity with ozone was observed also for anthraquinonic dye acid blue 40.

Minamitani *et al.* [180] used NMR analysis to evaluate the mechanism of indigo carmine degradation by the gas-phase pulsed corona discharge with the dye solution sprayed into the plasma. They have determined rapid decomposition of the chromogenic bond of the indigo dye; however, degradation of unsaturated bonds (i.e., double bonds of carbon in a benzene ring and the center of the formula of indigo carmine) took significantly longer. Ozone was determined to be the dominant species in the decomposing of the chromogenic bond, while the hydroxyl radical was the dominant species in breaking the unsaturated bonds. Pawlat and Hensel [181–184] and Wang *et al.* [185] studied decolorization of indigo blue and indigo carmine using discharge in foam and spraying DBD, respectively.

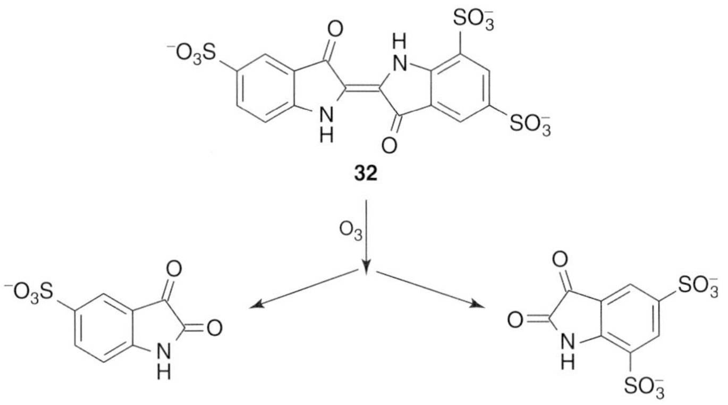

32

33a

33b

Figure 7.9 Ozone decomposition of indigo trisulfonate **32** to sulfonated isatin **33a, 33b**.

Other carbonyl dyes that were treated in aqueous solutions by electrical discharges include anthraquinonic dyes, acid green 25, alizarin S, alizarin red, reactive blue 49, acid blue 74, or quinone imine dye oxazin blue [3, 8, 175, 186–189]. For these dyes, no intermediates were reported, but in some cases, decrease in TOC, chemical oxygen demand (COD), and adsorbable organic halogens (AOX) were given. For example, for reactive blue 49 degradation by a diaphragm discharge in water, the TOC data indicated that partial mineralization of the dye simultaneously proceeded with the dye decolorization [189]. The rate of decrease of AOX was, however, significantly slower compared to the decrease of TOC, which was attributed to the high stability of cyanuric chloride as a reactive part of the dye molecule. Xue *et al.* [188] proposed a degradation mechanism of the alizarin red dye in a hybrid gas–liquid DBD plasma reactor based on evaluation of the chemical bond dissociation energies of the alizarin dye molecule. They suggested ozone attack as an initial step of the dye degradation, taking place most likely at the carbon bond adjacent to –C=O group of the central quinone ring.

7.3.2.3 **Aryl Carbonium Ion Dyes**

Aryl carbonium ion dyes are characterized by the presence of a methine (–CH=) group as the chromophoric unit. The essential structural feature of these dyes is the central (methine) carbon atom attached to either two or three aromatic rings. Commonly, these compounds are referred to as *diarylmethanes* and *triarylmethanes* (or *triphenylmethanes*). Auramine O (basic yellow 2) **34** and malachite green **35** are the typical examples of this class of dyes (Figure 7.10).

Conjugation of C=C and C=N double bonds in aromatic rings and methine group makes the central carbon atom of the molecule π electron-rich and an attractive center for electrophilic attack of these dyes. For the triphenylmethane dye phenol red (PR), which was treated by the pulsed corona discharge in water with O_2 bubbling, both ozone and hydroxyl radicals were suggested to contribute to the degradation through oxidative defragmentation of the dye molecule. 4-methoxy-acetophenone,

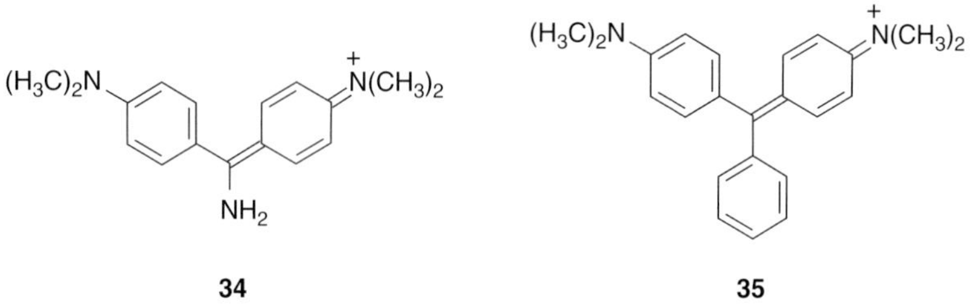

Figure 7.10 Auramine O **34** and malachite green **35** aryl carbonium ion dyes.

3,5-di-*tert*-butyl-4-hydroxybenzaldehyde, phenol, and hydroquinone were identified as aromatic by-products, along with the aliphatic products 2-*tert*-butoxyethanol, 3-methoxy-3-methyl-1-butanol, and 2-ethoxy-2methylpropane [159]. Other triphenylmethane dyes that were treated in aqueous solutions by electrical discharges include malachite green (MG), crystal violet, or brilliant green [174, 190–192].

For MG, it has been shown that the electrophilic attack of ozone (and $OH^\bullet$ radical) at central carbon atom may proceed through two degradation pathways (Figure 7.11) [211, 215, 216]. The attack on the central carbon (C^C) may take place either at the site bonded to the anilinic ring (C^N) or at the site bonded to the phenylic ring (C^P). The direction of the attack depends on the electronic character of the ring substituent(s), which alter electron density of the attached aromatic rings and the methine group. Electron-donating groups such as dimethylamine (-$N(CH_3)_2$) for MG (or hydroxyl (–OH) for PR) accelerate oxidative attack at the C^C–C^N

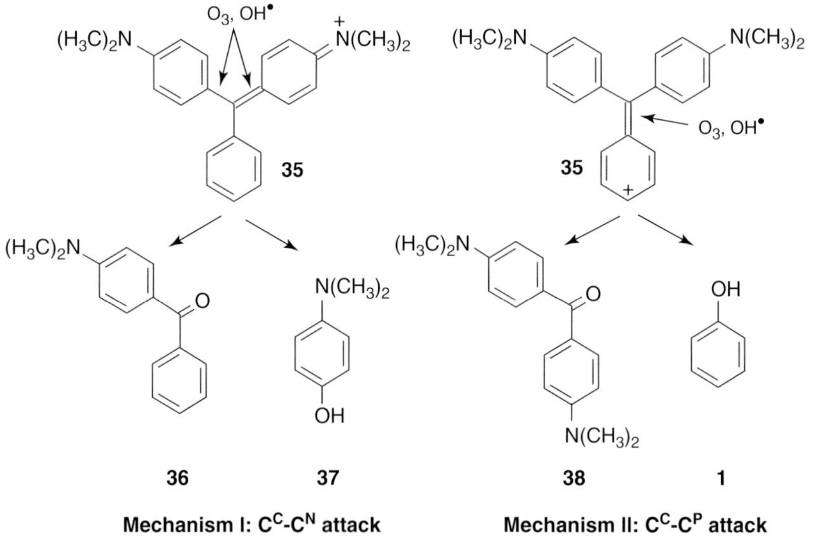

Figure 7.11 Scheme of oxidation of malachite green by electrophilic attack of ozone or $OH^\bullet$ radical at the central carbon atom.

Table 7.6 Heterocyclic Aryl Carbonium Ion Dyes and Their Aza Analogues.

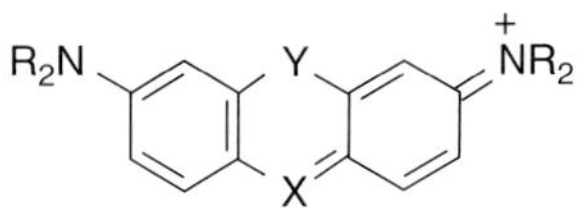

X-substituent	Y-substituent	Type of dye
–C(Ar)=	–O–	Xanthene
–C(Ar)=	–S–	Thioxanthene
–C(Ar)=	–NR–	Acridine
–N=	–O–	Oxazine
–N=	–S–	Thiazine
–N=	–NR–	Azine

Source: [209]. Reproduced by permission of the Royal Society of Chemistry.

bond, while electron-withdrawing groups such as the nitro group (-NO_2) activate the C^C-C^P carbon site. Thus, the oxidation of MG with ozone and $OH^{\bullet}$ radical attack at the central carbon atom proceeds preferentially via mechanism I, which gives *p*-benzyl-*N, N*-dimethylaniline **36**, and *N, N*-dimethylaminophenol **37** as the primary oxidation products (Figure 7.11). In addition, ozone may also react with *N, N*-dimethylamine substituents to form *N*-formyl-*N*-methylamino derivates [211].

Aryl carbonium dyes also include heterocyclic derivatives, which are derived from di- and triarylmethane dyes by bridging their aromatic rings at ortho–ortho positions to the central methine carbon atom with a heteroatom. These heterocyclic derivatives may be further categorized into xanthenes (oxygen bridged), thioxanthenes (sulfur bridged), acridines (nitrogen bridged), and their aza analogues oxazines, thiazines, and azines. Examples of the general structure of these dyes are illustrated in Table 7.6 [209].

From the heterocyclic aryl carbonium dyes, the thiazine type **39** methylene blue (MB) was the most studied dye in decolorization experiments with electrical discharges generated directly in water (with or without bubbling of O_2, O_3, and Ar and the addition of solid catalysts) [78, 95, 159, 193–195], in foams [182, 184, 196], or above a water surface (under air or oxygen atmosphere) [190, 197–199, 212]. Other dyes of the same class that were treated in aqueous solutions by electrical discharges include xanthene-type dyes rhodamine B, eosin, and fluorescein-4-isothiocyanate [66, 160, 161, 200–202], acridine-type dye acridine orange [203], or oxazine-type dye direct blue 106 [204–206]. Magureanu *et al.* [159, 194] have identified two aromatic products of MB (4-methoxy-acetophenone and 3, 5-di-tert-butyl-4-hydroxybenzaldehyde) and some aliphatic compounds (2-*tert*-butoxyethanol, 3-methoxy-3-methyl-1-butanol, and

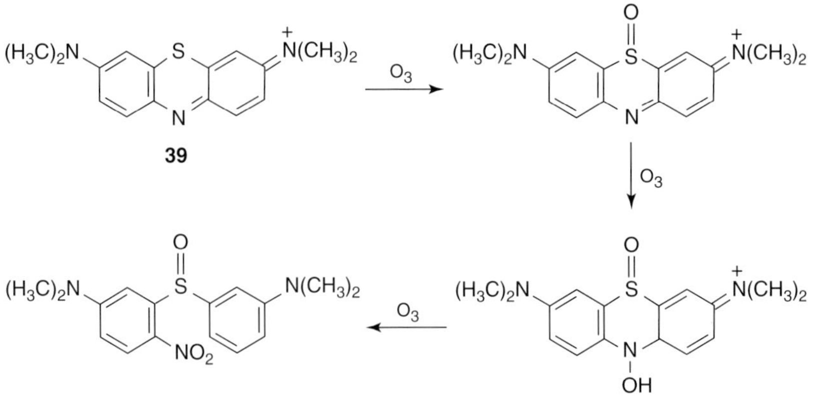

Figure 7.12 Proposed mechanism of methylene blue decolorization by ozone. (Source: Adapted from [24]. Copyright (1993), with permission of Taylor & Francis Group (http://www.informaworld.com).)

2-ethoxy-2methylpropane) during the dye degradation by the pulsed corona discharge in water bubbled with oxygen. Huang *et al.* [199] identified benzothiazole as the cleavage product of MB degradation by DBD air plasma above the water surface.

The significant role of ozone in MB removal from water by the discharge was demonstrated [95, 159, 193, 194, 212]. Benitez *et al.* [24] demonstrated that MB possesses relatively high reactivity with ozone. They determined a pH-dependent rate constant for the reaction of aqueous MB with ozone in the fast pseudo-0.5th order kinetic regime (see Table 7.5) and proposed a mechanism of MB decolorization by aqueous ozone through primary attack at C–S and C–N bonds at the central heterocyclic ring of the MB molecule (Figure 7.12). Huang *et al.* [199] evaluated the chemical bond dissociation energies in the MB molecule and proposed a different mechanism of MB degradation by primary attack of ozone (or $OH^{\bullet}$ radical) at the N-CH_3 bonds of *N, N*-dimethylamine substituents with subsequent rupture of C–S and C–N bonds at the central heterocyclic ring of the MB molecule.

Magureanu *et al.* [194] reported that when argon was bubbled through the solution, the rate of MB degradation by the discharge was much slower than with oxygen bubbling, suggesting that ozone contributed to MB degradation in addition to the dye oxidation by $OH^{\bullet}$ radicals. Grabowski *et al.* [212] observed that the addition of the $OH^{\bullet}$ scavenger *t*-butanol did not have a large effect on the degradation of MB using a pulsed corona discharge generated in air above the water surface. They estimated that $OH^{\bullet}$ radicals contributed only 20% to the total MB degradation by the discharge and showed that ozone was the main oxidizing agent.

On the other hand, Velikonja *et al.* [197] compared MB removal efficiency obtained by a DBD discharge generated in oxygen above a water surface with MB degradation obtained by direct ozonation. They supplied either dry or humid ozone into the chamber in the same concentrations as produced with the discharge.

Decolorization of aqueous solution of MB (prepared from deionized water) was three times faster in water treated by the discharge than by only ozonation. The authors concluded that MB oxidation by $OH^\bullet$ radicals produced by the discharge and possibly through Peroxone chemistry of ozone with hydrogen peroxide might be the reason for difference in MB degradation obtained by the discharge in comparison with the effect of direct ozonation (see Table 7.5 for the rate constants of $OH^\bullet$ radical and ozone with MB). Although the significant role of ozone in MB degradation has been reported and largely accepted, the understanding of plasmachemical processes participating in MB degradation (and generally of organic dyes) is limited.

7.3.3
Aliphatic Compounds

The plasmachemical degradation of aliphatic compounds in water has been studied to a much smaller extent than that of aromatic compounds (Sections 7.3.1 and 7.3.2). Among these species, alcohols have been studied more extensively because they are often used as scavengers of $OH^\bullet$ radicals in studies focused on fundamental mechanisms of plasmachemical processes in water (e.g., [80, 212, 217]) and also as substrates for generation of hydrogen gas on their decomposition by liquid-phase plasma [218–234]. Other aliphatic compounds, for example, dimethylsulfoxide (DMSO) and tetranitromethane (TNM), have been decomposed by electrical discharges in water and used for evaluation of aqueous-phase oxidative and reductive plasmachemical processes induced by the electrical discharges in water [44, 139, 235, 236]. Trichloroethylene was degraded by the corona discharge above and directly in water [36, 38, 237]. Methyl *tert*-butyl ether was decomposed using pulsed arc discharge in water [238] and by corona and glow discharge above water surface [237, 239]. In the following sections, we have selected methanol (Section 7.3.3.1), DMSO (Section 7.3.3.2), and TNM (Section 7.3.3.3) as examples of aliphatic organic compounds to demonstrate the main oxidative and reductive mechanisms involved in plasmachemical treatment of these compounds in water.

7.3.3.1 Methanol

Alcohols are frequently added to aqueous solutions treated by the electrical discharges to scavenge $OH^\bullet$ radicals, for example, to distinguish between the $OH^\bullet$ radical and ozone oxidative mechanisms induced in water by electrical discharges in gas–liquid environments (see Table 7.5 for the rate constants of $OH^\bullet$ radical and ozone with methanol) or to evaluate the formation mechanisms of hydrogen peroxide by discharge in water. Methanol, as a representative of simple aliphatic alcohols, reacts with the $OH^\bullet$ radical via hydrogen abstraction, predominantly from the carbon atom (Eq. (7.48)).

$$OH^\bullet + CH_3OH \longrightarrow H_2O + {}^\bullet CH_2OH \tag{7.48}$$

Methanol radicals both disproportionate and dimerize to form formaldehyde (Eq. (7.49)) and ethylene glycol (Eq. (7.50))

$$2^{\bullet}CH_2OH \longrightarrow CH_3OH + HCHO \quad (7.49)$$

$$2^{\bullet}CH_2OH \longrightarrow OH - CH_2CH_2 - OH \quad (7.50)$$

or in the presence of oxygen rapidly react with oxygen to form peroxy radicals (Eq. (7.51))

$$^{\bullet}CH_2OH + O_2 \longrightarrow {}^{\bullet}O_2CH_2OH \quad (7.51)$$

which generally disappear by a second-order process in acid or neutral solution with hydrogen peroxide and formaldehyde as products (Eqs. (7.52) and (7.53)).

$$2^{\bullet}O_2CH_2OH \longrightarrow 2\,HCHO + H_2O_2 + O_2 \quad (7.52)$$

$$^{\bullet}O_2CH_2OH + HO_2 \longrightarrow HCHO + H_2O_2 + O_2 \quad (7.53)$$

Several authors have investigated methanol decomposition in water using electrical discharges [218–221]. Liu *et al.* [219] reported ethylene glycol as the major product of methanol decomposition in aqueous solution treated by the corona discharge. Yan *et al.* [220] reported formaldehyde along with hydrogen, carbon monoxide, methane, ethane, and propane as the main products of the methanol decomposition induced by the glow discharge plasma electrolysis. Other authors using electrical discharge focused primarily on the production of molecular hydrogen from aqueous solutions or vapors of methanol (or other alcohols) [218–234].

Thagard *et al.* [218] studied the hydrogen production from varying methanol/water mixtures, which were treated by liquid-phase pulsed corona discharge. With the increase in methanol concentration in the mixture (varying from 20 vol.-% to 100 vol.-%) they have determined three times higher productions of hydrogen and higher energy efficiency. The major liquid and gas-phase products of methanol decomposition that they determined were methane, ethylene glycol, and CO_2. Chernyak *et al.* [223, 225] studied the formation of H_2 from ethanol/water mixtures using liquid-phase DC discharge generated between two point electrodes flowed with air. The H_2 yield increased with increasing discharge power, and the maximum was obtained in 50% ethanol/water mixtures. Burlica *et al.* [222] studied the formation of H_2 from methanol, propanol, and ethanol solutions exposed to a nonthermal pulsed plasma gliding arc reactor equipped with a spray nozzle. They have determined the highest H_2 energy yield formation for Ar carrier and pure methanol.

The main reactions in which molecular hydrogen is produced from methanol are with hydrogen radical by H-abstraction reaction (Eqs. (7.54) and (7.55)).

$$H^{\bullet} + CH_3OH \longrightarrow H_2 + {}^{\bullet}CH_2OH \quad (7.54)$$

$$H^{\bullet} + CH_3OH \longrightarrow H_2 + {}^{\bullet}CH_3O \quad (7.55)$$

In the solutions containing dissolved oxygen, however, methanol competes for hydrogen radical with oxygen (Eq. (7.9)). Considering the rate constants of $H^{\bullet}$ radical with oxygen and methanol (Table 7.5), reactions (Eq. (7.54) and (7.55)) occur only if the ratio $[CH_3OH]/[O_2]$ is greater than about 600 : 1 (i.e., if the concentration of methanol in air-saturated solutions is more than 0.15 mol l^{-1}, which is about 1 vol.-%). Consequently, in solutions with high content of methanol, the hydrogen radicals are produced by the liquid-phase plasma not only from water (Chapter 6) but also from methanol

$$CH_3OH + e^* \longrightarrow {}^{\bullet}CH_2OH + H + e' \qquad \Delta H = 401.7\,kJ\,mol^{-1} \quad (7.56)$$

$$CH_3OH + e^* \longrightarrow {}^{\bullet}CH_3O + H + e' \qquad \Delta H = 384.9\,kJ\,mol^{-1} \quad (7.57)$$

where e* and e' represent the excited electrons and electrons with less energy, respectively.

Additional reaction channels for H_2 formation from methanol were proposed by Han *et al.* [229]. The authors have used density functional theory (DFT) to investigate the mechanisms of methanol decomposition in nonthermal plasma and the feasibility of the production of various products, including hydrogen, carbon monoxide, and ethylene glycol from methanol. Among the pathways studied by DFT, the most probable involve direct dissociation of methanol to formaldehyde and its *cis*- and *trans*- hydroxycarbene isomers (HCOH) along with molecular hydrogen (Eqs. 7.58–7.60).

$$CH_3OH + e^* \longrightarrow CH_2O + H_2 + e' \quad (7.58)$$

$$CH_3OH + e^* \longrightarrow cis\text{-}HCOH + H_2 + e' \quad (7.59)$$

$$CH_3OH + e^* \longrightarrow trans\text{-}HCOH + H_2 + e' \quad (7.60)$$

HCOH rapidly rearranges to formaldehyde by hydrogen tunneling through the isomerization barrier [240]. Since formaldehyde is highly soluble in water and the addition of methanol increases the solubility [241], it might be expected that water and methanol–water solutions could sequester formaldehyde products in the liquid phase. Water–methanol solutions are commonly used to collect formaldehyde, which binds with the solvents to form glycols and oligomeric compounds [242]. Thus the chemistry of methanol and formaldehyde in aqueous solutions can be quite complex and requires further detailed studies in the case of liquid and gas–liquid electrical discharge plasma.

7.3.3.2 Dimethylsulfoxide

Dimethyl sulfoxide $(CH_3)_2SO$ is an organosulfur aliphatic compound and an effective scavenger of $OH^{\bullet}$ radicals similar to alcohols (Table 7.5). The plasmachemical degradation of DMSO in water was reported by several authors [139, 235, 236] using the pulsed corona discharge in water. Under these conditions, the DMSO was degraded mainly to methanesulfonate, formaldehyde, and sulfate ions. Lukes *et al.* [236] identified an additional product, dimethylsulfone $(CH_3)_2SO_2$, whose formation was attributed to the tungstate-catalyzed oxidation of DMSO by hydrogen peroxide (Section 7.4.3). The formation of methanesulfonate as the major degradation product of DMSO indicated that the degradation of the dimethyl sulfoxide in the discharge proceeded through $OH^{\bullet}$ radical attack. The reaction of the $OH^{\bullet}$ radical with DMSO is very fast (Table 7.5) and proceeds mainly by $OH^{\bullet}$ addition to form an OH• adduct, which rapidly decomposes into methanesulfinic acid, as the primary sulfur-containing intermediate of DMSO, and a methyl radical (Eq. (7.61)) [243, 244].

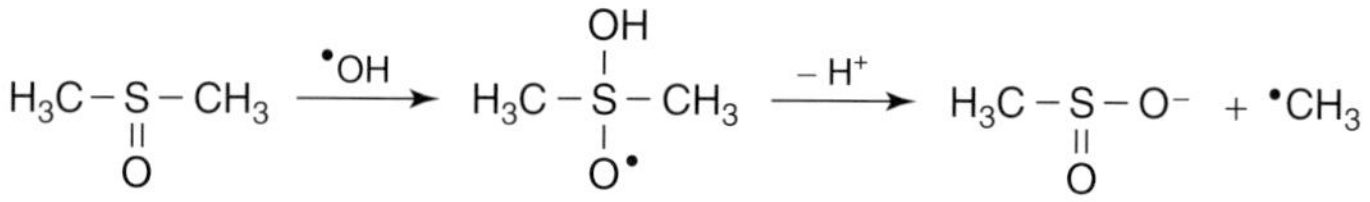

$$H_3C-S(=O)-CH_3 \xrightarrow{\bullet OH} H_3C-S(OH)(O^\bullet)-CH_3 \xrightarrow{-H^+} H_3C-S(=O)-O^- + {}^\bullet CH_3 \tag{7.61}$$

Methanesulfinic acid behaves in many respects like DMSO, that is, the OH• radical reacts preferentially by OH addition with subsequent fragmentation of the OH-adduct, giving rise mainly to the methanesulfonyl radical or the splitting off of a methyl radical along with the formation of bisulfite anion (Eq. (7.62)).

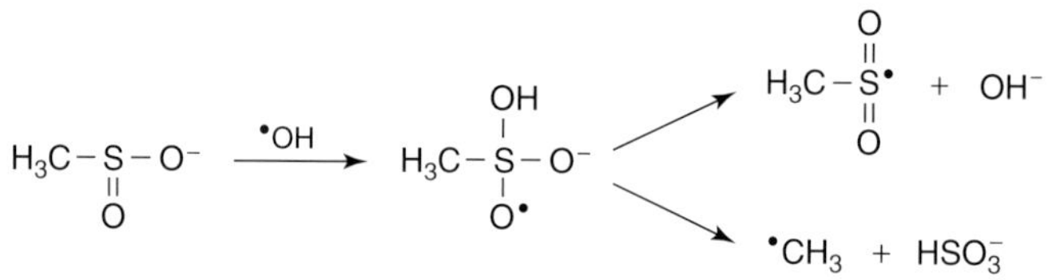

$$H_3C-S(=O)-O^- \xrightarrow{\bullet OH} H_3C-S(OH)(O^\bullet)-O^- \longrightarrow \begin{cases} H_3C-S(=O)_2^\bullet + OH^- \\ {}^\bullet CH_3 + HSO_3^- \end{cases} \tag{7.62}$$

In the presence of oxygen, the methanesulfonyl radical $CH_3S(O)O^\bullet$ initiates a rapid chain reaction between methylsulfonylperoxyl radical $CH_3S(O)_2O^\bullet$ (Eq. (7.63)) and methanesulfinate (Eq. (7.64)), generating methanesulfonic acid (Eq. (7.65)).

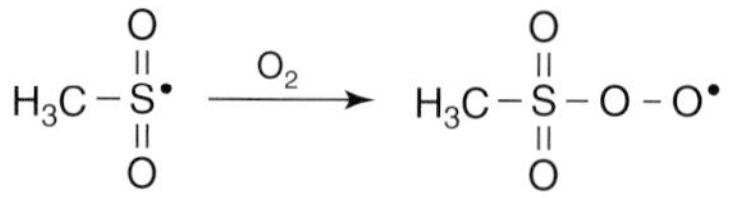

$$H_3C-S(=O)_2^\bullet \xrightarrow{O_2} H_3C-S(=O)_2-O-O^\bullet \tag{7.63}$$

$$H_3C-S(=O)_2-O-O^\bullet + H_3C-S(=O)-O^- \longrightarrow H_3C-S(=O)_2-O-O^- + H_3C-S(=O)_2^\bullet \tag{7.64}$$

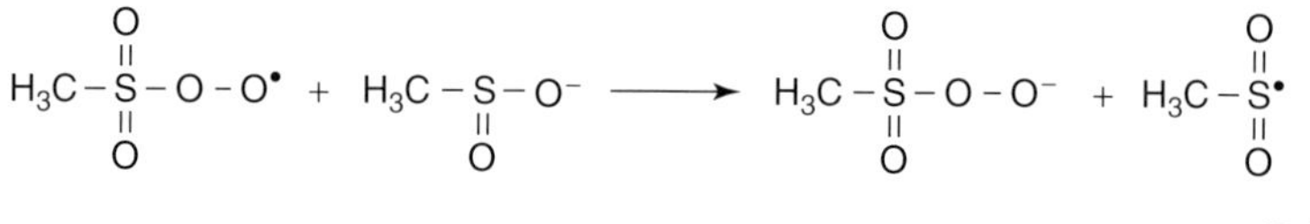

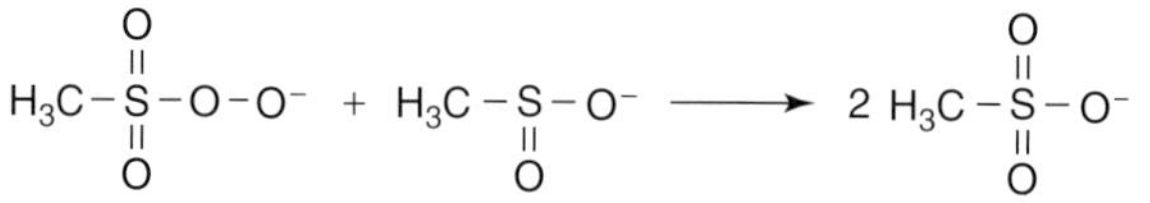

$$H_3C-S(=O)_2-O-O^- + H_3C-S(=O)-O^- \longrightarrow 2\, H_3C-S(=O)_2-O^- \tag{7.65}$$

Oxidation of methanesulfonate by OH• radical yields formaldehyde and the sulfate anion. Formaldehyde is also produced with the reaction of the methyl radical (formed in reactions (Eqs. (7.61) and (7.62)) with oxygen (Eq. (7.67)) through the formation of the methylperoxyl radical $CH_3O_2^\bullet$ intermediate (Eq. (7.66)).

$$^\bullet CH_3 + O_2 \longrightarrow CH_3O_2^\bullet \tag{7.66}$$

$$2\, CH_3O_2^\bullet + O_2 \longrightarrow 2\, HCHO + 2\, HO_2^\bullet \tag{7.67}$$

Formaldehyde can then be converted by the $OH^{\bullet}$ radical in the presence of oxygen to formate and the final product carbon dioxide.

Interesting results in the liquid-phase plasma-induced decomposition of DMSO were reported by Thagard *et al.* [218], where pure DMSO was used instead of an aqueous solution. During the discharge treatment, they observed the change in the color of the DMSO solution, which turned to yellow and then to dark red, accompanied with the formation of SO_2 and methane. They suggested that when the electrical discharge is generated in the DMSO liquid, the primary pyrolysis reaction of the DMSO takes place in the discharge plasma zone, which leads to splitting of the DMSO molecule into radicals (Eq. (7.68)).

$$H_3C-\underset{\underset{\displaystyle O}{\|}}{S}-CH_3 \longrightarrow H_3C-\underset{\underset{\displaystyle O}{\|}}{S^{\bullet}} + {}^{\bullet}CH_3 \qquad (7.68)$$

Methyl radicals may possibly get reduced to form carbon because they observed formation of diamondlike carbon films on the tip of the tungsten needle discharge electrode.

7.3.3.3 **Tetranitromethane**

Tetranitromethane $C(NO_2)_4$ reacts with a high reaction rate with reductive species such as the $H^{\bullet}$ radical (5.5×10^8 l (mol s)$^{-1}$) [22] and the $O_2^{\bullet-}$ radical (2×10^9 l (mol s)$^{-1}$) [49], and it is extensively used as a probe in free radical studies. TNM has been degraded by the pulsed corona discharge in water and used for the detection and quantification of reductive species ($H^{\bullet}$ and $O_2^{\bullet-}$ radicals) produced by the discharge in water (Section 7.2.3) [44]. The reaction of TNM with H and $O_2^{\bullet-}$ radicals proceeds via one-electron reduction, giving the products nitroform anion $C(NO_2)_3^-$ and nitrite radical (Eqs. (7.31) and (7.69)).

$$C(NO_2)_4 + H^{\bullet} \longrightarrow C(NO_2)_3^- + NO_2^{\bullet} + H^+ \qquad (7.69)$$

$$C(NO_2)_4 + O_2^{\bullet-} \longrightarrow C(NO_2)_3^- + NO_2^{\bullet} + O_2 \qquad (7.31)$$

The nitroform anion (NF^-) formation in TNM solution treated by the discharge was detected by UV-Vis spectrometry. The authors [44] evaluated the effects of various oxidative and reductive scavengers (chloroform, 2-propanol, and hydrogen peroxide) on TNM degradation and determined that the subsequent degradation of NF^- product is dominated by $OH^{\bullet}$ attack. They suggested that because of the volatile nature of TNM, a portion of the TNM degradation also occurred thermally in the discharge region or the surrounding high-temperature regions.

7.4
Aqueous-Phase Plasma-Catalytic Processes

A wide variety of homogeneous and heterogeneous catalysts have been introduced into systems with electrical discharges in water and gas–liquid environments in order to determine possible effects on plasmachemical processes. The evidence for plasma catalytic effects was initially revealed in connection with the erosion of high-voltage electrodes in plasma generated directly in the liquid phase [245–250]. Different materials either sputtered or dissolved from electrodes into the water were demonstrated to have significant effects on the plasmachemical processes in water. For example, high-voltage electrodes made from stainless steel (iron), platinum, or tungsten, decreased the yield of H_2O_2 produced by a pulsed corona discharge in water. At the same time, however, these materials, when used as high-voltage electrodes, increased the efficiency of plasmachemical removal of organic compounds from water [236, 249–253]. In addition, some electrode materials, such as silver and copper, have been shown to have bactericidal effects on microorganisms treated by the discharge in water, which were induced by metal ions and oxide nanoparticles that were released from the electrodes by the discharge [254–256]. Plasma-chemical activity of electrical discharges in water can also be enhanced by the addition of solid particles into the water. Several types of materials have been tested, such as activated carbon, silica gel, glass, alumina, titanium dioxide, and zeolites [92–95]. For example, UV radiation generated by the pulsed corona discharge in water has been demonstrated to cause plasma-induced photocatalytic processes with suspended TiO_2 particles [95–103, 154–157]. However, despite the observed synergistic effects of catalysts with plasma, there is limited knowledge about the role of these materials in the plasma-assisted processes in water. Heterogeneous catalyst materials can also indirectly affect the chemical activity of the discharge by changing electrical and plasma characteristics of the discharge. Homogeneous catalysts can affect solution properties such as pH and conductivity, and thereby change plasma properties. In this section, the main plasma catalytic processes induced by electrical discharges in aqueous environment are reviewed, with the main emphasis on the discharges generated directly in water.

7.4.1 Iron

The catalytic effects of iron on enhancing the efficiency of the plasmachemical removal of organic compounds from water is the most studied plasma catalytic process in underwater plasma. Pioneering work was published by Sharma *et al.* [84], who first demonstrated the positive effect of iron addition on degradation of phenol by pulsed corona discharge in water due to *Fenton's process* (Eq. (7.70)) involving plasmachemically produced hydrogen peroxide and ferrous ions (either directly added as a salt or released from an iron electrode).

$$Fe^{2+} + H_2O_2 \longrightarrow Fe^{3+} + OH^- + OH^\bullet \tag{7.70}$$

Since then, many papers have been published on this topic using iron either as homogeneous or as heterogeneous Fenton-type catalyst (i.e., ferrous/ferric salts or iron minerals/zeolites, respectively) [85–91, 127–130, 170–174].

Figure 7.13 shows a typical example of the enhancement of phenol degradation by the addition of an iron salt [20]. Line (1) demonstrates the negligible role of electrolysis in phenol degradation in this system. In these experiments, the applied

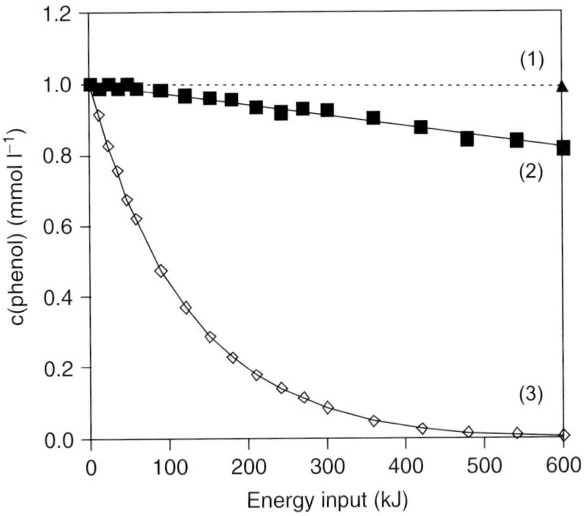

Figure 7.13 Phenol removal for (1) electrolysis only in 1 mmol l^{-1} NaCl, (2) corona discharge in 1 mmol l^{-1} NaCl, (3) corona discharge in 0.5 mmol l^{-1} $FeCl_2$ [20].

voltage was slightly below the inception value, and so, no discharge was generated. Line (2) corresponds to the corona discharge in the same type of solution (NaCl, without iron salt addition), and shows some slow degradation of phenol, caused apparently by the oxidative action of $OH^{\bullet}$ radicals produced by the discharge. Line (3) demonstrates a significant role of iron ions on phenol removal by the discharge with the addition of $FeCl_2$. Since hydrogen peroxide is produced by the discharge in water, the effect of iron addition is attributed to the increased yield of $OH^{\bullet}$ radicals from decomposition of H_2O_2 in Fenton-type reactions with iron ions (Eq. (7.70)).

The mechanism of the Fenton's process is well known and involves both ionic forms of iron [26, 257, 258]. In addition to ferrous ions (Eq. (7.70)), the decomposition of hydrogen peroxide is also catalyzed by ferric ions. In this process, Fe(III)-catalyzed decomposition of H_2O_2 leads to H_2O and O_2 and a steady-state concentration of ferrous ions is maintained during peroxide decomposition through reactions (Eqs. (7.71) and (7.72)).

$$Fe_{3+} + H_2O_2 \overset{-H^+}{\rightleftharpoons} Fe[OOH]^{2+} \rightleftharpoons Fe^{2+} + HO_2^{\bullet} \tag{7.71}$$

$$Fe^{3+} + HO_2^{\bullet} \longrightarrow Fe^{2+} + H^+ + O_2 \tag{7.72}$$

The oxidizing power of the Fenton-type systems can also be greatly enhanced by irradiation with UV or UV-visible light (e.g., emitted from the plasma) due to the photoreduction of hydroxylated ferric ion in aqueous solution (7.73), (7.74) [258].

$$Fe^{3+} + OH^- \rightleftharpoons Fe(OH)^{2+} \tag{7.73}$$

$$Fe(OH)^{2+} + h\nu \longrightarrow Fe^{2+} + OH^{\bullet} \tag{7.74}$$

Thus, the combined process (Eqs. 7.70, 7.73, and (7.74)), known as *the photo-Fenton reaction*, or *Haber-Weiss process* (Eqs. 7.70–7.72), may proceed in water, which produces a higher yield of $OH^{\bullet}$ radicals, and more importantly, iron cycling between the (II) and (III) oxidation states. However, in underwater plasma, these routes of iron cycling are limited (Section 7.4.1.1) and Fe^{3+} is the only stable ionic form of iron and Fe^{2+} ions are the dominant species involved in the decomposition of H_2O_2 via reaction (Eq. (7.70)). Nevertheless, depending on the solution composition (e.g., phenols, Section 7.4.1.1) or material of electrode (e.g., platinum, Section 7.4.2.1), other recovery mechanisms of ferrous ion involved in Fenton's process can take place.

7.4.1.1 **Catalytic Cycle of Iron in Plasmachemical Degradation of Phenol**

In the presence of phenol (or other hydroxylated aromatic compounds and aliphatic organic acids), additional reactions of Fe^{3+} (Eqs. (7.71)–(7.74)) can take place. These include (i) reduction of Fe^{3+} by phenol and formation of phenoxy radical (Eq. (7.75)) [104], (ii) reduction of Fe^{3+} by aromatic degradation products of phenol, that is, dihydroxybenzene and quinone intermediates (Eqs. (7.76)–(7.78)) (Section 8.31) [106], (iii) photolysis of Fe(III)–organic ligands complexes, for example, with organic acids formed during degradation of phenol (Eq. (7.79)) [259]. Catalytic cycle of iron between the (II) and (III) oxidation states is thus sustained in this case by reactions (Eq. (7.70)) and (Eqs. (7.75)–(7.79)).

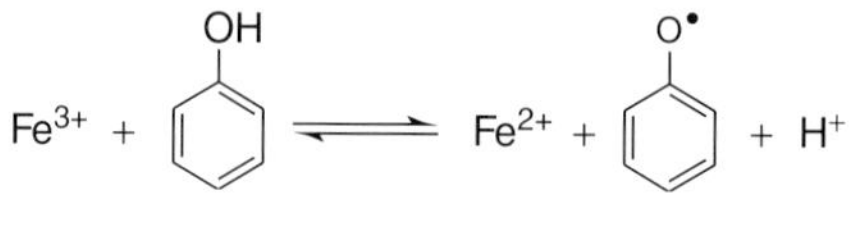

(7.75)

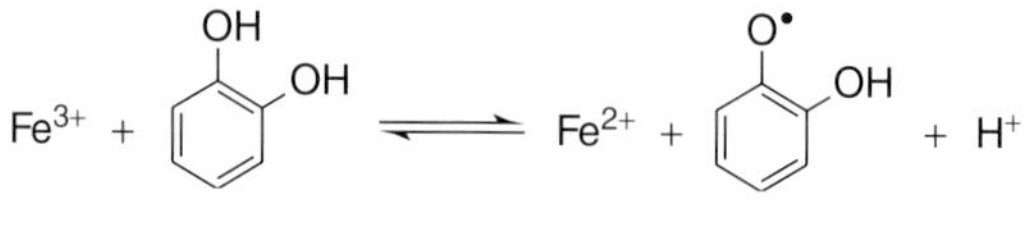

(7.76)

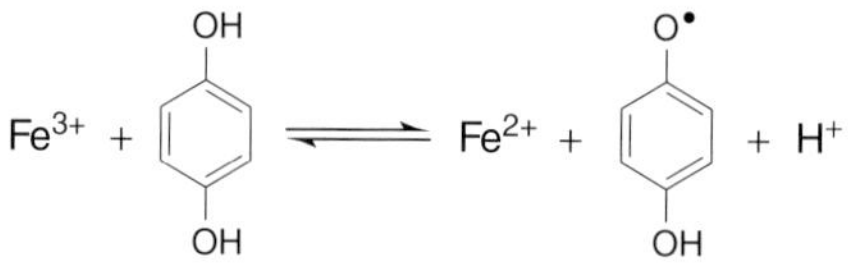

(7.77)

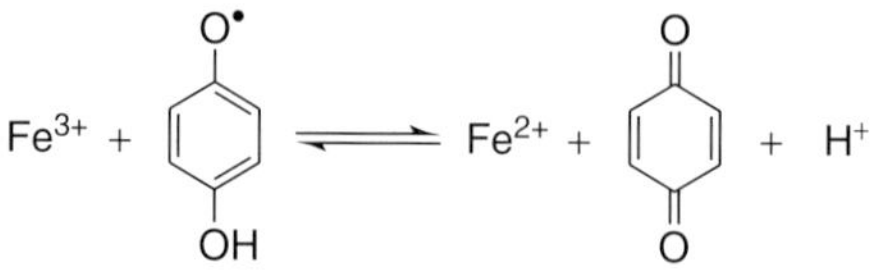

(7.78)

$$[Fe^{3+}(RCOO)^{-}]^{2+} + h\nu \longrightarrow Fe^{2+} + CO_2 + R^{\bullet} \qquad (7.79)$$

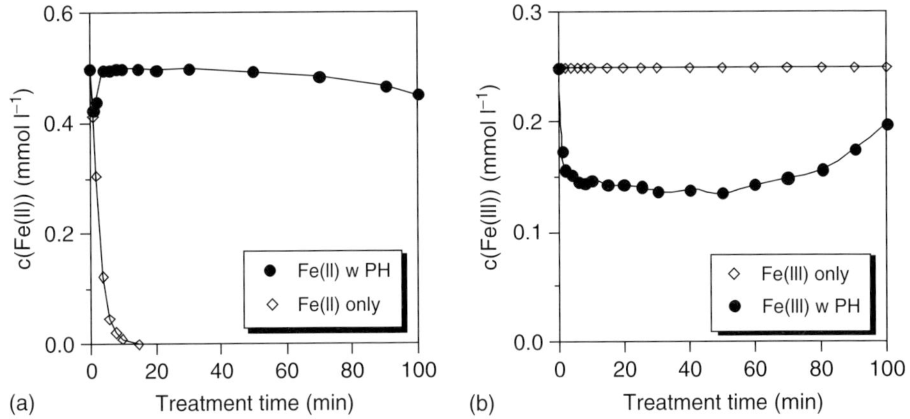

Figure 7.14 Evolution of iron ionic forms (ferrous and ferric) in the solution treated by corona discharge depending on the presence of 1 mM phenol (PH) (100 μS cm^{-1}, pH 3.4, 21 kV, 100 W). (a) 0.5 mmol l^{-1} $FeCl_2$, (b) 0.25 mmol l^{-1} $FeCl_3$ [20].

The above described reaction mechanism is demonstrated in Figure 7.14, which shows the evolution of ionic forms of iron (ferrous and ferric) in the solution treated by the corona discharge in the presence and absence of phenol. When a solution of $FeCl_2$ was treated by the discharge in the absence of phenol (Figure 7.14a), ferrous ions were completely oxidized to the ferric form in several minutes. In this case, ferrous ions were oxidized into ferric form by hydrogen peroxide in the Fenton process (Eq. (7.70)) and also by $OH^{\bullet}$ radicals (Eq. (7.80)) produced directly by the discharge from water and through the Fenton process (Eq. (7.70)). On the other hand, when the $FeCl_3$ solution was treated by the discharge, the ferric form of iron was stable in the solution (Figure 7.14b).

$$Fe^{2+} + OH^{\bullet} \longrightarrow Fe^{3+} + OH^{-} \tag{7.80}$$

A different situation occurred in the presence of phenol in the solution. When iron in the ferrous form was initially in the solution of phenol (Figure 7.14a), the concentration of ferrous ions (c_{Fe2+}) at first partly dropped, but then again recovered to the same initial value, and a near constant level of c_{Fe2+} was maintained during the presence of aromatic compounds in the solution.

In this case $OH^{\bullet}$ radicals produced by plasma and in Fenton's reaction (Eq. (7.70)) preferentially reacted with phenol rather than with Fe^{2+}, thus phenol served as an $OH^{\bullet}$ radical scavenger for ferrous ions (Table 7.5). The initial partial decline in c_{Fe2+} was caused by the consumption of Fe^{2+} in the reaction with H_2O_2 from the discharge (Eq. (7.70)). The subsequent recovery in c_{Fe2+} is attributable to the reduction of Fe^{3+} by organic intermediates of phenol (as dihydroxybenzenes and quinones) on reaching sufficient concentration in the phenol reaction mixture to participate in the iron(II) regeneration process (Eqs. (7.76)–(7.78)).

When iron in the ferric form was initially in the solution of phenol (Figure 7.14b), the ferrous–ferric ion equilibrium in the concentration ratio 2 : 3 was established soon after beginning of the discharge treatment. In this case, ferric ions were reduced by aromatic oxidation by-products of phenol to ferrous form (Eqs. (7.76)–(7.78)) and possibly also by photolysis of Fe(III) complexes of organic acids (Eq. (7.79)) induced by UV light emitted from the discharge in water. The subsequent decrease of c_{Fe2+} (Figure 7.14a) and slow recovery of c_{Fe3+} (Figure 7.14b), which were observed in later discharge treatment times, were caused by decreasing levels of phenol-reducing intermediates in the solution.

Mathematical modeling of the Fenton reaction system for direct discharge in water [91] fully supported the above mechanism, whereby H_2O_2 generated by the plasma in the liquid phase is the predominate source of $OH^\bullet$ radicals in the case of phenol decomposition. Similar kinetic modeling of gas–liquid hybrid reactors including gas-phase ozone formation and liquid-phase hydrogen peroxide generation [90, 93] also demonstrated the combined roles of $OH^\bullet$ radicals and ozone and the relative effects of these species on the primary target (phenol) and various reaction products.

7.4.2
Platinum

It has been demonstrated that platinum, when used as a high-voltage electrode, reduces the production rates of hydrogen, hydrogen peroxide, and oxygen by electrical discharge in water and enhances plasmachemical efficiency of pollutant removal from water in comparison to that with other electrode materials [251–253]. Specifically, it has been shown that Pt electrodes have a significant effect on removal of polychlorinated biphenyls and *s*-triazine by the discharge from water [37, 143]. It has been demonstrated that platinum particles could sputter from the solid platinum electrode, and thereafter cause heterogeneous catalytic reactions via a pH-dependent mechanism in which hydrogen, oxygen, and hydroxyl ions are adsorbed onto platinum metal particles, thereafter reacting with H_2O_2 molecule directly from the bulk liquid phase in an Eley–Rideal mechanism [253].

The first steps in the surface-catalyzed hydrogen peroxide decomposition involve the adsorption of molecular hydrogen (Eq. (7.81)), molecular oxygen (Eq. (7.82)), and hydroxyl ions (Eq. (7.83)) on the platinum surface.

$$H_2 + 2\,Pt \longrightarrow 2\,Pt - H \tag{7.81}$$

$$\tfrac{1}{2}\,O_2 + Pt \longrightarrow Pt - O \tag{7.82}$$

$$Pt + OH^- \longrightarrow Pt - (OH)_{ads} + e^- \tag{7.83}$$

Hydrogen and oxygen are adsorbed through dissociative adsorption in which hydrogen and oxygen atoms are directly bonded to the platinum. Hydroxyl ions are involved in a simple charge transfer reaction strongly dependent on pH.

$$Pt - H + H_2O_2 + e^- \longrightarrow Pt + H_2O + OH^- \tag{7.84}$$

$$Pt - (OH)_{ads} + H_2O_2 \longrightarrow Pt - (OOH)_{ads} + H_2O \tag{7.85}$$

Chemisorbed species further react with a molecule of hydrogen peroxide directly from the bulk liquid phase (Eqs. (7.84) and (7.85)) to form oxygen and water and recover the free platinum surface (Eqs. (7.86)–(7.88)).

$$Pt - (OOH)_{ads} + Pt - H \rightleftharpoons 2\,Pt + O_2 + H_2 \tag{7.86}$$

$$Pt - O + 2\,Pt - H \rightleftharpoons Pt - H_2O + 2\,Pt \tag{7.87}$$

$$Pt - H_2O \longrightarrow Pt + H_2O \tag{7.88}$$

Thus, two mechanisms are responsible for the catalytic hydrogen peroxide decomposition: (i) a hydrogen mechanism (Eqs. (7.81) and (7.84)), where the adsorbed hydrogen atoms decompose hydrogen peroxide and (ii) a pH mechanism (Eqs. (7.83) and (7.85)), which involves adsorption of hydroxyl ions. In acidic and moderate pH solutions, the first mechanism is dominant and hydrogen peroxide decomposition proceeds via hydrogen atom reduction (Pt–H). The second mechanism largely occurs only under alkaline conditions, causing additional decomposition of hydrogen peroxide and molecular hydrogen via Pt–OH.

7.4.2.1 The Role of Platinum as a Catalyst in Fenton's Reaction

It was found that platinum enhances the Fenton's reaction initiated by pulse electrical discharge in aqueous solutions containing ferrous/ferric salts [37, 143, 253]. It was shown that platinum particles emitted into the solution from the high-voltage Pt electrode can reduce ferrous iron to ferric iron by a catalytic reaction with plasma-generated H_2. Hydrogen gas adsorbs on the surface of platinum particles emitted into solution, where it dissociates forming hydrogen radicals (Eq. (7.81)), which then react with ferric ions as

$$Fe^{3+} + Pt{-}H \longrightarrow Fe^{2+} + Pt + H^+ \tag{7.89}$$

A catalytic cycle of iron between the ferrous and ferric forms is sustained through reactions (Eqs. (7.70),(7.81), and (7.89)), whereby the Fenton's reaction (Eq. (7.70)) is sustained to continuously produce hydroxyl radicals.

The reaction mechanism discussed above is demonstrated in Figure 7.15, which shows the evolution of ferrous ions in the solution treated by the corona discharge as a function of the electrode material (NiCr or Pt) and the initial form of iron in the solution (ferrous or ferric). Figure 7.15a shows that when ferrous ions were initially added to the reactor, c_{Fe2+} dropped to almost zero in the case of the NiCr electrode, while in the case of Pt electrode, c_{Fe2+} dropped to about one-fifth of its initial value and thereafter remained constant. When ferric ions were initially added to the reactor (Figure 7.15b), very small amounts of ferrous ions were formed in the case of NiCr. On the other hand, when Pt was used as the high-voltage electrode, there was a continuous increase in the c_{Fe2+}. These results demonstrate that Pt was directly involved in the reduction of the ferric ions to ferrous ions in the electrical discharge. The experiment with commercially available platinum particles suspended in solution showed results similar to the cases where Pt was emitted from the electrode, and these results confirmed that the reduction takes place on particles suspended in the bulk rather then on the high-voltage electrode.

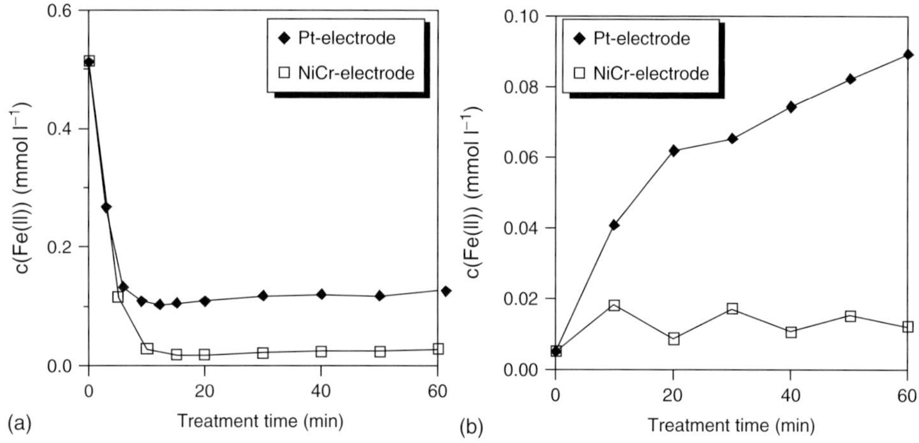

Figure 7.15 Ferrous concentration dependending on the material of electrode (NiCr or Pt) and ionic initial form of iron in the solution treated by corona discharge (130 μS cm^{-1}, 45 kV, 60 W). (a) Ferrous form (0.5 mmol l^{-1} $FeSO_4$) and (b) ferric form (0.185 mmol l^{-1} $FeCl_3$). (Source: Reprinted from [253]. Copyright (2007), with permission from Elsevier.)

7.4.3 Tungsten

Tungsten used as a high-voltage electrode has been demonstrated to significantly affect the plasmachemical activity of the discharge in water [236, 250]. Tungstate ions (WO_4^{2-}), and solid particles of tungsten, which were released into the water from the tungsten electrode by erosion in the discharge, were demonstrated to cause a catalytic decomposition of H_2O_2 produced by the corona discharge in water. While using DMSO as the probe organic compound, it was shown that released tungstates can also enhance the plasmachemical activity of the discharge due to tungstate-catalyzed oxidation of DMSO by H_2O_2 [236].

Reactions of H_2O_2 with tungsten (tungstate ions) have been demonstrated through comparison of the yields of H_2O_2 obtained using tungsten electrodes with those determined for titanium electrodes. Lower production of H_2O_2 was found using the tungsten electrode compared to the titanium electrode, and a subsequent decrease during the postdischarge period in the H_2O_2 concentration was observed in the solution treated using the W electrode after switching off the discharge [236]. Similar results were found in a gas–liquid discharge where Ti, Ni-Cr, Cu, stainless steel, tungsten-copper, and tungsten carbide had the same initial rates of formation, but the tungsten containing electrodes showed a drop in H_2O_2 at longer times [250].

A correlation between the amount of eroded tungsten material released into the solution and the chemical effects induced by the discharge was determined, and tungstate WO_4^{2-} ions, which involved up to 70% of the eroded tungsten electrode material released into the solution, were shown to play a dominant role in the decomposition of H_2O_2. Depending on the reaction conditions, the tungstate-catalyzed disproportionation of hydrogen peroxide proceeds mainly through the formation of the mono-, di-, and tetraperoxotungstate intermediates, $[WO_{4-n}(O_2)_n]^{2-}$ ($n = 1, 2, 4$), which are converted back into tungstate anion by reactions (Eqs. (7.90) and (7.91)) [260, 261].

$$WO_4^{2-} + nH_2O_2 \longrightarrow [WO_{4-n}(O_2)_n]^{2-} + nH_2O \quad (7.90)$$

$$[WO_{4-n}(O_2)_n]^{2-} + 2nH^+ + 2ne^- \longrightarrow WO_4^{2-} + nH_2O \quad (7.91)$$

The experiment with Na_2WO_4 added to the solution in the same amount as determined for the solution treated by the discharge showed a similar decrease in H_2O_2 concentration. In addition to WO_4^{2-}, tungsten particles sputtered from the W electrode were shown to contribute to the H_2O_2 decomposition.

Concerning the mechanism of the formation of tungstate ions by the discharge-induced erosion of tungsten electrodes, several possible pathways have been proposed. These include formation of tungsten oxide WO_3 either by the electrolytic anodic oxidation of tungsten metal (Eq. (7.92)) or by plasma sputtering of tungsten, followed by WO_3 dissolution into tungstate ions (Eq. (7.93)) [236].

$$W + 3H_2O = WO_3 + 6H^+ + 6e^- \quad (7.92)$$

$$WO_3 + OH^- \longrightarrow WO_4^{2-} + H^+ \quad (7.93)$$

Tungsten metal particles released from tungsten electrodes might be also oxidized in the solution by hydrogen peroxide produced by the discharge into WO_4^{2-} ions (Eq. (7.94)).

$$W + 3H_2O_2 \longrightarrow WO_4^{2-} + 2H^+ + 2H_2O \quad (7.94)$$

The effects of the tungsten electrode material and the catalytic role of tungstate ions in the plasmachemical activity of the corona discharge in water were also demonstrated by the decomposition of dimethylsulfoxide $(CH_3)_2SO$. A higher degradation of DMSO was determined in the tungsten electrode case compared to that with the Ti electrode, which was attributed to the combined action of DMSO oxidation by $OH^\bullet$ radicals (Section 7.3.3.2) and tungstate-catalyzed oxidation of DMSO by hydrogen peroxide (Eq. (7.95)).

$$(CH_3)_2SO + H_2O_2 \xrightarrow{WO_4^{2-}} (CH_3)_2SO_2 + H_2O \quad (7.95)$$

Such a mechanism was proved through the detection of the DMSO degradation by-products (methanesulfonate, sulfate, and dimethyl sulfone). Since the catalytic oxidation of DMSO in the WO_4^{2-}/H_2O_2 system proceeds almost exclusively through the formation of dimethylsulfone $(CH_3)_2SO_2$ (Eq. (7.95)) [262, 263], the detection of dimethylsulfone in the reaction mixture of DMSO provided strong evidence to the contribution of WO_4^{2-} ions in the degradation of DMSO by the discharge.

The catalytic role of tungstate ions on the plasmachemical activity of the discharge has also been observed to be affected by the pH of the aqueous solution. Decomposition of H_2O_2 and DMSO proceeded more slowly in the NaH_2PO_4 solution ($pH_0 = 5.5$) than in the H_3PO_4 solution ($pH_0 = 2.9$). Since tungstates form with decreasing pH (pH < 7.8) a number of different polynuclear species in an aqueous solution (e.g., $W_{12}O_{42}^{12-}$, $W_{12}O_{41}^{10-}$, $HW_6O_{21}^{5-}$, and $H_3W_6O_{21}^{3-}$), which have different reactivities with H_2O_2 [261, 262], these pH effects were attributed to the pH-dependent catalytic activity of tungstates in the disproportionation of H_2O_2.

7.4.4 Titanium Dioxide

Introducing photocatalytically active titanium dioxide solid particles into the liquid phase of the discharge plasma reactor (either in the form of powder or immobilized on glass beads or carbon fibers) has been shown to enhance plasmachemical processes in water. More efficient removal of organic compounds from aqueous solutions treated by pulsed corona discharge, gliding arc discharge, and DBD has been demonstrated with the presence of TiO_2 compared to the cases without TiO_2 in the solution (e.g., phenols, organic dyes, or halocarbons) [95–103, 125, 126, 154–157, 264]. At the same time, higher concentrations of hydrogen peroxide, $OH^\bullet$ radicals, and $O^\bullet$ radicals were determined in the liquid-phase plasma systems combined with TiO_2 catalyst [96–98]. Since electrical discharges produce significant quantities of UV light in the liquid phase [45], the TiO_2 enhancement was attributed mainly to the photocatalytic processes induced on the TiO_2 surface by UV radiation from plasma. Photoelectrocatalytic effects have not been studied within the context of high-voltage electrical discharge processes; however, at low voltage, TiO_2 photocatalysts can be activated using relatively low-voltage sources in water [265].

The photocatalytic activity of TiO_2 is derived from its ability to form charge carriers on its surface on illumination with the light of wavelength $\lambda < 390$ nm (i.e., with photons of greater energy than bandgap energy of TiO_2, $E_{bg} = 3.2$ eV). These charge carriers are the excited-state conduction band electrons (e^-_{cb}) and valence band holes (h^+_{vb})

$$TiO_2 + h\nu \longrightarrow e^-_{cb} + h^+_{vb} \tag{7.96}$$

In the presence of water and oxygen, these species produce highly reactive species on the TiO_2 surface such as $OH^\bullet$ and $O_2^{\bullet-}$ radicals

$$H_2O + h^+_{vb} \longrightarrow OH^\bullet + H^+ \tag{7.97}$$

$$OH^- + h^+_{vb} \longrightarrow OH^\bullet \tag{7.98}$$

$$O_2 + e^-_{cb} \longrightarrow O_2^{\bullet-} \tag{7.99}$$

Thus, in the case of a combined plasma/TiO_2 system, a higher yield of oxidative species can be generated and is available for the destruction of organic compounds in water. The oxidation by $OH^\bullet$ radical was found to be the dominant process in this case through analysis of the reaction intermediates formed during degradation of probe organic compounds in the liquid-phase plasma systems combined with the TiO_2 catalyst. For example, catechol, hydroquinone, and 1,4-benzoquinone were found to be the main by-products of phenol decomposition [96, 97], and

4-chloro-1,4-benzoquinone, hydroquinone, and 1,4-benzoquinone were the main by-products of 4-chlorophenol [126]. All these by-products correspond to the oxidation by $OH^\bullet$ radical via its electrophilic addition to the aromatic ring of phenol and 4-chlorophenol. The same reaction intermediates were reported during the photocatalytic decomposition of phenol and 4-chlorophenol in aqueous TiO_2 suspensions [266] as well as during plasmachemical degradation of these compounds by the discharge in water alone (Sections 7.3.1.1 and 7.3.1.2).

Wang *et al.* [98] demonstrated significantly higher relative emission intensities of $OH^\bullet$ (313 nm) and $O^\bullet$ radicals (777 nm) in pulsed corona discharge in water combined with TiO_2 catalyst than that in the discharge alone. While they used glass beads with immobilized TiO_2 as the photocatalyst introduced into the plasma reactor, the increase in relative emission intensities of $OH^\bullet$ and $O^\bullet$ radicals in the plasma/TiO_2 system could be partly associated with the change of plasma characteristics because of the presence of glass beads in the discharge. Nevertheless, experiments with $OH^\bullet$ scavengers (carbonates and *n*-butanol) performed by the same group [98, 100] showed a decrease of phenol removal with the increasing scavenging capacity of Na_2CO_3 and *n*-butanol (performed under constant solution conductivity) when a higher decrease in phenol degradation was determined in the plasma–photocatalytic system than in the plasma alone system. These results suggest a synergetic effect of combined pulsed discharge plasma with TiO_2 photocatalyst and that the $OH^\bullet$ radical is the most important species involved in enhancing the photocatalytic effect of TiO_2 induced in liquid-phase plasma/TiO_2 systems.

The enhancement of plasma-catalytic processes in the presence of TiO_2 through solution conductivity has been demonstrated with phenol degradation by pulsed corona discharge in water [96]. Since the intensity of UV radiation from the discharge in water increases with the solution conductivity [45], it was concluded that the effect of solution conductivity is most likely related to the increasing contribution of plasma-induced photocatalytic processes by TiO_2 on phenol degradation with higher solution conductivity.

Consequently, larger production of hydrogen peroxide was found in aqueous solutions treated in the plasma/TiO_2 systems with organic content compared to the yields of H_2O_2 produced by the discharge in water without the TiO_2 catalyst and organic content [96–98]. Since under photocatalysis molecular oxygen dissolved in water reacts on the TiO_2 surface with photogenerated electrons (Eq. (7.99)), hydrogen peroxide is produced from superoxide radical anion ($O_2^{\bullet-}$)

$$2\,O_2^{\bullet-} + 2\,H^+ \rightleftharpoons 2\,HO_2^\bullet \longrightarrow H_2O_2 + O_2 \quad (7.100)$$

Hydrogen peroxide is also formed by recombination of $OH^\bullet$ radicals produced in reactions (Eqs. (7.97) and (7.98)). The photocatalytic conversion of hydrogen peroxide by the conduction band electrons or valence band holes can than also occur on the TiO_2 surface:

$$H_2O_2 + e^-_{cb} \longrightarrow OH^\bullet + OH^- \quad (7.101)$$

$$H_2O_2 + h^+_{vb} \longrightarrow HO_2^\bullet + H^+ \quad (7.102)$$

In addition, the yield of photocatalytically formed H_2O_2 can also be increased by the addition of excessive amounts of a suitable organic or inorganic compound into the solution serving as an electron donor to the positive holes of the TiO_2 [266]. The higher production of H_2O_2 in the plasma/TiO_2 system with organic compounds in water can then be explained by the competition of these electron donors for valence band holes, with e^-_{cb}/h^+_{vb} recombination allowing conduction band electrons to react with molecular oxygen. In the absence of such species that are able to rapidly react with the photo-generated holes, the electron–hole recombination reaction becomes more efficient compared to molecular oxygen reduction, and the photocatalytic production of H_2O_2 is decreased.

7.4.5 Activated Carbon

It has been shown that activated carbon (AC) may function as both an adsorbent and a catalyst in the liquid-phase electrical discharge. Grymonpre *et al.* [91–93] have shown that, compared with pulsed corona discharge alone, the combination of pulsed corona discharge with suspended AC particles leads to enhancement of the overall removal of phenol from the aqueous solution. The enhancement has been attributed to phenol oxidation in the bulk liquid phase induced by pulsed corona discharge and physical adsorption of phenol on AC, as well as to putative surface-phase reactions on the AC induced by the electrical discharge. The combination of the AC and potassium salts, however, has also been shown to affect the power waveforms of the pulsed corona discharge in water. Oxygen bubbling through the high-voltage needle electrodes with addition of suspended AC into the discharge was shown to give the best phenol removal efficiency because of subsequent reactions of dissolved ozone and the adsorption and reaction effects on the AC. A mathematical model taking account an adsorption, mass transfer, and surface reaction on AC has been developed [93]. It was proposed that AC can participate in radical-based reactions under highly oxidative conditions in liquid-phase pulsed corona reactors, for example, as a catalyst to convert ozone into hydroxyl radicals.

Zhang *et al.* [163, 164] have shown that carbon chemical surface properties have significant effect on the plasma-induced decomposition of methyl orange by the pulsed electrical discharge in water combined with AC and ozone. They have observed an increase in the number of oxygen-containing functional groups of basic and acidic nature on the AC surface caused by its oxidation by plasma and ozone in the discharge. Acidic and basic carbon surfaces both accelerated decomposition of MO by the discharge in combination with AC and ozone. It was shown that spent carbon can be regenerated *in situ* by the plasma. When saturated AC was added into the reactor, its adsorption capacity was recovered after treatment with the discharge. The authors concluded that AC acts not only as adsorbent but also as catalyst promoting the decomposition of ozone into $OH^\bullet$ radicals and enhancing the MO degradation by the discharge.

Their conclusion is supported by work of Alvarez *et al.* [267] who reported that surface chemistry on AC plays a key role in promoting surface ozone decomposition by AC. They have demonstrated that basic and hydroxyl surface oxygen groups (SOG) of AC are the groups that most influence the ozone decomposition process. It was deduced that these groups (and also some types of acidic SOG) are the active sites for ozone transformation into $OH^{\bullet}$ radicals, which proceeds through the formation of H_2O_2 on surface of AC and its further dissociation via radical chain reaction initiated by OH^- and HO_2^- ions in the liquid bulk phase.

7.4.6 Silica Gel

Similar plasma-induced synergistic effects of solid-phase catalysts and ozone as in the case of AC (Section 7.4.5) have been reported for silica gel beads introduced into the pulse corona discharge directly in water [95]. A significant improvement in the rate of decomposition of phenol and MB in water was determined in a hybrid system including an underwater pulse corona discharge with silica gel and ozone addition. The positive effects of silica gel were explained by surface-mediated reactions between the adsorbed pollutant molecules and the chemically active species produced in close proximity to the solid surfaces. The occurrence of surface-mediated reactions has been demonstrated on the decolorization of blue beads with preadsorbed MB, which were treated with the corona discharge in the presence of ozone and distilled water flowing through the reactor. The observed transparent patches in partially regenerated beads were attributed to the higher concentration of chemically active species in those regions, which were produced by surface discharge at the contact points between the beads as a result of a higher electric field and the difference in permittivities of silica gel and water.

7.4.7 Zeolites

The presence of zeolite particles suspended in the liquid phase with direct discharge in the liquid and discharges in a combined gas–liquid environment has been shown to enhance the degradation of some organic dyes and phenol in water [89, 162, 172, 173]. Zeolites were found to affect the decomposition process in water differently depending on the zeolite type (namely NH_4ZSM5, FeZSM5, and HY) and the presence of ozone in water. Part of the enhancement in the case of HY zeolite was attributed to its acidic properties, which caused a decrease of pH on addition of the zeolite into water (pH 3.8) and which also affects dye decolorization. However, there was an additional effect of HY zeolite beyond that caused by the acid enhancement since a higher dye removal was obtained with HY zeolite in the discharge compared to that with dye removal, which was obtained by the discharge under acidic condition in the solution of the same pH without zeolite. Similar experiment performed with NH_4ZSM5 zeolite, which did not affect pH of the solution and gave even higher dye removal than that with the HY zeolite, implied that enhancement by zeolites can be caused by possible ozone–zeolite interactions induced by the discharge at the zeolite surface. This mechanism was further supported by experiments performed with phenol, which showed that the $OH^•$ radical produced by the discharge may be quenched and that ozone reactions may be enhanced by these zeolites. In the case of Fe-exchanged zeolite, FeZSM5, it has been demonstrated that the main enhancing effect of zeolite addition into the discharge is an additional generation of $OH^•$ radicals by Fenton reaction (Eq. (7.70)) between H_2O_2 produced by the discharge in water and Fe^{2+} ions doped in the zeolite framework as well as those leached from zeolite into the bulk solution.

7.5 Concluding Remarks

It has been shown in this chapter that electrical discharge plasma produce in water a number of chemical effects. The fundamentals of aqueous-phase processes induced by electrical discharges in water and gas–liquid interfaces were discussed and examples of their application in the plasmachemical destruction of various organic compounds in water were provided. However, there is still a need for much further work to improve our understanding of the mechanisms of plasmachemical processes induced in the liquid phase in order to optimize and enhance their efficiency since different compounds possess different sensitivity and susceptibility to the chemical effects induced by reactive chemical species formed by plasma in water and gas–liquid environments. It is also important to recognize that plasmachemical processes induced by various types of high-voltage electrical discharges considered in this chapter have many common features with other advanced oxidation technologies (AOTs) used for water pollutant abatement. Therefore, it is also necessary to answer a number of questions with regard to the prospects of using electrical discharges for potential environmental applications. For example, do electrical discharge processes in the liquid phase lead to fundamentally different reaction mechanisms and breakdown pathways than those of other AOTs? Most of the work on the liquid-phase organic compound degradation presented in this chapter considers hydroxyl radicals, formed directly in the water or indirectly through reactions from hydrogen peroxide, ozone, or peroxynitrite, as the primary reactive species for the degradation of organic compounds in water (in addition to the direct oxidation by ozone, peroxynitrite, or reduction by hydrogen or superoxide radicals). However, it is possible that other factors such as direct UV light, shockwaves, local supercritical conditions formed by the plasma, or even direct plasma contact may also contribute to the organic compound destruction. Further work at both the laboratory scale and the pilot scale is necessary to fully evaluate the potential of these types of processes for water pollution control. It is vitally important to develop data for the comparison of the various processes under similar conditions of contaminant concentration, solution pH, and conductivity, as well as residence time and energy density.

Furthermore, in addition to the use of plasma in liquids for chemical decontamination processes, plasma is also able to stimulate the processes in the opposite direction, that is, plasmachemical synthesis. These processes were not covered in this chapter; however, this field is of growing interest in the plasma-chemical applications of electrical discharges in the liquid and the gas–liquid environments. Pioneering work on aqueous-phase plasmachemical synthesis was performed by Miller [268] who demonstrated that using electric discharge in contact with water more complex bioorganic compounds, including many amino acids, could be produced. More recently, plasmas in liquids were used, for example, for the production of hydrogen gas from hydrocarbons (e.g., aliphatic alcohols, which were briefly reviewed in Section 7.3.3.1, see also Chapter 9), polymer surface functionalization [269–273], synthesis of polymers in aqueous solutions (e.g., polymethacrylate, polyacrylamide-co-acrylic acid hydrogels) [274–276], or production of nanoparticles in the liquid [277–280]. In this case, electrical discharges were generated, in addition to those in aqueous solutions, in organic solvents or ionic liquids, in which liquid-phase plasmas induce different type of plasma-chemical processes [218, 277, 281, 282]. Other applications of high-energy capacitor discharges, pulsed arcs, in water include simulation of underwater explosions, metal forming, rock fragmentation, shock wave lithotripsy, and biological and biomedical applications such as surgery and skin treatment [283]. Some of the last mentioned effects are evaluated more in the Chapter 8, which is devoted to the biological effects of electrical discharges in water and gas–liquid environments.

Acknowledgments

Lukes would like to acknowledge financial support from the Grant Agency of Academy of Sciences of the Czech Republic (project No. IAAX00430802). Locke would like to gratefully acknowledge financial support from the US National Science Foundation under grant number CBET-0932481.

References

1. Hoigne, J. and Bader, H. (1983) Rate constants of reactions of ozone with organic and inorganic compounds in water. I. Non-dissociating organic compounds. *Water Res.*, **17** (2), 173–183.
2. Hoigne, J. and Bader, H. (1983) Rate constants of reactions of ozone with organic and inorganic compounds in water. II. Dissociating organic compounds. *Water Res.*, **17** (2), 185–194.
3. Brisset, J.L., Benstaali, B., Moussa, D., Fanmoe, J., and Njoyim-Tamungang, E. (2011) Acidity control of plasma-chemical oxidation: Applications to dye removal, urban waste abatement and microbial inactivation. *Plasma Sources Sci. Technol.*, **20** (3), 034021.
4. Brisset, J.L., Prevot, F., Doubla, A., Lelievre, J., and Amouroux, J. (1990) Acid-base reactions induced by a plasma phase on liquid targets: Bronsted acidity and oxo synthesis. *J. Phys.*, **51** (18), C5245–C5252.
5. Brisset, J.L., Lelievre, J., Doubla, A., and Amouroux, J. (1990) Interactions with aqueous solutions of the air

corona products. *Rev. Phys. Appl.*, **25** (6), 535–543.
6. Goldman, A., Goldman, M., Sigmond, R.S., and Sigmond, T. (1989) Analysis of air corona products by means of their reactions in water, in *ISPC-9: 9th International Symposium on Plasma Chemistry* (ed. R.D. Agostino), University of Bari, Pugnochiuso, Italy, 1654.
7. Sharma, A.K., Camaioni, D.M., Josephson, G.B., Goheen, S.C., and Mong, G.M. (1997) Formation and measurement of ozone and nitric acid in a high voltage DC negative metallic point-to-aqueous-plane continuous corona reactor. *J. Adv. Oxid. Technol.*, **2** (1), 239–247.
8. Janca, J., Kuzmin, S., Maximov, A., Titova, J., and Czernichowski, A. (1999) Investigation of the chemical action of the gliding and "point" arcs between the metallic electrode and aqueous solution. *Plasma Chem. Plasma Process.*, **19** (1), 53–67.
9. Oehmigen, K., Hahnel, M., Brandenburg, R., Wilke, C., Weltmann, K.D., and von Woedtke, T. (2010) The role of acidification for antimicrobial activity of atmospheric pressure plasma in liquids. *Plasma Process. Polym.*, **7** (3-4), 250–257.
10. Benstaali, B., Moussa, D., Addou, A., and Brisset, J.L. (1998) Plasma treatment of aqueous solutes: Some chemical properties of a gliding arc in humid air. *Eur. Phys. J.: Appl. Phys*, **4** (2), 171–179.
11. Brisset, J.L., Moussa, D., Doubla, A., Hnatiuc, E., Hnatiuc, B., Youbi, G.K., Herry, J.M., Naitali, M., and Bellon-Fontaine, M.N. (2008) Chemical reactivity of discharges and temporal post-discharges in plasma treatment of aqueous media: examples of gliding discharge treated solutions. *Ind. Eng. Chem. Res.*, **47** (16), 5761–5781.
12. Moussa, D., Abdelmalek, F., Benstaali, B., Addou, A., Hnatiuc, E., and Brisset, J.L. (2005) Acidity control of the gliding arc treatments of aqueous solutions: Application to pollutant abatement and biodecontamination. *Eur. Phys. J.: Appl. Phys*, **29** (2), 189–199.
13. Moussa, D., Doubla, A., Kamgang-Youbi, G., and Brisset, J.L. (2007) Postdischarge long life reactive intermediates involved in the plasma chemical degradation of an azoic dye. *IEEE Trans. Plasma Sci.*, **35** (2), 444–453.
14. Burlica, R., Kirkpatrick, M.J., Finney, W.C., Clark, R.J., and Locke, B.R. (2004) Organic dye removal from aqueous solution by glidarc discharges. *J. Electrost.*, **62** (4), 309–321.
15. Burlica, R. and Locke, B.R. (2008) Pulsed plasma gliding-arc discharges with water spray. *IEEE Trans. Ind. Appl.*, **44** (2), 482–489.
16. Burlica, R., Kirkpatrick, M.J., and Locke, B.R. (2006) Formation of reactive species in gliding arc discharges with liquid water. *J. Electrost.*, **64** (1), 35–43.
17. Porter, D., Poplin, M.D., Holzer, F., Finney, W.C., and Locke, B.R. (2009) Formation of hydrogen peroxide, hydrogen, and oxygen in gliding arc electrical discharge reactors with water spray. *IEEE Trans. Ind. Appl.*, **45** (2), 623–629.
18. Burlica, R., Grim, R.G., Shih, K.Y., Balkwill, D., and Locke, B.R. (2010) Bacteria inactivation using low power pulsed gliding arc discharges with water spray. *Plasma Process. Polym.*, **7** (8), 640–649.
19. Shainsky, N., Dobrynin, D., Ercan, U., Joshi, S., Ji, H., Brooks, A., Fridman, G., Cho, Y., Fridman, A., and Friedman, G. (2011) Non-equilibrium plasma treatment of liquids, formation of plasma acid, *ISPC-20: 20th International Symposium on Plasma Chemistry*, A.J. Drexel Plasma Institute, Philadelphia, PA.
20. Lukes, P. (2002) *Water treatment by pulsed streamer corona discharge in water*. PhD Dissertation. Institute of Plasma Physics AS CR, Prague, Czech Republic.
21. Seinfeld, J.H. and Pandis, S.N. (2006) *Atmospheric Chemistry and Physics: From Air Pollution to Climate Change*, 2nd edn, John Wiley & Sons, Inc, Hoboken, NJ.

22. Buxton, G.V., Greenstock, C.L., Helman, W.P., and Ross, A.B. (1988) Critical review of rate constants for reactions of hydrated electrons, hydrogen atoms and hydroxyl radicals ($^{\bullet}OH/^{\bullet}O^{-}$) in aqueous solution. *J. Phys. Chem. Ref. Data*, **17** (2), 513–886.

23. Hoigne, J., Bader, H., Haag, W.R., and Staehelin, J. (1985) Rate constants of reactions of ozone with organic and inorganic compounds in water. III. Inorganic compounds and radicals. *Water Res.*, **19** (8), 993–1004.

24. Benitez, F.J., Beltranheredia, J., Gonzalez, T., and Pascual, A. (1993) Ozone treatment of methylene blue in aqueous solutions. *Chem. Eng. Commun.*, **119**, 151–165.

25. NDRL/NIST Solution Kinetics Database on the Web. (2011) NIST standard reference database 40, *http://kinetics.nist.gov/solution*.

26. Tarr, M.A. (2003) *Chemical Degradation Methods for Wastes and Pollutants: Environmental and Industrial Applications*, Marcel Dekker, Inc., New York.

27. Hoigne, J. (1998) Chemistry of aqueous ozone and transformation of pollutants by ozonation and advanced oxidation processes, in *The Handbook of Environmental Chemistry, Vol. 5, Part C, Quality and Treatment of Drinking Water II* (ed. J. Hrubec), Springer-Verlag, Berlin Heidelberg, pp. 83–141.

28. Hoigne, J. and Bader, H. (1976) Role of hydroxyl radical reactions in ozonation processes in aqueous solutions. *Water Res.*, **10** (5), 377–386.

29. Staehelin, J. and Hoigne, J. (1982) Decomposition of ozone in water - rate of initiation by hydroxide ions and hydrogen peroxide. *Environ. Sci. Technol.*, **16** (10), 676–681.

30. Beckman, J.S., Beckman, T.W., Chen, J., Marshall, P.A., and Freeman, B.A. (1990) Apparent hydroxyl radical production by peroxynitrite - implications for endothelial injury from nitric oxide and superoxide. *Proc. Natl. Acad. Sci. U.S.A.*, **87** (4), 1620–1624.

31. Beckman, J.S. and Koppenol, W.H. (1996) Nitric oxide, superoxide, and peroxynitrite: the good, the bad, and the ugly. *Am. J. Physiol. Cell Physiol.*, **271** (5), C1424–C1437.

32. Goldstein, S., Squadrito, G.L., Pryor, W.A., and Czapski, G. (1996) Direct and indirect oxidations by peroxynitrite, neither involving the hydroxyl radical. *Free Radical Biol. Med.*, **21** (7), 965–974.

33. Squadrito, G.L. and Pryor, W.A. (1998) Oxidative chemistry of nitric oxide: the roles of superoxide, peroxynitrite, and carbon dioxide. *Free Radical Biol. Med.*, **25** ((4–5), 392–403.

34. Peyton, G.R., Girin, E., and Lefaivre, M.H. (1995) Reductive destruction of water contaminants during treatment with hydroxyl radical processes. *Environ. Sci. Technol.*, **29** (6), 1710–1712.

35. Watts, R.J., Bottenberg, B.C., Hess, T.F., Jensen, M.D., and Teel, A.L. (1999) Role of reductants in the enhanced desorption and transformation of chloroaliphatic compounds by modified Fenton's reactions. *Environ. Sci. Technol.*, **33** (19), 3432–3437.

36. Sahni, M., Finney, W.C., Clark, R.J., Landing, W., and Locke, B.R. (2002) Degradation of aqueous phase trichloroethylene using pulsed corona discharge, in *Proceedings of the 8th International Symposium High Pressure, Low Temperature Plasma Chemistry (HAKONE 8)*, University of Tartu, Puhajarve, Estonia.

37. Sahni, M., Finney, W.C., and Locke, B.R. (2005) Degradation of aqueous phase polychlorinated biphenyls (PCB) using pulsed corona discharges. *J. Adv. Oxid. Technol.*, **8** (1), 105–111.

38. Sahni, M. (2006) Analysis of chemical reactions in pulsed streamer discharges: an experimental study. PhD Dissertation. Florida State University, Tallahassee, FL.

39. Sunka, P., Babicky, V., Clupek, M., Lukes, P., Simek, M., Schmidt, J., and Cernak, M. (1999) Generation of chemically active species by electrical discharges in water. *Plasma Sources Sci. Technol.*, **8**, 258–265.

40. Sun, B., Sato, M., and Clements, J.D. (1997) Optical study of active species produced by a pulsed streamer corona discharge in water. *J. Electrost.*, **39**, 189–202.

41. Namihira, T., Sakai, S., Yamaguchi, T., Yamamoto, K., Yamada, C., Kiyan, T., Sakugawa, T., Katsuki, S., and Akiyama, H. (2007) Electron temperature and electron density of underwater pulsed discharge plasma produced by solid-state pulsed-power generator. *IEEE Trans. Plasma Sci.*, **35** (3), 614–618.
42. An, W., Baumung, K., and Bluhm, H. (2007) Underwater streamer propagation analyzed from detailed measurements of pressure release. *J. Appl. Phys.*, **101** (5), 053302.
43. Kirkpatrick, M.J. and Locke, B.R. (2005) Hydrogen, oxygen, and hydrogen peroxide formation in aqueous phase pulsed corona electrical discharge. *Ind. Eng. Chem. Res.*, **44** (12), 4243–4248.
44. Sahni, M. and Locke, B.R. (2006) Quantification of reductive species produced by high voltage electrical discharges in water. *Plasma Process. Polym.*, **3** (4-5), 342–354.
45. Lukes, P., Clupek, M., Babicky, V., and Sunka, P. (2008) Ultraviolet radiation from the pulsed corona discharge in water. *Plasma Sources Sci. Technol.*, **17** (2), 024012.
46. Joshi, A.A., Locke, B.R., Arce, P., and Finney, W.C. (1995) Formation of hydroxyl radicals, hydrogen peroxide and aqueous electrons by pulsed streamer corona discharge in aqueous solution. *J. Hazard. Mater.*, **41**, 3–30.
47. Thagard, S.M., Takashima, K., and Mizuno, A. (2009) Chemistry of the positive and negative electrical discharges formed in liquid water and above a gas-liquid surface. *Plasma Chem. Plasma Process.*, **29** (6), 455–473.
48. Spinks, J.W.T. and Woods, R.J. (1990) *An Introduction to Radiation Chemistry*, 3rd edn, John Wiley & Sons, Inc, New York.
49. Bielski, B.H.J., Cabelli, D.E., Arudi, R.L., and Ross, A.B. (1985) Reactivity of HO_2/O_2^- radicals in aqueous solution. *J. Phys. Chem. Ref. Data*, **14** (4), 1041–1100.
50. Legrini, O., Oliveros, E., and Braun, A.M. (1993) Photochemical processes for water treatment. *Chem. Rev.*, **93** (2), 671–698.
51. Bolton, J. and Cater, S. (1994) Homogenous photodegradation of pollutants in contaminated water: An introduction, in *Aquatic and Surface Photochemistry* (eds G. Heiz, R. Zepp, and D. Crosby), Lewis Publishers, Boca Raton, FL, pp. 467–490.
52. Peyton, G.R. and Glaze, W.H. (1988) Destruction of pollutants in water with ozone in combination with ultraviolet-radiation. 3.Photolysis of aqueous ozone. *Environ. Sci. Technol.*, **22** (7), 761–767.
53. Hunt, J.P. and Taube, H. (1952) The photochemical decomposition of hydrogen peroxide - quantum yields, tracer and fractionation effects. *J. Am. Chem. Soc.*, **74** (23), 5999–6002.
54. Baxendale, J.H. and Wilson, J.A. (1957) The photolysis of hydrogen peroxide at high light intensities. *Trans. Faraday Soc.*, **53** (3), 344–356.
55. Malik, M.A. (2010) Water purification by plasmas: which reactors are most energy efficient? *Plasma Chem. Plasma Process.*, **30** (1), 21–31.
56. Locke, B.R. and Shih, K.Y. (2011) Review of the methods to form hydrogen peroxide in electrical discharge plasma with liquid water. *Plasma Sources Sci. Technol.*, **20** (3), 034006.
57. Dors, M., Metel, E., and Mizeraczyk, J. (2006) Influence of iron ions on the phenol oxidation. *Czech. J. Phys.*, **56**, B1271–B1276.
58. Dang, T.H., Denat, A., Lesaint, O., and Teissedre, G. (2009) Pulsed electrical discharges in water for removal of organic pollutants: a comparative study. *Eur. Phys. J.: Appl. Phys*, **47** (2), 22818.
59. Shi, J.W., Bian, W.J., and Yin, X.L. (2009) Organic contaminants removal by the technique of pulsed high-voltage discharge in water. *J. Hazard. Mater.*, **171** (1-3), 924–931.
60. Hoeben, W., van Velduizen, E.M., Rutgers, W.R., and Kroesen, G.M.W. (1999) Gas phase corona discharges for oxidation of phenol in an aqueous

solution. *J. Phys. D: Appl. Phys.*, **32** (24), L133–L137.

61. Hoeben, W., van Veldhuizen, E.M., Rutgers, W.R., Cramers, C., and Kroesen, G.M.W. (2000) The degradation of aqueous phenol solutions by pulsed positive corona discharges. *Plasma Sources Sci. Technol.*, **9** (3), 361–369.
62. Hayashi, D., Hoeben, W., Dooms, G., van Veldhuizen, E.M., Rutgers, W.R., and Kroesen, G.M.W. (2000) Influence of gaseous atmosphere on corona-induced degradation of aqueous phenol. *J. Phys. D: Appl. Phys.*, **33** (21), 2769–2774.
63. Hayashi, D., Hoeben, W.F.L.M., Dooms, G., van Veldhuizen, E.M., Rutgers, W.R., and Kroesen, G.M.W. (2000) LIF diagnostic for pulsed-corona-induced degradation of phenol in aqueous solution. *J. Phys. D: Appl. Phys.*, **33**, 1484–1486.
64. Hayashi, D., Hoeben, W., Dooms, G., van Veldhuizen, E., Rutgers, W., and Kroesen, G. (2001) Laser-induced fluorescence spectroscopy for phenol and intermediate products in aqueous solutions degraded by pulsed corona discharges above water. *Appl. Opt.*, **40** (6), 986–993.
65. Sugiarto, A.T. and Sato, M. (2001) Pulsed plasma processing of organic compounds in aqueous solution. *Thin Solid Films*, **386** (2), 295–299.
66. Sano, N., Kawashima, T., Fujikawa, J., Fujimoto, T., Kitai, T., Kanki, T., and Toyoda, A. (2002) Decomposition of organic compounds in water by direct contact of gas corona discharge: Influence of discharge conditions. *Ind. Eng. Chem. Res.*, **41** (24), 5906–5911.
67. Lukes, P. and Locke, B.R. (2005) Plasmachemical oxidation processes in a hybrid gas-liquid electrical discharge reactor. *J. Phys. D: Appl. Phys.*, **38** (22), 4074–4081.
68. Grabowski, L.R., van Veldhuizen, E.M., and Rutgers, W.R. (2005) Removal of phenol from water: a comparison of energization methods. *J. Adv. Oxid. Technol.*, **8** (2), 142–149.
69. Grabowski, L.R., van Veldhuizen, E.M., Pemen, A.J.M., and Rutgers, W.R. (2006) Corona above water reactor for systematic study of aqueous phenol degradation. *Plasma Chem. Plasma Process.*, **26** (1), 3–17.
70. Sato, M., Tokutake, T., Ohshima, T., and Sugiarto, A.T. (2008) Aqueous phenol decomposition by pulsed discharges on the water surface. *IEEE Trans. Ind. Appl.*, **44** (5), 1397–1402.
71. Li, J., Sato, M., and Ohshima, T. (2007) Degradation of phenol in water using a gas-liquid phase pulsed discharge plasma reactor. *Thin Solid Films*, **515** (9), 4283–4288.
72. Yan, J.H., Du, C.M., Li, X.D., Sun, X.D., Ni, M.J., Cen, K.F., and Cheron, B. (2005) Plasma chemical degradation of phenol in solution by gas-liquid gliding arc discharge. *Plasma Sources Sci. Technol.*, **14** (4), 637–644.
73. Piskarev, I.M. (1999) Phenol oxidation by OH, H, O, and O_3 species formed in electric discharge. *Kinet. Catal.*, **40** (4), 452–458.
74. Tezuka, M. and Iwasaki, M. (1997) Oxidative degradation of phenols by contact glow discharge electrolysis. *Denki Kagaku*, **65** (12), 1057–1060.
75. Sun, B., Sato, M., and Clements, J.S. (1999) Use of a pulsed high-voltage discharge for removal of organic compounds in aqueous solution. *J. Phys. D: Appl. Phys.*, **32** (15), 1908–1915.
76. Sun, B., Sato, M., and Clements, J.S. (2000) Oxidative processes occurring when pulsed high voltage discharges degrade phenol in aqueous solution. *Environ. Sci. Technol.*, **34** (3), 509–513.
77. Sano, N., Fujimoto, T., Kawashima, T., Yamamoto, D., Kanki, T., and Toyoda, A. (2004) Influence of dissolved inorganic additives on decomposition of phenol and acetic acid in water by direct contact of gas corona discharge. *Sep. Purif. Technol.*, **37** (2), 169–175.
78. Shin, W.T., Yiacoumi, S., Tsouris, C., and Dai, S. (2000) A pulseless corona-discharge process for the oxidation of organic compounds in water. *Ind. Eng. Chem. Res.*, **39** (11), 4408–4414.
79. Pokryvailo, A., Wolf, M., Yankelevich, Y., Wald, S., Grabowski, L.R., van Veldhuizen, E.M., Rutgers, W.R.,

Reiser, M., Glocker, B., Eckhardt, T., Kempenaers, P., and Welleman, A. (2006) High-power pulsed corona for treatment of pollutants in heterogeneous media. *IEEE Trans. Plasma Sci.*, **34** (5), 1731–1743.

80. Polyakov, O.V., Badalyan, A.M., and Bakhturova, L.F. (2004) Relative contributions of plasma pyrolysis and liquid-phase reactions in the anode microspark treatment of aqueous phenol solutions. *High Energy Chem.*, **38** (2), 131–133.
81. Shen, Y.J., Lei, L.C., Zhang, X.W., Zhou, M.H., and Zhang, Y. (2008) Effect of various gases and chemical catalysts on phenol degradation pathways by pulsed electrical discharges. *J. Hazard. Mater.*, **150** (3), 713–722.
82. Miyazaki, Y., Satoh, K., and Itoh, H. (2011) Pulsed discharge purification of water containing nondegradable hazardous substances. *Electr. Eng. Jpn.*, **174** (2), 1–8.
83. Marotta, E., Ceriani, E., Shapoval, V., Schiorlin, M., Ceretta, C., Rea, M., and Paradisi, C. (2011) Characterization of plasma-induced phenol advanced oxidation process in a DBD reactor. *Eur. Phys. J.: Appl. Phys.*, **55** (1), 13811.
84. Sharma, A.K., Locke, B.R., Arce, P., and Finney, W.C. (1993) A preliminary-study of pulsed streamer corona discharge for the degradation of phenol in aqueous-solutions. *Hazard. Waste Hazard. Mater.*, **10** (2), 209–219.
85. Lukes, P., Clupek, M., Sunka, P., Babicky, V., and Janda, V. (2002) Effect of ceramic composition on pulse discharge induced processes in water using ceramic-coated wire to cylinder electrode system. *Czech. J. Phys.*, **52**, 800–806.
86. Chen, Y.S., Zhang, X.S., Dai, Y.C., and Yuan, W.K. (2004) Pulsed high-voltage discharge plasma for degradation of phenol in aqueous solution. *Sep. Purif. Technol.*, **34** (1-3), 5–12.
87. Liu, Y.J. and Jiang, X.Z. (2005) Phenol degradation by a nonpulsed diaphragm glow discharge in an aqueous solution. *Environ. Sci. Technol.*, **39** (21), 8512–8517.
88. Yan, J.H., Du, C.M., Li, X.D., Cheron, B.G., Ni, M.J., and Cen, K.F. (2006) Degradation of phenol in aqueous solutions by gas-liquid gliding arc discharges. *Plasma Chem. Plasma Process.*, **26** (1), 31–41.
89. Kusic, H., Koprivanac, N., and Locke, B.R. (2005) Decomposition of phenol by hybrid gas/liquid electrical discharge reactors with zeolite catalysts. *J. Hazard. Mater.*, **125** (1-3), 190–200.
90. Grymonpre, D.R., Finney, W.C., Clark, R.J., and Locke, B.R. (2004) Hybrid gas-liquid electrical discharge reactors for organic compound degradation. *Ind. Eng. Chem. Res.*, **43** (9), 1975–1989.
91. Grymonpre, D.R., Sharma, A.K., Finney, W.C., and Locke, B.R. (2001) The role of Fenton's reaction in aqueous phase pulsed streamer corona reactors. *Chem. Eng. J.*, **82** (1-3), 189–207.
92. Grymonpre, D.R., Finney, W.C., and Locke, B.R. (1999) Aqueous-phase pulsed streamer corona reactor using suspended activated carbon particles for phenol oxidation: model-data comparison. *Chem. Eng. Sci.*, **54** (15-16), 3095–3105.
93. Grymonpre, D.R., Finney, W.C., Clark, R.J., and Locke, B.R. (2003) Suspended activated carbon particles and ozone formation in aqueous-phase pulsed corona discharge reactors. *Ind. Eng. Chem. Res.*, **42** (21), 5117–5134.
94. Sano, N., Yamane, Y., Hori, Y., Akatsuka, T., and Tamon, H. (2011) Application of multiwalled carbon nanotubes in a wetted-wall corona-discharge reactor to enhance phenol decomposition in water. *Ind. Eng. Chem. Res.*, **50** (17), 9901–9909.
95. Malik, M.A. (2003) Synergistic effect of plasmacatalyst and ozone in a pulsed corona discharge reactor on the decomposition of organic pollutants in water. *Plasma Sources Sci. Technol.*, **12** (4), S26–S32.
96. Lukes, P., Clupek, M., Sunka, P., Peterka, F., Sano, T., Negishi, N., Matsuzawa, S., and Takeuchi, K. (2005) Degradation of phenol by underwater

pulsed corona discharge in combination with TiO_2 photocatalysis. *Res. Chem. Intermed.*, **31** (4-6), 285–294.
97. Wang, H.J., Li, J., Quan, X., Wu, Y., Li, G.F., and Wang, F.Z. (2007) Formation of hydrogen peroxide and degradation of phenol in synergistic system of pulsed corona discharge combined with TiO_2 photocatalysis. *J. Hazard. Mater.*, **141** (1), 336–343.
98. Wang, H.J., Li, J., Quan, X., and Wu, Y. (2008) Enhanced generation of oxidative species and phenol degradation in a discharge plasma system coupled with TiO_2 photocatalysis. *Appl. Catal. B: Environ.*, **83** (1-2), 72–77.
99. Wang, H.J., Chu, J.Y., Ou, H.X., Zhao, R.J., and Han, J.G. (2009) Analysis of TiO_2 photocatalysis in a pulsed discharge system for phenol degradation. *J. Electrost.*, **67** (6), 886–889.
100. Wang, H.J. and Chen, X.Y. (2011) Kinetic analysis and energy efficiency of phenol degradation in a plasma-photocatalysis system. *J. Hazard. Mater.*, **186** (2-3), 1888–1892.
101. Sano, N., Yamamoto, T., Takemori, I., Kim, S.I., Eiad-ua, A., Yamamoto, D., and Nakaiwa, M. (2006) Degradation of phenol by simultaneous use of gas-phase corona discharge and catalyst-supported mesoporous carbon gels. *Ind. Eng. Chem. Res.*, **45** (8), 2897–2900.
102. Bubnov, A.G., Burova, E.Y., Grinevich, V.I., Rybkin, V.V., Kim, J.K., and Choi, H.S. (2007) Comparative actions of NiO and TiO_2 catalysts on the destruction of phenol and its derivatives in a dielectric barrier discharge. *Plasma Chem. Plasma Process.*, **27** (2), 177–187.
103. Bubnov, A.G., Burova, E.Y., Grinevich, V.I., Rybkin, V.V., Kim, J.K., and Choi, H.S. (2006) Plasma-catalytic decomposition of phenols in atmospheric pressure dielectric barrier discharge. *Plasma Chem. Plasma Process.*, **26** (1), 19–30.
104. Mihailovic, M.L. and Cekovic, Z. (1971) Oxidation and reduction of phenols, in *The Chemistry of the Hydroxyl Group, Part 2* (ed. S. Patai), John Wiley & Sons, Ltd, London, pp. 505–592.
105. Pan, X.M., Schuchmann, M.N., and Vonsonntag, C. (1993) Oxidation of benzene by the OH radical - a product and pulse-radiolysis study in oxygenated aqueous solution. *J. Chem. Soc., Perkin Trans. 2*, (3), 289–297.
106. Chen, R.Z. and Pignatello, J.J. (1997) Role of quinone intermediates as electron shuttles in Fenton and photoassisted Fenton oxidations of aromatic compounds. *Environ. Sci. Technol.*, **31** (8), 2399–2406.
107. Bailey, P.S. (1982) *Ozonation in Organic Chemistry: Nonolefinic Compounds*, vol. 2, Academic Press, New York.
108. Singer, P.C. and Gurol, M.D. (1983) Dynamics of the ozonation of phenol. 1. Experimental observations. *Water Res.*, **17** (9), 1163–1171.
109. Yamamoto, Y., Niki, E., Shiokawa, H., and Kamiya, Y. (1979) Ozonation of organic compounds. 2. Ozonation of phenol in water. *J. Org. Chem.*, **44** (13), 2137–2141.
110. Vione, D., Belmondo, S., and Carnino, L. (2004) A kinetic study of phenol nitration and nitrosation with nitrous acid in the dark. *Environ. Chem. Lett.*, **2** (3), 135–139.
111. Yenes, S. and Messeguer, A. (1999) A study of the reaction of different phenol substrates with nitric oxide and peroxynitrite. *Tetrahedron*, **55** (49), 14111–14122.
112. Challis, B.C. and Lawson, A.J. (1971) The chemistry of nitroso compounds. II. The nitrosation of phenol and anisole. *J. Chem. Soc. B*, (4), 770–775.
113. Daiber, A., Mehl, M., and Ullrich, V. (1998) New aspects in the reaction mechanism of phenol with peroxynitrite: the role of phenoxy radicals. *Nitric Oxide: Biol. Chem.*, **2** (4), 259–269.
114. Lemercier, J.N., Padmaja, S., Cueto, R., Squadrito, G.L., Uppu, R.M., and Pryor, W.A. (1997) Carbon dioxide modulation of hydroxylation and nitration of phenol by peroxynitrite. *Arch. Biochem. Biophys.*, **345** (1), 160–170.
115. Uppu, R.M., Lemercier, J.N., Squadrito, G.L., Zhang, H.W., Bolzan, R.M., and Pryor, W.A. (1998) Nitrosation by peroxynitrite: use of phenol as a probe. *Arch. Biochem. Biophys.*, **358** (1), 1–16.

116. Lukes, P., Clupek, M., Babicky, V., Sunka, P., Winterova, G., and Janda, V. (2003) Non-thermal plasma induced decomposition of 2-chlorophenol in water. *Acta Phys. Slovaca*, **53** (6), 423–428.
117. Lukes, P. and Locke, B.R. (2005) Degradation of substituted phenols in a hybrid gas-liquid electrical discharge reactor. *Ind. Eng. Chem. Res.*, **44** (9), 2921–2930.
118. Wen, Y.Z., Jiang, X.Z., and Liu, W.P. (2002) Degradation of 4-chlorophenol by high-voltage pulse corona discharges combined with ozone. *Plasma Chem. Plasma Process.*, **22** (1), 175–185.
119. Willberg, D.M., Lang, P.S., Hochemer, R.H., Kratel, A., and Hoffmann, M.R. (1996) Degradation of 4-chlorophenol, 3,4-dichloroaniline, and 2,4,6-trinitrotoluene in an electrohydraulic discharge reactor. *Environ. Sci. Technol.*, **30** (8), 2526–2534.
120. Zhang, Y., Zhou, M.H., Hao, X.L., and Lei, L.C. (2007) Degradation mechanisms of 4-chlorophenol in a novel gas-liquid hybrid discharge reactor by pulsed high voltage system with oxygen or nitrogen bubbling. *Chemosphere*, **67** (4), 702–711.
121. Zhang, Y., Zhou, M.H., and Lei, L.C. (2007) Degradation of 4-chlorophenol in different gas-liquid electrical discharge reactors. *Chem. Eng. J.*, **132** (1-3), 325–333.
122. Lei, L.C., Zhang, Y., Zhang, X.W., Du, Y.X., Dai, Q.Z., and Han, S. (2007) Degradation performance of 4-chlorophenol as a typical organic pollutant by a pulsed high voltage discharge system. *Ind. Eng. Chem. Res.*, **46** (17), 5469–5477.
123. Bian, W.J., Ying, X.L., and Shi, J.W. (2009) Enhanced degradation of p-chlorophenol in a novel pulsed high voltage discharge reactor. *J. Hazard. Mater.*, **162** (2-3), 906–912.
124. Yin, X.L., Bian, W.J., and Shi, J.W. (2009) 4-chlorophenol degradation by pulsed high voltage discharge coupling internal electrolysis. *J. Hazard. Mater.*, **166** (2-3), 1474–1479.
125. Hao, X.L., Zhou, M.H., Zhang, Y., and Lei, L.C. (2006) Enhanced degradation of organic pollutant 4-chlorophenol in water by non-thermal plasma process with TiO_2. *Plasma Chem. Plasma Process.*, **26** (5), 455–468.
126. Hao, X.L., Zhou, M.H., and Lei, L.C. (2007) Non-thermal plasma-induced photocatalytic degradation of 4-chlorophenol in water. *J. Hazard. Mater.*, **141** (3), 475–482.
127. Du, C.M., Yan, J.H., and Cheron, B.G. (2007) Degradation of 4-chlorophenol using a gas-liquid gliding arc discharge plasma reactor. *Plasma Chem. Plasma Process.*, **27** (5), 635–646.
128. Hao, X.L., Zhou, M.H., Xin, Q., and Lei, L.C. (2007) Pulsed discharge plasma induced Fenton-like reactions for the enhancement of the degradation of 4-chlorophenol in water. *Chemosphere*, **66** (11), 2185–2192.
129. Gao, J.Z., Pu, L.M., Yang, W., Yu, J., and Li, Y. (2004) Oxidative degradation of nitrophenols in aqueous solution induced by plasma with submersed glow discharge electrolysis. *Plasma Process. Polym.*, **1** (2), 171–176.
130. Liu, Y.J. (2009) Aqueous p-chloronitrobenzene decomposition induced by contact glow discharge electrolysis. *J. Hazard. Mater.*, **166** (2-3), 1495–1499.
131. Tezuka, M. and Iwasaki, M. (1998) Plasma induced degradation of chlorophenols in an aqueous solution. *Thin Solid Films*, **316** (1-2), 123–127.
132. Tezuka, M. and Iwasaki, M. (1999) Liquid-phase reactions induced by gaseous plasma. Decomposition of benzoic acids in aqueous solution. *Plasmas Ions*, **2** (1), 23–26.
133. Tezuka, M. and Iwasaki, M. (2001) Plasma-induced degradation of aniline in aqueous solution. *Thin Solid Films*, **386** (2), 204–207.
134. Gai, K. (2009) Anodic oxidation with platinum electrodes for degradation of p-xylene in aqueous solution. *J. Electrost.*, **67** (4), 554–557.
135. Manolache, S., Shamamian, V., and Denes, F. (2004) Dense medium plasma-enhanced decontamination of water of aromatic compounds. *J. Environ. Eng. ASCE*, **130** (1), 17–25.

136. Wen, Y. and Jiang, X. (2000) Degradation of acetophenone in water by pulsed corona discharges. *Plasma Chem. Plasma Process.*, **20** (3), 343–351.
137. Wen, Y.Z. and Jiang, X.Z. (2001) Pulsed corona discharge-induced reactions of acetophenone in water. *Plasma Chem. Plasma Process.*, **21** (3), 345–354.
138. Tomizawa, S. and Tezuka, M. (2006) Oxidative degradation of aqueous cresols induced by gaseous plasma with contact glow discharge electrolysis. *Plasma Chem. Plasma Process.*, **26** (1), 43–52.
139. Sahni, M. and Locke, B.R. (2006) Quantification of hydroxyl radicals produced in aqueous phase pulsed electrical discharge reactors. *Ind. Eng. Chem. Res.*, **45** (17), 5819–5825.
140. Brisset, J.L. (1997) Air corona removal of phenols. *J. Appl. Electrochem.*, **27** (2), 179–183.
141. Sharma, A.K., Josephson, G.B., Camaioni, D.M., and Goheen, S.C. (2000) Destruction of pentachlorophenol using glow discharge plasma process. *Environ. Sci. Technol.*, **34** (11), 2267–2272.
142. Mededovic, S. and Locke, B.R. (2007) Side-chain degradation of atrazine by pulsed electrical discharge in water. *Ind. Eng. Chem. Res.*, **46** (9), 2702–2709.
143. Mededovic, S., Finney, W.C., and Locke, B.R. (2007) Aqueous-phase mineralization of s-triazine using pulsed electrical discharge. *Int. J. Plasma Environ. Sci. Technol.*, **1** (1), 82–90.
144. Wang, L., Jiang, X.Z., and Liu, Y.J. (2008) Degradation of bisphenol A and formation of hydrogen peroxide induced by glow discharge plasma in aqueous solutions. *J. Hazard. Mater.*, **154** (1-3), 1106–1114.
145. Abdelmalek, F., Torres, R.A., Combet, E., Petrier, C., Pulgarin, C., and Addou, A. (2008) Gliding arc discharge (GAD) assisted catalytic degradation of bisphenol A in solution with ferrous ions. *Sep. Purif. Technol.*, **63** (1), 30–37.
146. Gerrity, D., Stanford, B.D., Trenholm, R.A., and Snyder, S.A. (2010) An evaluation of a pilot-scale nonthermal plasma advanced oxidation process for trace organic compound degradation. *Water Res.*, **44** (2), 493–504.
147. Krause, H., Schweiger, B., Schuhmacher, J., Scholl, S., and Steinfeld, U. (2009) Degradation of the endocrine disrupting chemicals (EDCs) carbamazepine, clofibric acid, and iopromide by corona discharge over water. *Chemosphere*, **75** (2), 163–168.
148. Krause, H., Schweiger, B., Prinz, E., Kim, J., and Steinfeld, U. (2011) Degradation of persistent pharmaceuticals in aqueous solutions by a positive dielectric barrier discharge treatment. *J. Electrost.*, **69** (4), 333–338.
149. Magureanu, M., Piroi, D., Mandache, N.B., David, V., Medvedovici, A., and Parvulescu, V.I. (2010) Degradation of pharmaceutical compound pentoxifylline in water by non-thermal plasma treatment. *Water Res.*, **44** (11), 3445–3453.
150. Magureanu, M., Piroi, D., Mandache, N.B., David, V., Medvedovici, A., Bradu, C., and Parvulescu, V.I. (2011) Degradation of antibiotics in water by non-thermal plasma treatment. *Water Res.*, **45** (11), 3407–3416.
151. Yan, J.H., Liu, Y.N., Bo, Z., Li, X.D., and Cen, K.F. (2008) Degradation of gas-liquid gliding arc discharge on acid orange II. *J. Hazard. Mater.*, **157** (2-3), 441–447.
152. Du, C.M., Shi, T.H., Sun, Y.W., and Zhuang, X.F. (2008) Decolorization of acid orange 7 solution by gas-liquid gliding arc discharge plasma. *J. Hazard. Mater.*, **154** (1-3), 1192–1197.
153. Du, C.M., Zhang, L.L., Wang, J., Zhang, C.R., Li, H.X., and Xiong, Y. (2010) Degradation of acid orange 7 by gliding arc discharge plasma in combination with advanced Fenton catalysis. *Plasma Chem. Plasma Process.*, **30** (6), 855–871.
154. Du, C.M., Xiong, Y., Zhang, L.L., Wang, J., Jia, S.G., Chan, C.Y., and Shi, T.H. (2011) Degradation of acid orange 7 solution by air-liquid gliding arc discharge in combination with TiO_2

catalyst. *J. Adv. Oxid. Technol.*, **14** (1), 17–22.

155. Mok, Y.S., Jo, J.O., and Whitehead, J.C. (2008) Degradation of an azo dye orange II using a gas phase dielectric barrier discharge reactor submerged in water. *Chem. Eng. J.*, **142** (1), 56–64.
156. Li, J., Wang, H.J., Li, G.F., Wu, Y., Quan, X., and Liu, Z.G. (2007) Synergistic decolouration of azo dye by pulsed streamer discharge immobilized TiO_2 photocatalysis. *Plasma Sci. Technol.*, **9** (4), 469–473.
157. Li, J., Zhou, Z.G., Wang, H.J., Li, G.F., and Wu, Y. (2007) Research on decoloration of dye wastewater by combination of pulsed discharge plasma and TiO_2 nanoparticles. *Desalination*, **212** (1-3), 123–128.
158. Jin, X.L., Bai, H., Wang, F., Wang, X.C., Wang, X.Y., and Ren, H.X. (2011) Plasma degradation of acid orange 7 with contact glow discharge electrolysis. *IEEE Trans. Plasma Sci.*, **39** (4), 1099–1103.
159. Magureanu, M., Mandache, N.B., and Parvulescu, V.I. (2007) Degradation of organic dyes in water by electrical discharges. *Plasma Chem. Plasma Process.*, **27** (5), 589–598.
160. Sugiarto, A.T., Ohshima, T., and Sato, M. (2002) Advanced oxidation processes using pulsed streamer corona discharge in water. *Thin Solid Films*, **407** (1-2), 174–178.
161. Sugiarto, A.T., Ito, S., Ohshima, T., Sato, M., and Skalny, J.D. (2003) Oxidative decoloration of dyes by pulsed discharge plasma in water. *J. Electrost.*, **58** (1-2), 135–145.
162. Vujevic, D., Koprivanac, N., Bozic, A.L., and Locke, B.R. (2004) The removal of direct orange 39 by pulsed corona discharge from model wastewater. *Environ. Technol.*, **25** (7), 791–800.
163. Zhang, Y.Z., Zheng, J.T., Qu, X.F., and Chen, H.G. (2007) Effect of granular activated carbon on degradation of methyl orange when applied in combination with high-voltage pulse discharge. *J. Colloid Interface Sci.*, **316** (2), 523–530.
164. Zhang, Y.Z., Sun, B.Y., Deng, S.H., Wang, Y.J., Peng, H., Li, Y.W., and Zhang, X.H. (2010) Methyl orange degradation by pulsed discharge in the presence of activated carbon fibers. *Chem. Eng. J.*, **159** (1-3), 47–52.
165. Gong, J.Y. and Cai, W.M. (2007) Degradation of methyl orange in water by contact glow discharge electrolysis. *Plasma Sci. Technol.*, **9** (2), 190–193.
166. Gong, J.Y., Wang, J., Xie, W.J., and Cai, W.M. (2008) Enhanced degradation of aqueous methyl orange by contact glow discharge electrolysis using Fe^{2+} as catalyst. *J. Appl. Electrochem.*, **38** (12), 1749–1755.
167. Doubla, A., Bello, L.B., Fotso, M., and Brisset, J.L. (2008) Plasmachemical decolourisation of Bromothymol Blue by gliding electric discharge at atmospheric pressure. *Dyes Pigment.*, **77** (1), 118–124.
168. Li, J., Wang, T.C., Lu, N., Zhang, D.D., Wu, Y., Wang, T.W., and Sato, M. (2011) Degradation of dyes by active species injected from a gas phase surface discharge. *Plasma Sources Sci. Technol.*, **20** (3), 034019.
169. Kozakova, Z., Nejezchleb, M., Krcma, F., Halamova, I., Caslavsky, J., and Dolinova, J. (2010) Removal of organic dye direct red 79 from water solutions by DC diaphragm discharge: analysis of decomposition products. *Desalination*, **258** (1-3), 93–99.
170. Trifi, B., Cavadias, S., and Bellakhal, N. (2011) Decoloration of methyl red by gliding arc discharge. *Desalin. Water Treat.*, **25** (1-3), 65–70.
171. Sunka, P., Babicky, V., Clupek, M., Lukes, P., and Balcarova, J. (2003) Modified pinhole discharge for water treatment, in *PPC-2003: 14th IEEE International Pulsed Power Conference, Digest of Technical Papers*, Vols 1 and 2 (eds M. Giesselmann and A. Neuber), IEEE, New York, pp. 229–231.
172. Kusic, H., Koprivanac, N., Peternel, I., and Locke, B.R. (2005) Hybrid gas/liquid electrical discharge reactors with zeolites for colored wastewater degradation. *J. Adv. Oxid. Technol.*, **8** (2), 172–181.
173. Peternel, I., Kusic, H., Koprivanac, N., and Locke, B.R. (2006) The roles of ozone and zeolite on reactive dye

degradation in electrical discharge reactors. *Environ. Technol.*, **27** (5), 545–557.

174. Abdelmalek, F., Ghezzar, M.R., Belhadj, M., Addou, A., and Brisset, J.L. (2006) Bleaching and degradation of textile dyes by nonthermal plasma process at atmospheric pressure. *Ind. Eng. Chem. Res.*, **45** (1), 23–29.
175. Stara, Z., Krcma, F., Nejezchleb, M., and Skalny, J.D. (2009) Organic dye decomposition by DC diaphragm discharge in water: effect of solution properties on dye removal. *Desalination*, **239** (1-3), 283–294.
176. Stara, Z., Krcma, F., Nejezchleb, M., and Skalny, J.D. (2008) Influence of solution composition and chemical structure of dye on removal of organic dye by DC diaphragm discharge in water solutions. *J. Adv. Oxid. Technol.*, **11** (1), 155–162.
177. Lu, D.L., Chen, J.R., Gao, A.H., Zissis, G., Hu, S.B., and Lu, Z.G. (2010) Decolorization of aqueous acid red B solution during the cathode process in abnormal glow discharge. *IEEE Trans. Plasma Sci.*, **38** (10), 2854–2859.
178. Sato, M., Kon-no, D., Ohshima, T., and Sugiarto, A.T. (2005) Decoloration of organic dye in water by pulsed discharge plasma generated simultaneously in gas and liquid media. *J. Adv. Oxid. Technol.*, **8** (2), 198–204.
179. Clements, J.S., Sato, M., and Davis, R.H. (1987) Preliminary investigation of prebreakdown phenomena and chemical-reactions using a pulsed high-voltage discharge in water. *IEEE Trans. Ind. Appl.*, **23** (2), 224–235.
180. Minamitani, Y., Shoji, S., Ohba, Y., and Higashiyama, Y. (2008) Decomposition of dye in water solution by pulsed power discharge in a water droplet spray. *IEEE Trans. Plasma Sci.*, **36** (5), 2586–2591.
181. Pawlat, J. and Hensel, K. (2004) Discoloration of the solutions in the foaming environment. *Czech. J. Phys.*, **54**, C964–C969.
182. Pawlat, J., Ihara, S., Yamabe, C., and Pollo, I. (2005) Oxidant formation and the decomposition of organic compounds in foaming systems. *Plasma Process. Polym.*, **2** (3), 218–221.
183. Pawlat, J., Hensel, K., and Ihara, S. (2006) Generation of oxidants and removal of indigo blue by pulsed power in bubbling and foaming systems. *Czech. J. Phys.*, **56**, B1174–B1178.
184. Pawlat, J. and Ihara, S. (2007) Removal of color caused by various chemical compounds using electrical discharges in a foaming column. *Plasma Process. Polym.*, **4** (7-8), 753–759.
185. Wang, Z.H., Xu, D.X., Chen, Y., Hao, C.X., and Zhang, X.Y. (2008) Plasma decoloration of dye using dielectric barrier discharges with earthed spraying water electrodes. *J. Electrost.*, **66** (9-10), 476–481.
186. Ghezzar, M.R., Abdelmalek, F., Belhadj, M., Benderdouche, N., and Addou, A. (2007) Gliding arc plasma assisted photocatalytic degradation of anthraquinonic acid green 25 in solution with TiO_2. *Appl. Catal. B: Environ.*, **72** (3-4), 304–313.
187. Gao, J.Z., Yu, J., Lu, Q.F., He, X.Y., Yang, W., Li, Y., Pu, L.M., and Yang, Z.M. (2008) Decoloration of alizarin red S in aqueous solution by glow discharge electrolysis. *Dyes Pigment.*, **76** (1), 47–52.
188. Xue, J., Chen, L., and Wang, H.L. (2008) Degradation mechanism of alizarin red in hybrid gas-liquid phase dielectric barrier discharge plasmas: Experimental and theoretical examination. *Chem. Eng. J.*, **138** (1-3), 120–127.
189. Bozic, A.L., Koprivanac, N., Sunka, P., Clupek, M., and Babicky, V. (2004) Organic synthetic dye degradation by modified pinhole discharge. *Czech. J. Phys.*, **54**, C958–C963.
190. Goheen, S.C., Durham, D.E., McCulloch, M., and Heath, W.O. (1992) The degradation of organic dyes by corona discharge, in *Proceedings of the 2nd International Symposium on Chemical Oxidation: Technology for the Nineties* (eds W.W. Eckenfelder, A.R. Bowers, and J.A. Roth), Vanderbilt University, Nashville, TN, pp. 356–367.

191. Hoeben, W.F.L.M. (2000) Pulsed corona-induced degradation of organic materials in water. PhD Dissertation. Technische Universiteit Eindhoven, Eindhoven, NL.

192. Gao, J.Z., Yu, J., Li, Y., He, X.Y., Bo, L.L., Pu, L.M., Yang, W., Lu, Q.F., and Yang, Z.M. (2006) Decoloration of aqueous brilliant green by using glow discharge electrolysis. *J. Haz. Mater.*, **137** (1), 431–436.

193. Malik, M.A., ur Rehman, U., Ghaffar, A., and Ahmed, K. (2002) Synergistic effect of pulsed corona discharges and ozonation on decolourization of methylene blue in water. *Plasma Sources Sci. Technol.*, **11** (3), 236–240.

194. Magureanu, M., Piroi, D., Gherendi, F., Mandache, N.B., and Parvulescu, V. (2008) Decomposition of methylene blue in water by corona discharges. *Plasma Chem. Plasma Process.*, **28** (6), 677–688.

195. Maehara, T., Miyamoto, I., Kurokawa, K., Hashimoto, Y., Iwamae, A., Kuramoto, M., Yamashita, H., Mukasa, S., Toyota, H., Nomura, S., and Kawashima, A. (2008) Degradation of methylene blue by RF plasma in water. *Plasma Chem. Plasma Process.*, **28** (4), 467–482.

196. Pawlat, J., Hensel, K., and Ihara, S. (2005) Decomposition of humic acid and methylene blue by electric discharge in foam. *Acta Phys. Slovaca*, **55** (5), 479–485.

197. Velikonja, J., Bergougnou, M.A., Castle, G.S.P., Cairns, W.L., and Inculet, I. (2002) Ozone dissolution vs. aqueous methylene blue degradation in semi-batch reactors with dielectric barrier discharge over the water surface. *Ozone Sci. Eng.*, **24** (3), 159–170.

198. Magureanu, M., Piroi, D., Mandache, N.B., and Parvulescu, V. (2008) Decomposition of methylene blue in water using a dielectric barrier discharge: Optimization of the operating parameters. *J. Appl. Phys.*, **104** (10), 103306.

199. Huang, F.M., Chen, L., Wang, H.L., and Yan, Z.C. (2010) Analysis of the degradation mechanism of methylene blue by atmospheric pressure dielectric barrier discharge plasma. *Chem. Eng. J.*, **162** (1), 250–256.

200. Baroch, P., Anita, V., Saito, N., and Takai, O. (2008) Bipolar pulsed electrical discharge for decomposition of organic compounds in water. *J. Electrost.*, **66** (5-6), 294–299.

201. Baroch, P., Saito, N., and Takai, O. (2008) Special type of plasma dielectric barrier discharge reactor for direct ozonization of water and degradation of organic pollution. *J. Phys. D: Appl. Phys.*, **41** (8), 085207.

202. Gao, J.Z., Ma, D.P., Guo, X., Wang, A.X., Fu, Y., Wu, J.L., and Yang, W. (2008) Degradation of anionic dye eosin by glow discharge electrolysis plasma. *Plasma Sci. Technol.*, **10** (4), 422–427.

203. Gao, J.Z., Hu, Z.G., Wang, X.Y., Hou, J.G., Lu, X.Q., and Kang, J.W. (2001) Oxidative degradation of acridine orange induced by plasma with contact glow discharge electrolysis. *Thin Solid Films*, **390** (1-2), 154–158.

204. Nemcova, L., Nikiforov, A., Leys, C., and Krcma, F. (2011) Chemical efficiency of H_2O_2 production and decomposition of organic compounds under action of DC underwater discharge in gas bubbles. *IEEE Trans. Plasma Sci.*, **39** (3), 865–870.

205. Nikiforov, A.Y., Leys, C., Li, L., Nemcova, L., and Krcma, F. (2011) Physical properties and chemical efficiency of an underwater dc discharge generated in He, Ar, N_2 and air bubbles. *Plasma Sources Sci. Technol.*, **20** (3), 034008.

206. Krcma, F., Stara, Z., and Prochazkova, J. (2010) Diaphragm discharge in liquids: Fundamentals and applications. *J. Phys.: Conf. Ser.*, **207**, 012010.

207. Abdelmalek, F., Gharbi, S., Benstaali, B., Addou, A., and Brisset, J.L. (2004) Plasmachemical degradation of azo dyes by humid air plasma: Yellow Supranol 4 GL, Scarlet Red Nylosan F3 GL and industrial waste. *Water Res.*, **38** (9), 2339–2347.

208. Ghezzar, M.R., Abdelmalek, F., Belhadj, M., Benderdouche, N., and Addou, A. (2009) Enhancement of the bleaching and degradation of textile

wastewaters by gliding arc discharge plasma in the presence of TiO_2 catalyst. *J. Hazard. Mater.*, **164** (2-3), 1266–1274.
209. Christie, R.M. (2001) *Colour Chemistry*, Royal Society of Chemistry, Cambridge, London.
210. Konstantinou, I.K. and Albanis, T.A. (2004) TiO_2-assisted photocatalytic degradation of azo dyes in aqueous solution: kinetic and mechanistic investigations - A review. *Appl. Catal. B: Environ.*, **49** (1), 1–14.
211. Matsui, M. (1996) Ozonation, in *Environmental Chemistry of Dyes and Pigments* (eds A. Reife and H.S. Freeman), John Wiley & Sons, Inc, New York. Chapter 3.
212. Grabowski, L.R., van Veldhuizen, E.M., Pemen, A.J.M., and Rutgers, W.R. (2007) Breakdown of methylene blue and methyl orange by pulsed corona discharge. *Plasma Sources Sci. Technol.*, **16** (2), 226–232.
213. Bader, H. and Hoigne, J. (1981) Determination of ozone in water by the indigo method. *Water Res.*, **15** (4), 449–456.
214. Bader, H. and Hoigne, J. (1982) Determination of ozone in water by the indigo method - a submitted standard method. *Ozone Sci. Eng.*, **4** (4), 169–176.
215. Oturan, M.A., Guivarch, E., Oturan, N., and Sires, I. (2008) Oxidation pathways of malachite green by Fe^{3+}-catalyzed electro-Fenton process. *Appl. Catal. B: Environ.*, **82** (3-4), 244–254.
216. Kusvuran, E., Gulnaz, O., Samil, A., and Yildirim, O. (2011) Decolorization of malachite green, decolorization kinetics and stoichiometry of ozone-malachite green and removal of antibacterial activity with ozonation processes. *J. Hazard. Mater.*, **186** (1), 133–143.
217. Sato, M., Ohgiyama, T., and Clements, J.S. (1996) Formation of chemical species and their effects on microorganisms using a pulsed high-voltage discharge in water. *IEEE Trans. Ind. Appl.*, **32** (1), 106–112.
218. Thagard, S.M., Takashima, K., and Mizuno, A. (2009) Electrical discharges in polar organic liquids. *Plasma Process. Polym.*, **6** (11), 741–750.
219. Liu, X.Z., Liu, C.J., and Eliasson, B. (2003) Hydrogen production from methanol using corona discharges. *Chin. Chem. Lett.*, **14** (6), 631–633.
220. Yan, Z.C., Chen, L., and Wang, H.L. (2007) Glow discharge plasma electrolysis of methanol solutions. *Acta Phys. Chim. Sin.*, **23** (6), 835–840.
221. Yan, Z.C., Li, C., and Lin, W.H. (2009) Hydrogen generation by glow discharge plasma electrolysis of methanol solutions. *Int. J. Hydrog. Energy*, **34** (1), 48–55.
222. Burlica, R., Shih, K.Y., Hnatiuc, B., and Locke, B.R. (2011) Hydrogen generation by pulsed gliding arc discharge plasma with sprays of alcohol solutions. *Ind. Eng. Chem. Res.*, **50** (15), 9466–9470.
223. Chernyak, V.Y., Olszewski, S.V., Yukhymenko, V., Solomenko, E.V., Prysiazhnevych, I.V., Naumov, V.V., Levko, D.S., Shchedrin, A.I., Ryabtsev, A.V., Demchina, V.P., Kudryavtsev, V.S., Martysh, E.V., and Verovchuck, M.A. (2008) Plasma-assisted reforming of ethanol in dynamic plasma-liquid system: experiments and modeling. *IEEE Trans. Plasma Sci.*, **36** (6), 2933–2939.
224. Yan, Z.C., Chen, L., and Wang, H.L. (2008) Hydrogen generation by glow discharge plasma electrolysis of ethanol solutions. *J. Phys. D: Appl. Phys.*, **41** (15), 155205.
225. Levko, D., Shchedrin, A., Chernyak, V., Olszewski, S., and Nedybaliuk, O. (2011) Plasma kinetics in ethanol/water/air mixture in a 'tornado'-type electrical discharge. *J. Phys. D: Appl. Phys.*, **44** (14), 145206.
226. Kabashima, H., Einaga, H., and Futamura, S. (2003) Hydrogen generation from water, methane, and methanol with nonthermal plasma. *IEEE Trans. Ind. Appl.*, **39** (2), 340–345.
227. Wang, B.W., Lu, Y.J., Zhang, X., and Hu, S.H. (2011) Hydrogen generation from steam reforming of ethanol in dielectric barrier discharge. *J. Nat. Gas Chem.*, **20** (2), 151–154.

228. Aubry, O., Met, C., Khacef, A., and Cormier, J.M. (2005) On the use of a non-thermal plasma reactor for ethanol steam reforming. *Chem. Eng. J.*, **106** (3), 241–247.

229. Han, Y., Wang, J.G., Cheng, D.G., and Liu, C.J. (2006) Density functional theory study of methanol conversion via cold plasmas. *Ind. Eng. Chem. Res.*, **45** (10), 3460–3467.

230. Palo, D.R., Dagle, R.A., and Holladay, J.D. (2007) Methanol steam reforming for hydrogen production. *Chem. Rev.*, **107** (10), 3992–4021.

231. Petitpas, G., Rollier, J.D., Darmon, A., Gonzalez-Aguilar, J., Metkemeijer, R., and Fulcheri, L. (2007) A comparative study of non-thermal plasma assisted reforming technologies. *Int. J. Hydrog. Energy*, **32** (14), 2848–2867.

232. Jimenez, M., Yubero, C., and Calzada, M.D. (2008) Study on the reforming of alcohols in a surface wave discharge (SWD) at atmospheric pressure. *J. Phys. D: Appl. Phys.*, **41** (17), 175201.

233. Deminsky, M., Jivotov, V., Potapkin, B., and Rusanov, V. (2002) Plasma-assisted production of hydrogen from hydrocarbons. *Pure Appl. Chem.*, **74** (3), 413–418.

234. Sarmiento, B., Brey, J.J., Viera, I.G., Gonzalez-Elipe, A.R., Cotrino, J., and Rico, V.J. (2007) Hydrogen production by reforming of hydrocarbons and alcohols in a dielectric barrier discharge. *J. Power Sources*, **169** (1), 140–143.

235. Lee, C., Lee, Y., and Yoon, J. (2006) Oxidative degradation of dimethylsulfoxide by locally concentrated hydroxyl radicals in streamer corona discharge process. *Chemosphere*, **65** (7), 1163–1170.

236. Lukes, P., Clupek, M., Babicky, V., Sisrova, I., and Janda, V. (2011) The catalytic role of tungsten electrode material in the plasmachemical activity of a pulsed corona discharge in water. *Plasma Sources Sci. Technol.*, **20** (3), 034011.

237. Even-Ezra, I., Mizrahi, A., Gerrity, D., Snyder, S., Salveson, A., and Lahav, O. (2009) Application of a novel plasma-based advanced oxidation process for efficient and cost-effective destruction of refractory organics in tertiary effluents and contaminated groundwater. *Desalin. Water Treat.*, **11** (1-3), 236–244.

238. Johnson, D.C., Shamamian, V.A., Callahan, J.H., Denes, F.S., Manolache, S.O., and Dandy, D.S. (2003) Treatment of methyl tert-butyl ether contaminated water using a dense medium plasma reactor: A mechanistic and kinetic investigation. *Environ. Sci. Technol.*, **37** (20), 4804–4810.

239. Tong, S.P., Ni, Y.Y., Shen, C.S., Wen, Y.Z., and Jiang, X.Z. (2011) Degradation of methyl tert-butyl ether (MTBE) in water by glow discharge plasma. *Water Sci. Technol.*, **63** (12), 2814–2819.

240. Schreiner, P.R., Reisenauer, H.P., Pickard, F.C., Simmonett, A.C., Allen, W.D., Matyus, E., and Csaszar, A.G. (2008) Capture of hydroxymethylene and its fast disappearance through tunnelling. *Nature*, **453** (7197), 906–909.

241. Grutzner, T. and Hasse, H. (2004) Solubility of formaldehyde and trioxane in aqueous solutions. *J. Chem. Eng. Data*, **49** (3), 642–646.

242. Maiwald, M., Fischer, H.H., Ott, M., Peschla, R., Kuhnert, C., Kreiter, C.G., Maurer, G., and Hasse, H. (2003) Quantitative NMR spectroscopy of complex liquid mixtures: Methods and results for chemical equilibria in formaldehyde-water-methanol at temperatures up to 383 K. *Ind. Eng. Chem. Res.*, **42** (2), 259–266.

243. Flyunt, R., Makogon, O., Schuchmann, M.N., Asmus, K.D., and von Sonntag, C. (2001) OH-Radical-induced oxidation of methanesulfinic acid. The reactions of the methanesulfonyl radical in the absence and presence of dioxygen. *J. Chem. Soc., Perkin Trans. 2*, (5), 787–792.

244. Alfassi, Z.B. (1999) *The Chemistry of Free Radicals: S-Centered Radicals*, John Wiley & Sons, Ltd, Chichester.

245. Goryachev, V.L., Rutberg, F.G., and Fedyukovich, V.N. (1996) Some properties of pulse-periodic discharge in water with energy per pulse of similar to 1 J for water purification. *High Temp.*, **34** (5), 746–749.

246. Blokhin, V.I., Vysikailo, F.I., Dmitriev, K.I., and Efremov, N.M. (1999) Systems with different electrode materials for treatment of water by a pulsed electric discharge. *High Temp.*, **37** (6), 963–965.
247. Goryachev, V.L., Ufimtsev, A.A., and Khodakovskii, A.M. (1997) Mechanism of electrode erosion in pulsed discharges in water with a pulse energy of similar to 1 J. *Tech. Phys. Lett.*, **23** (5), 386–387.
248. Potocky, S., Saito, N., and Takai, O. (2009) Needle electrode erosion in water plasma discharge. *Thin Solid Films*, **518** (3), 918–923.
249. Lukes, P., Clupek, M., Babicky, V., Sunka, P., Skalny, J.D., Stefecka, M., Novak, J., and Malkova, Z. (2006) Erosion of needle electrodes in pulsed corona discharge in water. *Czech. J. Phys.*, **56**, B916–B924.
250. Holzer, F. and Locke, B.R. (2008) Influence of high voltage needle electrode material on hydrogen peroxide formation and electrode erosion in a hybrid gas-liquid series electrical discharge reactor. *Plasma Chem. Plasma Process.*, **28** (1), 1–13.
251. Kirkpatrick, M.J. and Locke, B.R. (2006) Effects of platinum electrode on hydrogen, oxygen, and hydrogen peroxide formation in aqueous phase pulsed corona electrical discharge. *Ind. Eng. Chem. Res.*, **45** (6), 2138–2142.
252. Mededovic, S. and Locke, B.R. (2006) Platinum catalysed decomposition of hydrogen peroxide in aqueous-phase pulsed corona electrical discharge. *Appl. Catal. B: Environ.*, **67** (3-4), 149–159.
253. Mededovic, S. and Locke, B.R. (2007) The role of platinum as the high voltage electrode in the enhancement of Fenton's reaction in liquid phase electrical discharge. *Appl. Catal. B: Environ.*, **72** (3-4), 342–350.
254. Edebo, L., Holme, T., and Selin, I. (1968) Microbicidal action of compounds generated by transient electric arcs in aqueous systems. *J. Gen. Microbiol.*, **53**, 1–7.
255. Efremov, N.M., Adamiak, B.Y., Blochin, V.I., Dadashev, S.J., Dmitriev, K.I., Semjonov, V.N., Levashov, V.F., and Jusbashev, V.F. (2000) Experimental investigation of the action of pulsed electrical discharges in liquids on biological objects. *IEEE Trans. Plasma Sci.*, **28** (1), 224–229.
256. Rutberg, P.G., Kolikov, V.A., Kurochkin, V.E., Panina, L.K., and Rutberg, A.P. (2007) Electric discharges and the prolonged microbial resistance of water. *IEEE Trans. Plasma Sci.*, **35** (4), 1111–1118.
257. Walling, C. (1975) Fentons reagent revisited. *Acc. Chem. Res.*, **8** (4), 125–131.
258. Pignatello, J.J., Oliveros, E., and MacKay, A. (2006) Advanced oxidation processes for organic contaminant destruction based on the Fenton reaction and related chemistry. *Crit. Rev. Environ. Sci. Technol.*, **36** (1), 1–84.
259. Balzani, V. and Carassiti, V. (1970) *Photochemistry of Coordination Compounds*, Academic Press, London.
260. Nardello, V., Marko, J., Vermeersch, G., and Aubry, J.M. (1998) W-183 NMR study of peroxotungstates involved in the disproportionation of hydrogen peroxide into singlet oxygen ($^1O_2$, $^1\Delta_g$) catalyzed by sodium tungstate in neutral and alkaline water. *Inorg. Chem.*, **37** (21), 5418–5423.
261. Howarth, O.W. (2004) Oxygen-17 NMR study of aqueous peroxotungstates. *Dalton Trans.*, (3), 476–481.
262. Noyori, R., Aoki, M., and Sato, K. (2003) Green oxidation with aqueous hydrogen peroxide. *Chem. Commun.*, (16), 1977–1986.
263. Ogata, Y. and Tanaka, K. (1981) Kinetics of the oxidation of dimethylsulfoxide with aqueous hydrogen-peroxide catalyzed by sodium tungstate. *Can. J. Chem. Rev. Can. Chim.*, **59** (4), 718–722.
264. Maroulf-Khelifa, K., Abdelmalek, F., Khelifa, A., and Addou, A. (2008) TiO_2-assisted degradation of a perfluorinated surfactant in aqueous solutions treated by gliding arc discharge. *Chemosphere*, **70** (11), 1995–2001.
265. Egerton, T. and Christensen, P. (2004) Photoelectrocatalysis processes, in *Advanced Oxidation Processes for Water and*

Wastewater Treatment (ed. S. Parson), IWA Publishing, London.
266. Hoffmann, M.R., Martin, S.T., Choi, W.Y., and Bahnemann, D.W. (1995) Environmental applications of semiconductor photocatalysis. *Chem. Rev.*, **95** (1), 69–96.
267. Alvarez, P.M., Garcia-Araya, J.F., Beltran, F.J., Giraldez, I., Jaramillo, J., and Gomez-Serrano, V. (2006) The influence of various factors on aqueous ozone decomposition by granular activated carbons and the development of a mechanistic approach. *Carbon*, **44** (14), 3102–3112.
268. Miller, S.L. (1955) Production of some organic compounds under possible primitive earth conditions. *J. Am. Chem. Soc.*, **77** (9), 2351–2361.
269. Brablec, A., Slavicek, P., Stahel, P., Cizmar, T., Trunec, D., Simor, M., and Cernak, M. (2002) Underwater pulse electrical diaphragm discharges for surface treatment of fibrous polymeric materials. *Czech. J. Phys.*, **52**, 491–500.
270. Nikiforov, A.Y. and Leys, C. (2006) Surface treatment of cotton yarn by underwater capillary electrical discharge. *Plasma Chem. Plasma Process.*, **26** (4), 415–423.
271. Joshi, R., Schulze, R.D., Meyer-Plath, A., and Friedrich, J.F. (2008) Selective surface modification of poly(propylene) with OH and COOH groups using liquid-plasma systems. *Plasma Process. Polym.*, **5** (7), 695–707.
272. Joshi, R., Schulze, R.D., Meyer-Plath, A., Wagner, M.H., and Friedrich, J.F. (2009) Selective surface modification of polypropylene using underwater plasma technique or underwater capillary discharge. *Plasma Process. Polym.*, **6**, S218–S222.
273. Joshi, R., Friedrich, J., and Wagner, M. (2011) Role of hydrogen peroxide in selective OH group functionalization of polypropylene surfaces using underwater capillary discharge. *J. Adhes. Sci. Technol.*, **25** (1-3), 283–305.
274. Sengupta, S.K., Sandhir, U., and Misra, N. (2001) A study on acrylamide polymerization by anodic contact glow-discharge electrolysis: A novel tool. *J. Polym. Sci. A: Polym. Chem.*, **39** (10), 1584–1588.
275. Malik, M.A., Ahmed, M., Ejaz ur, R., Naheed, R., and Ghaffar, A. (2003) Synthesis of superabsorbent copolymers by pulsed corona discharges in water. *Plasmas Polym.*, **8** (4), 271–279.
276. Wang, A.X., Gao, J.Z., Yuan, L., and Yang, W. (2009) Synthesis and characterization of polymethylmethacrylate by using glow discharge electrolysis plasma. *Plasma Chem. Plasma Process.*, **29** (5), 387–398.
277. Hatakeyama, R., Kaneko, T., Kato, T., and Li, Y.F. (2011) Plasma-synthesized single-walled carbon nanotubes and their applications. *J. Phys. D: Appl. Phys.*, **44** (17), 174004.
278. Graham, W.G. and Stalder, K.R. (2011) Plasmas in liquids and some of their applications in nanoscience. *J. Phys. D: Appl. Phys.*, **44** (17), 174037.
279. Takai, O. (2008) Solution plasma processing (SPP). *Pure Appl. Chem.*, **80** (9), 2003–2011.
280. Meiss, S.A., Rohnke, M., Kienle, L., El Abedin, S.Z., Endres, F., and Janek, J. (2007) Employing plasmas as gaseous electrodes at the free surface of ionic liquids: Deposition of nanocrystalline silver particles. *Chem. Phys. Chem.*, **8** (1), 50–53.
281. Nomura, S., Toyota, H., Mukasa, S., Yamashita, H., and Maehara, T. (2006) Microwave plasma in hydrocarbon liquids. *Appl. Phys. Lett.*, **88** (21), 211503.
282. Malik, M.A. and Ahmed, M. (2008) Preliminary studies on formation of carbonaceous products by pulsed spark discharges in liquid hydrocarbons. *J. Electrost.*, **66** (11-12), 574–577.
283. Locke, B.R., Sunka, P., Sato, M., Hoffmann, M., and Chang, J.S. (2006) Electrohydraulic discharge and non thermal plasma for water treatment. *Ind. Eng. Chem. Res.*, **45**, 882–905.

8
Biological Effects of Electrical Discharge Plasma in Water and in Gas–Liquid Environments

Petr Lukes, Jean-Louis Brisset, and Bruce R. Locke

8.1
Introduction

Plasma interactions with biologically derived compounds and materials as well as with live organisms, including proteins, DNA, infectious agents, single microbial cells and viruses, eukaryotic cells, biofilms, and cellular tissue within living animals, has recently generated considerable interest for potential contamination control and medical applications [1–19]. The control of bacteria and other disease-carrying agents by plasma processes has been investigated for sterilization in food safety and medical equipment, and, more recently, directly to the human body for wound healing [20, 21] and dental operations [22, 23]. Other direct applications of plasma to the human body that do not involve microbial disinfection include cancer treatment [24–29], lithotripsy [30–33], and electrosurgery [34, 35].

In this chapter, we are specifically concerned with the life science applications of plasma processes. While this is a very large and growing field, ranging from surgery to microbial inactivation as mentioned above, there are some common aspects with regard to the basic chemical and physical processes that are important in many of these applications. Detailed discussions of both the chemical and physical processes induced by electrical discharges in gases, liquids, and gas–liquid environments are covered in Chapters 6 and 7 of this book. In this chapter, the chemical and physical processes that specifically affect biological materials is discussed, and particular emphasis is given to the interactions of plasma with bacteria since that is the most commonly studied cell type.

Methods for inactivating and killing microorganisms, including spores, are generally based on chemical and physical (ultraviolet (UV) and gamma irradiation), mechanical (pressure), and thermal processes. Many of the conventional processes suffer from disadvantages of cost, difficulty of user-acceptance, formation of residues on surfaces, formation of disinfection byproducts, changes to surface properties, and acquired microbial resistances, and there is a need to develop more effective technologies. Reviews on atmospheric plasma [1–4, 36–43] show that inactivation of microorganisms in such systems have five major mechanisms for action. These are thermal, electric field, UV radiation, direct chemical reactions of

Plasma Chemistry and Catalysis in Gases and Liquids, First Edition.
Edited by Vasile I. Parvulescu, Monica Magureanu, and Petr Lukes.

neutral reactive species, and interactions of charged particles [12, 44] with cellular components. Among these mechanisms, the reactive species (e.g., atomic oxygen, metastable oxygen molecules, ozone, and $OH^{\bullet}$ radicals) are generally accepted to play the dominant role in the inactivation process in the nonthermal atmospheric pressure plasma systems [3, 10]. In general, these oxidizing species are referred to as reactive oxygen species (ROS), and they have been commonly utilized for antimicrobial activity in many fields including the food and pharmaceutical industries and in medical environments. In the case of plasmas in gas–liquid environments, the presence of water adds more complexity to the system. For example, in underwater discharges, it is expected that in addition to the chemical effects (largely attributed to $OH^{\bullet}$ radicals and hydrogen peroxide), the physical processes such as high electric field, UV radiation, and shock waves may significantly contribute to the inactivation of microorganisms [45–53]. There are also possible synergistic effects of the above mentioned processes on the destructive efficiency of discharge in water. In addition to the production of ROS, electric discharge plasma can produce reactive nitrogen species (RNS) under atmospheric conditions with suitable nitrogen sources such as with air or nitrogen carrier gases. As with ROS, the RNS [54–56] can also directly lead to DNA damage and some enzymes can protect against RNS [57]. These effects are described in the following sections of this chapter. We first focus on the chemical interactions with charged and reactive species since much evidence given below points to their primary importance in plasma disinfection processes (Section 8.3). Physical mechanisms of plasma interactions with biological materials are discussed in Section 8.4. Examples of treated microorganisms using plasmas in water and in gas–liquid environments are presented in Section 8.2.

8.2
Microbial Inactivation by Nonthermal Plasma

Microbial inactivation by plasma can be classified by the plasma environment into (i) dry gas plasma, (ii) humid gas plasma, (iii) gas plasma in contact with liquids, and (iv) plasma formed directly in the liquid phase. Chapter 1 discusses the major aspects related to category (i) and Chapters 6 and 7 highlight some of the basic physics and chemistry for categories (ii) through (iv). One of the first plasma sterilization techniques utilized radio frequency (RF) argon gas plasma [58], and subsequently many research activities as well as some commercial devices have been developed for dry gas plasma sterilization. Electric discharge plasma processes for disinfection applications directly in liquids stem from early studies of both pulsed electric fields (PEFs) in nonplasma-generating conditions [59–67] and plasma generation directly in liquid water [45–53]. PEFs in nonplasma-generating situations are widely used in biological applications, including as a laboratory technique to deliver DNA into or out of the cell, as well as in other applications [68–72]. Discharges in gas-phase plasmas over liquids are also of particular importance to many disinfection applications since the microorganisms are typically located on hydrated surfaces, and in some cases, the desired objective is to purify water that contains suspended microbes. Many example plasmas under atmospheric pressure conditions have been used in such sterilization applications [36, 37, 73–75] including corona discharges [76–79], dielectric barrier discharges (DBD) [80–82], glow discharge [83–88], contact glow discharge electrolysis, gliding arc discharges [89], microwave, and RF discharges [90, 91]. In the following sections, we provide a brief overview of microbial inactivation by plasma classified according to the plasma environment utilized.

8.2.1 Dry Gas Plasma

A wide range of microbes, including spore forming and nonspore forming, have been inactivated by various dry-gas plasmas as shown in Tables 8.1, 8.2, and 8.3 for microwave plasma [92–97], DBD, and glow discharge, respectively. Additional studies with glow discharge in dry gases have shown inactivation of a variety of bacteria *Staphylococcus aureus, Escherichia coli, Bacillus stearothermophilus, Bacillus subtilis, Bacillus pumilus (spores), Bacillus subtilis niger (spores), Pseudomonas aeruginosa, Deinococcus radiodurans,* the yeasts *Saccharomyces cerevisiae* and *Candida albicans,* and the viruses bacteriophage phi X174 [83–86].

Corona discharge in various gases including air, oxygen, argon, and nitrogen showed 3–4 orders of magnitude decreases in colony forming unit (CFU) of a population of sporulated *B. stearothermophillus* within 1 h [108], and RF plasma has been used for killing bacteria [78, 79] and destroying spores [109]. Pointu *et al.* [110] found 5 CFU log units with an improved pulse device, and Ekem *et al.* [111] inactivated 3.5 CFU log units of *St. aureus* within 12 min of treatment. The RF discharge in air to make the plasma needle or pencil has effectively inactivated *E. coli* [90, 112], *Streptococcus mutans* [91]. *B. subtilis* spores have also been extensively inactivated with DBD [81] and RF O_2 plasma [113].

Table 8.1 Bacterial inactivation by microwave plasma.

Microorganism	Gas	(log CFU)//(min)	References
E. coli	O_2/N_2	13 // 25	[98]
E. coli	N_2O	> 5 // 2	[99]
B. subtilis (spores)	O_2/CF_4	5 // 7.5	[100]
B. subtilis (spores)	Ar	2 // 40	[99]
B. subtilis (spores)	Ar/O_2	7 // 40	[101]
B. subtilis (spores)	air	6 // 3	[96]
B. pumilus (spores)	Ne	3 // 30	[101]
B. pumilus (spores)	air	4 // 30	[101]
B. stearothermophillus (spores)	N_2O	> 5 // 20	[92]
Pseudomonas fluorescens	N_2O	> 5 // 15	[92]
Proteus vulgaris	N_2O	> 5 // 5	[92]

Table 8.2 Bacterial inactivation by DBD plasma.

Microorganism	Gas	(log CFU)//(min)	References
E.coli	air	7 // 20	[102]
E. coli	air	5 // 1.5	[103]
E. coli	air	4 // 1	[82]
E.coli	He	7 // 2	[104]
E. coli	Ar	5 // 1	[105]
B. subtilis (spores)	air	5 // 2	[103]
B. subtilis (spores)	air	1 // 0.3	[3]
B. subtilis (spores)	N_2; N_2O	4 // 10	[106]
B. subtilis (spores)	N_2	4 // 20	[106]
P. aeruginosa	He; air	5 // 15	[1]
St. aureus	air	5 // 1.5	[103]
St. aureus	Ar	5 // 1.5	[105]
C. albicans	air	5 // 0.3	[107]

Table 8.3 Bacterial inactivation by glow discharge plasma.

Microorganism	Gas	(log CFU)//(min)	References
E. coli	air	7 // 25	[83]
E. coli	air	6 // 0.5	[84]
B. subtilis	air	5 // 5	[88]
St. aureus	air	4 // 10	[84]
St. aureus	air	6 // 0.5	[84]
Virus phiX174	air	3 // 10	[83]
B. strearothermophillus	air	5 // 7	[88]

These data generally show that significant amounts and many kinds of bacteria and other organisms (yeast and viruses) can be inactivated in relatively short time. While a range of plasma types have been studied, many studies suggest that the etching mechanism or chemical reactions to destroy the surface of the cells are the key reasons for dry gas plasma inactivation of bacteria and spores [40, 100, 105–107, 114].

Reactive neutral species (e.g., atomic oxygen, metastable oxygen, ozone, and OH radicals) are generally accepted to play dominant roles in cellular inactivation processes in nonthermal atmospheric pressure plasma systems, but the specific species formed strongly depend on the composition of the carrier gas [5, 8, 36, 37]. For example, plasma with flowing Ar gas and additional UV light destroyed cell membranes and DNA [76]. Emissions data on oxygen species in plasma has been reported [115], and it has been demonstrated that the chemical etching by oxygen radicals on the surface of membranes leads to microorganism inactivation

[116]. Plasma with nitrogen [117] and with nitrogen–oxygen mixtures [118] have demonstrated inactivation of dental bacteria [23] and *E. coli* [98, 117, 119, 120]. The densities of N and O atoms in such plasmas were measured [119], and these species (rather than UV) were shown to play a major role in microorganism destruction [3].

Alternatively, and depending on how the bacteria are exposed to the plasma, stable chemical species can be responsible for sterilization. For example, DBD are commonly used to generate ozone in dry air or oxygen [121], and the main mechanisms for sterilization in such systems is by the reactions of ozone [82, 122, 123].

8.2.2 Humid Gas Plasma

Humidified gases have been found to inactivate *E. coli* with RF discharge [124] and bacteria, algae, and protozoa with DBD [80]. Table 8.4 shows examples of bacterial inactivation using humid gas gliding arc discharge. These data generally indicate relatively fast inactivation by many orders of magnitude for a range of bacterial species. In contrast to dry gas plasma systems, the addition of water, either as a vapor or as a contacting liquid phase, can dramatically change the reaction products, and therefore the mechanisms of bacterial inactivation. For example, humidified gas [124] with a high-frequency air plasma led to significant *E. coli* inactivation, but complicating the interpretation, the role of water vapor was found to be optimal at an intermediate relative humidity. Discharges with water, as discussed in Chapter 6, also produce $OH^{\bullet}$ radicals and hydrogen peroxide.

8.2.3 Gas Plasma in Contact with Liquids

8.2.3.1 Discharge over Water and Hydrated Surfaces

Plasma discharges have been formed in the gas phase over liquid water solutions containing suspended bacteria or over hydrated (agar) surfaces. A wide variety of such configurations are shown in Figure 6.1 (see Chapter 6). By way of an

Table 8.4 Bacterial inactivation by gliding arc discharge plasma in humid air.

Microorganism	Gas	(log CFU)//(min)	References
E. coli	air	8 // 8	[39]
E. cartovora	air	10 // 5	[125]
St. epidermidis	air	7 // 5	[126]
H. alvei	air	7 // 9	[39, 127]
S. cerevisiae	air	4 // 20	[39, 127]
L. mesenteroides	air	4 // 20	[39, 127]

example, the gliding arc plasma discharge can be formed above a liquid solution. In such a case, the high-velocity gas carries reactive species from the plasma to the liquid surface, and in some cases, the plasma plume can directly impinge on the liquid surface, causing the reactive species to be formed in the liquid or at the liquid–gas interface [128, 129]. Such plasma has been shown to inactivate bacteria including *Erwinia spp.* [125, 130] and *Hafnia alvei* [89]. Very fast, within seconds, inactivation by 4–5 orders of magnitude of *E. coli* in agar dishes below a plasma needle discharge (RF discharge) have been reported [79, 90]. AC discharge over the liquid phase has been shown to inactivate *E. coli, S. aureus*, and yeast [131]. DC discharge over water can inactivate *E. coli and P. aeruginosa* [132], and a bipolar pulsed discharge inactivates *E. coli* [133] and *Microcystis aeruginosa* [134]. Atmospheric glow plasma over a water solution inactivates *E. coli K12, E. coli O157 : H7, Salmonella typhimurium*, and *P. aeruginosa* [135].

8.2.3.2 Discharge with Water Spray

A very efficient gliding arc reactor incorporating water spray and pulsed low power has also been developed and demonstrated to have very high energy yields for hydrogen peroxide generation and for inactivation of *E. coli* downstream of the plasma spray [136–140]. Part of this enhancement may be due to the increased surface area or contact area caused by the liquid spray. However, it is also very likely that spraying the liquid through the plasma region affects the type and distribution of active species formed in the plasma by the presence of the condensed water droplets affecting the radical and reaction quenching processes. DC discharges with the bacteria contained within the water droplets also showed high degree of bacterial inactivation [141, 142].

8.2.3.3 Gas Discharge in Bubbles

Further studies of microorganism inactivation in gas–liquid systems involve injection of bubbles into the liquid phase and/or generation of the discharge within the bubbles. Details on these reactor configurations and chemical processes are discussed in Chapter 6; however, it can be noted that the general motivation to use such configurations is to increase energy efficiency for plasma generation (in gas bubbles verses directly in the liquid) and to change plasma chemistry by, for example, the addition of oxygen. Pulsed plasma discharges within bubbles immersed in the liquid have led to inactivation of *Campylobacter jejuni, E. coli, Listeria monocytogenes, S. aureus*, and *Salmonella* [143, 144]. With gas bubble, injection into the liquid with a DC discharge generated with a rotating electrode has led to inactivation of *Bacillus, Corynebacterium, Enterobacter, Klebsiella, Micrococcus, Proteus, Pseudomonas, Shigella*, and *Staphylococcus* [145].

8.2.4 Plasma Directly in Water

Electrical discharges directly in water have been shown to destroy a wide variety of microorganisms (including bacteria and yeasts, as well as viruses) through combinations of chemical and electrical effects. Pulsed electric discharges directly in the liquid phase with energies in the range of Joules per pulse have been shown to inactivate a wide range of bacteria including *E. coli, S. aureus, S. enterititus, Microcystis aeruginosa, bacilli, Pseudomonas putida,* food pathogens, and others [45–53, 77, 146–152]. Electrohydraulic discharges generally utilize kiloJoules per pulse and have been shown to inactivate bacteria directly in solution [49, 153, 154]. Microwave discharges in water have also inactivated *E. coli* [155]. Contributions from the high electric fields, UV radiation, and shock waves may also contribute to the inactivation of microorganisms by discharge directly in water [47–53]. Electric discharge in water should be clearly distinguished from PEFs where the electric field is below the threshold for plasma discharge formation [60, 67, 77].

8.2.5 Kinetics of Microbial Inactivation

Analysis of the rates of bacterial inactivation by plasma are generally reported in survival diagrams showing the number, N, of CFU per unit volume as a function of the time of exposure to the discharge. These plots are usually composed of linear sections on semilogarithmic scales [156–165]. Some survival curves have a single-slope exponential decay (Chick's law [156]), while many others exhibit multiexponential decay (Figure 8.1). A more generalized semiempirical formula for such decay can be written as

$$\frac{\mathrm{d}N}{\mathrm{d}t} = -kmN^{x}C^{n}t^{m-1} \tag{8.1}$$

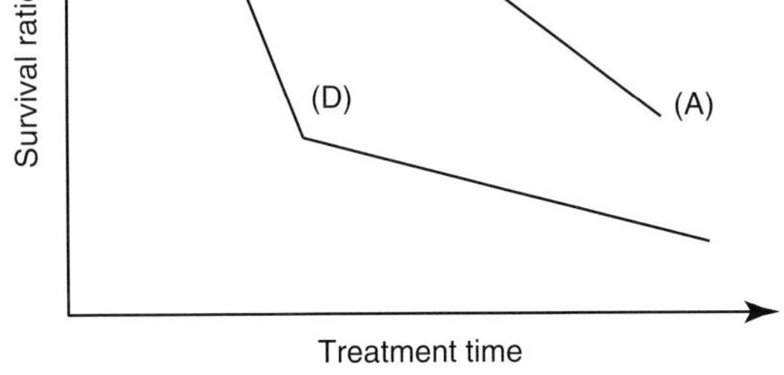

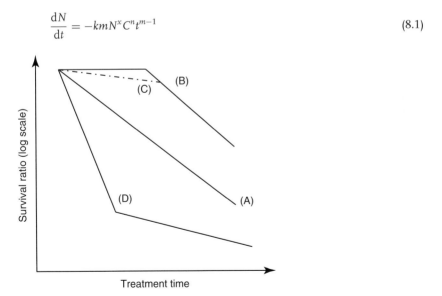

Figure 8.1 Typical first-order and biphasic inactivation of a bacterial population: (A) first-order kinetics, (B,C) shouldering period followed by first-order kinetics, (D) rapid decay followed by slower regime.

where t is contact time, C is concentration of disinfectant, and m, n, and x are empirical constants [162]. Chick [156] utilized the chemical kinetic mechanisms whereby

$$xN + nC \longrightarrow p\ (\text{inactive microbe with rate constant } k) \tag{8.2}$$

For excess disinfectant, this reduces to a first-order decay known as *Chick's law* [156, 162]. Watson utilized an empirical function to account for the amount of disinfectant that combines with the microbe and utilizing a first-order decay of the disinfectant concentration reduces to [157, 162]

$$\frac{dN}{dt} = -kNC_0^n e^{-k'tn} \tag{8.3}$$

More details are given in the literature for a variety of proposed disinfection mechanisms and extensions to the Chick–Waston approach [164], which reproduce a range of types of observed behaviors like those shown in Figure 8.1 [162].

In general, the characteristics of the plasma inactivation kinetics depend on the type and state of the bacteria, the type of liquid medium, as well as the type and working conditions of the electrical discharge, including the method of exposure of the liquid (direct or indirect). Most of the inactivation processes induced by electric discharges in liquid and gas–liquid environments obey biexponential behavior with either a larger initial slope followed by a region with lower slope (Figure 8.1, line D) or a smaller initial slope (often erroneous referred to as a *lag phase*) followed by more steeply decreasing second slope (Figure 8.1, lines B and C). This two-step inactivation process may be processed automatically with the help of the GinaFiT program [166], which combines the two kinetic steps, that is, a zero-order step followed by a first-order step. Other situations such as the "temporal post-discharge reactions" (TPDR), which develop after the discharge is switched off, are discussed in Section 8.3.4.

8.2.5.1 **Comments on Sterilization and Viability Tests**

The basic protocol to assess bacterial inactivation is to grow cells in a suitable medium after they have been exposed to the plasma inactivation process and thereafter to count the bacterial population using the classical colony counting methodology [167]. While care must be taken to establish proper control experiments, measurement of the number of CFUs of the surviving population has been considered a satisfactory method to determine the efficacy of the inactivation process. However, more detailed analysis of cell metabolism and viability using molecular probes can be utilized.

Other tests of bacterial metabolism involve fluorescent indicators (life/death tests) that are able to pass through the membrane of living cells [168–170]. Clearly, bacteria that can remain viable but not culturable (VBNC) can present problems in analysis [171, 172]. In addition, the use of such molecular probes in such life/death tests should be carefully conducted in plasma systems because the chemical reagents may be degraded by the active species produced by the plasma if the reagent is incorporated into the medium before exposure or during post-discharge situations.

It is important to note that much work in plasma and other sterilization methods utilize the concept of sterilization. Since complete sterilization implies no remaining microbes, it is generally linked to the measurement method and sometimes to the assumption that microorganism inactivation follows exponential kinetics, which can be extrapolated indefinitely. von Woedtke *et al.* [173] have defined the concept of sterilization assurance level (SAL) based on four microbiological safety levels dependent on the intended products. The first level, related to heat-resistant species, applies to pharmaceutical sterilization and is set at 10^{-6} or "the probability of not more than one viable microorganism in one million sterilized items of the final product." The other four levels connote different degrees of sterilization, and details can be found in Ref. [173].

8.3 Chemical Mechanisms of Electrical Discharge Plasma Interactions with Bacteria in Water

As already described in Chapters 6 and 7, the type and quantity of the reactive species formed in the discharge depend on the nature and composition of the ambient gas and, in the case of liquid water contacting the plasma, the properties of the solution. The amount of energy delivered by the discharge, other parameters, (e.g., utilization of a separator or a barrier), and the nature of the working electrodes can be important. Air is typically used in discharges for environmental applications because of the nature of the formed species and the low operating costs. Water is present at many material surfaces exposed to the atmosphere and is the most common medium containing wastes and microorganism. Microorganisms generally grow in aqueous media or on wet material surfaces in humid atmospheres where a thin water layer generally coats the surface of the substrate. In some cases, very dry environments may be of importance. It is thus reasonable to utilize solution chemistry in water in order to analyze many decontamination phenomena occurring under atmospheric pressure plasma that contact liquid water (detailed consideration of the chemical aspects of plasma with liquids is covered in Chapters 6 and 7). In addition, the contribution of physical processes such as the direct impact of plasma species on the external membranes of the microorganisms or the effects induced by strong electric fields, heat, UV light, and shockwaves have to be considered, particularly for direct discharge in liquid water (Section 8.4).

The interaction of chemical species formed in the plasma with bacteria and other cells begins at the surface of the cell where chemical destruction of the cell wall and membrane and associated components can occur. This is followed, in some cases, by transport of active species into the cell where internal cell damage can occur through destruction of DNA, proteins, and other internal components of the cell. Figure 8.2 shows SEM images (scanning electron micrograph) of morphology change of the bacterial membrane of *Erwinia spp.* after treatment by a gliding arc discharge generated in humid air above the water surface [174]. The

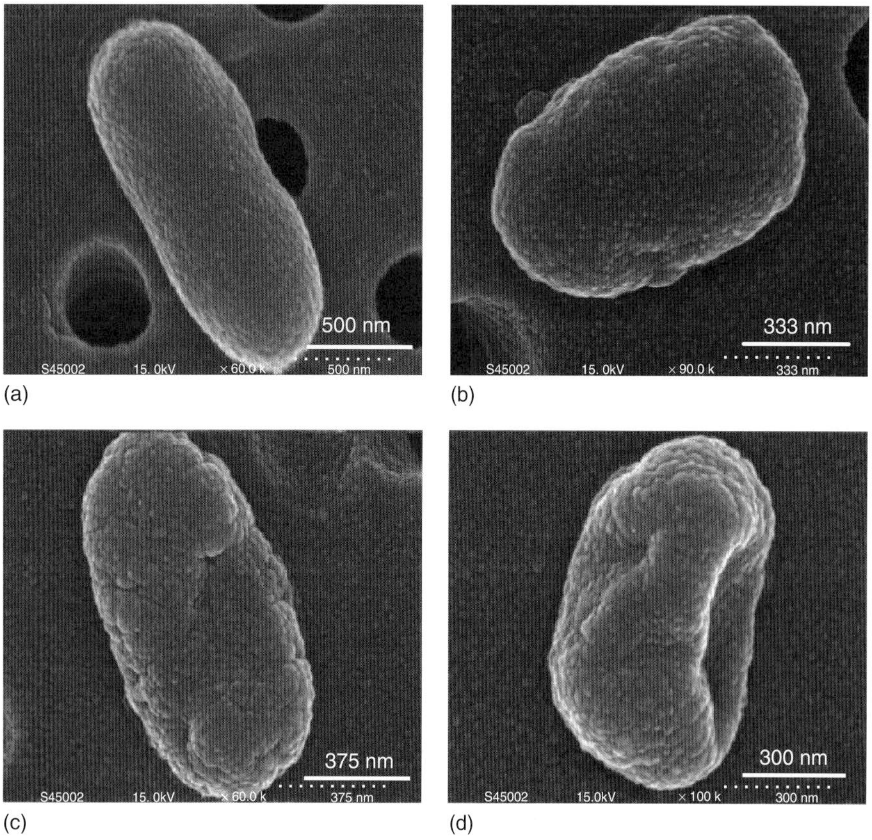

Figure 8.2 SEM images of the effects of gliding arc discharge in air on the morphology of *Erwinia carotovora*. (a) Nontreated sample, (b–d) bacteria exposed for 1, 2.5, and 5 min to the discharge. (Source: Reproduced from [174] with permission from American Society for Microbiology.)

corresponding electrophoresis experiments showed an increase in the quantity of extracted proteins, with slightly reduced molecular masses. This result may be accounted for by an attack at the lipoproteins present on the bacterial surface [175–177].

In order to penetrate into the cell or liquid environment, some of the plasma species must be soluble in water, while for cases where the bacteria directly contacts the gas, as in biofilms or with dry spores, other plasma species may react at the gas–cell interface. Therefore, discussion of biocidal effects of plasma at gas–liquid interfaces suggests humid air as one of the most effective atmospheres. The oxidizing properties of the species formed in such discharges are largely known (Chapters 6 and 7). The role of water in the inactivation processes is now also commonly acknowledged, and this role is well established in the case of underwater plasma. It should be noted that although electrical discharges directly in water and in gas–liquid environments were shown to be efficient tools for microbial inactivation, we are not yet in a position to mathematically model the discharge or the associated lethal effects for many bacteria and viruses because of the complexity of the chemical processes. However, the major chemical effects have been identified and may be applied to other living matter, for example, other cell types and tissues.

As discussed in Chapters 6 and 7, depending on the type of plasma contacting with the liquid and the chemical composition of the gas and liquid environments, the main reactive species formed in an electric discharge in humid air are ROS and RNS. The main reactive species formed in a discharge burning in humid air are hydroxyl radical, ozone, and nitric oxide. The main reactive species formed by discharges directly in liquid water are hydroxyl radical, hydrogen peroxide, and peroxynitrite. In the following sections, the main properties of these species in terms of their biological effects are presented with brief description of the chemical structure of the cell membrane of bacteria as the primary target of plasma.

8.3.1 Bacterial Structure

All cells have some type of barrier to control interactions with the environment. In the case of bacteria, the barrier includes phospholipid membranes with integral proteins and peptidoglycan layers that make up the cell wall. The thickness and composition of this barrier vary with the type of the bacteria. Gram-positive bacteria have much thicker cell walls (10–80 nm) than gram-negative bacteria (10 nm) [178]. The external membrane of gram-positive bacteria consists mainly of peptidoglycan, teichoic acids (i.e., polymerized ribitol and glycerol phosphates), and surface proteins. The membrane of gram-negative bacteria is more complex, because its external surface contains phospholipids and lipopolysaccharides over a peptidoglycan layer (periplasmic space), with the most common cell lipid being phosphatidylethanolamine. The membrane of gram-negative bacteria is richer in lipids and lipopolysaccharides compared to that of gram-positive bacteria. The murein or peptidoglycan is made of alternate chains of *N-acetylmuramic acid* (NAM) and *N-acetylglycosamine* (NAG) with linking oligopeptides at the NAM molecules. The bacterial cell wall maintains the rod, spherical, or helical shape and provides mechanical protection against osmotic pressure effects and other physical and chemical stresses.

Peptidoglycan and the surrounding protein-containing coatings (exosporium, spore coat) are present around the bacterial spores. The main components of exosporium are proteins (52%), polysaccharides (20%), and lipids (18%) [179]. Peptidoglycan is a cross-linked complex of polysaccharide and peptides with C–N, C–C, and C–O bonds largely weaker than the C–H bond (bonding enthalpy $413\,\mathrm{kJ\,mol^{-1}}$), which makes them easily broken by direct impact of the plasma species.

The outer surface of the cells or spores contacts the plasma species and their products. For example, the degradation of bacterial outer structures by plasma species may involve attack on the various functional groups. It is well known that sugar and alcohol functional groups (OH) in NAM and NAG are easily oxidized. In addition, chemical species can pass through the external membrane either through lipopolysaccharides and phospholipids (hydrophobic path) or through proteins and porines (hydrophilic path). Once inside the cell, for example, peroxynitrite is reported to attack the RNA bases (e.g., formation of nitrotyrosine). Peptides, either inside the cell or on the cell surface, can be targets for peroxynitrite.

Examples of microbial inactivation by plasma shown in Section 8.2 illustrate that electric discharges are efficient in initiating lethal effects for both gram-positive and gram-negative bacteria in gases, on surfaces, and dispersed in aqueous media, at temperatures close to ambient, thereby suggesting that chemical degradation pathways are more important than thermal effects. Most plasma bacterial degradation is probably related to chemical effects, since highly oxidizing compounds are present in plasma formed in dry and humid gas phases and directly in the liquid phase, that is, $OH^{\bullet}$ radicals, H_2O_2, and ONOOH. These species attack the membrane and degrade specific components such as the lipidic part of lipopolysaccharide (LPS) or phospholipids. The membrane is weakened, and the inner fluid may leak out, which prevents the bacteria from stimulating the antistress and repairing processes. An "etching process" was thus postulated.

8.3.2 Reactive Oxygen Species

8.3.2.1 Hydroxyl Radical

The hydroxyl radical $OH^{\bullet}$ is the major ROS produced in electric discharges with water. The $OH^{\bullet}$ radical has a high oxidizing power ($E^0 = 2.85\ V/SHE$), and it is the strongest oxidant that can exist in an aqueous environment. It reacts with most organic compounds with rates that approach diffusion-controlled limits (Chapters 6 and 7). In the case of microbial cells, the primary target of the $OH^{\bullet}$ radical is the outer cell wall, including the cell membrane. The cell membrane, composed largely of organic compounds such as lipids, proteins, and polysaccharides, is susceptible to the $OH^{\bullet}$ radical attack via the reaction mechanisms described in Chapter 7. Lipids are macromolecules of the cell membrane that are most vulnerable to oxidation. Lipid reactions with $OH^{\bullet}$ radicals proceed mainly via H-abstraction from the unsaturated carbon bonds of fatty acids, which in the presence of oxygen causes lipid peroxidation (Figure 8.3) [180]. The final product of this process is malondialdehyde, which can be quantified spectrophotometrically by its reaction with thiobarbituric acid. Therefore, malondialdehyde can be used as an indicator of the peroxidation of membrane lipids because of interaction of the ROS generated in the plasma with the cell membrane. Malondialdehyde was positively analyzed, for example, during decontamination of water containing bacteria *S. typhimurium and Bacillus cereus* and yeast *Saccharomyces cervisiae* using water spray DC discharge, proving the occurrence of such an inactivation route in gas–liquid plasmas [141, 142].

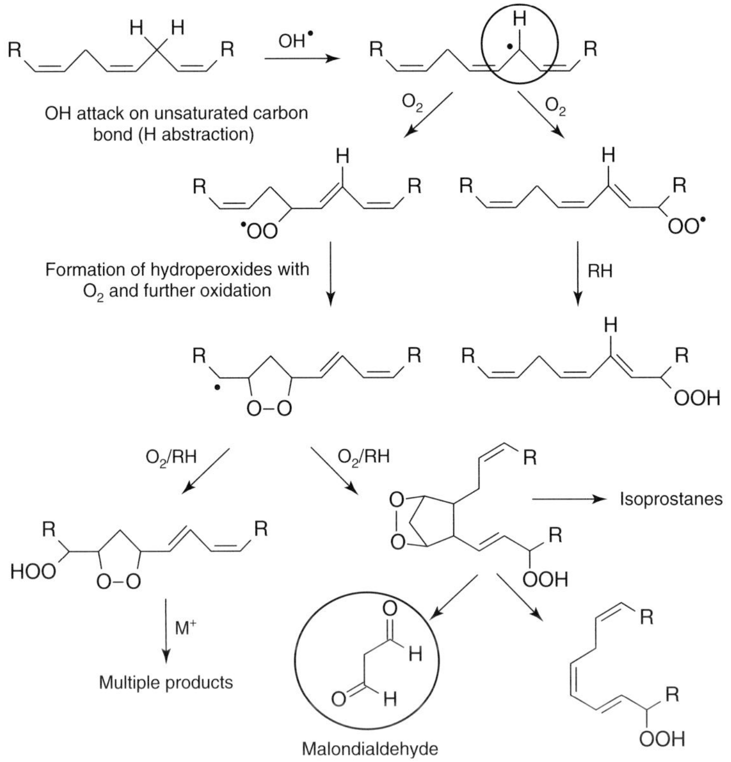

Figure 8.3 Scheme of OH radical attack on lipids. (Source: Modified from Ref. [180]. Copyright (1999), with permission from Elsevier.)

Similarly, $OH^•$ radicals can damage membrane proteins by H abstraction from α-carbon of peptide bonds –CO–NH– between chain peptide-linked amino acids. $OH^•$ radical attack leads then to peroxidation and backbone cleavage of proteins (Figure 8.4) [181].

In addition to reaction with $OH^•$ radicals, lipids and proteins are sensitive to oxidation by other ROS such as atomic oxygen and ozone. The reactivities of ozone and atomic oxygen are, however, more selective compared to the reactivity of the $OH^•$ radical. Ozone is less reactive with saturated and aliphatic hydrocarbons, and its reactivity in water is strongly pH dependent as discussed in Chapter 7.

8.3.2.2 Hydrogen Peroxide

Hydrogen peroxide is a very important species, which increases the collective oxidizing power of the plasma, especially in the case of underwater plasmas, in which hydrogen peroxide is the most abundant long-lived plasmachemical product. The antimicrobial properties of H_2O_2 are well recognized, and a variety of applications have been suggested and developed. H_2O_2 is able to react in the

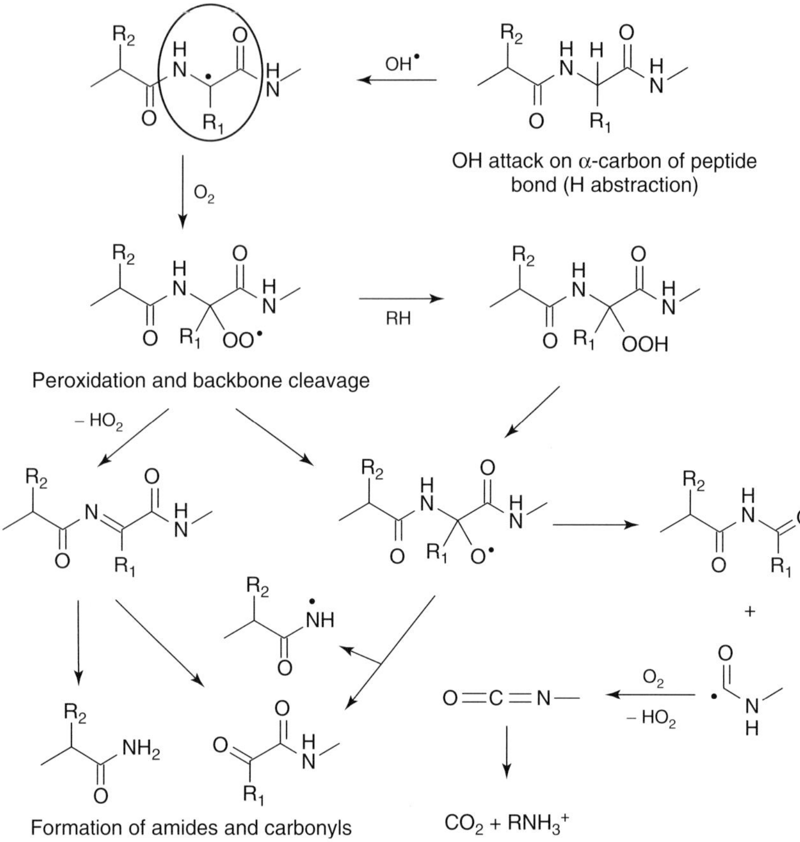

Figure 8.4 Scheme of OH radical attack on proteins. (Source: Modified from Ref. [181].)

gas phase, at the liquid surface, and in the bulk target solution with solutes since it is completely soluble in water and can readily transfer across cell membranes. Hydrogen peroxide is a strong oxidizing agent able to oxidize many organic compounds. This property explains why hydrogen peroxide is of common use for disinfecting superficial wounds. It is thus reasonable to consider hydrogen peroxide in the plasma processes of microbial abatement. Its action may take place both in direct exposure of the target to the discharge and in post-discharge conditions (Section 8.3.4).

The principal mechanism of the H_2O_2 cytotoxicity involves penetration into cells and the generation of hydroxyl radicals through interactions of H_2O_2 with intracellular transition metals (Cu^+ and/or Fe^{2+}) by the Fenton's reaction (Chapter 7):

$$H_2O_2 + Fe^{2+} + H^+ \longrightarrow OH^\bullet + Fe^{3+} + H_2O \quad (8.4)$$

In a bacterial cell, this process might be initiated with the reduction of ions Fe^{3+} to Fe^{2+} liberated from bacterial iron storage proteins, ferritin or bacterioferritin, by superoxide radical $O_2^{\bullet -}$. The released Fe^{2+} ions can then react with H_2O_2 outside the proteins or close to their surface to yield hydroxyl radicals that might cause oxidative damage in bacterial cells and stimulate double-strand breaks in DNA [182].

Such mechanisms, however, require relatively large concentrations of hydrogen peroxide to penetrate into the cell since it is known that bacterial cells possess an adaptive response to low concentrations of H_2O_2 and that they may protect themselves against its oxidant toxicity with defense proteins and enzymes, and that they even produce H_2O_2 [183–189]. For example, *E. coli* cells generate intracellular H_2O_2 at a rate of about 10^{-20} mol s^{-1} of H_2O_2 by a single cell [190]. Experiments performed with *E. coli* in pulsed corona discharge in water, which produces H_2O_2 with the rate of formation typically of the order of 10^{-6} mol s^{-1} have shown that even concentrations of H_2O_2 of the order of millimoles per liter were not sufficient to cause significant inactivation effect of bacteria [191]. Sato *et al.* [48], on the other hand, have shown H_2O_2 to be the key chemical agent responsible for inactivation of yeast cells caused by the same type of discharge through comparison of the lethal effect of water containing H_2O_2 generated by the discharge with nonplasma solutions containing H_2O_2. However, Imlay *et al.* [192] found that H_2O_2 concentrations between 1.5 and 2.5 mmol l^{-1} were optimal for *E. coli* killing in solution (in absence of electric discharge).

The protective mechanisms of bacterial cells against hydrogen peroxide might be strongly reduced under the influence of high electric field because of polarization and a subsequent increase of the cell membrane permeability. Nakamura *et al.* [193] determined electric field intensity of 320 kV cm^{-1} at the streamer tip of a pulsed streamer discharge in water. Such high-intensity electric field may significantly increase the flux of H_2O_2 into the cell, causing the intracellular H_2O_2 concentration to overwhelm the basal level of the scavenging capacity of the cell, giving rise to oxidative stress conditions. In this respect, there is the possibility that the high electric field induced by the discharge can enhance lethal activity of the H_2O_2 formed by the discharge. Such synergy was documented in the inactivation of *E. coli* performed by means of a homogeneous PEF generated in water using a plate-to-plate electrode configuration with addition of H_2O_2 (Figure 8.5). The PEF caused a decrease in the survival ratio of *E. coli*, which is in agreement with the lethal effect of PEF reported in a number of papers [60, 64, 67, 175]. Consequently, the effect of the PEF was further improved by the addition of 1 mmol l^{-1} H_2O_2 to the bacterial suspension (the value for which only a negligible effect of H_2O_2 on viability of *E. coli* was determined without the presence of PEF) [191]. Abou-Ghazala *et al.* [50] compared *E. coli* decontamination by pulsed corona discharge in water with the effect of PEF with the conclusion that the decontamination efficiency of the discharge method was slightly higher than that of the PEF.

Owing to relatively high chemical stability of hydrogen peroxide in water, it can act on microorganisms dispersed in solutions treated by the plasma and also in the postdischarge reactions through peroxynitrite chemistry (Sections 8.3.3.1 and 8.3.4).

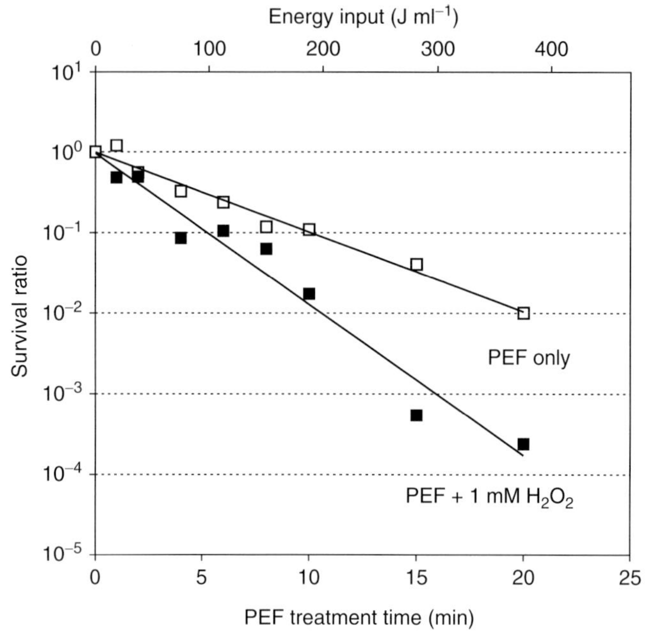

Figure 8.5 Inactivation of *E. coli* by pulsed electric field and with addition of 1 mmol l^{-1} H_2O_2 as a function of applied energy input (E = 39 kV cm^{-1}, FWHM = 200 ns). (Source: © 2007 IEEE. Reprinted, with permission, from [191].)

8.3.3 Reactive Nitrogen Species

The main RNS to be considered are nitric oxide NO, the primary species formed by an electric discharge in air, and its derivatives formed with water, including nitrites NO_2^-, nitrates NO_3^-, and peroxynitrites $O{=}NOO^-$. NO is the reducing agent of the NO^+/NO system [$E^0(NO^+/NO) = 1.21$ V/SHE at pH = 7] [194], whose oxidizer, the nitrosonium ion NO^+, is involved in the selective nitrosation reaction of primary and secondary amines and in bacterial inactivation [195]. NO participates in a number of reactions, including the formation of nitrous acid HNO_2

$$OH^{\bullet} + NO + M \longrightarrow HONO + M \quad (8.5)$$

Nitrous acid is a medium acid in water ($pK_a = 3.3$) and is not thermodynamically stable at pH < 6 since it slowly disproportionates into NO and NO_3^-. Nitric oxide also rapidly reacts with oxygen in air and yields NO_2, which is transformed into nitrite. The pH of the solution is then reduced, which favors the disproportionation of NO_2^- to NO and NO_3^-. Thus, NO is transformed into nitrates in water by the sequence (Eq. (8.6), see Chapters 6 and 7)

$$NO \longrightarrow NO_2 \longrightarrow NO_2^- \longrightarrow ONO_2^- \longleftrightarrow NO_3^- \quad (8.6)$$

The resulting pH change in the liquid is very important in the lethal effects of atmospheric discharges since acidity significantly contributes to some bacterial inactivation.

8.3.3.1 Peroxynitrite

Formation of nitrate may also proceed via the formation of peroxynitrite, which is a very strong oxidizer. The oxidant reactivity of peroxynitrite is highly pH dependent and both anionic ($O{=}N{-}OO^-$, peroxynitrite anion) and protonated forms ($O{=}N{-}OOH$, peroxynitrous acid) ($pK_a = 6.8$) can participate in oxidation reactions [$E^0(ONOOH/NO_2) = 2.02\,V/SHE$; $E^0(ONOO^-/NO_2) = 2.44\,V/SHE$]. It can be formed by (i) the reaction of nitric oxide and superoxide anion radicals, (ii) the reaction of the nitrite anion with hydrogen peroxide, or (iii) the reaction of $NO_2^\bullet$ with an $OH^\bullet$ radical (for more about peroxynitrite chemistry, see Chapter 7). The main chemical properties of peroxynitrite/peroxynitrous acid are well established, and their occurrence in stress phenomena of living matter is now largely accepted. For more details, see review [196, 197]. Koppenol [195] and Pryor and Squadrito [198] demonstrated that peroxynitrite was a key agent in oxidative stress and is involved in various diseases (e.g., HIV, Alzheimer's disease, arteriosclerosis, gut inflammation) [199–201].

Peroxynitrite oxidizes organic molecules directly or through H^+ or CO_2-catalyzed homolysis, yielding nitrogen dioxide radical $NO_2^\bullet$, $OH^\bullet$ radical, or carbonate anion radical $CO_3^{\bullet-}$. Only a few chemical groups directly react with peroxynitrite, which favors selective reactions with key moieties in proteins, such as thiols, iron/sulfur centers, and zinc fingers [202]. The half-life of peroxynitrite is short ($\sim$10–20 ms), but sufficient to allow it to cross cell membranes and diffuse quite far on a cellular scale ($\sim$one to two cell diameters) [203] and to allow for significant interactions with most critical biomolecules [198]. This is a significant difference between the reactivity and biocidal effects of hydroxyl radical in the cell [202, 204]. The hydroxyl radical is intracellularly formed by a rather slow reaction via the reaction of ferrous iron with hydrogen peroxide (Section 8.3.2.2), but it is so reactive that it can only diffuse about the diameter of a typical protein [205]. In contrast, peroxynitrite is formed each time superoxide and $NO^\bullet$ collide, but reacts slowly enough to react more selectively throughout the cell. Thus, peroxynitrite can have more subtle and specific actions on cells [206].

The proton-catalyzed decomposition of peroxynitrite to form $OH^\bullet$ and $NO_2^\bullet$ radicals may become relevant in hydrophobic phases, resulting in the initiation of lipid oxidation, which may lead not only to lipid peroxidation but also to the formation of nitrated products [207]. Figure 8.6 summarizes the peroxynitrite reactivity with lipids. ONOOH homolyzes to $NO_2^\bullet$ and $OH^\bullet$, and these species abstract H from unsaturated carbon bonds of lipids to form a carbon-centered radical, which in the presence of oxygen leads to the formation of peroxyl radical and lipid peroxidation (Section 8.3.2.1). Nitrated products would arise from $NO_2^\bullet$ reaction with a carbon-centered radical, potentially via a caged radical rearrangement of unstable alkyl peroxynitrite intermediates. Alternatively, the $NO_2{}^\bullet$ radical can mediate nitration of unsaturated fatty acids through homolytic attack at the double bond,

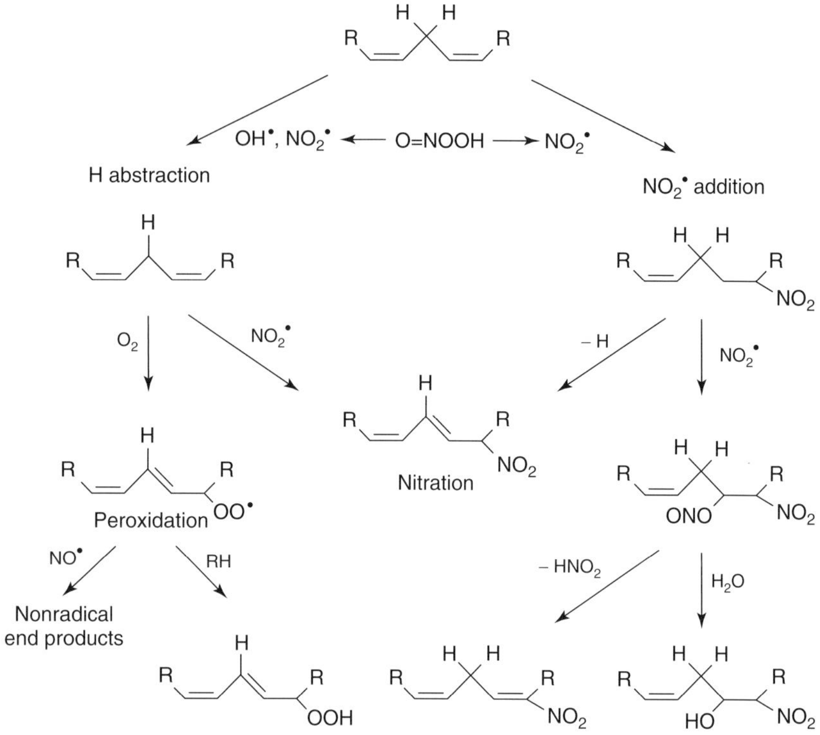

Figure 8.6 Overview of peroxynitrite reactivity with lipids. (Source: Modified from Ref. [207]. Copyright (2009), with permission from Elsevier.)

yielding a β-nitroalkyl radical that, at low oxygen content, combines with a second $NO_2^{\bullet}$ radical to form nitro/nitrite intermediates. Nitroalkenes are formed from these intermediates because of the loss of nitrous acid HNO_2, while its hydrolysis yields nitroalcohols. Since $NO_2^{\bullet}$ can also initiate lipid peroxidation reactions, yields of nitration versus oxidation/peroxidation will depend on O_2 levels: at low O_2 concentrations, nitrated products formation will predominate, while during aerobic conditions, lipid oxidation processes should be favored. Lipid oxidation is highly sensitive to inhibition by $NO^{\bullet}$, which is able to react with unsaturated lipid reactive species such as alkyl ($R^{\bullet}$), epoxyallylic ($R(O^{\bullet})$), alkoxyl ($RO^{\bullet}$), or peroxyl ($ROO^{\bullet}$) radicals, yielding nitrogen-containing lipid derivatives that rearrange or further react to form nonradical end products of lipid oxidation [207]. Note also the action of CO_2, which reacts with peroxynitrous acid and yields active nitrosoperoxycarbonate.

Owing to its weak chemical instability, the peroxynitrite is difficult to measure; however, the occurrence of peroxynitrite acid in plasma-treated solutions was recently identified in aqueous solutions exposed to a gliding arc discharge in a medium of suitable acidity [208]. The presence of peroxynitrite in plasma-treated solutions may also be related to the observed failing ion balance in conductivity measurements (Section 7.2) [209, 210]. Peroxynitrite attack at the transmembrane proteins was suggested in the lethal effects of the gliding arc discharge on many gram-positive and gram-negative bacteria [19, 43, 89, 127].

8.3.4 Post-discharge Phenomena in Bacterial Inactivation

Prolonged antibacterial activity of plasma-treated aqueous solutions was recently reported for electrical discharges generated directly in water and in gas–liquid environments. An apparent biocidal effect on microorganisms present in water was observed even several days after exposure of the solution to the discharge [19, 39, 43, 89, 127, 211–214]. This so-called *temporal post-discharge reaction* phenomenon can be generally defined as the chemical reactions that initiate or continue after the plasma discharge is switched off and in the absence of any external energy source, and they probably involve the presence of long-lived reactive species. TPDR was also observed in chemical degradation of organic compounds in water by a gliding arc discharge air plasma [208, 215]; for example, the TPDR degradation of slaughterhouse effluents that occurred over at least 9 days [215]. In the case of microbial inactivation, this phenomenon was known by different names, for example, *plasma-activated water* (PAW), *plasma acid, plasma pharmacology*, or *prolonged microbial resistance of water* [211–214]. For solutions that were treated by air-liquid-phase plasmas, TPDR was tentatively attributed to the effect of low pH and the remaining H_2O_2 and peroxynitrite present in noticeable concentrations in the solution. Note the similarity with the antimicrobial properties of the so-called *electrolyzed oxidizing water* in which the TPDR effect was attributed to the chlorine and oxychlorine ions produced from diluted salt solutions by electrolysis [216]. The TPDR phenomena in water treated by direct liquid-phase discharge was attributed to the effects of ions and oxide nanoparticles of heavy metals (e.g., Ag, Cu) that were released from the electrodes by erosion in the discharge [148, 211].

Figure 8.7 demonstrates an example of postdischarge reactions for *H. alvei*, which was exposed to gliding arc discharge generated in humid air above an aqueous solution for different discharge treatment times t^* (2, 3, 4, and 5 min) [89]. After switching off the discharge, the cells were left in the aqueous media without neutralization in order to reveal destruction by temporal postdischarge. Under these conditions, the treatment time was equal to the plasma discharge time t^* plus the post-discharge time t_p. Neutralization was performed after the post-discharge time. The efficiency of TPDR was clearly affected by the duration of discharge treatment. For longer time t^*, the decay was faster and the lag time was shorter.

Figure 8.8 shows the postdischarge behavior of *H. alvei* without precontact with the discharge. A NaCl solution was exposed to the discharge for 5 min, and the bacteria were then immediately placed in contact with the plasma-treated solution (referred to as *plasma-activated water*) for time t_c before neutralization. The effect of PAW was compared to that obtained for exposure of bacteria to direct discharge for

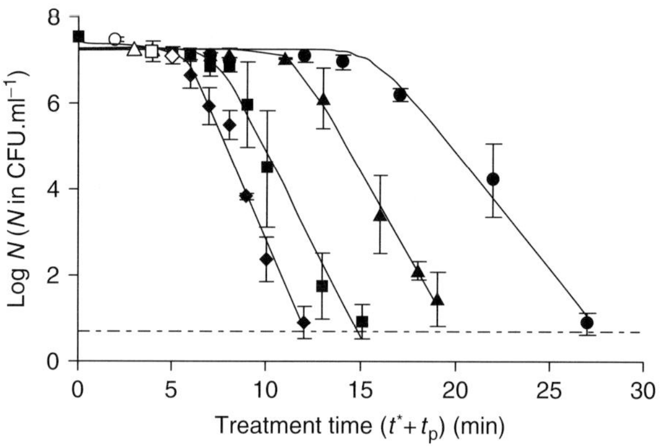

Figure 8.7 Inactivation of *Hafnia alvei* in TPDR conditions t_p followed after various gliding arc discharge treatment times t^*: 5 min (diamonds), 4 min (squares), 3 min (triangles), and 2 min (dots). Empty symbols indicate the beginning of TPDR. (Source: Reproduced from [89] with permission from American Society for Microbiology.)

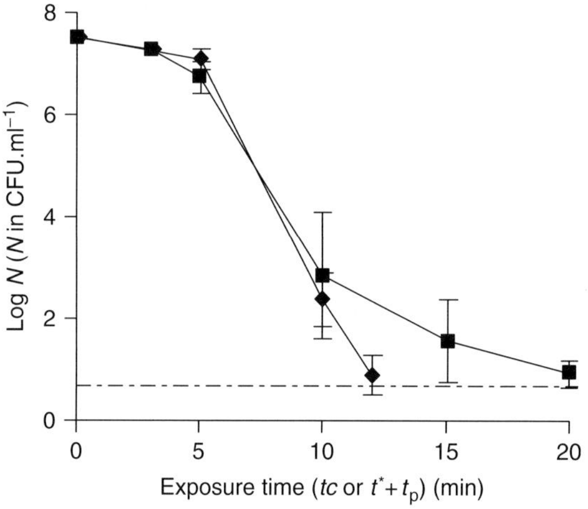

Figure 8.8 Inactivation of *Hafnia alvei* by PAW without precontact of bacteria with the discharge (squares). PAW was prepared from 0.15 mol l^{-1} NaCl solution exposed for 5 min by the gliding arc discharge in humid air. Line with diamonds shows inactivation of bacteria obtained directly by discharge after 5 min treatment. (Source: Reproduced from [89] with permission from American Society for Microbiology.)

5 min. Even if comparable, inactivation was slightly less efficient when cells were not subjected to the discharge. However, the reduction of more than 5-logs was obtained with PAW.

Other microorganisms, *Leuconostoc mesenteroides, Staphylococcus epidermidis, and S. cerevisiae,* showed similar results to TPDR [212]. Figure 8.9 shows changes in the

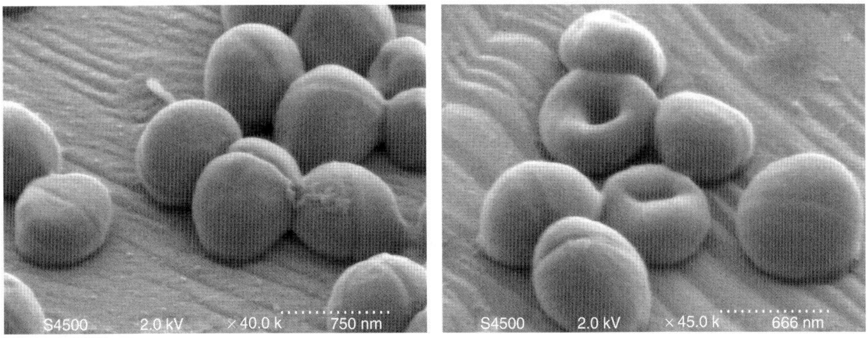

(a) (b)

Figure 8.9 SEM pictures of adherent *S. epidermidis* in contact with PAW for (a) 0 min and (b) 30 min. PAW was prepared from 0.15 mol l^{-1} NaCl solution exposed for 5 min to the gliding arc discharge in humid air [126].

structure of adherent bacteria *S. epidermidis* on exposure to PAW prepared from NaCl solution treated by gliding arc discharge in humid air.

Naitali *et al.* [19, 43] investigated the roles of ROS and RNS produced by the gliding arc discharge in TPDR inactivation of *H. alvei*. Table 8.5 shows the inactivation of *H. alvei* by treatment with PAW in comparison to acidified solutions (by HCl) containing nitrites, nitrates, or hydrogen peroxide and their mixtures at

Table 8.5 Inactivation of *Hafnia alvei* in various solutions.

Solution	pH	log (CFU_0)[a]	log (CFU_t)[a]			
			5 min	10 min	20 min	30 min
PAW[b]	3	7.9	6.8	5.7	2.5	<2
H_2O	3	8.0	7.9	7.7	7.6	7.6
H_2O_2		8.0	7.8	7.8	7.7	7.6
NO_3^-	3	8.0	7.8	7.7	7.6	7.5
NO_2^-	3	7.9	7.6	7.1	5.5	4.0
PAW + sulfamic acid[c]	2.1	7.9	7.8	7.7	7.4	6.9
$NO_2^- + H_2O_2$	3	7.9	7.5	7.0	5.0	3.4
$NO_2^- + NO_3^-$	3	7.9	7.4	6.8	4.7	2.9
$NO_2^- + NO_3^- + H_2O_2$	3	7.9	7.3	6.7	3.8	2.2
PAW[d]	6	7.9	7.8	7.8	7.8	7.7
PAW[e]	6	7.8	7.8	7.7	7.6	7.6

[a] CFU_0 and CFU_t, the number of cultivable cells before and after treatment, respectively.
[b] PAW was prepared from sterile distilled H_2O exposed by gliding arc discharge in air.
[c] Sulfamic acid was added to water before plasma exposure to trap NO_2^-.
[d] PAW was neutralized by NaOH.
[e] Distilled H_2O was buffered before exposure by gliding arc discharge in air.
Source: Reproduced from [19] with permission from American Society for Microbiology.

the concentrations found in PAW [19]. The largest inactivation was by treatment with PAW. However, based on the results shown in the Table 8.5, there are some contributions to inactivation by the combination of acidified nitrites, nitrates, and H_2O_2. Buffered and neutralized PAW was ineffective at inactivating the microbes. Acidified nitrites were the only compounds that caused a significant lethal effect when used alone. When nitrite formation was prevented by the use of sulfamic acid, no lethal effect of PAW was noted. Acidified nitrates and H_2O_2 were not lethal when utilized alone; however, their addition to nitrites enhanced the lethal effect (about 60% of the bacterial reduction achieved by PAW). Additional experiments showed that PAW slowly looses its efficiency with increasing storage time but remains active 24 h after it was prepared [89]. These results indicate the important role of peroxynitrite chemistry involved in the inactivation process by air plasma as discussed above.

Because rather complex peroxynitrite chemistry is involved in TPDR process, it is understandable that the comparison of the effects of externally added nitrites or nitrates and pH of the solutions might not necessary correlate with experimentally observed phenomena of the antibacterial effects of plasma-treated water. While the pH-dependent peroxynitrite chemistry might be a reason to explain the prolonged antibacterial phenomena of plasma-treated water, it is not possible to describe the exact mechanism of this process. Since the peroxysystem is reversible, intracellular cytotoxicity of the peroxynitrite system might be driven not only by its decomposition into $OH^\bullet$ and $NO_2^\bullet$ radicals but also through superoxide and nitric oxide radicals (i.e., precursors of $O{=}N{-}OO^-$ anion), which might be formed through reversible dissociation of peroxynitrite, when pH becomes more alkaline and thus more favorable for the reverse reaction of peroxynitrite (Eq. (8.7)) [206].

$$O_2^{\bullet -} + NO^\bullet \longleftrightarrow O{=}NOO^- + H^+ \longleftrightarrow O{=}NOOH \longrightarrow OH^\bullet + NO_2^\bullet \qquad (8.7)$$

It was also reported that peroxynitrite might be preserved under alkaline conditions at $-18^\circ C$ for several weeks [217]. All these features indicate a very important role of peroxynitrites in the discharge and in postdischarge processes induced in water by plasma, which clearly needs more study.

8.4
Physical Mechanisms of Electrical Discharge Plasma Interactions with Living Matter

It is not easy to distinguish physical and chemical properties in complex phenomena such nonthermal plasma and direct discharges in water. Thermal effects from the plasma interacting with the liquid water depend strongly on the plasma energy. In low-energy (Joules per pulse) corona-type discharges in and over water, small thermal effects are found as is the case in gliding arc discharges developed over the liquid. However, high-power kiloJoule per pulse discharges and other cases discussed in Chapter 6 thermal effects and shock waves can be important in bacterial disinfection. Solution conductivity as well as input power can affect the formation and intensity of UV light penetrating into the liquid phase. The formation of charged species close to or at the liquid surface can also increase the ionic strength of the liquid and thus form a significant osmotic pressure increase, which may damage bacterial membranes and induce lethal effects. In addition, discharges are generally associated with high electric fields that may affect bacterial inactivation by plasma in liquid. In the following sections, the biological effects of UV radiation (Section 8.4.1), shock waves (Section 8.4.3), and pulsed electric field (Section 8.4.5) in relation to their contributions to the electrical discharge plasma interactions with living matter are briefly reviewed.

8.4.1 UV Radiation

Unlike nonequilibrium air plasmas where UV radiation does not play a significant direct role in the sterilization process (radiation intensity of 50 $\mu W\,cm^{-2}$) [103], in the case of plasma formed by direct discharge in water, the role of UV photolysis on microbial inactivation has to be considered. UV radiation in the 200–300 nm wavelength range with doses of several miliwatts per square centimeters is known to cause lethal damage of cells. Among UV effects on bacteria is the dimerization of thymine bases in their DNA strands (Figure 8.10). This inhibits the ability of bacteria to replicate properly.

In pulsed high-current/high-voltage arc discharges [49, 218, 219], a large part of the energy is consumed in the formation of a high-temperature plasma channel (several tens of thousands of Kelvin). The discharge channel thus functions as a blackbody radiation source with a significant portion of the radiation in the UV region of the spectrum [220]. For underwater plasmas initiated by exploding fine wires, it was shown that up to 28% of the energy transferred into the plasma (1.5 kJ per pulse) is converted to UV radiation with a peak radiant power of 200 MW [221]. For the pulsed arc discharge generated in the plate-rod reactor, it was found that the UV light energy has a 3.2% total energy injected into the reactor [222]. Thus, UV radiation is one of the principal forms of energy, which is dissipated from arc electrical discharges in water. This radiation, especially vacuum UV (75–185 nm), could cause photolysis, resulting in the dissociation of water molecules into the primary radicals, hydrogen atoms, and hydroxyl radicals. UV light with a wavelength greater than 185 nm penetrates into the bulk of the solution and can inactivate microorganisms or degrade photolytically labile chemical compounds dissolved in water. Willberg *et al.* [219] reported that direct photolysis might be the primary

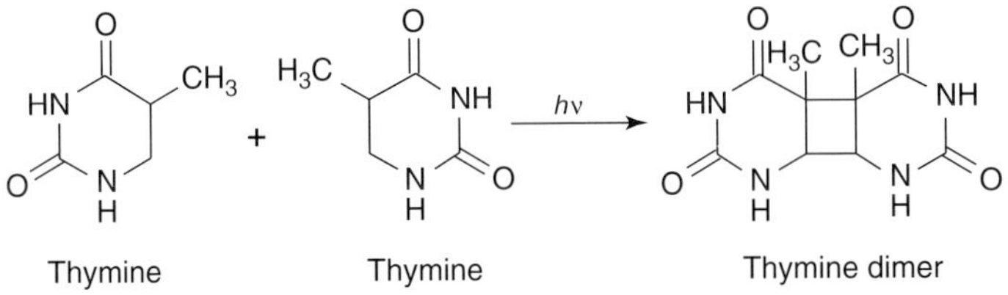

Figure 8.10 Thymine DNA base photolytic change.

mechanism of degradation of 4-chlorophenol in the electrohydraulic discharge reactor. Ching *et al.* [49] estimated the UV light intensity generated by the pulsed arc discharge of the order of 10^6 W cm^{-2} and determined the UV radiation as the main disinfection process in the discharge.

The emission of UV light from the pulsed corona discharge in water using point-plate electrode geometry has been demonstrated experimentally in [53]. Measurements using chemical actinometry revealed that significant UV emission from the discharge occurs with increasing solution conductivity and the pulse radiant power of the emitted UV radiation could reach levels of the order of tens to hundreds of watts per pulse in the range of solution conductivity of 100–500 $\mu S\,cm^{-1}$. This radiant power corresponds to UV radiation intensity of the order 0.1–10 $mW\,cm^{-2}$. This is a significantly higher intensity level than reported for nonequilibrium air plasma. It was estimated that the UV radiation from the underwater corona discharge contributes about 30% to the overall inactivation of bacteria *E. coli* [53].

Combined thermal and UV effects on spores of *B. atrophaeus* in a dry gas of N_2/O_2 flowing plasma afterglow showed some synergistic effects of heat and UV on increasing the inactivation of the spores placed on a polystyrene petri dish [223]. Further studies were conducted with gas-phase high-frequency plasma specifically to generate significant VUV light intensity without leading to chemical etching that often affects both the bacteria and the underlying surface [224].

8.4.2
X-Ray Emission

X-ray emission in the energy range 10–42 eV has recently been demonstrated [225] when a wire-plate corona discharge was operated in ambient air. The feature is related with the occurrence of streamers in pulse discharges. No biochemical or biophysical application has been reported, but living tissue is well known to be sensitive to X-rays. Further investigation of these effects may be of interest.

8.4.3
Shockwaves

In underwater discharges (especially pulsed electrohydraulic sparks/arcs with the pulse energy of the order of kiloJoules per pulse), a rather significant amount of discharged energy is transformed into the formation of shockwaves in water. Gilliland and Speck [226, 227] investigated bactericidal action produced by electrohydraulic shock generated by submerged high-voltage spark discharge in the 1960s. They concluded that there were no indications of death caused by mechanical disruption of cellular integrity. Later studies, however, demonstrated the significant role of shockwaves in electrohydraulic treatment of water when used for decontamination of sludge, killing of bacteria, and removal of zebra mussels from the intake pipes of water treatment facilities [154]. Li *et al.* [51, 52] observed mechanical rupture of intracellular gaseous vacuoles of cyanobacteria cells exposed to underwater streamer

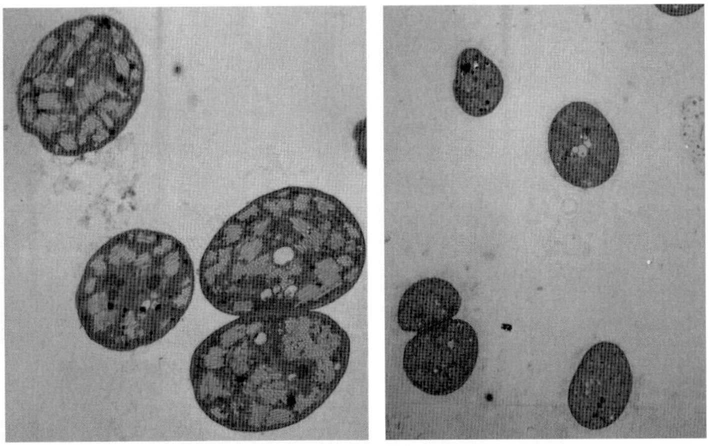

(a) (b)

Figure 8.11 TEM micrographs showing rupture of gas vesicles of cyanobacteria caused by underwater plasma: (a) before and (b) after. (Source: Reprinted from [52]. Copyright (2008), with permission of Taylor & Francis Group (http://www.informaworld.com).)

discharge, which was caused presumably by the physical effects of shockwaves induced by the discharge (Figure 8.11).

Underwater shock waves are also used in extracorporeal shockwave lithotripsy (ESWL) to noninvasively disintegrate kidney stones and gallstones [30, 31, 33]. In fact, ESWL is one of the most successful applications of shock waves in medicine. Since the mid-1980s, lithotripsy has been in clinical use worldwide for treating kidney stones and continues to be the favored method for uncomplicated, upper urinary tract calculi, even with the advent of percutaneous surgical methods. The treatment involves focusing shock waves generated by an ESWL device (lithotripter) outside the patient's body to disintegrate the stone at a depth in tissue. In electrohydraulic lithotripters, the shock wave is generated by an underwater high-current spark discharge between a pair of electrodes placed at the focus of an ellipsoidal reflector. The targeted stone is located at the second focus of the ellipsoid, and several hundreds to thousands of shock wave pulses are delivered to comminute the stone. Acoustic coupling with the patient is achieved through water baths or liquid-filled pillows. Stone debris, following urinary stone treatment, passes down the urinary tract or, in the case of gallstones, is chemically dissolved. A typical pressure waveform at the lithotripter focus in water consists of a leading shock front (compressive wave) with a peak positive pressure in the range of 30–150 MPa and a phase duration of 0.5–3 μs, followed by a tensile wave with a peak negative pressure of 20 MPa and a duration of 2–20 μs [30, 33, 52].

The success of the ESWL stimulated research on applications of focused shock waves in other branches of medicine. Great attention has been given to the role of cavitation in ESWL-induced biological effects on soft tissues and possible treatment of some types of cancers. This subject was initially studied in order to identify possible side effects of the lithotripsy therapy such as haematuria, renal colic, perirenal and intrarenal haematomas, pancreatitis, and arrhythmias [30]. The evidence suggests that microinhomogeneities or microbubbles in the fluid (e.g., water or urine) that are located in the vicinity of the focal zone may produce cavitations, which may interact not only with the stone but also with the cells/tissue [30]. To accelerate stone comminution and/or reduce/enhance tissue damage, special generators of shock waves are being designed to modify the cavitation field and to control cavitation [228–231]. Several authors have reported controlling bubble growth and collapse by the so-called tandem shock waves, which intensify the collapse of cavitation bubbles by sending a second shock wave at the moment before the bubbles produced by the first shock wave start to collapse [33, 232–234]. Most of these experimental studies demonstrated a significant effect on stone comminution and/or tissue injury, attributed in large part to the modified cavitation field. Shock waves have been shown to cause significant cytotoxic effects in tumor cells both *in vitro* and *in vivo*. Furthermore, it was discovered that shock waves induce necroses in tumors, which delay tumor growth. However, the optimal shock wave profile and pulse combination has not been found [235].

One reason is that there is a fundamental difference in the acoustic impedance of kidney stones and tissue. A kidney stone represents relatively strong acoustical nonhomogeneity in comparison with the surrounding liquid and soft tissues. The nonhomogeneity localizes the action of the shock wave because the shock propagates through soft tissues and liquid with a small attenuation almost without interacting with them. In the case of cancer tissue, no acoustical nonhomogeneities exist between cancer and healthy tissues. The localized action of the shock waves in an acoustically homogeneous medium (soft tissue) is attributed to the cavitations produced by focused shock waves. Collapsing cavitation creates strong secondary shock waves of nanosecond duration (tens of micrometers in scale) that can interact with cell-scale structures. Local thermal effects (of the order of micrometer dimensions) accompanying cavitation collapse (sonoluminiscence) and the production of chemical radicals may also play a role in cell damage. In general, the intensity of the secondary shock wave and the velocity of the microjet are assumed to depend on the initial bubble radius. The larger the bubbles grow (>1 mm), the more violent their collapse, which may be more suitable for stone or tissue damage. Small bubble sizes ($<200\,\mu m$) that are comparable to the size of cells may be used to enhance shock-wave-mediated drug delivery and gene transfer. In the case of tandem shock waves, the second wave can interact with the cavitation produced by the first wave in a different phase of its evolution. Thus, by using tandem shock waves, inertial bubble interactions may be tailored for different biomedical applications [231].

8.4.4
Thermal Effects and Electrosurgical Plasmas

Plasma generation in saline solutions for electrosurgery applications involves significant generation of thermal effects in addition to the chemical species formed

[34, 35, 236–240]. Local heating near the electrodes leads to the formation of vapor layers, where plasma is generated [237]. The radicals and ions formed in the plasma zone, coupled with possible thermal effects, lead to reactions with proteins such as collagen and tissue cutting and cauterization. Since the denaturing of tissue protein due to thermal necrosis initiates at temperatures above 43°C (requiring only 0.1 s contact with tissue at 60°C) [31], it is possible that local thermal effects as well as the chemical effects may participate in the tissue ablation, and clearly, thermal effects can generally increase the rates of the chemical reactions. Future work is needed to account for the combination of chemical and thermal effects in the wide range of discharges in the liquid phase.

8.4.5
Electric Field Effects and Bioelectrics

Existence of high electric fields in plasma discharges can lead to some contribution to the plasma inactivation of cells, especially in the underwater discharges (Section 8.3.2.2). Nakamura *et al.* [193] determined at the streamer tip of a pulsed streamer intensity of discharge in water of electric field of 320 kV cm^{-1}. Such a high-intensity electric field may induce significant lethal effects to the cells present in the liquid close to the discharge zone. Pulsed electric fields are known to induce changes in morphology and functions of cells depending on the strength of the field and pulse duration. Electrical pulses with duration in excess of microseconds caused transient increases in the permeability of the cell membranes because of the formation of pores [59, 60]. This process, known as *electroporation*, has been widely utilized in biological research and applications [68, 241–243]. The physical mechanism of electroporation is based on the charging of membranes through capacitive coupling of the voltage pulse to the cell, forming potential difference across the cell membrane. For applied electric fields with magnitudes that reach the voltage across the membrane of approximately 1 V, the membrane becomes permeable to large molecules. This effect may be reversible if the size of the electrically generated pore(s) in the outer membrane stays below a critical diameter, or it may be irreversible, leading to cell death [244]. Consequently, the size of the pores and the fate of the cell depend not only on the applied electric field but also on the duration of the electric field pulse. Typical pulses range from tens of milliseconds with amplitudes of several 100 V cm^{-1} to pulses of a few microseconds and several kV cm^{-1} [245].

In the area of disinfection, lethal effects of pulsed electric fields have been used for inactivation of a wide range of microorganisms in water [67] including bacteria and yeast [60], viruses [48, 62], bacteria and fungi [246], bacteria, yeast, and spores [63, 65, 247, 248], and food pathogens [249]. There is a multitude of medical applications for electroporation from gene therapy to electrochemotherapy [250–253]. The most advanced of these applications is the electrochemotherapy for cancer treatment, where clinical studies are underway. In electrochemotherapy, a chemotherapeutic drug is injected into the tumor or intravenously and the tumor is then treated with pulsed electric fields of 10–100 μs duration and amplitudes on the order of 1 kV cm^{-1}. This is done by means of an electrode array that is inserted into the affected tissue [250, 253].

Advances in pulsed power technology that allow generation of extremely short high-intensity electrical pulses in the nanosecond to subnanosecond range and of the high frequency over tens of megahertz range provide new opportunities for other applications in biomedicine and lead to development of new research directions in the field of bioelectrics [245, 250, 254–261]. Generally, the electrical pulses do not affect the intracellular membranes because the outer membrane shields the interior from the influence of electric fields. However, if the pulse duration becomes very short and consequently the cutoff frequency of the Fourier spectrum of the pulse becomes very high, the electric field can penetrate the outer membrane and affect subcellular structures [250]. This allows control and stimulation of cell functions such as to induce apoptosis in cancer cells without permanent damage to outer cell membrane (i.e., programmed cell death –a process that is inhibited in cancer cells, leading to uncontrolled cell proliferation) [244, 245, 255, 261, 262] and, thus, opening up new possibilities for using pulsed electric field in medical applications (e.g., in tumor therapy) [29, 263, 264]. Even with these new developments, however, application of pulsed electrical field relies on the electrode systems that can be brought close to the tumors. Accordingly, targets that can be treated have to be located close to the skin surface or they will require invasive surgery. Thus, a new idea involves focusing strong electric fields in the picosecond range into a patient by using ultrawideband antennas in somewhat similar approach as in the case of focused shock waves [265–267].

8.5
Concluding Remarks

Electrical discharge plasma in dry and humid gases as well as over and directly inside liquid water can clearly effectively inactivate a wide range of microorganisms over many orders of magnitudes by chemical and physical means. The chemical interactions of ROS and RNS with cellular components, including membranes, proteins, and nucleic acids, are responsible for cellular inactivation in both dry and humid environments. Plasma generated directly in a dry noble gas phase, for example, can lead to etching of the surface of cellular structures by relatively nonspecific chemical pathways somewhat akin to plasma etching of inorganic materials in microelectronics processing. The addition of water vapor, oxygen, and/or nitrogen carriers leads to the formation of ROS and/or RNS, which react at relatively high rates with specific functional groups on the various organic compounds that make up the cellular surface, and in some cases ROS and RNS can transport into the interior of the cell and disrupt chemical metabolism. These ROS and RNS can be molecular compounds such as ozone, hydrogen peroxide, and peroxonitrous acid as well as radicals such as $OH^{\bullet}$, $HO_2^{\bullet}$, $NO^{\bullet}$, and $NO_2^{\bullet}$. When plasma is generated directly in the liquid phase, the input power can strongly affect the formation of UV emissions and the generation of shockwaves, particularly at high power (>kiloJoules per pulse) and high solution conductivity, and both these physical factors, as well as high electric fields, can disrupt cellular structure. While our knowledge of many of the chemical pathways and the various physical factors that affect cellular structures are known, there remain many challenges to develop predictive tools and mathematical analysis to quantify these effects and to determine how much of a given inactivation is due to the specific factor. Beyond microbial inactivation, current and future efforts in biomedicine involve plasma interaction with eukaryotic cells and, in particular, with human cells from stem cells to cancer cells. These processes were not covered in this chapter; however, the research on biomedical applications of plasmas, also called *plasma medicine* [11, 13, 15], is a rapidly developing field. A wide range of studies in the field of plasma medicine include the treatment of skin diseases, wound sterilization, treatment of dental cavities, plasma stimulation of blood coagulation [11, 268, 269], treatment of skin cancer [27, 28], as well as inactivation of prion, protein, and pyrogenic substances [270, 271]. Such developments clearly require the combination of new plasma technologies (e.g., new power supplies, and methods to deliver and control the plasma) [272] with advanced understanding of the biochemical and physiological responses. We can also note that the large field of plasma surface modification on polymers can have significant effects on human cells as, for example, recently reported for mesenchymal stem cell differentiation [273].

Acknowledgments

Lukes would like to acknowledge financial support from the Grant Agency of Academy of Sciences of the Czech Republic (project No. IAAX00430802). J.-L.Brisset is largely indepted to his microbiology colleagues Prof. M.-N. Bellon-Fontaine, Dr. M. Naitali and Dr. J.-M. Herry (AgroParitech UMR 732-BHM) for driving his attention to the role of peroxynitrite in microbiological inactivation phenomena. Locke would like to gratefully acknowledge financial support from the US National Science Foundation under grant number CBET-0932481.

References

1. Laroussi, M. (2000) Biological decontamination by nonthermal plasma. *IEEE Trans. Plasma Sci.*, **28**, 184–188.
2. Moisan, M., Barbeau, J., Moreau, S., Pelletier, J., Tabrizian, M., and Yahia, L.H. (2001) Low-temperature sterilization using gas plasmas: a review of the experiments and an analysis of the inactivation mechanisms. *Int. J. Pharm.*, **226** (1–2), 1–21.
3. Laroussi, M. and Leipold, F. (2004) Evaluation of the roles of reactive species, heat, and UV radiation in the inactivation of bacterial cells by air plasmas at atmospheric pressure. *Int. J. Mass Spectrom.*, **233** (1–3), 81–86.
4. Laroussi, M. (2009) Low-temperature plasmas for medicine? *IEEE Trans. Plasma Sci.*, **37** (6), 714–725.
5. Laroussi, M. (1996) Sterilization of contaminated matter with an atmospheric pressure plasma. *IEEE Trans. Plasma Sci.*, **24**, 1188–1191.
6. Sunka, P. (2001) Pulse electrical discharges in water and their applications. *Phys. Plasmas*, **8**, 2587–2594.
7. Montie, T.C., Kelly-Wintenberg, K., and Roth, J.R. (2000) An overview of

research using the one atmosphere uniform glow discharge plasma (OAUGDP) for sterilization of surfaces and materials. *IEEE Trans. Plasma Sci.*, **28** (1), 41–50.

8. Moisan, M., Barbeau, J., Crevier, M.C., Pelletier, J., Philip, N., and Saoudi, B. (2002) Plasma sterilization. Methods mechanisms. *Pure Appl. Chem.*, **74** (3), 349–358.

9. Odic, E., Goldman, A., Goldman, M., Delaveau, S., and Le Hegarat, F. (2002) Plasma sterilization technologies and processes. *High Temp. Mater. Process.*, **6** (3), 385–396.

10. Gaunt, L.F., Beggs, C.B., and Georghiou, G.E. (2006) Bactericidal action of the reactive species produced by gas-discharge nonthermal plasma at atmospheric pressure: A review. *IEEE Trans. Plasma Sci.*, **34** (4), 1257–1269.

11. Fridman, G., Friedman, G., Gutsol, A., Shekhter, A.B., Vasilets, V.N., and Fridman, A. (2008) Applied plasma medicine. *Plasma Proc. Polym.*, **5** (6), 503–533.

12. Stoffels, E., Sakiyama, Y., and Graves, D.B. (2008) Cold atmospheric plasma: Charged species and their interactions with cells and tissues. *IEEE Trans. Plasma Sci.*, **36** (4), 1441–1457.

13. Kong, M.G., Kroesen, G., Morfill, G., Nosenko, T., Shimizu, T., van Dijk, J., and Zimmermann, J.L. (2009) Plasma medicine: an introductory review. *New J. Phys.*, **11**, 115 012.

14. Laroussi, M. (2008) The biomedical applications of plasma: a brief history of the development of a new field of research. *IEEE Trans. Plasma Sci.*, **36** (4), 1612–1614.

15. von Woedtke, T., Junger, M., Kocher, T., Kramer, A., Lademann, J., Lindequist, U., and Weltmann, K.D. (2009) Plasma medicine - therapeutic application of physical plasmas, in *World Congress on Medical Physics and Biomedical Engineering*, IFMBE Proceedings, vol. 25 (eds O. Dossel and W.C. Schlegel), Springer, Munich, Germany, pp. 82–83.

16. Weltmann, K.D., Kindel, E., von Woedtke, T., Hahnel, M., Stieber, M., and Brandenburg, R. (2010) Atmospheric-pressure plasma sources: Prospective tools for plasma medicine. *Pure Appl. Chem.*, **82** (6), 1223–1237.

17. Wende, K., Landsberg, K., Lindequist, U., Weltmann, K.D., and von Woedtke, T. (2010) Distinctive activity of a non-thermal atmospheric-pressure plasma jet on eukaryotic and prokaryotic cells in a cocultivation approach of keratinocytes and microorganisms. *IEEE Trans. Plasma Sci.*, **38** (9), 2479–2485.

18. Weltmann, K.D. and von Woedtke, T. (2011) Basic requirements for plasma sources in medicine. *Eur. Phys. J. Appl. Phys.*, **55** (1), 13 807.

19. Naitali, M., Kamgang-Youbi, G., Herry, J.M., Bellon-Fontaine, M.N., and Brisset, J.L. (2010) Combined effects of long living chemical species during microbial inactivation using atmospheric plasma treated water. *Appl. Environ. Microbiol.*, **76** (22), 7662–7664.

20. Daeschlein, G., von Woedtke, T., Kindel, E., Brandenburg, R., Weltmann, K.D., and Junger, M. (2010) Antibacterial activity of an atmospheric pressure plasma jet against relevant wound pathogens *in vitro* on a simulated wound environment. *Plasma Proc. Polym.*, **7** (3–4), 224–230.

21. Lloyd, G., Friedman, G., Jafri, S., Schultz, G., Friedman, A., and Harding, K. (2010) Gas plasma: Medical uses and developments in wound care. *Plasma Proc. Polym.*, 7, 194–211.

22. Sladek, R.E.J., Stoffels, E., Walraven, R., Tielbeek, P.J.A., and Koolhoven, R.A. (2004) Plasma treatment of dental cavities: A feasibility study. *IEEE Trans. Plasma Sci.*, **32** (4), 1540–1543.

23. Villeger, S., Ricard, A., and Sixou, M. (2004) Sterilization of dental bacteria in a N_2-O_2 microwaves post-discharge, at low pressure: influence of temperature. *Eur. Phys. J. Appl. Phys.*, **26** (3), 203–208.

24. Benes, J., Sunka, P., Kralova, J., Kaspar, J., and Pouckova, P. (2007) Biological effects of two successive shock waves focused on liver tissues and melanoma cells. *Physiol. Res.*, **56**, S1–S4.

25. Sunka, P., Babicky, V., Clupek, M., Benes, J., and Pouckova, P. (2004)

Localized damage of tissues induced by focused shock waves. *IEEE Trans. Plasma Sci.*, **32** (4), 1609–1613.

26. Sunka, P., Stelmashuk, V., Benes, J., Pouckova, P., and Kralova, J. (2007) Potential applications of tandem shock waves in cancers therapy, *2007 IEEE Pulsed Power Conference*, **1–4**, IEEE, Albuquerque, NM, pp. 1074–1077.
27. Sensenig, R., Kalghatgi, S., Cerchar, E., Fridman, G., Shereshevsky, A., Torabi, B., Arjunan, K.P., Podolsky, E., Fridman, A., Friedman, G., Azizkhan-Clifford, J., and Brooks, A.D. (2011) Non-thermal plasma induces apoptosis in melanoma cells via production of intracellular reactive oxygen species. *Ann. Biomed. Eng.*, **39** (2), 674–687.
28. Fridman, G., Shereshevsky, A., Jost, M.M., Brooks, A.D., Fridman, A., Gutsol, A., Vasilets, V., and Friedman, G. (2007) Floating electrode dielectric barrier discharge plasma in air promoting apoptotic behavior in melanoma skin cancer cell lines. *Plasma Chem. Plasma Process.*, **27** (2), 163–176.
29. Hall, E.H., Schoenbach, K.H., and Beebe, S.J. (2007) Nanosecond pulsed electric fields induce apoptosis in p53-wildtype and p53-null HCT116 colon carcinoma cells. *Apoptosis*, **12** (9), 1721–1731.
30. Coleman, A.J. and Saunders, J.E. (1993) A review of the physical properties and biological effects of the high amplitude acoustic fields used in extracorporeal lithotripsy. *Ultrasonics*, **31**, 75–89.
31. Bailey, M.R., Khokhlova, V.A., Sapoznikov, O.A., Kargl, S.G., and Crum, L.A. (2003) Physical mechanisms of the therapeutic effect of ultrasound (a review). *Acoust. Phys.*, **49**, 369–388.
32. Sunka, P., Babicky, V., Clupek, M., Fuciman, M., Lukes, P., Simek, M., Benes, J., Locke, B., and Majcherova, Z. (2004) Potential applications of pulse electrical discharges in water. *Acta Phys. Slovaca*, **54** (2), 135–145.
33. Loske, A.M. (2010) *New Trends in Shock Wave Applications to Medicine and Biotechnology*, Research Signpost, Kerala, India.
34. Stalder, K.R., McMillen, D.F., and Woloszko, J. (2005) Electrosurgical plasmas. *J. Phys. D.: Appl. Phys.*, **38** (11), 1728–1738.
35. Stalder, K.R. and Woloszko, J. (2007) Some physics and chemistry of electrosurgical plasma discharges. *Contrib. Plasma Phys.*, **47** (1–2), 64–71.
36. Laroussi, M. (2002) Nonthermal decontamination of biological media by atmospheric-pressure plasma: review, analysis and prospects. *IEEE Trans. Plasma Sci.*, **30** (4), 1409–1415.
37. Laroussi, M. (2005) Low temperature plasma-based sterilization: overview and state-of-the-art. *Plasma Proc. Polym.*, **2** (5), 391–400.
38. Laroussi, M., Minayeva, O., Dobbs, F.C., and Woods, J. (2006) Spores survivability after exposure to low-temperature plasmas. *IEEE Trans. Plasma Sci.*, **34** (4), 1253–1256.
39. Brisset, J.L., Moussa, D., Doubla, A., Hnatiuc, E., Hnatiuc, B., Youbi, G.K., Herry, J.M., Naitali, M., and Bellon-Fontaine, M.N. (2008) Chemical reactivity of discharges and temporal post-discharges in plasma treatment of aqueous media: examples of gliding discharge treated solutions. *Ind. Eng. Chem. Res.*, **47** (16), 5761–5781.
40. Deng, X.T., Shi, J.J., and Kong, M.G. (2006) Physical mechanisms of inactivation of Bacillus subtilis spores using cold atmospheric plasmas. *IEEE Trans. Plasma Sci.*, **34** (4), 1310–1316.
41. Kalghatgi, S., Dobrynin, D., Fridman, G., Cooper, M., Nagaraj, G., Peddinghaus, L., Balasubramanian, M., Barbee, K., Brooks, A., Vasilets, V., Gutsol, A., Fridman, A., and Friedman, G. (2008) Applications of non thermal atmospheric pressure plasma in medicine, in *Plasma Assisted Decontamination of Biological and Chemical Agents*, NATO Science for Peace and Security Series A-Chemistry and Biology (eds S. Guceri and A. Fridman), Springer-Verlag, pp. 173–181.
42. Dobrynin, D., Fridman, G., Friedman, G., and Fridman, A. (2009) Physical and biological mechanisms of direct

plasma interaction with living tissue. *New J. Phys.*, **11**, 115 020.
43. Naitali, M., Hnatiuc, B., Herry, J.M., Hnatiuc, E., Bellon-Fontaine, M.N., and Brisset, J.L. (2009) Decontamination of chemical and microbial targets using gliding electric discharge, in *Biological and Environmental Applications of Gas Discharge Plasmas* (ed. G. Brelles-Marino), Nova Science Publishers Inc, New York, pp. 189–236.
44. Fridman, G., Brooks, A.D., Balasubramanian, M., Fridman, A., Gutsol, A., Vasilets, V.N., Ayan, H., and Friedman, G. (2007) Comparison of direct and indirect effects of non-thermal atmospheric-pressure plasma on bacteria. *Plasma Proc. Polym.*, **4** (4), 370–375.
45. Edebo, L., Holme, T., and Selin, I. (1968) Microbicidal action of compounds generated by transient electric arcs in aqueous systems. *J. Gen. Microbiol.*, **53**, 1–7.
46. Edebo, L., Holme, T., and Selin, I. (1969) Influence of conductivity of discharge liquid on microbicidal effect of transient electric arcs in aqueous systems. *Appl. Microbiol.*, **17** (1), 59–62.
47. Edebo, L. and Selin, I. (1968) Effect of pressure shock wave and some electrical quantities in microbicidal effect of transient electric arcs in aqueous systems. *J. Gen. Microbiol.*, **50**, 253–259.
48. Sato, M., Ohgiyama, T., and Clements, J.S. (1996) Formation of chemical species and their effects on microorganisms using a pulsed high-voltage discharge in water. *IEEE Trans. Ind. Appl.*, **32**, 106–112.
49. Ching, W.K., Colussi, A.J., Sun, H.J., Nealson, K.H., and Hoffmann, M.R. (2001) Escherichia coli disinfection by electrohydraulic discharges. *Environ. Sci. Technol.*, **35** (20), 4139–4144.
50. Abou-Ghazala, A., Katsuki, S., Schoenbach, K.H., Dobbs, F.C., and Moreira, K.R. (2002) Bacterial decontamination of water by means of pulsed-corona discharges. *IEEE Trans. Plasma Sci.*, **30**, 1449–1453.
51. Li, Z., Sakai, S., Yamada, C., Wang, D., Chung, S., Lin, X., Namihira, T., Katsuki, S., and Akiyama, H. (2006) The effects of pulsed streamerlike discharge on cyanobacteria cells. *IEEE Trans. Plasma Sci.*, **34**, 1719–1725.
52. Li, Z., Ohno, T., Sato, H., Sakugawa, T., Akiyama, H., Kunitomo, S., Sasaki, K., Ayukawa, M., and Fujiwara, H. (2008) A method of water-bloom prevention using underwater pulsed streamer discharge. *J. Environ. Sci. Health, Part A*, **43**, 1209–1214.
53. Lukes, P., Clupek, M., Babicky, V., and Sunka, P. (2008) Ultraviolet radiation from the pulsed corona discharge in water. *Plasma Sources Sci. Technol.*, **17** (2), 024 012.
54. Fang, F.C. (1997) Mechanisms of nitric oxide related antimicrobial activity. *J. Clin. Invest.*, **100** (12), S43–S50.
55. Kawanishi, S., Hiraku, Y., and Inoue, S. (1999) DNA damage induced by Salmonella test-negative carcinogens through the formation of oxygen and nitrogen-derived reactive species (review). *Int. J. Mol. Med.*, **3** (2), 169–174.
56. Mukhopadhyay, P., Zheng, M., Bedzyk, L.A., LaRossa, R.A., and Storz, G. (2004) Prominent roles of the NorR and Fur regulators in the Escherichia coli transcriptional response to reactive nitrogen species. *Proc. Natl. Acad. Sci. U.S.A.*, **101** (3), 745–750.
57. Bryk, R., Griffin, P., and Nathan, C. (2000) Peroxynitrite reductase activity of bacterial peroxiredoxins. *Nature*, **407** (6801), 211–215.
58. Menashi, W. (1968) Surface treatments. US patent 3383163, filed Jan. 24, 1964 and issued May 14, 1968.
59. Hamilton, W.A. and Sale, A.J.H. (1967) Effects of high electric fields on microorganisms II. Mechanism of action of the lethal effect. *Biochim. Biophys. Acta*, **148**, 789–800.
60. Sale, A.J.H. and Hamilton, W.A. (1967) Effects of high electric fields on microorganisms I. Killing of bacteria and yeasts. *Biochim. Biophys. Acta*, **148**, 781–788.
61. Mizuno, A. and Hori, Y. (1988) Destruction of living cells by pulsed

high-voltage application. *IEEE Trans. Ind. Appl.*, **24**, 387–394.
62. Mizuno, A., Inoue, T., Yamaguchi, S., Sakamoto, K., Saeki, T., Matsumoto, Y., and Minamiyama, K. (1990) Inactivation of viruses using pulsed high electric fields, in *Conference Record of the 1990 IEEE Industry Applications Society Annual Meeting*, IEEE, Seattle, WA, pp. 77–83.
63. Grahl, T. and Markl, H. (1996) Killing of microorganisms by pulsed electric fields. *Appl. Microbiol. Biotechnol.*, **45**, 148–157.
64. Schoenbach, K.H., Peterkin, F.E., Alden, R.W., and Beebe, S.J. (1997) The effect of pulsed electric fields on biological cells: experiments and applications. *IEEE Trans. Plasma Sci.*, **25**, 284–292.
65. Qin, B.L., Barbosa-Canovas, V., Swanson, B.G., Pedrow, P.D., and Olsen, R.G. (1998) Inactivating microorganisms using a pulsed electric field continuous treatment system. *IEEE Trans. Ind. Appl.*, **34**, 43–50.
66. Abou-Ghazala, A. and Schoenbach, K.H. (2000) Biofouling prevention with pulsed electric fields. *IEEE Trans. Plasma Sci.*, **28**, 115–121.
67. Schoenbach, K.H., Joshi, R.P., Stark, R.H., Dobbs, F.C., and Beebe, S.J. (2000) Bacterial decontamination of liquids with pulsed electric fields. *IEEE Trans. Dielectr. Electr. Insul.*, **7**, 637–645.
68. Neumann, E., Sowers, A.E., and Jordan, C.A. (1989) *Electroporation and Electrofusion in Cell Biology*, Plenum Press, New York.
69. Zimmermann, U. (1983) Electrofusion of cells: principles and industrial potential. *Trends Biotechnol.*, **1**, 149–155.
70. Ohshima, T., Ono, T., and Sato, M. (1999) Decomposition of nucleic acid molecules in pulsed electric field and its release from recombinant Escherichia coli. *J. Electrostatics*, **46**, 163–170.
71. Ohshima, T., Hama, Y., and Sato, M. (2000) Releasing profiles of gene products from recombinant Escherichia coli in a high-voltage pulsed electric field. *Biochem. Eng. J.*, **5**, 149–155.
72. Hall, E.H., Schoenbach, K.H., and Beebe, S.J. (2007) Nanosecond pulsed electric fields have differential effects on cells in the S-phase. *DNA Cell Biol.*, **26** (3), 160–171.
73. Laroussi, M. and Lu, X. (2005) Room-temperature atmospheric pressure plasma plume for biomedical applications. *Appl. Phys. Lett.*, **87** (11), 113 902.
74. Rossi, F., Kylian, O., and Hasiwa, M. (2006) Decontamination of surfaces by low pressure plasma discharges. *Plasma Proc. Polym.*, **3** (6–7), 431–442.
75. Foest, R., Schmidt, M., and Becker, K. (2006) Microplasmas, an emerging field of low-temperature plasma science and technology. *Int. J. Mass Spectrom.*, **248** (3), 87–102.
76. Sato, T., Miyahara, T., Doi, A., Ochiai, S., Urayama, T., and Nakatani, T. (2006) Sterilization mechanism for Escherichia coli by plasma flow at atmospheric pressure. *Appl. Phys. Lett.*, **89** (7), 073 902.
77. Sato, M. (2008) Environmental and biotechnological applications of high-voltage pulsed discharges in water. *Plasma Sources Sci. Technol.*, **17** (2), 024 021.
78. Sharma, A., Pruden, A., Stan, O., and Collins, G.J. (2006) Bacterial inactivation using an RF-powered atmospheric pressure plasma. *IEEE Trans. Plasma Sci.*, **34** (4), 1290–1296.
79. Sharma, A., Pruden, A., Yu, Z.Q., and Collins, G.J. (2005) Bacterial inactivation in open air by the afterglow plume emitted from a grounded hollow slot electrode. *Environ. Sci. Technol.*, **39** (1), 339–344.
80. Bai, X.Y., Zhang, Z.T., Bai, M.D., Yang, B., and Bai, M.B. (2005) Killing of invasive species of ship's ballast water in 20t/h system using hydroxyl radicals. *Plasma Chem. Plasma Process.*, **25** (1), 41–54.
81. Stryczewska, H.D., Ebihara, K., Takayama, M., Gyoutoku, Y., and Tachibana, M. (2005) Non-thermal plasma-based technology for soil treatment. *Plasma Proc. Polym.*, **2** (3), 238–245.

82. Choi, J.H., Han, I., Baik, H.K., Lee, M.H., Han, D.W., Park, J.C., Lee, I.S., Song, K.M., and Lim, Y.S. (2006) Analysis of sterilization effect by pulsed dielectric barrier discharge. *J. Electrostatics*, **64** (1), 17–22.
83. Ben Gadri, R., Roth, J.R., Montie, T.C., Kelly-Wintenberg, K., Tsai, P.P.Y., Helfritch, D.J., Feldman, P., Sherman, D.M., Karakaya, F., and Chen, Z.Y. (2000) Sterilization and plasma processing of room temperature surfaces with a one atmosphere uniform glow discharge plasma (OAUGDP). *Surf. Coat. Technol.*, **131** (1–3), 528–542.
84. Kelly-Wintenberg, K., Montie, T.C., Brickman, C., Roth, J.R., Carr, A.K., Sorge, K., Wadsworth, L.C., and Tsai, P.P.Y. (1998) Room temperature sterilization of surfaces and fabrics with a one atmosphere uniform glow discharge plasma. *J. Ind. Microbiol. Biotechnol.*, **20** (1), 69–74.
85. Kelly-Wintenberg, K., Hodge, A., Montie, T.C., Deleanu, L., Sherman, D., Roth, J.R., Tsai, P., and Wadsworth, L. (1999) Use of a one atmosphere uniform glow discharge plasma to kill a broad spectrum of microorganisms. *J. Vac. Sci. Technol., A*, **17** (4), 1539–1544.
86. Vujosevic, D., Mozetic, M., Cvelbar, U., Krstulovic, N., and Milosevic, S. (2007) Optical emission spectroscopy characterization of oxygen plasma during degradation of Escherichia coli. *J. Appl. Phys.*, **101** (10), 103 305.
87. Ohkawa, H., Akitsu, T., Tsuji, M., Kimura, H., and Fukushima, K. (2005) Initiation and microbial disinfection characteristics of wide-gap atmospheric pressure glow discharge using soft X-ray ionization. *Plasma Proc. Polym.*, **2** (2), 120–126.
88. Akitsu, T., Ohkawa, H., Tsuji, M., Kimura, H., and Kogoma, M. (2005) Plasma sterilization using glow discharge at atmospheric pressure. *Surf. Coat. Technol.*, **193** (1–3), 29–34.
89. Kamgang-Youbi, G., Herry, J.M., Bellon-Fontaine, M.N., Brisset, J.L., Doubla, A., and Naitali, M. (2007) Evidence of temporal postdischarge decontamination of bacteria by gliding electric discharges: Application to Hafnia alvei. *Appl. Environ. Microbiol.*, **73** (15), 4791–4796.
90. Sladek, R.E.J. and Stoffels, E. (2005) Deactivation of Escherichia coli by the plasma needle. *J. Phys. D.: Appl. Phys.*, **38** (11), 1716–1721.
91. Sladek, R.E.J., Baede, T.A., and Stoffels, E. (2006) Plasma-needle treatment of substrates with respect to wettability and growth of Escherichia coli and Streptococcus mutans. *IEEE Trans. Plasma Sci.*, **34** (4), 1325–1330.
92. Chau, T.T., Kao, K.C., Blank, G., and Madrid, F. (1996) Microwave plasmas for low-temperature dry sterilization. *Biomaterials*, **17** (13), 1273–1277.
93. Lerouge, S., Fozza, A.C., Wertheimer, M.R., Marchand, R., Tabrizian, M., and Yahia, L. (1999) Plasma sterilization: spore destruction by microwave plasmas, in *Plasma Deposition and Treatment of Polymers*, MRS Proceedings, vol. 544 (eds W.W. Lee, R. Dagostino, and M.R. Wertheimer), Cambridge University Press, UK, pp. 33–37.
94. Purevdorj, D., Igura, N., Shimoda, M., Ariyada, O., and Hayakawa, I. (2001) Kinetics of inactivation of Bacillus spores using low temperature argon plasma at different microwave power densities. *Acta Biotechnol.*, **21** (4), 333–342.
95. Purevdorj, D., Igura, N., Hayakawa, I., and Ariyada, O. (2002) Inactivation of Escherichia coli by microwave induced low temperature argon plasma treatments. *J. Food Eng.*, **53** (4), 341–346.
96. Feichtinger, J., Schulz, A., Walker, M., and Schumacher, U. (2003) Sterilisation with low-pressure microwave plasmas. *Surf. Coat. Technol.*, **174**, 564–569.
97. Schneider, J., Baumgartner, K.M., Feichtinger, J., Kruger, J., Muranyi, P., Schulz, A., Walker, M., Wunderlich, J., and Schumacher, U. (2005) Investigation of the practicability of low-pressure microwave plasmas in the sterilisation of food packaging materials at industrial level. *Surf. Coat. Technol.*, **200** (1–4), 962–966.

98. Villeger, S., Cousty, S., Ricard, A., and Sixou, M. (2003) Sterilization of E. coli bacterium in a flowing N_2-O_2 post-discharge reactor. *J. Phys. D.: Appl. Phys.*, **36** (13), L60–L62.
99. Moreau, S., Tabrizian, M., Barbeau, J., Moisan, M., Leduc, A., Pelletier, J., Lagarde, T., Rohr, M., Desor, F., Vidal, D., and Yahia, L.H. (1999) Essential parameters for plasma sterilization, in *ICPIG-24: Proceedings of the 24th International Conference on Phenomena in Ionized Gases*, vol. 1 (eds P. Pisarczyk, T. Pisarczyk, and J. Wolowski), Warsaw, Poland, pp. 51–52.
100. Lerouge, S., Wertheimer, M.R., Marchand, R., Tabrizian, M., and Yahia, L. (2000) Effect of gas composition on spore mortality and etching during low-pressure plasma sterilization. *J. Biomed. Mater. Res.*, **51** (1), 128–135.
101. Purevdorj, D., Igura, N., Ariyada, O., and Hayakawa, I. (2003) Effect of feed gas composition of gas discharge plasmas on Bacillus pumilus spore mortality. *Lett. Appl. Microbiol.*, **37** (1), 31–34.
102. Maeda, Y., Igura, N., Shimoda, M., and Hayakawa, I. (2003) Bactericidal effect of atmospheric gas plasma on Escherichia coli K12. *Int. J. Food Sci. Technol.*, **38** (8), 889–892.
103. Shi, X.M., Yuan, Y.K., Sun, Y.Z., Yuan, W., Peng, F.L., and Qiu, Y.H. (2006) Experimental research of inactivation effect of low-temperature plasma on bacteria. *Plasma Sci. Technol.*, **8** (5), 569–572.
104. Yu, H., Perni, S., Shi, J.J., Wang, D.Z., Kong, M.G., and Shama, G. (2006) Effects of cell surface loading and phase of growth in cold atmospheric gas plasma inactivation of Escherichia coli K12. *J. Appl. Microbiol.*, **101** (6), 1323–1330.
105. Xu, G.M., Zhang, G.J., Shi, X.M., Ma, Y., Wang, N., and Li, Y. (2009) Bacteria inactivation using DBD plasma jet in atmospheric pressure argon. *Plasma Sci. Technol.*, **11** (1), 83–88.
106. Boudam, M.K., Moisan, M., Saoudi, B., Popovici, C., Gherardi, N., and Massines, F. (2006) Bacterial spore inactivation by atmospheric-pressure plasmas in the presence or absence of UV photons as obtained with the same gas mixture. *J. Phys. D.: Appl. Phys.*, **39** (16), 3494–3507.
107. Shi, X.M., Zhang, G.J., Yuan, Y.K., Ma, Y., Xu, G.M., and Yang, Y. (2008) Research on the inactivation effect of low-temperature plasma on Candida albicans. *IEEE Trans. Plasma Sci.*, **36** (2), 498–503.
108. Kuzmitchev, A., Soloshenko, I., Tsoilkop, V., Kryzhanovsky, V., Bazhenov, V., Mikhno, I., and Khomich, V. (2000) Features of sterilization by different types of atmospheric pressure discharges in *Hakone VII: Proc. Int. Symp. High Pressure Low Temp. Plasma Chem.*, (eds. J.B.H. Wagner, G. Babucke), Greifswald, Germany, pp. 402–406.
109. Lassen, K.S., Nordby, B., and Grun, R. (2005) The dependence of the sporicidal effects on the power and pressure of RF-generated plasma processes. *J. Biomed. Mater. Res. B: Appl. Biomater.*, **74B** (1), 553–559.
110. Pointu, A.M., Ricard, A., Dodet, N., Odic, E., Larbre, J., and Ganciu, M. (2005) Production of active species in N_2-O_2 flowing post-discharges at atmospheric pressure for sterilization. *J. Phys. D.: Appl. Phys.*, **38** (12), 1905–1909.
111. Ekem, N., Akan, T., Akgun, Y., Kiremitci, A., Pat, S., and Musa, G. (2006) Sterilization of Staphylococcus aureus by atmospheric pressure pulsed plasma. *Surf. Coat. Technol.*, **201** (3–4), 993–997.
112. Laroussi, M., Tendero, C., Lu, X., Alla, S., and Hynes, W.L. (2006) Inactivation of bacteria by the plasma pencil. *Plasma Proc. Polym.*, **3** (6–7), 470–473.
113. Cvelbar, U., Vujosevic, D., Vratnica, Z., and Mozetic, M. (2006) The influence of substrate material on bacteria sterilization in an oxygen plasma glow discharge. *J. Phys. D.: Appl. Phys.*, **39** (16), 3487–3493.
114. Lerouge, S., Wertheimer, M.R., Marchand, R., Tabrizian, M., and Yahia, L.H. (2004) Effects of gas composition on spore mortality and etching

during low pressure plasma sterilization. *J. Biomed. Mater. Res.*, **51**, 128–135.

115. Sharma, S.P., Cruden, B.A., Rao, M.V.V.S., and Bolshakov, A.A. (2004) Analysis of emission data from O_2 plasmas used for microbe sterilization. *J. Appl. Phys.*, **95** (7), 3324–3333.

116. Nagatsu, M., Terashita, F., Nonaka, H., Xu, L., Nagata, T., and Koide, Y. (2005) Effects of oxygen radicals in low-pressure surface-wave plasma on sterilization. *Appl. Phys. Lett.*, **86** (21), 211 502.

117. Cousty, S., Villeger, S., Sarette, J.P., Ricard, A., and Sixou, M. (2006) Inactivation of Escherichia coli in the flowing afterglow of an N_2 discharge at reduce pressure: study of the destruction mechanisms of bacteria and hydrodynamics of the afterglow flow. *Eur. Phys. J. Appl. Phys.*, **34** (2), 143–146.

118. Villeger, S., Sarrette, J.P., and Ricard, A. (2005) Synergy between N and O atom action and substrate surface temperature in a sterilization process using a flowing N_2-O_2 microwave post discharge. *Plasma Proc. Polym.*, **2** (9), 709–714.

119. Henriques, J., Villeger, S., Levaton, J., Nagai, J., Santana, S., Amorim, J., and Ricard, A. (2005) Densities of N- and O-atoms in N_2-O_2 flowing glow discharges by actinometry. *Surf. Coat. Technol.*, **200** (1–4), 814–817.

120. Pintassilgo, C.D., Loureiro, J., and Guerra, V. (2005) Modelling of a N_2-O_2 flowing afterglow for plasma sterilization. *J. Phys. D.: Appl. Phys.*, **38** (3), 417–430.

121. Kogelschatz, U. and Eliasson, B. (1995) Ozone generation and applications, in *Handbook of Electrostatic Processes* (eds J.S. Chang, A.J. Kelly, and J.M. Crowley), Marcel Dekker, Inc., New York, pp. 581–605. Chapter 26.

122. Ohkawa, H., Akitsu, T., Ohtsuki, K., Ohnishi, M., and Tsuji, M. (2004) High-grade disinfection using high-density ozone. *J. Adv. Oxid. Technol.*, **7** (2), 154–160.

123. Ohkawa, H., Tsuji, M., Ohtsuki, K., Ohnishi, M., and Akitsu, T. (2005) High density ozone disinfection of medical care materials for dental surgery. *Plasma Proc. Polym.*, **2** (2), 112–119.

124. Maeda, Y., Igura, N., Shimoda, M., and Hayakawa, I. (2003) Inactivation of Escherichia coli K12 using atmospheric gas plasma produced from humidified working gas. *Acta Biotechnol.*, **23** (4), 389–395.

125. Moreau, M., Feuilloley, M.G.J., Orange, N., and Brisset, J.L. (2005) Lethal effect of the gliding arc discharges on Erwinia spp. *J. Appl. Microbiol.*, **98** (5), 1039–1046.

126. Kamgang, J.O., Briandet, R., Herry, J.M., Brisset, J.L., and Naitali, M. (2007) Destruction of planktonic, adherent and biofilm cells of Staphylococcus epidermidis using a gliding discharge in humid air. *J. Appl. Microbiol.*, **103** (3), 621–628.

127. Kamgang-Youbi, G., Herry, J.M., Brisset, J.L., Bellon-Fontaine, M.N., Doubla, A., and Naitali, M. (2008) Impact on disinfection efficiency of cell load and of planktonic/adherent/detached state: case of Hafnia alvei inactivation by Plasma Activated Water. *Appl. Microbiol. Biotechnol.*, **81** (3), 449–457.

128. Benstaali, B., Moussa, D., Addou, A., and Brisset, J.L. (1998) Plasma treatment of aqueous solutes: some chemical properties of a gliding arc in humid air. *Eur. Phys. J. Appl. Phys.*, **4** (2), 171–179.

129. Moussa, D. and Brisset, J.L. (1996) Gliding arc treatment of aqueous solutions: acid and oxidizing effects of an humid air treatment, in *Hakone V: 5th International symposium on High Pressure Low Temperature Plasma Chemistry*, Masaryk University, Brno, Czech Republic, pp. 165–169.

130. Moreau, M., Orange, N., and Brisset, J.L. (2005) Application of electric discharges at atmospheric pressure and ambient temperature for bio-decontamination. *Ozone Sci. Eng.*, **27** (6), 469–473.

131. Chen, C.W., Lee, H.M., and Chang, M.B. (2008) Inactivation of aquatic microorganisms by low-frequency AC

discharges. *IEEE Trans. Plasma Sci.*, **36** (1), 215–219.
132. Shmelev, V.M., Evtyukhin, N.V., and Che, D.O. (1996) Water sterilization by pulse surface discharge. *Chem. Phys. Rep.*, **15** (3), 463–468.
133. Zhang, R., Wang, L., Wu, Y., Guan, Z., and Jia, Z. (2006) Bacterial decontamination of water by bipolar pulsed discharge in a gas-liquid-solid three-phase discharge reactor. *IEEE Trans. Plasma Sci.*, **34** (4), 1370–1374.
134. Wang, C.H., Li, G.F., Wu, Y., Wang, Y., Li, J., Li, D., and Wang, N.H. (2007) Role of bipolar pulsed DBD on the growth of Microcystis aeruginosa in three-phase discharge plasma reactor. *Plasma Chem. Plasma Process.*, **27** (1), 65–83.
135. Maeda, Y., Igura, N., Shimoda, M., and Hayakawa, I. (2003) Inactivation of vegetative bacteria in a liquid medium by gas plasma under atmospheric pressure. *J. Fac. Agric. Kyushu Univ.*, **48** (1–2), 159–166.
136. Burlica, R. and Locke, B.R. (2008) Pulsed plasma gliding arc discharges with water spray. *IEEE Trans. Ind. Appl.*, **44** (2), 482–489.
137. Burlica, R., Grim, R.G., Shih, K.Y., Balkwill, D., and Locke, B.R. (2010) Bacteria inactivation using low power pulsed gliding arc discharges with water spray. *Plasma Proc. Polym.*, **7** (8), 640–649.
138. Locke, B.R. and Shih, K.Y. (2011) Review of the methods to form hydrogen peroxide in electrical discharge plasma with liquid water. *Plasma Sources Sci. Technol.*, **20** (3), 034 006.
139. Burlica, R., Kirkpatrick, M., and Locke, B.R. (2006) The formation of reactive species in gliding arc discharges with liquid water. *J. Electrostatics*, **64** (1), 35–43.
140. Burlica, R., Kirkpatrick, M., Finney, W.C., Clark, R., and Locke, B.R. (2004) Organic dye removal from aqueous solution by glidarc discharges. *J. Electrostatics*, **62/4**, 309–321.
141. Machala, Z., Jedlovsky, I., Chladekova, L., Pongrac, B., Giertl, D., Janda, M., Sikurova, L., and Polcic, P. (2009) DC discharges in atmospheric air for bio-decontamination - spectroscopic methods for mechanism identification. *Eur. Phys. J. D*, **54** (2), 195–204.
142. Machala, Z., Chladekova, L., and Pelach, M. (2010) Plasma agents in bio-decontamination by dc discharges in atmospheric air. *J. Phys. D.: Appl. Phys.*, **43**, 222 001.
143. Rowan, N.J., Espie, S., Harrower, J., Farrell, H., Marsili, L., Anderson, J.G., and MacGregor, S.J. (2008) Evidence of lethal and sublethal injury in food-borne bacterial pathogens exposed to high-intensity pulsed-plasma gas discharges. *Lett. Appl. Microbiol.*, **46** (1), 80–86.
144. Rowan, N.J., Espie, S., Harrower, J., Anderson, J.G., Marsili, L., and MacGregor, S.J. (2007) Pulsed-plasma gas-discharge inactivation of microbial pathogens in chilled poultry wash water. *J. Food Prot.*, **70** (12), 2805–2810.
145. Manolache, S., Somers, E.B., Wong, A.C.L., Shamamian, V., and Denes, F. (2001) Dense medium plasma environments: A new approach for the disinfection of water. *Environ. Sci. Technol.*, **35**, 3780–3785.
146. Sato, M., Tokita, K., Sadakata, M., Sakai, T., and Nakanishi, K. (1990) Sterilization of microorganisms by a high-voltage, pulsed discharge under water. *Int. Chem. Eng.*, **30**, 695–698.
147. Marsili, L., Espie, S., Anderson, J.G., and MacGregor, S.J. (2002) Plasma inactivation of food-related microorganisms in liquids. *Radiat. Phys. Chem.*, **65**, 507–513.
148. Efremov, N.M., Adamiak, B.Y., Blochin, V.I., Dadshev, S.J., Dmitriev, K.I., Semjonov, V.N., Levashov, V.F., and Jusbashev, V.F. (2000) Experimental investigation of the action of pulsed electrical discharges in liquids on biological objects. *IEEE Trans. Plasma Sci.*, **28**, 224–228.
149. Bogomaz, A.A., Goryachev, V.L., Remennyi, A.S., and Rutberg, F.G. (1991) The effectiveness of a pulsed electrical discharge in decontaminating water. *Sov. Tech. Phys. Lett.*, **17**, 448–449.
150. Fudamoto, T., Namihira, T., Katsuki, S., Akiyama, H., Imakubo, T., and

Majima, T. (2008) Sterilization of E. coli by underwater pulsed streamer discharges in a continuous flow system. *Electr. Eng. Jpn.*, **164** (1), 1–7.

151. Gupta, S.B. and Bluhm, H. (2008) The potential of pulsed underwater streamer discharges as a disinfection technique. *IEEE Trans. Plasma Sci.*, **36** (4), 1621–1632.

152. Anpilov, A.M., Barkhudarov, E.M., Christofi, N., Kop'ev, V.A., Kossyi, I.A., Taktakishvili, M.I., and Zadiraka, Y. (2002) Pulsed high voltage electric discharge disinfection of microbially contaminated liquids. *Lett. Appl. Microbiol.*, **35** (1), 90–94.

153. Takeda, T., Chang, J.S., Ishizaki, T., Saito, N., and Takai, O. (2008) Morphology of high-frequency electrohydraulic discharge for liquid-solution plasmas. *IEEE Trans. Plasma Sci.*, **36** (4), 1158–1159.

154. Zastawny, H.Z., Romat, H., Karpel vel Leitner, N., and Chang, J.S. (2004) Pulsed arc discharges for water treatment and disinfection, *Electrostatics 2003*, Institute of Physics Conference Series, vol. 137, IOP Publishers, Bristol, 325–330. 178.

155. Sakiyama, Y., Tomai, T., Miyano, M., and Graves, D.B. (2009) Disinfection of E. coli by nonthermal microplasma electrolysis in normal saline solution. *Appl. Phys. Lett.*, **94** (16), 161 501.

156. Chick, H. (1908) An investigation of the laws of disinfection. *J. Hyg.*, **8** (1), 92–158.

157. Watson, H.E. (1908) A note on the variation of the rate of disinfection with change in the concentration of the disinfectant. *J. Hyg.*, **8** (4), 536–542.

158. Hiatt, C.W. (1964) Kinetics of inactivation of viruses. *Bacteriol. Rev.*, **28** (2), 150–163.

159. Severin, B.F., Suidan, M.T., and Engelbrecht, R.S. (1983) Kinetic modeling of UV disinfection of water. *Water Res.*, **17** (11), 1669–1678.

160. Haas, C.N. and Karra, S.B. (1984) Kinetics of microbial inactivation by chlorine. I. Review of results in demand-free systems. *Water Res.*, **18** (11), 1443–1449.

161. Meulemans, C.C.E. (1987) The basic principles of UV disinfection of water. *Ozone Sci. Eng.*, **9** (4), 299–313.

162. Gyurek, L.L. and Finch, G.R. (1998) Modeling water treatment chemical disinfection kinetics. *J. Environ. Eng. ASCE*, **124** (9), 783–793.

163. Hijnen, W.A.M. and Medema, G.J. (2005) Inactivation of viruses, bacteria, spores and protozoa by ultraviolet irradiation in drinking water practice: A review. *Water Sci. Technol.*, **5** (5), 93–99.

164. Jensen, J.N. (2010) Disinfection model based on excess inactivation sites: Implications for linear disinfection curves and the Chick-Watson dilution coefficient. *Environ. Sci. Technol.*, **44** (21), 8162–8168.

165. Morent, R. and de Geyter, N. (2011) Inactivation of bacteria by non-thermal plasma, in *Biomedical Engineering - Frontiers and Challenges* (ed. R. Fazel-Rezai), InTech, pp. 25–50.

166. Geeraerd, A.H., Valdramidis, V.P., and Van Impe, J.F. (2005) GinaFiT, a freeware tool to assess non-linear microbial survivor curves. *Int. J. Food Microbiol.*, **102**, 95–105.

167. Greenberg, A.E., Clesceri, L.S., and Eaton, A.D. (1992) *Standard methods for the examination of water and wastewater*, American Public Health Association, American Water Works Association, Water Environment Federation, Washington, DC.

168. Byrd, J.J., Xu, H.S., and Colwell, R.R. (1991) Viable but nonculturable bacteria in drinking water. *Appl. Environ. Microbiol.*, **57** (3), 875–878.

169. Schaule, G., Flemming, H.C., and Ridgway, H.F. (1993) Use of 5-cyano-2,3-ditoyl tetrazolium chloride for quantifying planktonic and sessile respiring bacteria in drinking water. *Appl. Environ. Microbiol.*, **59** (11), 3850–3857.

170. Defives, C., Guyard, S., Oulare, M.M., Mary, P., and Hornez, J.P. (1999) Total counts, culturable and viable, and non-culturable microflora of a French mineral water: a case study. *J. Appl. Microbiol.*, **86** (6), 1033–1038.

171. Kogure, K., Simidu, U., and Taga, N. (1979) Tentative direct microscopic method for counting living marine bacteria. *Can. J. Microbiol.*, **25** (3), 415–420.
172. Xu, H.S., Roberts, N., Singleton, F.L., Attwell, R.W., Grimes, D.J., and Colwell, R.R. (1982) Survival and viability of nonculturable Escherichia coli and Vibriu cholerae in the estuarine and marine environment. *Microb. Ecol.*, **8** (4), 313–323.
173. von Woedtke, T., Kramer, A., and Weltmann, K.D. (2008) Plasma sterilization: what are the conditions to meet this claim? *Plasma Proc. Polym.*, **5** (6), 534–539.
174. Moreau, M., Feuilloley, M.G.J., Veron, W., Meylheuc, T., Chevalier, S., Brisset, J.L., and Orange, N. (2007) Gliding arc discharge in the potato pathogen Erwinia carotovora subsp atroseptica: mechanism of lethal action and effect on membrane-associated molecules. *Appl. Environ. Microbiol.*, **73** (18), 5904–5910.
175. Ohshima, T., Sato, M., and Saito, M. (1995) Selective release of intracellular protein using pulsed electric field. *J. Electrostatics*, **35**, 103–112.
176. Dodet, B., Beggar, D., Odic, E., Salamitou, S., Le Hegarat, F., Leblon, G., Goldman, A., and Goldman, M. (2004) Covalent cross-linking of proteins by a non-thermal plasma process. *High Temp. Mater. Process.*, **8** (2), 321–332.
177. Bae, P.H., Hwang, Y.J., Jo, H.J., Kim, H.J., Lee, Y., Park, Y.K., Kim, J.G., and Jung, J. (2006) Size removal on polyester fabrics by plasma source ion implantation device. *Chemosphere*, **63** (6), 1041–1047.
178. Salton, M.R.J. and Kim, K.S. (1996) Structure, in *Medical Microbiology* (ed. S. Baron), 4th edn, University of Texas Medical Branch at Galveston, Galveston, TX. Chapter 2.
179. Matz, L.L., Cabrera Beaman, T., and Gerhardt, P. (1970) Chemical composition of exosporium from spores of Bacillus cereus. *J. Bacteriol.*, **101**, 196–201.
180. Marnett, L.J. (1999) Lipid peroxidation - DNA damage by malondialdehyde. *Mutat. Res. Fundam. Mol. Mech. Mugag.*, **424** (1–2), 83–95.
181. Davies, M.J. (2003) *Protein oxidations: concepts, mechanisms and new insights, http://www.sfrbm.org/frs/Davies2003.pdf* (accessed 2011).
182. Keyer, K., Gort, A.S., and Imlay, J.A. (1995) Superoxide and the production of oxidative DNA damage. *J. Bacteriol.*, **177** (23), 6782–6790.
183. Pericone, C.D., Park, S., Imlay, J.A., and Weiser, J.N. (2003) Factors contributing to hydrogen peroxide resistance in Streptococcus pneumoniae include pyruvate oxidase (SpxB) and avoidance of the toxic effects of the Fenton reaction. *J. Bacteriol.*, **185** (23), 6815–6825.
184. LeBlanc, J.J., Davidson, R.J., and Hoffman, P.S. (2006) Compensatory functions of two alkyl hydroperoxide reductases in the oxidative defense system of Legionella pneumophila. *J. Bacteriol.*, **188** (17), 6235–6244.
185. Sabri, M., Leveille, S., and Dozois, C.M. (2006) A SitABCD homologue from an avian pathogenic Escherichia coli strain mediates transport of iron and manganese and resistance to hydrogen peroxide. *Microbiology*, **152**, 745–758.
186. Lu, S., Killoran, P.B., Fang, F.C., and Riley, L.W. (2002) The global regulator ArcA controls resistance to reactive nitrogen and oxygen intermediates in Salmonella enterica serovar enteritidis. *Infect. Immun.*, **70** (2), 451–461.
187. Elgrably-Weiss, M., Park, S., Schlosser-Silverman, E., Rosenshine, I., Imlay, J., and Altuvia, S. (2002) A Salmonella enterica serovar typhimurium hemA mutant is highly susceptible to oxidative DNA damage. *J. Bacteriol.*, **184** (14), 3774–3784.
188. Horsburgh, M.J., Wharton, S.J., Karavolos, M., and Foster, S.J. (2002) Manganese: elemental defence for a life with oxygen? *Trends Microbiol.*, **10** (11), 496–501.
189. Kehres, D.G., Janakiraman, A., Slauch, J.M., and Maguire, M.E. (2002) Regulation of Salmonella enterica serovar

typhimurium mntH transcription by H_2O_2, Fe^{2+}, and Mn^{2+}. *J. Bacteriol.*, **184** (12), 3151–3158.

190. Seaver, L.C. and Imlay, J.A. (2001) Hydrogen peroxide fluxes and compartmentalization inside growing Escherichia coli. *J. Bacteriol.*, **183** (24), 7182–7189.

191. Lukes, P., Clupek, M., Babicky, V., and Vykouk, T. (2007) Bacterial inactivation by pulsed corona discharge in water, in *2007 IEEE Pulsed Power Conference*, **1–4**, IEEE, Albuquerque, NM, pp. 320–323. doi: 10.1109/ppps.2007.4651849.

192. Imlay, J.A., Chin, S.M., and Linn, S. (1988) Toxic DNA damage by hydrogen peroxide through the Fenton reaction *in vivo* and *in vitro*. *Science*, **240** (4852), 640–642.

193. Nakamura, S., Minamitani, Y., Handa, T., Katsuki, S., Namihira, T., and Akiyama, H. (2009) Optical measurements of the electric field of pulsed streamer discharges in water. *IEEE Trans. Dielectr. Electr. Insul.*, **16** (4), 1117–1123.

194. Katsumura, Y. (1998) NO_2 and NO_3 radicals in radiolysis of nitric acid solutions, in *The Chemistry of Free Radicals: N-Centered Radicals* (ed. Z.B. Alfassi), John Wiley & Sons, Ltd, Chichester, pp. 393–412.

195. Koppenol, W. (1998) The basic chemistry of nitrogen monoxide and peroxynitrite. *Free Radical Biol. Med.*, **25**, 385–391.

196. Edwards, J.O. and Plumb, R.C. (1994) The chemistry of peroxonitrite. *Prog. Inorg. Chem.*, **41**, 599.

197. Ray, J.D. (1962) Heat of isomerisation of peroxynitrite to nitrate and kinetics of isomerization of peroxynitrous to nitric acid. *J. Inorg. Nucl. Chem.*, **24**, 1159–1162.

198. Pryor, W. and Squadrito, G. (1995) The chemistry of peroxynitrite: a product from the reaction of nitric oxide with superoxide. *Am. J. Physiol.*, **268**, L699–L722.

199. Viera, L., Ye, Y.Z., Estevez, A., and Beckman, J.S. (1999) Immunohistochemical methods to detect nitrotyrosine. *Methods Enzymol.*, **301**, 373–381.

200. Laskin, D.L., (2010) *Oxidative/Nitrosative Stress and Disease*, Annals of the New York Academy of Science., John Wiley & Sons, Inc, New York. vol. **1203**.

201. Seago, N., Clark, D., and Miller, M. (1995) Role of inducible oxide synthase (NOS) and peroxynitrite in gut inflammation. *Inflam. Res.*, **44**, S153–S154.

202. Beckman, J.S. and Koppenol, W.H. (1996) Nitric oxide, superoxide, and peroxynitrite: the good, the bad, and the ugly. *Am. J. Physiol. Cell Physiol.*, **271** (5), C1424–C1437.

203. Denicola, A., Souza, J.M., and Radi, R. (1998) Diffusion of peroxynitrite across erythrocyte membranes. *Proc. Natl. Acad. Sci. U.S.A.*, **1998** (95), 3566–3571.

204. Beckman, J.S. (1994) Peroxynitrite versus hydroxyl radical: the role of nitric oxide in superoxide-dependent cerebral injury. *Ann. N.Y. Acad. Sci.*, **738**, 69–75.

205. Hutchinson, F. (1957) The distance that a radical formed by ionizing radiation can diffuse in a yeast cell. *Radiat. Res.*, **7** (5), 473–483.

206. Pacher, P., Beckman, J.S., and Liaudet, L. (2007) Nitric oxide and peroxynitrite in health and disease. *Physiol. Rev.*, **87**, 315–424.

207. Rubbo, H., Trostchansky, A., and O'Donnell, V.B. (2009) Peroxynitrite-mediated lipid oxidation and nitration: Mechanisms and consequences. *Arch. Biochem. Biophys.*, **484**, 167–172.

208. Moussa, D., Doubla, A., Kamgang-Youbi, G., and Brisset, J.L. (2007) Postdischarge long life reactive intermediates involved in the plasma chemical degradation of an azoic dye. *IEEE Trans. Plasma Sci.*, **35** (2), 444–453.

209. Porter, D., Poplin, M.D., Holzer, F., Finney, W.C., and Locke, B.R. (2009) Formation of hydrogen peroxide, hydrogen, and oxygen in gliding arc electrical discharge reactors with water spray. *IEEE Trans. Ind. Appl.*, **45** (2), 623–629.

210. Brisset, J.L., Doubla, A., Amouroux, J., Lelievre, J., and Goldmann, M. (1989) Chemical reactivity of the gaseous species in a plasma discharge in air: an acid-base study. *Appl. Surf. Sci.*, **36** (1–4), 530–538.

211. Rutberg, P.G., Kolikov, V.A., Kurochkin, V.E., Panina, L.K., and Rutberg, A.P. (2007) Electric discharges and the prolonged microbial resistance of water. *IEEE Trans. Plasma Sci.*, **35** (4), 1111–1118.

212. Kamgang-Youbi, G., Herry, J.M., Meylheuc, T., Brisset, J.L., Bellon-Fontaine, M.N., Doubla, A., and Naitali, M. (2009) Microbial inactivation using plasma-activated water obtained by gliding electric discharges. *Lett. Appl. Microbiol.*, **48** (1), 13–18.

213. Friedman, G. (2010) Plasma pharmacology. *ICPM-3: 3rd International Conference on Plasma Medicine.*, Greifswald, Germany.

214. Shainsky, N., Dobrynin, D., Ercan, U., Joshi, S., Ji, H., Brooks, A., Fridman, G., Cho, Y., Fridman, A., and Friedman, G. (2011) Non-equilibrium plasma treatment of liquids, formation of plasma acid, in *ISPC-20: 20th International Symposium on Plasma Chemistry*, A. J. Drexel Plasma Institute, Philadelphia, PA.

215. Gnokam, F., Doubla, A., and Brisset, J.L. (2010) Temporal post discharge in plasma-chemical degradation of slaughterhouse effluents. *Chem. Eng. Commun.*, **198**, 483–493.

216. Huang, Y.R., Hung, Y.C., Hsu, S.Y., Huang, Y.W., and Hwang, D.F. (2008) Application of electrolyzed water in the food industry (Review). *Food Control*, **19**, 329–345.

217. Beckman, J.S., Chen, J., Ischiropoulos, H., and Crow, J.P. (1994) Oxidative chemistry of peroxynitrite. *Methods Enzymol.*, **233**, 229–240.

218. Lang, P.S., Ching, W.K., Willberg, D.M., and Hoffmann, M.R. (1998) Oxidative degradation of 2,4,6-trinitrotoluene by ozone in an electrohydraulic discharge reactor. *Environ. Sci. Technol.*, **32**, 3142–3148.

219. Willberg, D.M., Lang, P.S., Hochemer, R.H., Kratel, A., and Hoffmann, M.R. (1996) Degradation of 4-chlorophenol, 3,4-dichloroanilin, and 2,4,6-trinitrotoluene in an electrohydraulic discharge. *Environ. Sci. Technol.*, **30**, 2526–2534.

220. Martin, E.A. (1960) Experimental investigation of a high-energy density, high-pressure arc plasma. *J. Appl. Phys.*, **31**, 255–267.

221. Robinson, J.W., Ham, M., and Balaster, A.N. (1973) Ultraviolet radiation from electrical discharges in water. *J. Appl. Phys.*, **44**, 72–75.

222. Hori, H., Nagaoka, Y., Yamamoto, A., Sano, T., Yamashita, N., Taniyasu, S., Kutsuna, S., Osaka, I., and Arakawa, R. (2006) Efficient decomposition of environmentally persistent perfluorooctanesulfonate and related fluorochemicals using zerovalent iron in subcritical water. *Environ. Sci. Technol.*, **40** (3), 1049–1054.

223. Boudam, M.K. and Moisan, M. (2010) Synergy effect of heat and UV photons on bacterial-spore inactivation in an N_2-O_2 plasma-afterglow sterilizer. *J. Phys. D.: Appl. Phys.*, **43** (29), 295 202.

224. Pollak, J., Moisan, M., Keroack, D., and Boudam, M.K. (2008) Low-temperature low-damage sterilization based on UV radiation through plasma immersion. *J. Phys. D.: Appl. Phys.*, **41** (13), 135 212.

225. Nguyen, C.V., van Deursen, A.P.J., van Heesch, E.J.M., Winands, G.J.J., and Pemen, A.J.M. (2010) X-ray emission in streamer-corona plasma. *J. Phys. D.: Appl. Phys.*, **43** (2), 025 202.

226. Gilliland, S.E. and Speck, M.L. (1967) Mechanism of the bactericidal action produced by electrohydraulic shock. *Appl. Microbiol.*, **15**, 1038–1044.

227. Gilliland, S.E. and Speck, M.L. (1967) Inactivation of microorganisms by electrohydraulic shock. *Appl. Microbiol.*, **15**, 1031–1037.

228. Bailey, M.R., Blackstock, D.T., Cleveland, R.O., and Crum, L.A. (1999) Comparison of electrohydraulic lithotripters with rigid and pressure-release ellipsoidal reflectors. II. Cavitation fields. *J. Acoust. Soc. Am.*, **106**, 1149–1160.

229. Huber, P., Debus, J., Jochle, K., Simiantonakis, I., Jenne, J., Rastert,

R., Spoo, J., Lorenz, W.J., and Wannenmacher, M. (1999) Control of cavitation activity by different shock-wave pulsing regimes. *Phys. Med. Biol.*, **44**, 1427–1437.

230. Sokolov, D.L., Bailey, M.R., and Crum, L.A. (2001) Use of a dual-pulse lithotripter to generate a localized and intensified cavitation field. *J. Acoust. Soc. Am.*, **110** (3), 1685–1695.

231. Zhong, P., Lin, H.F., Xi, X.F., Zhu, S.L., and Bhogte, E.S. (1999) Shock wave-inertial microbubble interaction: Methodology, physical characterization, and bioeffect study. *J. Acoust. Soc. Am.*, **105** (3), 1997–2009.

232. Loske, A.M., Prieto, F.E., Fernandez, F., and van Cauwelaert, J. (2002) Tandem shock wave cavitation enhancement for extracorporeal lithotripsy. *Phys. Med. Biol.*, **47**, 3945–3957.

233. Sunka, P., Stelmashuk, V., Babicky, V., Clupek, M., Benes, J., Pouckova, P., Kaspar, J., and Bodnar, M. (2006) Generation of two successive shock waves focused to a common focal point. *IEEE Trans. Plasma Sci.*, **34** (4), 1382–1385.

234. Alvarez, U.M., Ramirez, A., Fernandez, F., Mendez, A., and Loske, A.M. (2008) The influence of single-pulse and tandem shock waves on bacteria. *Shock Waves*, **17** (6), 441–447.

235. Huber, P.E. and Debus, J. (2001) Tumor cytotoxicity *in vivo* and radical formation *in vitro* depend on the shock wave-induced cavitation dose. *Radiat. Res.*, **156** (3), 301–309.

236. Stalder, K.R., Woloszko, J., Brown, I.G., and Smith, C.D. (2001) Repetitive plasma discharges in saline solutions. *Appl. Phys. Lett.*, **79**, 4503–4505.

237. Woloszko, J., Stalder, K.R., and Brown, I.G. (2002) Plasma characteristics of repetitively-pulsed electrical discharges in saline solutions used for surgical procedures. *IEEE Trans. Plasma Sci.*, **30**, 1376–1383.

238. Stalder, K.R., Nersisyan, G., and Graham, W.G. (2006) Spatial and temporal variation of repetitive plasma discharges in saline solutions. *J. Phys. D.: Appl. Phys.*, **39** (16), 3457–3460.

239. Schaper, L., Graham, W.G., and Stalder, K.R. (2011) Vapour layer formation by electrical discharges through electrically conducting liquids-modelling and experiment. *Plasma Sources Sci. Technol.*, **20** (3), 034 003.

240. Schaper, L., Stalder, K.R., and Graham, W.G. (2011) Plasma production in electrically conducting liquids. *Plasma Sources Sci. Technol.*, **20** (3), 034 004.

241. Needham, D. and Rockmuth, R.M. (1989) Electro-mechanical permeabilization of lipid vesicles. *Biophys. J.*, **55**, 1001–1009.

242. Garner, A.L., Chen, G., Chen, N., Sridhara, V., Kolb, J.F., Swanson, R.J., Beebe, S.J., Joshi, R.P., and Schoenbach, K.H. (2007) Ultrashort electric pulse induced changes in cellular dielectric properties. *Biochem. Biophys. Res. Commun.*, **362** (1), 139–144.

243. Pakhomov, A.G., Shevin, R., White, J.A., Kolb, J.F., Pakhomova, O.N., Joshi, R.P., and Schoenbach, K.H. (2007) Membrane permeabilization and cell damage by ultrashort electric field shocks. *Arch. Biochem. Biophys.*, **465** (1), 109–118.

244. Schoenbach, K.H., Beebe, S.J., and Buescher, E.S. (2001) Intracellular effect of ultrashort electrical pulses. *Bioelectromagnetics*, **22** (6), 440–448.

245. Schoenbach, K.H., Hargrave, B., Joshi, R.P., Kolb, J.F., Nuccitelli, R., Osgood, C., Pakhomov, A., Stacey, M., Swanson, R.J., White, J.A., Xiao, S., Zhang, J., Beebe, S.J., Blackmore, P.F., and Buescher, E.S. (2007) Bioelectric effects of intense nanosecond pulses. *IEEE Trans. Dielectr. Electr. Insul.*, **14** (5), 1088–1109.

246. Mazurek, B., Lubicki, P., and Staroniewicz, Z. (1995) Effect of short HV pulses on bacteria and fungi. *IEEE Trans. Dielectr. Electr. Insul.*, **2**, 418–425.

247. Lubicki, P. and Jayaram, S. (1997) High voltage pulse application for the destruction of the Gram-negative bacterium Yersinia enterocolitica. *Bioelectrochem. Bioenerg.*, **43**, 135–141.

248. Ohshima, T., Sato, K., Terauchi, H., and Sato, M. (1997) Physical and chemical modifications of high-voltage pulse sterilization. *J. Electrostatics*, **42**, 159–166.

249. MacGregor, S.J., Farish, O., Fouracre, R., Rowan, N.J., and Anderson, J.G. (2000) Inactivation of pathogenic and spoilage microorganisms in a test liquid using pulsed electric fields. *IEEE Trans. Plasma Sci.*, **28**, 144–149.

250. Schoenbach, K.H., Katsuki, S., Stark, R.H., Buescher, E.S., and Beebe, S.J. (2002) Bioelectrics - new applications for pulsed power technology. *IEEE Trans. Plasma Sci.*, **30** (1), 293–300.

251. Nanda, G.S., Sun, F.X., Hofmann, G.A., Hoffman, R.M., and Dev, S.B. (1998) Electroporation enhances therapeutic efficacy of anticancer drugs: treatment of human pancreatic tumor in animal model. *Anticancer Res.*, **18** (3A), 1361–1366.

252. Nanda, G.S., Sun, F.X., Hofmann, G.A., Hoffman, R.M., and Dev, S.B. (1998) Electroporation therapy of human larynx tumors HEp-2 implanted in nude mice. *Anticancer Res.*, **18** (2A), 999–1004.

253. Dev, S.B., Nanda, G.S., Zn, Z., Wang, X., Hofman, R.M., and Hofmann, G.A. (1997) Effective electroporation therapy of human pancreatic tumors implanted in nude mice. *Drug Deliv.*, **4**, 293–299.

254. Zhang, J., Blackmore, P.F., Hargrave, B.Y., Xiao, S., Beebe, S.J., and Schoenbach, K.H. (2008) Nanosecond pulse electric field (nanopulse): a novel non-ligand agonist for platelet activation. *Arch. Biochem. Biophys.*, **471** (2), 240–248.

255. Schoenbach, K.H., Xiao, S., Joshi, R.P., Camp, J.T., Heeren, T., Kolb, J.F., and Beebe, S.J. (2008) The effect of intense subnanosecond electrical pulses on biological cells. *IEEE Trans. Plasma Sci.*, **36** (2), 414–422.

256. Xiao, S., Guo, S.Q., Nesin, V., Heller, R., and Schoenbach, K.H. (2011) Subnanosecond electric pulses cause membrane permeabilization and cell death. *IEEE Trans. Biomed. Eng.*, **58** (5), 1239–1245.

257. White, J.A., Pliquett, U., Blackmore, P.F., Joshi, R.P., Schoenbach, K.H., and Kolb, J.F. (2011) Plasma membrane charging of Jurkat cells by nanosecond pulsed electric fields. *Eur. Biophys. J. Biophys. Lett.*, **40** (8), 947–957.

258. Long, G., Shires, P.K., Plescia, D., Beebe, S.J., Kolb, J.F., and Schoenbach, K.H. (2011) Targeted tissue ablation with nanosecond pulses. *IEEE Trans. Biomed. Eng.*, **58** (8), 2161–2167.

259. Nomura, N., Yano, M., Katsuki, S., Akiyama, H., Abe, K., and Abe, S.I. (2009) Intracellular DNA damage induced by non-thermal, intense narrowband electric fields. *IEEE Trans. Dielectr. Electr. Insul.*, **16** (5), 1288–1293.

260. Katsuki, S., Nomura, N., Koga, H., Akiyama, H., Uchida, I., and Abe, S.I. (2007) Biological effects of narrow band pulsed electric fields. *IEEE Trans. Dielectr. Electr. Insul.*, **14** (3), 663–668.

261. Katsuki, S., Mitsutake, K., Yano, M., Akiyama, H., Kai, H., and Shuto, T. (2010) Non-thermal and transient thermal effects of burst 100 MHz sinusoidal electric fields on apoptotic activity in HeLa cells. *IEEE Trans. Dielectr. Electr. Insul.*, **17** (3), 678–684.

262. Chen, X.H., Kolb, J.F., Swanson, R.J., Schoenbach, K.H., and Beebe, S.J. (2010) Apoptosis initiation and angiogenesis inhibition: melanoma targets for nanosecond pulsed electric fields. *Pigment Cell Melanoma Res.*, **23** (4), 554–563.

263. Chen, X.H., Swanson, R.J., Kolb, J.F., Nuccitelli, R., and Schoenbach, K.H. (2009) Histopathology of normal skin and melanomas after nanosecond pulsed electric field treatment. *Melanoma Res.*, **19** (6), 361–371.

264. Kolb, J.F. (2012) Subcellular biological effects of nanosecond pulsed electric fields, in *Plasma for Bio-Decontamination, Medicine and Food Security*, NATO Science for Peace and Security Series A: Chemistry and Biology (eds Z. Machala, K. Hensel, and Y. Akishev), Springer.

265. Kumar, P., Baum, C.E., Altunc, S., Buchenauer, J., Xiao, S., Christodoulou, C.G., Schamiloglu, E., and Schoenbach,

K.H. (2011) A hyperband antenna to launch and focus fast high-voltage pulses onto biological targets. *IEEE Trans. Microw. Theory Tech.*, **59** (4), 1090–1101.

266. Bajracharya, C., Xiao, S., Baum, C.E., and Schoenbach, K.H. (2011) Target detection with impulse radiating antenna. *IEEE Antennas Wire. Propag. Lett.*, **10**, 496–499.

267. Xiao, S., Altunc, S., Kumar, P., Baum, C.E., and Schoenbach, K.H. (2010) A reflector antenna for focusing subnanosecond pulses in the near field. *IEEE Antennas Wire. Propag. Lett.*, **9**, 12–15.

268. Fridman, G., Peddinghaus, M., Ayan, H., Fridman, A., Balasubramanian, M., Gutsol, A., Brooks, A., and Friedman, G. (2006) Blood coagulation and living tissue sterilization by floating-electrode dielectric barrier discharge in air. *Plasma Chem. Plasma Process.*, **26** (4), 425–442.

269. Vasilets, V.N., Gutsol, A., Shekhter, A.B., and Fridman, A. (2009) Plasma medicine. *High Energy Chem.*, **43** (3), 229–233.

270. Bernard, C., Leduc, A., Barbeau, J., Saoudi, B., Yahia, L., and De Crescenzo, G. (2006) Validation of cold plasma treatment for protein inactivation: a surface plasmon resonance-based biosensor study. *J. Phys. D.: Appl. Phys.*, **39** (16), 3470–3478.

271. Hasiwa, M., Kylian, O., Hartung, T., and Rossi, F. (2008) Removal of immune-stimulatory components from surfaces by plasma discharges. *Innate Immun.*, **14** (2), 89–97.

272. Ayan, H., Fridman, G., Gutsol, A.F., Vasilets, V.N., Fridman, A., and Friedman, G. (2008) Nanosecond-pulsed uniform dielectric-barrier discharge. *IEEE Trans. Plasma Sci.*, **36** (2), 504–508.

273. Mwale, F., Wang, H.T., Nelea, V., Luo, L., Antoniou, J., and Wertheimer, M.R. (2006) The effect of glow discharge plasma surface modification of polymers on the osteogenic differentiation of committed human mesenchymal stem cells. *Biomaterials*, **27** (10), 2258–2264.

9
Hydrogen and Syngas Production from Hydrocarbons

Moritz Heintze

9.1
Introduction: Plasma Catalysis

The application of electric gas discharges in chemical synthesis has been under investigation for a long time. The ability of plasma to enable unconventional chemical routes and products has been shown as early as 1857 with ozone synthesis in a dielectric barrier discharge (DBD) [1]. Probably, the most striking evidence for unconventional plasma reactions may be Miller's synthesis of amino acids in an atmosphere simulating earth's early development [2]. Several books have been published in the twentieth century, summarizing the results at their time; some examples are the books by Drost [3], McTaggart [4], and Eremin [5]. The latter may be the most relevant to assess earlier work in the current context, presenting an empirical but detailed kinetic discussion of hydrocarbon plasma chemistry. A recent comprehensive review of plasma chemistry can be found in Fridman's book [6].

In spite of the potential that plasmas appear to offer in chemical synthesis, their application for the industrial production of bulk chemicals remained an exception. The cause of this is the basic nature of plasmas. Thermal plasmas are an efficient but expensive way to heat a gas to extremely high temperatures of several thousands degrees, thus enabling the production of compounds that are stable at those temperatures. However, they may be less economic than chemical heating, for example, in flames. Nonthermal plasmas provide highly excited species such as radicals and ions or electronically excited states up to several tens of electron volts in a relatively cold environment (< 1000 K). These will react to the unconventional products mentioned above; however, the high-energy intermediates tend to react to a wide range of products with a poor selectivity.

In technical chemical synthesis, the method of choice for achieving the best product yields and selectivities is the application of catalysis; a catalyst is, unlike the reactants, not consumed in the process, but steers the reaction in the desired direction either by lowering the activation energy or by increasing the probability of the desired path by steric control.

Plasma Chemistry and Catalysis in Gases and Liquids, First Edition.
Edited by Vasile I. Parvulescu, Monica Magureanu, and Petr Lukes.

The idea of combining the two worlds of plasma chemistry and catalysis has since quite some time spurred considerable research interest, and several reviews were published [7–9]. There are, however, two fundamentally different notations of what plasma catalysis can refer to: (i) the use of plasma-excited species to initiate a chemical reaction, as discussed in the work done at the Kurchatov Institute in several examples of hydrocarbon conversion (see [6] for a review): the plasma contributes only a small amount to the total energy input needed for the reaction, but leads, similar to a catalyst, to a considerable increase in conversion. (ii) The plasma is applied to initiate a gas-phase reaction, either in the presence of or followed by a solid (heterogeneous) catalyst bed, which controls the product distribution and yield. A review of the possible interactions of plasma-excited species and solid catalyst surfaces in plasma-catalytic hydrocarbon conversion was recently published [10], although with detailed examples of the actual processes not being considered. A third research field, the preparation or activation of catalysts by plasma before their use, has recently received wide interest; but this is beyond the scope of the present review.

Since, so far, the conversion of hydrocarbons remains a challenge in many potentially important applications, it is not surprising that the interest in research into plasma catalysis persists, in spite of the limited success in the past decades as far as the introduction of commercial processes is concerned.

9.2 Current State of Hydrogen Production, Applications, and Technical Requirements

At present, hydrogen is perceived as a potentially important energy carrier for the future. Currently, however, hydrogen production plays a relatively small role compared to the world's energy supply: the total energy content of annually produced hydrogen amounts to only 1.4% of the world's energy demand [11]. Two factors will almost inevitably lead to an increased use of hydrogen in the energy supply chain, even if timescale and extent are matters of conjecture: (i) the decline of crude oil resources and (ii) the need to respond to global warming. CO_2-neutral primary energy sources are emerging, and their increased share is unanimously predicted; however, they have in common that the major part of the energy is harvested as electricity. The need for transportation fuel and means of energy storage remains an unresolved challenge. Obviously, the production and use of hydrogen constitute one of the most promising alternatives. The options and perspectives for hydrogen production have been reviewed recently [12]. Not being biased toward plasma technology, this review sees its scope only in small-scale applications.

As an intermediate on the way to a sustainable energy supply, natural gas plays a key role in future development. First, it is more favorable in terms of CO_2 balance than coal or crude oil, and second, the world's natural gas resources are by far greater than those of crude oil. Finally, in a renewable energy scenario, technologies developed for natural gas utilization may eventually be used for biogas processing.

Already today, the conversion of natural gas to liquid hydrocarbon fuels (GTL) has become a commercial reality with Shell's synthetic fuel from the Qatar plant being part of high-quality diesel or the test flight in 2009 by a commercial airline using 50% GTL fuel [13].

The most obvious source of hydrogen is water. In all reforming processes employing oxidants, if hydrogen is the desired final product, the carbon monoxide must be converted to additional hydrogen in the water gas shift (WGS) reaction:

$$CO + H_2O \longrightarrow CO_2 + H_2 \quad \Delta H = -41.1 \text{ kJ mol}^{-1} \quad (-0.43 \text{ eV}) \tag{9.1}$$

Even in steam reforming (SR) of natural gas combined with water gas shifting, half of the hydrogen originates from the water employed in the reforming and WGS: in summary, $CH_4 + 2\,H_2O \rightarrow 4\,H_2 + CO_2$. Water being abundant and not a cost factor, the production methods are categorized and named after the hydrocarbon used or effectively by the energy source. This may be electricity in the case of electrolysis or coal in the case of coal gasification. Methane contains 25% hydrogen by weight, whereas coal still contains about 5%; so the choice of fuel in hydrocarbon reforming does not make a fundamental difference to the process materials' balance.

In order to judge the relevance and scope for development of potential future hydrogen production methods, it is important to bear in mind the present day hydrogen production and use. About 90% of hydrogen is produced by SR from natural gas and hydrocarbons, mainly heavy oil. Half of this is used for the production of ammonia, which is further processed to fertilizers. The other half of hydrogen is produced and used in oil refineries for efficiently processing crude oil into a range of fuels and other hydrocarbons (e.g., lubricants, wax, ethylene, and other monomers).

This work reviews recent research into the use of electric gas discharges or plasma for the production of hydrogen and syngas, the latter being an intermediate for producing other liquid fuels. Fossil hydrocarbons are also the primary feedstock in the chemical industry, but here, the same applies as to the hydrogen demand: by far the largest market is energy supply. Hence, future innovative plasma processes for hydrogen conversion will have to compete with and outperform current technology. This need not be in terms of production cost alone; a range of applications is conceivable where other factors become dominant, such as dynamic response and small-scale operation.

The work published on plasma reforming and hydrocarbon conversion may, in principle, be categorized by the process, feedstock, or type of plasma used. Since the reactions available and the product composition are practically the most prominent features of a process, the latter is the main classification criterion in this review.

9.2.1
Steam Reforming: SR

SR refers to the conversion of hydrocarbons with steam into syngas. The reaction is endothermic; in the case of methane, the reaction may be summarized as follows:

$$CH_4 + H_2O \longrightarrow 3\,H_2 + CO \quad \Delta H = 206.2 \text{ kJ mol}^{-1} \quad (2.14 \text{ eV}) \tag{9.2}$$

This is currently the most cost-effective method of hydrogen production, and high-volume production plants with a capacity up to 100 000 $m^3 h^{-1}$ are in widespread use. Owing to its endothermic nature, the reaction is carried out at relatively high temperatures (700–1100 °C) over a Ni catalyst. The reactor is heated by external combustion of part of the hydrocarbon fuel. The advantage is the ready availability of the reactants in concentrated form, even if desulfurization of the hydrocarbon is mandatory to prevent catalyst deactivation. Owing to the relatively low productivity per volume GHSV (gas hourly space velocity), the required preprocessing, and the need of indirect heating at high temperatures, the economic advantage of SR can only be realized in large-scale installations.

If the syngas is to be used for Fischer–Tropsch synthesis (FTS), the molar ratio H_2/CO must be decreased from 3 to values between 1 and 2, depending on the details of the subsequent FTS synthesis [14]. The same applies for the processing of syngas to oxygenates such as methanol or dimethyl ether.

9.2.2 Partial Oxidation: POX

The second path is the partial oxidation (POX) of hydrocarbon fuel in an oxygen-deficient atmosphere

$$CH_4 + \frac{1}{2}O_2 \longrightarrow 2\,H_2 + CO \quad \Delta H = -43.6\ \text{kJ mol}^{-1} \quad (-0.45\ \text{eV}) \quad (9.3)$$

The reaction is slightly exothermal; thus, it will readily proceed without the need of externally heating the reactor. Thus a much simpler and more compact reactor design may be achieved. The second potential advantage lies in the different stoichiometry: the molar product ratio is better suited for FTS. One disadvantage is the need to produce pure oxygen by air separation, which contributes considerably to the overall process cost. Since in FTS the gaseous reactants are only partly converted, they need to be run in a cycle, leading to nitrogen accumulation, if air is used as oxidant in POX.

The product yield in POX is usually limited by the formation of water and carbon, the latter being particularly critical in a catalytic reactor, where it leads to catalyst deactivation. POX is currently used in the valorization of heavy oil residues that would rapidly poison an SR catalyst. POX of natural gas is not yet employed at a large industrial scale, but it has recently received considerable attention in the context of realizing small and compact fuel reformers for powering fuel cells in mobile applications. It is particularly attractive in mobile applications, where carrying a water supply in addition to the fuel is out of the question. As shown in the following sections, this is a field where the application of plasma looks most promising and hence receives a significant amount of attention.

The two processes SR and POX may be combined by supplying both steam and air to the reformer and adjusting the ratio of oxygen and steam to tune the thermal balance of the reactor. In principle, the reaction enthalpy can be set to zero; this is why the process is called *autothermal reforming*. In fact, a direct temperature regulation of the reactor by the reactant ratio is conceivable. In addition to thermal management, autothermal reforming also offers control of the molar product ratio, which makes it attractive as a source of syngas for FTS.

9.2.3 Dry Carbon Dioxide Reforming: CDR

Hydrocarbons and, in particular, methane can also be brought to reaction with carbon dioxide as oxidizer.

$$CH_4 + CO_2 \longrightarrow 2H_2 + 2\,CO \quad \Delta H = 247.9\ \text{kJ mol}^{-1} \quad (2.57\ \text{eV}) \tag{9.4}$$

The main incentive for methane dry reforming simply lies in the fact that natural gas contains large amounts of CO_2 that may well exceed 50%. The same applies to gas generated in anaerobic fermentation of organic material, that is, biogas or off-gas from municipal waste. Carbon dioxide reforming (CDR) may also be an effective way to utilize this CO_2. CDR could also be an important step in carbon dioxide sequestration or reuse in a carbon dioxide neutral fuel cycle. A further advantage of dry reforming consists in its capability to provide a CO-rich syngas with the composition approximately at optimum for FTS and oxygenate synthesis.

Consequently, in recent years, CDR has received considerable attention, both catalytic or plasma induced. In catalytic CDR, the catalyst is particularly prone to coking and deactivation. Therefore, plasma might be a suitable means for obtaining a stable process. A general review of different catalytic processes, including but not restricted to plasma activation, was published recently [15]. The paper provides a detailed overview of the catalysts and catalyst supports used.

9.2.4 Pyrolysis

Pyrolysis refers to the decomposition of hydrocarbons upon heat or plasma exposure without the addition of oxygen or an oxygen-containing reactant. In the case of methane, the pyrolysis reaction is endothermic and methane will decompose readily to hydrogen and solid carbon at sufficiently high temperatures:

$$CH_4 \longrightarrow 2H_2 + C\,(s) \quad \Delta H = 74.81\ \text{kJ mol}^{-1} \quad (0.78\ \text{eV}) \tag{9.5}$$

Methane plasma pyrolysis has received considerable attention for a long time [5]. Currently, it is proposed as a path to hydrogen from natural gas without producing additional CO_2, and it was argued that the process is economically highly attractive if the carbon soot is additionally used as a raw material. At this point, however, a word of caution is in place: carbon black is a material with a very complex microscopic and mesoscopic structure. Depending on this and on the impurity content, it can be a valuable industrial feedstock or just waste, which must be disposed of at high cost. An example of the former is the filler for polymers and rubber, and that for the latter is the carbon by-product in acetylene flame synthesis. A wide range of energy sources to provide the energy needed for pyrolysis have been investigated recently, ranging from nuclear to solar power [16]. This search for options demonstrates the interest in finding novel carbon-dioxide-free hydrogen sources.

Methane plasma pyrolysis has also been thoroughly considered for plasma synthesis of higher hydrocarbons, in particular, acetylene.

$$2\,CH_4 \longrightarrow C_2H_2 + 3H_2 \quad \Delta H = 376.62\ \text{kJ mol}^{-1} \quad (3.9\ \text{eV}) \tag{9.6}$$

The Hüls process [17], which for some time was an industrially important method for acetylene production, is one of the very few plasma-driven bulk chemical synthesis processes ever employed at an industrial scale. It was established commercially in 1939 and is still in use for producing 1,4-butanediol [18].

9.3
Description and Evaluation of the Process

9.3.1
Materials Balance: Conversion, Yield, and Selectivity

The conversion of each reactant describes the fraction that is converted to any of the products. The measurement of product or reactant streams leaving the reactor is based on concentration measurements. The conversion may be expressed by the concentrations at the reactor inlet and outlet:

$$\text{Conv.}\ (CH_4) = 1 - \frac{[CH_4]_{out} \bullet \text{Total flow out}}{[CH_4]_{in} \bullet \text{Total flow in}} \tag{9.7}$$

Only in the cases when the number of molecules and, hence, the total volume stream is not changed by the reaction, is the stream of any compound directly related to its concentration. For reactions under consideration in this review, this is generally not the case; in fact, the number of molecules and the total volume stream doubles during methane reforming (Eqs. (9.2)–(9.6)). Therefore, if the change in total gas stream is neglected, the calculated conversion will be too high. One method of determining the total stream at the reactor output consists in adding a known flow of a gas not itself being involved in the process, conveniently N_2 or Ar. In a catalytic reactor, such substances will not react or affect the process if added in relatively small concentrations. In a plasma reactor, they should, however, be introduced at the outlet to determine the total product stream.

The yield of a product describes the actual amount produced in relation to the theoretical amount if all reactants are fully reacting only to the product in question; in SR, a CO yield of 100% is achieved if the hydrocarbon is completely converted and no CO_2 is formed. The yield of carbon-containing species is usually calculated on the basis of available carbon in the reactant and for H_2, by analogy, on the basis of available hydrogen.

$$Y\left(C_xH_y\right) = \frac{\frac{1}{x}\left[C_xH_y\right]_{out} \bullet \text{Total flow out}}{[CH_4]_{in} \bullet \text{Total flow in}}$$
$$Y\left(H_2\right) = \frac{[H_2]_{out} \bullet \text{Total flow out}}{\left(\frac{1}{2}[CH_4]_{in} + [H_2O]_{in}\right) \bullet \text{Total flow in}} \tag{9.8}$$

The selectivity to a particular product describes the amount of reactant that was converted to this particular product.

$$S\left(C_xH_y\right) = \frac{\frac{1}{x}\left[C_xH_y\right]_{out} \bullet \text{Total flow out}}{[CH_4]_{in} \bullet \text{Total flow in} - [CH_4]_{out} \bullet \text{Total flow out}} \tag{9.9}$$

These magnitudes are interrelated as $Y = \text{Conv.} \times S$. Thus, a high selectivity may not necessarily imply a high yield, but a high yield requires both a high selectivity and conversion. The respective by-products, water and soot, are comparatively difficult to measure and, hence, not determined in several studies. In this case, they are sometimes neglected in the calculation of Y and S; under these circumstances, substantial errors of the values quoted can arise.

The carbon balance is the ratio of total carbon in the products analyzed to the carbon in the reactants. It may be used as an assessment of the carbon deposited in the reactor or, if the latter is measured, for example, by weighing after a defined time of operation, as an indication of the total measurement error.

9.3.2
Energy Balance: Energy Requirement and Efficiency

The specific energy input (SEI) or input energy in a plasma chemical reactor is the electrical energy used to operate the plasma in relation to the reactant stream.

$$\text{SEI} = \frac{\text{Plasma power}}{[CH_4]_{in} \bullet \text{Total flow in}} \tag{9.10}$$

The electric energy required to convert the reactant(s) is termed as *conversion energy*:

$$E_{\text{conv.}} = \frac{\text{Plasma power}}{[CH_4]_{in} \bullet \text{Total flow in} - [CH_4]_{out} \bullet \text{Total flow out}} \tag{9.11}$$

The specific energy requirement (SER), on the other hand, describes the electrical energy required per amount of product. For syngas this is

$$\text{SER} = \frac{\text{Plasma power}}{\left([H_2]_{out} + [CO]_{out}\right) \bullet \text{Total flow out}} \tag{9.12}$$

Since the energy of combustion is very similar for H_2 and CO, it is justified to quote the energy requirement per mole of syngas produced.

Two different definitions are used for the reformer efficiency; they should always be based on the lower heating values (LHVs), that is, the heat available when using a fuel, not taking into account the condensation of the water formed, since it normally cannot be recuperated

$$\eta = \frac{\text{LHV of products}}{\left(\text{Plasma power} + \text{LHV of reactants}\right)} \tag{9.13}$$

This value is always less than 1 in an exothermal reforming reaction and could approach 1 in an ideal endothermal reaction if all electrical energy was converted into chemical energy (a case closely approached in electrolysis). The second definition is the ratio of enthalpy change during the reforming reaction to electric power. This value could also approach 1 in an ideal endothermal reaction but is zero in an autothermal process irrespective of the SER and even negative under exothermal conditions such as POX. To distinguish these, the first was termed *fuel production efficiency* and the second *chemical energy efficiency* [19]. In the following discussion, efficiencies will be quoted according to Eq. (9.13). In some cases, where rather high reformer efficiencies around 80% are quoted, this is simply due to neglecting the plasma power when calculating efficiency.

A characteristic magnitude describing the productivity of a reactor is the product stream per reactor volume. In heterogeneously catalyzed gas-phase reactions, this is termed *gas hourly space velocity* (GHSV), referring to the hourly product volume per catalyst volume.

9.4 Plasma-Assisted Reforming

9.4.1 Steam Reforming

Being a successful industrial process run at high temperature, SR has been investigated by many research groups with the objective of achieving a good dynamic response in small-scale units. In order to achieve satisfactory reformer efficiency, some of the energy supplied to the plasma must be recovered as chemical energy in the products.

9.4.1.1 Conversion of Methane

The notion of plasma catalysis in terms of supplying a small amount of the energy required for the reaction by a plasma was demonstrated at the Kurchatov Institute [6]. The researchers supplied most of the energy required by preheating the reactants to 500–570 °C and achieved a marked increase in hydrogen production by adding a relatively small (5–10%) amount of power in the form of a pulsed microwave plasma. The degree to which the nonthermal plasma can control SR was investigated at GREMI [20]. Two different configurations (R + and R−) of the gliding arc (GA) reactor are compared. In the R+ configuration, the current is higher and the voltage lower, indicating more thermal plasma, whereas R− provides a higher degree of nonequilibrium plasma. In this configuration, methane conversion and hydrogen concentration are very similar to the results of thermal equilibrium calculations, with the nonthermal plasma (R−) yielding a better methane conversion and carbon selectivity to CO. The energy cost of hydrogen production is, however, the lowest at $CH_4/H_2O = 4$, that is, under conditions where the reaction proceeds mostly as pyrolysis.

The energy balance of plasma-SR in a DBD was analyzed at Siemens, Erlangen [21, 22]. Their reactor is thermally insulated, and the temperatures at the inlet and outlet are measured. They found that only 3% of plasma power is converted to chemical enthalpy of the (endothermal) SR. Conversion of steam is much less than methane conversion, indicating a competition between SR and pyrolysis. From numeric simulations, the authors concluded that most energy supplied by the plasma leads to vibrational excitation and thus eventually gas heating. The most important reaction is the electron-induced hydrogen abstraction

$$CH_4 + e^- \longrightarrow CH_3 + H + e^- \quad \Delta H = 421\ \text{kJ mol}^{-1} \quad (4.36\ \text{eV}) \qquad (9.14)$$

Most (80%) radicals are lost by recombination with methane, and atomic hydrogen reacts with methane to form H_2 and more CH_3 radicals. Hydroxyl radicals from water dissociations also contribute to the formation of CH_3 radicals. In their DBD reactor, the authors observe a higher yield of C_2-hydrocarbons than H_2.

In a more recent work using a DBD [23] and running relatively low conversion values of 10–20%, the main products were also C_2H_6 and H_2, with selectivity to CO being <5%. Work using a microwave plasma source also shows the competition between CO and carbon formation [24]. The selectivity to CO was higher than the selectivity to carbon only at $H_2O/CH_4 = 3$.

A more successful approach might be the combination of the gas discharge with a solid catalyst. A study at the Tokyo Institute of Technology investigated barrier discharge and catalytic hybrid reactor [25]. The effect of the catalyst + discharge was compared with both, catalyst and discharge, alone. The discharge led to a temperature-independent methane conversion with high selectivities to CO_2 and higher hydrocarbons, whereas the catalyst was active only above 600 °C, with H_2 and CO being the main products. The selectivities to CO and CO_2 then were found to approach the equilibrium values. The combination of plasma and catalyst led to a considerable synergistic enhancement of the methane conversion.

In later studies [26, 27], Ni/Al_2O_3 catalyst was used, and the temperature of the external electrode (controlled by a heater) and the catalyst bed (heated by the discharge and reaction enthalpy) were monitored by an infrared camera. Methane conversion and product selectivity was also compared with thermodynamic equilibrium calculations. At comparatively high GHSV, the equilibrium conversion was only reached with catalyst + plasma. The equilibrium nature of the concentrations was proved by adding $CO_2 + H_2$ and observing the backward reaction to CH_4. Under conditions, where equilibrium was not reached, the rate of CH_4 dissociation was analyzed as a function of temperature as shown in Figure 9.1 for different steam-to-carbon ratios (S/C). The reaction-limited (low T) and diffusion-limited (high T) regions were identified, and the discharge was shown to enhance the preexponential factors, but not the activation energies. This was tentatively explained by the population of vibrational excited states through electron impact. Under equilibrium conditions, the plasma was found to act only as a source of heat.

One serious problem with the approach of combining plasma and catalysts in SR may arise from the rapid deactivation of the catalyst by the carbon formed.

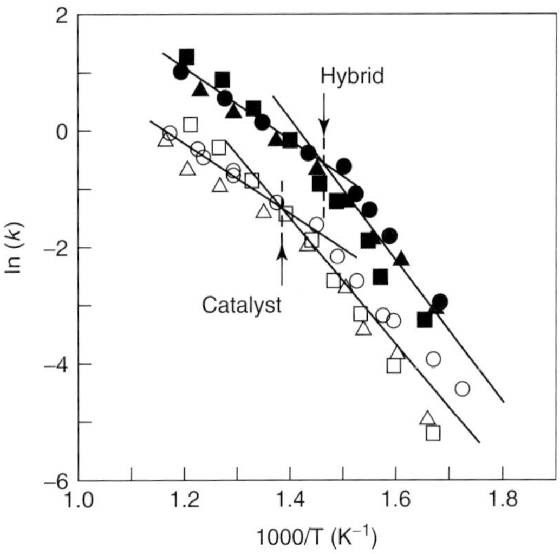

Figure 9.1 Arrhenius plot of the forward CH_4 reaction rate constant [27]. ○ : GHSV = 18 000 per h, S/C = 1; □ : GHSV = 18 000 per h, S/C = 3; △ : GHSV = 10 800 per h, S/C = 1.

9.4.1.2 **Conversion of Higher Hydrocarbons**

The conversion of higher hydrocarbons to hydrogen had attracted considerable attention around the year 2000 owing to its potential application in future fuel-cell-powered vehicles. In a transition from combustion engines to electric cars powered by battery or hydrogen fuel cells, the conversion of gasoline into hydrogen followed by generation of electricity in fuel cells were, for some time, regarded as an attractive intermediate technology. Probably, the slow dynamic response and recent advances in lithium battery development have stalled these activities.

At present, the standby power in heavy vehicles is generated at low efficiency by the idling engine. Diesel reforming and fuel cells could be an environmentally friendly solution. Catalytic reforming is difficult because of the sulfur content and decomposition near the boiling point. Therefore, GA plasma reformers have been developed up to the prototype scale [28]. A reformer efficiency of 92.9% was stated for SR as compared to 74.1% for POX. In principle, the application of plasma could have two advantages over catalysts: immediate activity on start-up and lack of deactivation due to carbon deposition. Another application discussed for plasma-SR of heavy hydrocarbons is the valorization of municipal solid waste [29] or of refining residues [30].

SR of propane was investigated at GREMI in a sliding discharge reactor [31]. The results show some characteristic differences of higher hydrocarbons when compared to methane. The selectivity to SR (amount of carbon in CO/total carbon-containing products) was found to be >60% for H_2O/C_3H_8 up to 19; therefore cracking to smaller hydrocarbons and soot is a relevant reaction path even at high H_2O dilutions. The data were correlated with thermodynamic equilibrium calculations, showing good agreement of the propane conversion and product distribution for an equilibrium temperature of 2000 K. The only free fitting parameter used reflects the fact that the sliding discharge does not fill the reactor space homogeneously, so that only 40–55% of the inlet gas is processed by the plasma.

Microwave plasma [32] was found to yield a good product selectivity in the SR of hexane, with very little carbon produced or unreacted water at O/C = 1. The authors also compared SR of hexane and isooctane [33] and obtained the following results: experimental production rates fall short of thermodynamic equilibrium calculations, hexane can be converted at slightly better energy efficiency, and the energy conversion efficiency is rather low at about 50%, both experimentally and using calculations. Consequently, the total energy balance of a system comprising the plasma reformer and fuel cell generator is negative, even when several optimistic assumptions regarding other system losses (e.g., magnetron efficiency, energy need for water gas shift) are made.

Toluene was investigated as a model substance for reforming tars that are generated in the gasification of biomass [34]. A spark gap followed by a Ni/SiO_2 solid catalyst bed is used. The temperature was fixed at 500 °C. The authors compare all four configurations: (i) empty tube, (ii) catalyst only, (iii) spark gap only, and (iv) spark gap above catalyst. The toluene conversion increased with catalyst to the same extent as with discharge, but syngas formation appeared to be enhanced to a higher extent by the discharge.

9.4.1.3 Conversion of Oxygenates

Oxygenates are an option for the synthesis of hydrogen in small mobile fuel cell power generators. Some of them, such as methanol, may be directly pyrolyzed to syngas, which is discussed in Section 9.4.4.3. With the addition of steam, one might attempt to increase hydrogen yield and to shift the product distribution toward CO_2 by adding water gas shift. Methanol reforming was found to proceed at high conversion and good selectivity to H_2 in a ferroelectric bed DBD using $BaTiO_3$ [35]. When adding steam, the conversion was raised to 100% and the CO could be partly converted to CO_2. Steam addition was found to be beneficial with and without the presence of a Cu–Mn catalyst to the reactor.

The reforming of ethanol is more of a technical interest, since it can be produced directly from biomass. Several research groups use a novel type of reactor in ethanol reforming in which the liquid is in direct contact with the plasma, and the plasma serves two purposes such as evaporation and initiation of the reforming reaction. One arrangement consists in a discharge between two pointed electrodes immersed in the liquid, where the current path is focused by a narrow diaphragm between the electrodes. With this arrangement, ethanol reforming efficiency was found to improve for small pinhole diameters [36]. In a 10-mm spark operated at the liquid surface of ethanol/water mixtures [37], the simultaneous reforming to H_2, CO, and CO_2 was observed. However, undesirable by-products such as methane, ethylene, and acetylene were also observed. The plasma power was mainly consumed for reactant evaporation and to a lesser extent for providing the enthalpy driving the (endothermic) reforming reaction.

Another process with direct plasma formation in the liquid reactant mixture was termed *glow discharge plasma electrolysis*. In the U-shaped glass tube shown in Figure 9.2, two electrodes were inserted, one stainless steel plate and one tungsten needle. Methanol [38] or ethanol [39] was used for reforming, and the alcohol was made conductive by adding alkaline electrolyte. At a breakdown voltage of about 300–500 V, a gas envelope formed around the tungsten electrode, within which the plasma was ignited. The addition of water favored CO_2 formation; however, the main reaction products were formaldehyde (in the case of methanol) and acetaldchyde (in the case of ethanol).

Coevaporated ethanol and water were used in a reactor with a spark gap inside a catalyst bed at 420–520 °C [40]. Three catalysts were tested: Pd, Rh, and Pt on CeO_2. The activity of the catalysts under thermal conditions was very low, with conversions <10%. Comparatively low voltages (130–600 V) and power levels (0.6–2 W) were required to increase the conversion to 60–70%. The authors maintain that the enhancement is not due to a discharge but solely caused by the electric field.

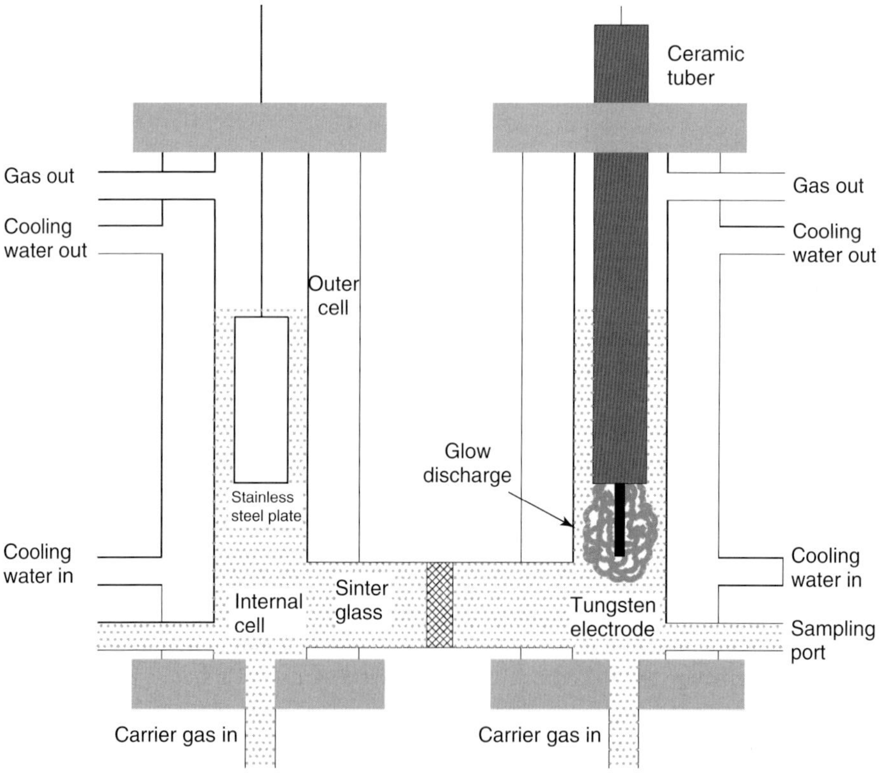

Figure 9.2 Reactor for the glow discharge plasma electrolysis of methanol and ethanol [39].

However, it has to be borne in mind that such spark gaps can form very small and poorly visible corona discharges or just hot spots at the tips.

Recently, the SR of ethanol was reported in a modified point-to-plate pulsed corona directly from liquid precursor [41]. Reformer efficiencies (Eq. (9.13)) up to 65% were reported based on H_2 and CO in the reformate. The efficiency could even be improved by raising the gas pressure up to 2 bar (absolute).

9.4.2 Partial Oxidation

The POX of hydrocarbons in a plasma-assisted reactor has been intensively investigated over the past decades because of the possibility of realizing compact systems with fast dynamic responses for mobile applications. Being an exothermal reaction, the chances are high that both requirements might be met in a plasma reactor that produces active radicals instantly on start-up. Extensive research was carried out at the MIT, beginning in the 1990s with a thermal plasma torch, the plasmatron. Results until about 2005 have been reviewed in Fridman's book [6]. In later work, the initial arc plasma design, which essentially provides a thermal plasma source, was abandoned in favor of GA varieties, where a nonequilibrium arc is maintained by an appropriate control of the gas flow.

Technical applications range from the hydrogen production for fuel cells to the enhancement of the operation of internal combustion engines by providing hydrogen as a fuel additive. The latter application was proposed at the MIT, and the performance enhancement was also shown in recent work [42]. A further application might be cleaning of diesel particulate filters by a hydrogen-rich reformate.

9.4.2.1 Conversion of Methane

The POX of methane was investigated partly as a model substance for the conversion of liquid hydrocarbons to hydrogen or simply for the use of natural gas. Depending on the reaction conditions, however, higher hydrocarbons or soot are formed in varying quantities. In an RF-plasma reactor at reduced pressure, a wide range of polycyclic aromatic hydrocarbons were formed [43] or, in another recent example, methanol, formaldehyde, and formic acid were produced in a microplasma reactor [44] with high selectivity to the oxygenates of up to 80%.

The simplest system investigated uses methane and oxygen in a reactor without catalyst. In a microwave plasma torch at a slightly reduced pressure of 933 mbar, maximum selectivities to CO and H_2 near 100% were reached at a stoichiometric feed gas ratio [45]. Lower O/C ratios favor acetylene, and higher ratios favor CO_2 by-product formation. As an alternative to pure oxygen, air was studied as oxidant, thus reflecting conditions more closely related to practical applications. Plasma sources used include spark gaps [46, 47], or GAs [48, 49]. Increasing the number of reactor stages from one to four enhanced the conversion of CH_4 and O_2 from about 20 to 70%, however, at the penalty of 50% increase in power consumption per amount of hydrogen produced [49]. Low O/C ratios favor low energy cost for hydrogen production and low CO_2 formation, however, usually at the cost of incomplete methane conversion.

Several studies focus on the combination of plasma and a catalyst. A spark plasma with Pt–Rh commercial monolith catalyst yielded thermal reformer efficiencies of about 70%, which could be increased to 77% by applying heat insulation [50]. The contribution of the catalyst to the conversion was not investigated separately.

In a DBD, the activity of a Ni catalyst on γ-Al_2O_3 was studied by comparing the reaction with reduced and oxidized catalyst as well as with the alumina support only [51]. Temperature-programed reduction/oxidation results (TPR/O) of the catalyst were correlated with the temperature of activity onset. The result is shown in Figures 9.3 and 9.4. The activity of the catalyst was found to depend on its ability to be reduced and oxidized: The temperature at which the catalyst becomes active and induces CO_2 formation (Figure 9.3) corresponds to the peak in TPR/O (Figure 9.4) between 300 and 400 °C. It has to be born in mind that at 400 °C, the equilibrium concentration of CO_2 is high and that only above 700 °C CO is favored [25]. A detailed study in the same reactor with Ni–Mg amnesite clay catalyst showed that the discharge leads to some methane conversion below the catalyst extinction temperature, but that the product composition is determined by the chemical equilibrium, when the catalyst is active. Some increase in methane conversion with increasing plasma power is observed, but this can be explained by a local overheating of the catalyst surface by ~30–40 °K [52].

In a GA plasma reactor followed by a Ni–γ-Al_2O_3 catalyst bed, the catalyst was found to increase H_2 concentration, CH_4 conversion, and reformer efficiency by 20, 31, and 25%, respectively [53]. Methane conversion was about 70% and the reformer thermal efficiency about 50%.

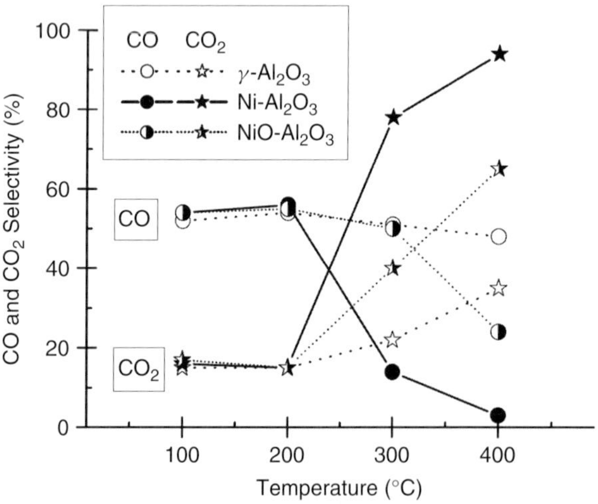

Figure 9.3 Selectivity to CO and CO_2 over γ-Al_2O_3, Ni/γ-Al_2O_3, and NiO/γ-Al_2O_3 in the presence of the discharge [51].

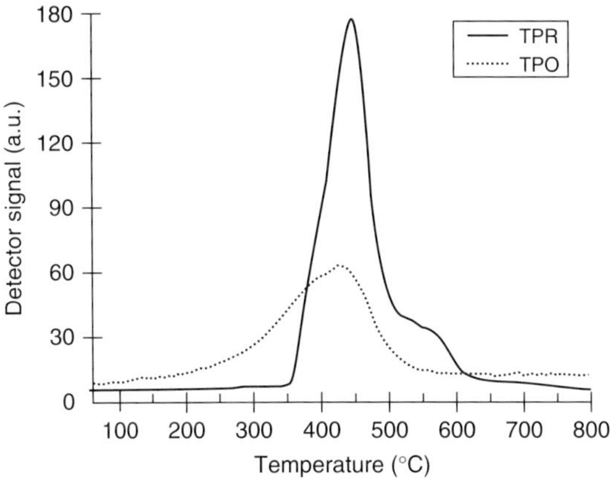

Figure 9.4 TPR profile of $NiO/\gamma\text{-}Al_2O_3$ (solid line) and TPO profile of $Ni/\gamma\text{-}Al_2O_3$ (dotted line) [51].

9.4.2.2 Conversion of Higher Hydrocarbons

Propane was studied in a spark reactor followed by a Pt–Rh commercial catalyst [54]. Fuel conversion, hydrogen yield, and thermal efficiency of the reactor were significantly dependent on reformate gas temperature, this being in close correspondence with equilibrium calculations. A good reformer thermal efficiency was achieved for methane above 750 °C at 72% as shown in Figure 9.5. With propane,

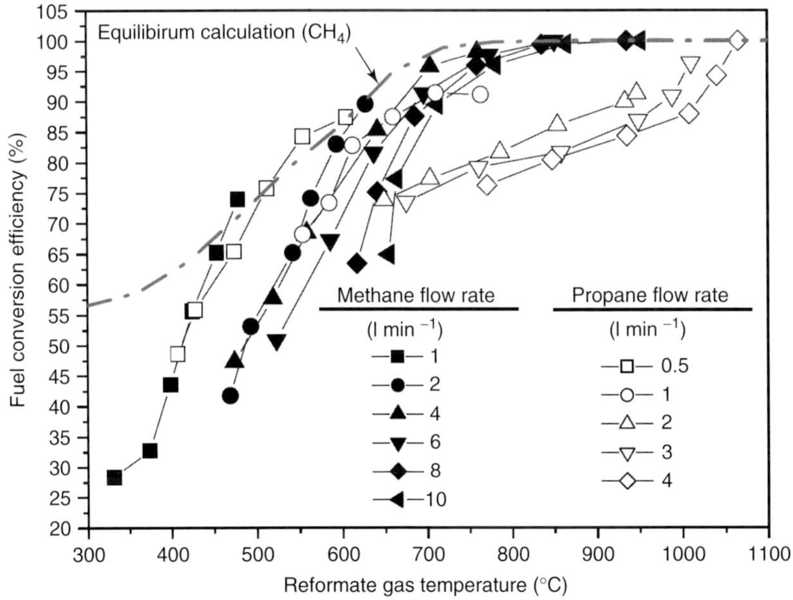

Figure 9.5 Conversion efficiencies of methane and propane compared to equilibrium values (dashed line) [54].

the results were less favorable, with 58% efficiency at 850 °C. A comparison of different catalysts shows that a high Pt/Rh ratio and a high density of catalyst cells favors good conversion. The selectivity to H_2 and CO, however, tends to be better with catalysts yielding less conversion [55].

Since Fridman's review, the work on plasma reforming of diesel fuel at Drexel University has led to a well-characterized GA reformer [56]. Two different concepts to stabilize the nonthermal plasma were used: the reverse vortex flow gliding arc reactor (RVF-GA) and the GA-plasmatron. Both reactors performed equally well. The product yield was collated with equilibrium calculations in a 400 °C adiabatic flame: *n*-tetradecane was used as diesel surrogate and owing to its composition, the highest CO and H_2 yields were expected at an O/C ratio near one. While under equilibrium conditions yields of almost 100% are expected, experimental values reached about 60% for the CO yield and 40% for hydrogen. The missing products were identified as light hydrocarbons from incomplete reforming. The plasma power was set to only 2–5% of the total chemical energy of the processed fuel. If the light hydrocarbons are included in the product balance, the reformer thermal efficiency, shown in Figure 9.6, is 80–90%. Water formation was higher than predicted for the chemical equilibrium. This and the observed hydrocarbons were explained by the short dwell time in the active reaction zone. It is known that the POX proceeds in two steps: the first is rapid oxidation until all free oxygen is

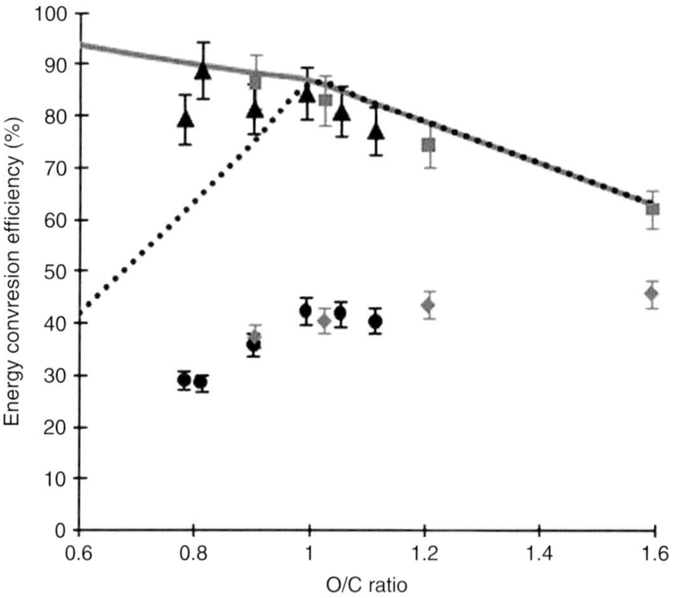

Figure 9.6 Thermodynamic equilibrium and experimental energy conversion efficiency of $H_2 + CO$ (dotted line, ●, ♦) and $H_2 + CO +$ light HCs (solid line, ▲, ■); black symbols: RVF-GA, gray symbols: GA plasmatron [56].

consumed and the second is the actual reforming by the water formed as a result of oxidation. In conclusion, there is a scope for optimizing the reactor design, but as yet, the reformate is well suited to fuel solid oxide fuel cells (SOFCs), which can handle methane and CO. At the Kurchatov Institute, it was shown that microwave plasma can be more efficient than thermal energy in kerosene POX [57].

The research is currently followed by prototype development from several research groups and commercial enterprises [28, 58, 59]. During the AIChE 2009, spring meeting session #17 was addressing plasma reformers and fuel cell systems [60]. One current main goal is the development of a shipborne electricity supply [59].

Oxygenates were also subjected to reforming by POX. In a spark discharge reactor, dimethyl ether was converted at an efficiency of only 4.2% [61]. In a GA reformer, a mixture of glycerol and natural gas was converted to syngas using oxygen-enriched air [62]. An energetic reformer efficiency of 64% was obtained with a reformate output (LHV) of 2.8 kW at 50 W plasma power.

9.4.3 Carbon Dioxide Dry Reforming

9.4.3.1 Reforming of Methane to Syngas

The research in CDR in nonthermal plasmas was reviewed some years ago, with the emphasis on DBD, corona, and especially on the combination of the plasma with catalysts [63]. These discharges often lead to considerable oxidative methane coupling and are discussed in the next section. The characteristic of most work is a much lower conversion of CO_2 as compared to CH_4. Equal selectivities to H_2 and CO were found, for example, in an investigation using a spatially homogeneous atmospheric pressure pulsed glow discharge [64]. In this configuration, the conversion was about 25% at stoichiometric composition and chemical energy efficiency (enthalpy gain/power input) was about 20%. As a reason for the efficient CO_2 activation, the authors discuss the high vibrational excitation available in the short (tens of nanoseconds) plasma pulses. In a later study [19], they showed that the efficiency is improved by using lower pulse energy, raising the chemical energy efficiency to over 30%. A C_2 selectivity of about 20% was reported for this system.

In a study at Bochum University, CO_2 and CH_4 were reacted in a microwave plasma at a scale large enough for allowing an assessment of economic feasibility [65–67]. The authors used a CYRANNUS® plasma source that supplies microwave power at 2.45 GHz to a cylindrical resonator through a slotted annular waveguide. A fairly uniform plasma with a diameter of about 13 cm can be produced using this source at power levels around 5 kW. The reactor, without the annular waveguide around the plasma, is shown in Figure 9.7.

Gas temperatures were relatively high, that is, up to 800 °C. Maximum conversions and syngas yields around 95% were obtained at O/C near one, when 70% Ar was added to the process gas. Soot formation is a problem that limits the range of O/C ratios, which permit a stable process. Ar was reduced to 6% of the total gas

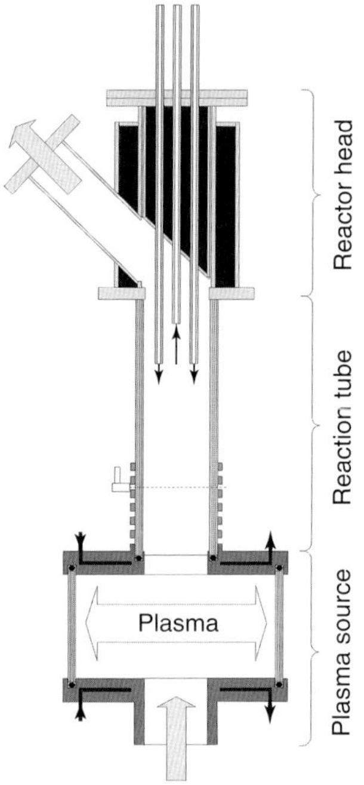

Figure 9.7 Technical scale plasma reactor [67]. The diameter of the plasma chamber is 130 mm.

flow, although at increased soot formation. Emission spectroscopy revealed C_2, C_3, and C_4 species as intermediates in the plasma; they were, however, absent in the product stream. The energy requirement was much lower than values reported for a DBD, meaning that about 60% of electrical energy was recovered as chemical enthalpy. The authors presented a concept study for the technical implementation of microwave plasma reforming [68]. Process control with optical emission spectroscopy was demonstrated. The appearance of emission from H_α and C_4 clusters was shown to be a good indication of soot formation, which should be avoided in a stable process by adjusting the reactant ratio. Cost estimation was carried out taking into account both operating costs and capital depreciation. On the basis of a possible scale-up to 75 kW sources running at 915 MHz, the cost was still considered prohibitively high for profitable industrial application; however, it was out only by about a factor of two.

In a cold plasma jet with a gas temperature of 630 °C [69] and followed by a Ni/γ-Al_2O_3 catalyst, similarly high efficiencies were reported, with an SER of 150 kJ mol^{-1} syngas. The plasma jet was operated at an AC frequency of 20 kHz and 50% dilution of the reactants in N_2.

A thermal arc also appears to work rather well in methane CDR with a chemical efficiency of about 55% [70]. The authors compare the efficiencies reported in several studies, confirming the better results in homogeneous discharges such as pulsed microwave plasma or abnormal glow. Thermodynamic equilibrium calculations confirm the significance of sufficiently high temperatures for methane dry reforming [71]. However, the model does not include the formation of higher hydrocarbons or acetylene. For achieving almost complete conversion, 800 °C was required, and at lower temperatures, an increased reaction to water and solid carbon was predicted. Also, as shown in Figure 9.8, the CH_4/CO_2 ratio should not exceed 1 for the highest hydrogen yields and to prevent water formation.

The synergistic effect of a plasma combined with a catalyst may be expressed as the ratio of $\text{conversion}_{(\text{catalyst+plasma})}/\text{conversion}_{(\text{catalyst})} + \text{conversion}_{(\text{plasma})}$. This was analyzed for a DBD reactor using a Ni/Al_2O_3 catalyst [72, 73]. The ratio was found to be about 1, ranging from 0.5 to about 1.5. The catalytic reaction gradually started above 300 °C, attaining conversions of about 40% for CH_4 and 20% for CO_2 at 550 °C. As an alternative to the packed bed, a fluidized bed reactor was studied, although without any significant improvement of the catalyst activity. Fluidized bed catalyst reactors are well established in chemical processing, and although they seem, in principle, well suited for plasma–catalyst combination, they have received only little attention so far. Another example is mentioned in Section 9.4.4.2.

In a study aimed at biogas reforming, small traces of H_2S were shown to be effectively decomposed to sulfur and to actually enhance the CDR reaction. This might turn out to be an interesting alternative to desulfurizing and separating CO_2 from biogas before use [74, 75].

A review on methane CDR in different cold and thermal plasmas was published at the time of writing this chapter [76]. The authors present a compilation of the

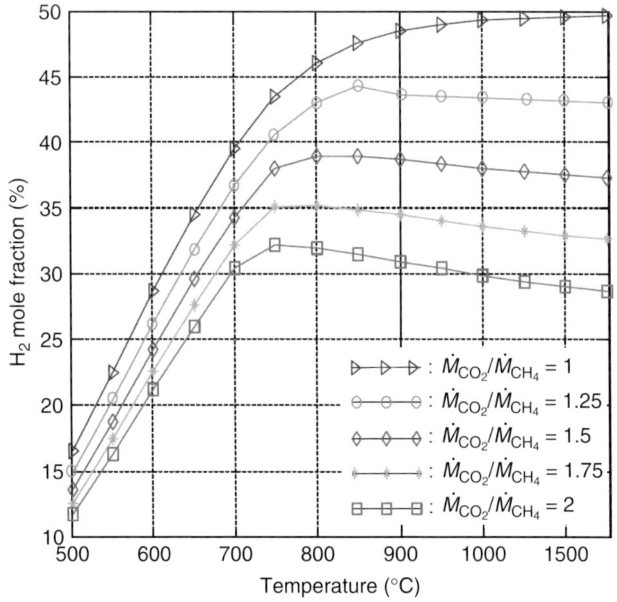

Figure 9.8 H_2 molar fraction in chemical equilibrium calculations of CDR. The missing hydrogen at $CO_2/CH_4 > 1$ is found in H_2O [71].

SER and reformer efficiencies obtained in 13 different studies. From this, it may be concluded that some typical cold plasmas such as DBD or corona discharges are much less effective in CDR than thermal plasmas. The most successful plasmas, however, are dense nonequilibrium plasmas.

9.4.3.2 **Coupling to Higher Hydrocarbons**

While in POX usually only a small percentage of the reactants are converted into higher hydrocarbons, selectivities to higher hydrocarbons and oxygenates in the range of 20% were reported by several authors in CDR. This may appear as a highly desirable situation; however, one has to be careful in distinguishing between a valuable product and undesired by-products. Often the cost of product separation counteracts the success of a seemingly attractive direct synthesis method. Plasma chemical processes are particularly prone to an unspecific product distribution because of the high reactivities of the plasma radicals. In fact, in catalysis research, the effort during the past years was directed toward achieving selectivities to the desired products close to 100%, quite far from what can be seen in most work on plasma catalysis.

It is, however, interesting that in DBD and corona-CDR, the carbon chain growth is favored over the dehydrogenation of the C_2 hydrocarbons in the first C–C coupling step. In the earlier work at ABB, Switzerland, the selectivity to acetylene was only 1–2%, and to C_4 and higher hydrocarbons, almost 50% [77]. In a computational study, the authors identified hydrogen attachment to unsaturated species as part of the main chain growth mechanism [78]. More recent research confirmed the preferential chain growth. In a study comparing a plain Al_2O_3 carrier in a DBD with noble metal catalysts [79], the catalyst had only little effect on the conversion and product selectivities between 120 and 300 °C. In a DBD, in a cordierite monolith with Ni/La_2O_3 catalyst on an alumina washcoat, the conversion was only found to double on increasing the temperature from 20 to 500 °C and the selectivity to higher hydrocarbons was almost as high as that to hydrogen [80]. This was probably due to the catalyst not being active in the temperature range investigated. During CDR in a GA discharge, on the other hand, dehydrogenation was found to be more prominent [81]: The selectivity to acetylene was always higher (up to 37% at $CH_4/CO_2 = 2$) than to ethylene (up to 20%), and no noticeable ethane was produced.

9.4.3.3 **Reforming of Higher Hydrocarbons**

Only limited research effort focuses on the CDR of higher hydrocarbons. At AIST, Ibaraki, Japan, the reactivity of methane, propane, and neopentane was compared in a ferroelectric packed bed reactor [82]. Hydrocarbon conversion increased in the sequence $CH_4 \ll C_3H_8 \ll C_5H_{12}$, reflecting the different covalent bond strengths. In addition, the conversion was found to increase with temperature, especially in the case of methane. Hydrogen yield was much higher for propane and neopentane than for methane. The latter apparently led to water formation, since the carbon balance was better (less soot produced) for methane than for higher hydrocarbons. These results may, however, not be representative of practical conditions, since the process gases were highly diluted in N_2. Unlike a normal DBD, the ferroelectric packed bed reactor only leads to rather limited C–C coupling of methane, and hydrogen abstraction is regarded as a dominating step [83].

The oxidative dehydrogenation of ethane was investigated in a pulsed corona over rare earth catalysts with the objective to upgrade ethane as the main constituent of natural gas [84]. The selectivity to acetylene was higher than to ethylene; however, an energy input of 1500 kJ mol^{-1} was necessary to achieve 72% conversion, which approximately corresponds to the heat of ethane combustion.

9.4.4 Plasma Pyrolysis

Plasma pyrolysis of methane is probably one of the best investigated plasma chemical processes. A detailed macrokinetic analysis was made on the basis of results obtained about 50 years ago [5] Already at that time, two distinct regimes in the methane plasma chemistry were recognized: (i) low SEI with relatively low conversion and a high selectivity to alkane or alkene chain growth and (ii) the conditions of high SEI, where hydrogen abstraction dominates and the main products are acetylene or soot.

Low pressure is the obvious way to generate cold and homogeneous plasma. Under such conditions, higher hydrocarbons are formed as intermediates to solid polymer formation. In an inductively coupled plasma, the above-mentioned dependence of methane conversion on SEI was also shown [85]: At an SEI below 10 eV, conversion depends on electron impact and is proportional to the SEI, and above 10 eV, the conversion is boosted by secondary reactions of plasma-generated radicals.

9.4.4.1 Methane Pyrolysis to Hydrogen and Carbon

The production of hydrogen and carbon black with a thermal plasma torch was developed in the 1980s by Kvaerner Engineering S.A., Norway [86]. However, after tests with a 3 MW pilot plant, the activities have apparently been abandoned. Simultaneous production of carbon black and hydrogen was studied using the CYRANNUS microwave plasma source mentioned in Section 9.4.3.1 [87]. The plasma was operated at reduced pressure of 100–300 mbar, and noble metal catalysts were introduced for enhancing the conversion efficiency. The carbon black was characterized by high-resolution TEM, specific surface measurement, and thermal and electrical analysis. TEM results are shown in Figure 9.9. The properties and particle size were found comparable to classic furnace black. Nevertheless, the energy requirement was not analyzed.

In a high flow rate microwave plasma source (quartz cylinder through waveguide), methane was decomposed into hydrogen and carbon at 99% conversion and 100% selectivity to H_2 [88, 89]. Microwave power of 577 g H_2 per kWh of is reported, which amounts to an SER of 12.4 eV. Given the endothermic nature of the reaction (Eq. (9.5)), this leaves the source of the reaction enthalpy unclear. Recently, another Norwegian company named GasPlas [90] promotes hydrogen and carbon black

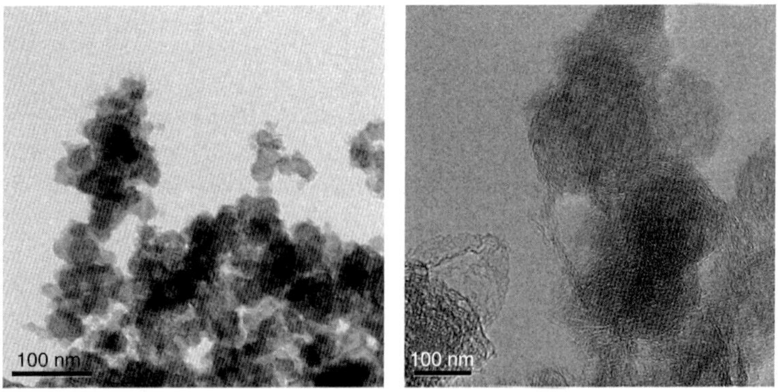

(a) (b)

Figure 9.9 (a) TEM micrograph of plasma-produced carbon black. The magnified image (b) shows the concentric structure of the particles [87].

production from natural gas in a cold microwave discharge, but no results were published so far.

Yet another possible future application of methane pyrolysis consists in closed cycle oxygen supply in future manned space missions. Oxygen can be recuperated from CO_2 by methane formation in the Sabatier reaction ($CO_2 + 4H_2 \rightarrow CH_4 + 2H_2O$). If the latter is pyrolyzed to hydrogen, the only waste will be the resulting carbon. To this end, a microwave plasma was investigated [91]. Plasma is an attractive alternative to the catalytic route in this case, since a compact reactor is needed for which long-term reliability is more important than energy efficiency.

9.4.4.2 **Production of Acetylene**

As in CDR (Section 9.4.3.2), cold plasma pyrolysis tends to lead to hydrocarbon chain growth. A large number of papers, even as per today [92], report alkanes, alkenes, and acetylene in varying concentrations. However, it has to be pointed out clearly that (i) the C_2–C_6 alkanes are common constituents of natural gas or crude oil and (ii) the cost of recovering desired compounds, for example, alkenes from a typical product mixture may be prohibitive. Therefore, this work is not reviewed here in further detail. Acetylene, on the other hand, may be obtained with comparatively small amounts of by-products.

Since the commercialization of the Hüls process, there is a clear benchmark for the production of acetylene from methane [17] with 70% conversion and 75% selectivity at an SEI of 3 eV and an SER of 11 eV. In the 1990s, the Kurchatov Institute claimed a lower SER of 6 eV in nonequilibrium microwave plasma [93], which, however, has not been confirmed so far. At RITE, Kyoto, a high-frequency pulsed plasma was developed, and the maximum efficiency for acetylene and hydrogen coproduction was found in a point-to-point spark discharge [94]. Acetylene is obtained at 9.2 kWh kg^{-1}, which amounts to 8.9 eV, although at a rather low methane conversion at 23.5%. Comparing their process with the Hüls arc and thermal synthesis through POX (at present the most widespread method), the authors conclude that the high-frequency pulsed plasma may be the favorable method [95]. This would require recycling of unreacted methane, given a rather low methane conversion of 39% and a high acetylene selectivity of 83%.

In a pulsed microwave plasma operated at a reduced pressure of 30 mbar, acetylene formation was optimized [96]. In this study, both mass spectrometry and gas chromatography were used for analysis, intended at minimizing the risk of measurement errors. As shown in Figure 9.10, it was found that a high SEI and, consequently, high methane conversion favor acetylene formation at low energy cost.

This is in agreement with results mentioned above and also correlates with the gas-phase temperature as measured by the rotational distribution in the C_2 (Swan) and H_2 (Fulcher) bands [97]. In a reactor with 1200 W peak power, the SER could be reduced from 900 eV to about 20 eV by increasing the SEI from 1.6 to about 7 eV. Time-resolved OES during the microwave plasma pulses showed the instant presence of CH emission from CH_4 excitation and the gradual increase of emission from reaction products (H_α and C_2).

The experiments were also scaled up to a 30 mm diameter quartz reactor running at 1500 W continuous power [98]. A reduction of the SER for C_2H_2 to 10 eV was achieved at a pressure around 120 mbar. This translates to about 3.3 eV for the

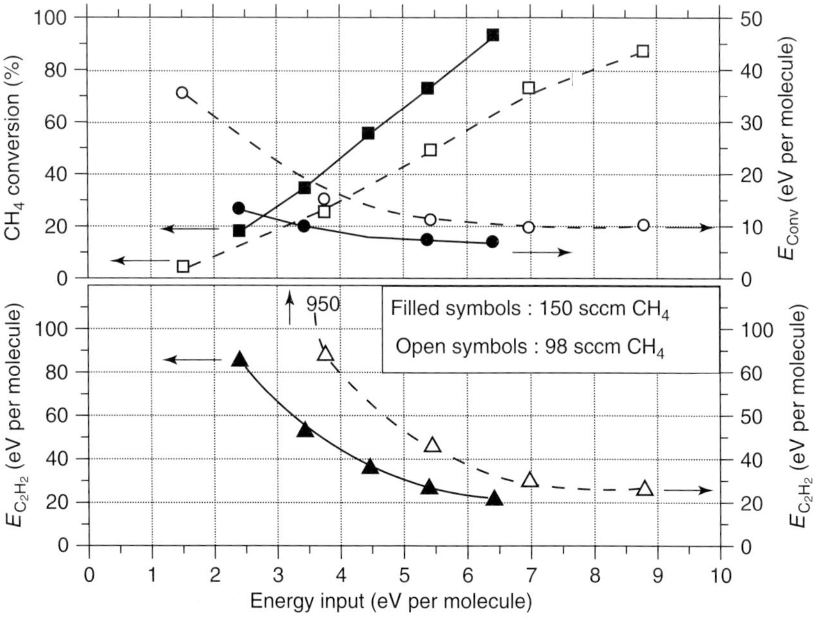

Figure 9.10 Methane conversion, energy requirement for conversion (E_{conv}.), and for acetylene formation (SER) as a function of the SEI in a pulsed microwave discharge [96].

coproduced hydrogen. From the blackbody radiation of the soot particles formed, the gas temperature was determined to be 2600–2700 °C. Soot formation could be suppressed by applying a pulsed mode.

A potentially interesting reactor for the combination of plasma and catalyst is the operation of a GA plasma in a spouted bed of catalyst particles shown in Figure 9.11. The local injection of the process gas at the bottom of the reactor between the electrodes will lead to a good control of the GA and will agitate the particle bed.

A study using different supported metal catalysts found some suppression of soot formation by the spouted bed [99]. Noteworthy is the reduced selectivity to acetylene in favor of ethylene and ethane in the presence of Pt catalyst. Pt is known to rapidly absorb and recombine atomic hydrogen, thus reducing the hydrogen abstraction to acetylene. Recently, similar experiments in a spouted bed were reported [100]. The authors attribute the decreased acetylene formation in the presence of noble metal catalysts to a hydrogenation reaction.

Further alternative products from methane dehydrogenation might be aromatic compounds, which themselves could be used as liquid fuels. With a suitable combination of microwave plasma excitation and surface catalysis [101], the selectivity to benzene was found to be up to about 100 times higher than to aliphatic C_6 compounds. The formation of aromatics was catalyzed by the carbon deposit formed

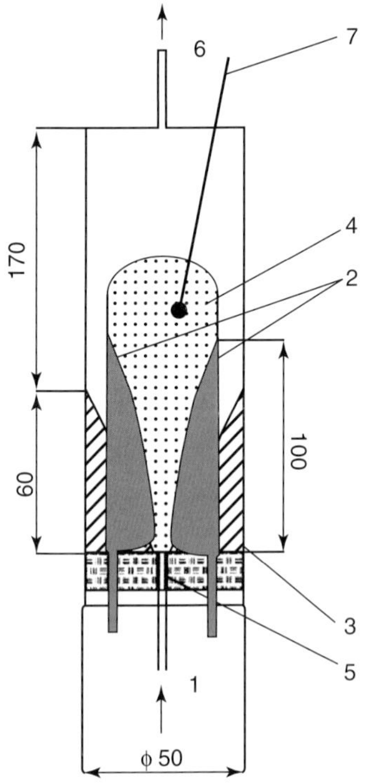

Figure 9.11 Gliding arc plasma reactor with spouted bed [99]. 1, gas inlet; 2, electrodes; 3, ceramic lining; 4, spouted bed of catalyst; 5, nozzle; 6, gas outlet; 7, thermocouple.

during the first minutes of operating the discharge. A similar product distribution was reported in a streamer discharge combined with a HZSM-5 catalyst [102].

9.4.4.3 Pyrolysis of Oxygenates

In certain cases, oxygenates may be desirable sources of hydrogen in small mobile applications. Thus, they may be used as liquid fuels for transportation or for replacing batteries, when recharging from an external electrical source is not feasible. Usually, they are easier to convert into hydrogen than C_xH_y hydrocarbons owing to their better reactivity. The simplest example is methanol, which directly can be split into syngas

$$CH_3OH \longrightarrow CO + 2H_2 \quad \Delta H = 90.5\ \text{kJ mol}^{-1} \quad (0.94\ \text{eV}) \qquad (9.15)$$

Methanol, however, is itself produced from syngas and therefore must only be considered an intermediate energy store. Even if it were energetically quite attractive, there would still remain considerable environmental problems in the use of methanol as automotive fuel, since it will readily mix with and contaminate ground water in the case of accidental spilling. It is therefore not surprising that work published on methanol is rather scarce.

Methanol pyrolysis was studied in a ferroelectric packed bed DBD, with and without the addition of a Cu–Mn catalyst [35]. Maximum conversion was 96 and 92%, respectively. The catalyst, however, led to some water formation, reducing the H_2 yield from 100 to 80%. In a later study, the authors studied the conversion in three different DBD arrangements: (i) empty discharge gap, (ii) alumina packed bed, and (iii) barium titanate (ferroelectric) packed bed [103]. The highest conversion of 92% was achieved in alumina, although at an SER of 15 eV, which could be lowered to 8.3 eV by decreasing the conversion to 76%.

Glycerol, which is currently available as a by-product of biofuel fabrication, is also a potential candidate for an intermediate liquid fuel for fuel cells. Plasma reforming of glycerol was demonstrated in a point-to-plate corona reactor, featuring direct liquid injection in the needle [104]. Pure glycerol could be converted to syngas at 80–90% conversion and 15% selectivity to C_2. The addition of 8% water reduced the conversion to 60–80% at about 10% selectivity to C_2. The SER of syngas production ($H_2 + CO$) was about 3 eV, corresponding to a reformer efficiency of about 50%.

9.4.5
Combined Processes

The introduction of a third reactant evidently leads to an increased complexity of the reaction pathways, and it becomes considerably more difficult to determine the relative contribution of the different reactions. With two oxygen-containing species, a careful analysis may be needed to correctly determine which species is consumed and which is just a by-product.

9.4.5.1 Autothermal Reforming of Methane

The addition of oxygen to SR was investigated in a sliding discharge [105]. However, an increased methane conversion was not observed, and the oxygen was found to cause increased water formation. The addition of steam to POX was studied in arc plasma, and an augmented hydrogen yield was reported [106]. However, $H_2/CO = 6.6$ in the reformate was rather high at 99% methane conversion, and other carbon-containing products (CO_2, C_2H_2) were seen only in small amounts, indicating that there must have been considerable soot formation.

The combined oxygen-SR was examined in a DBD with or without Ni catalyst [107]. In plasma reforming without catalyst, the steam concentration at the reactor output was higher than at the entrance. Therefore, H_2O has to be regarded as a product rather than as a reactant. Over Ni catalyst, the steam increased the hydrogen yield through WGS, provided the temperature was sufficiently high for the catalyst to become active, and full oxygen conversion was achieved. At lower temperatures, the addition of steam decreased the conversion of methane and oxygen. In this study, some critical points in operating a reactor with combined plasma and catalyst are also described: depending on the amount of catalyst used, the plasma can ignite the reaction locally, leading to the formation of a hot spot, where the reaction will proceed even if the plasma is shut off. Also, after some time onstream, the active catalyst volume may be reduced by local coke deposition.

One method to analyze chemical reactions in more detail consists in using isotopes in the reactants. This was done by combining CH_4 with deuterated water D_2O in SR and combined reforming [108]. In SR with $S/C = 2$, 67% H_2 and 25% HD were found, whereas in the combined reforming at $CH_4/O_2 = 2$ and higher, no significant HD was formed, confirming the missing steam conversion in combined reforming. The DBD was operated at 130 °C under three conditions: gas phase, alumina beads, and Ni/Al_2O_3 catalyst, and no significant effect of the solid bed or catalyst was reported. Kinetic modeling and a sensitivity analysis showed that C_1 oxygenates are important intermediates in plasma-activated POX.

9.4.5.2 Autothermal Reforming of Liquid Fuels

The work on plasma reforming for mobile applications published until 2006 was reviewed at the CEP, Paris [109]. The processes include SR, POX, and combinations thereof. Thermal and nonthermal reactors of some research groups were reviewed, and a compilation of the SERs, fuel conversions, and reformer efficiencies was given. The graphic summary of SER values reported is reproduced in Figure 9.12. Numerical modeling supporting the experimental work was also reviewed. In conclusion, dense nonthermal plasmas were considered the most efficient.

Following their review, the CEP published the results of an extensive study in gasoline reforming carried out in cooperation with Renault [110]. For the experimental work, a compact reactor with a nonthermal plasma torch followed by a reaction zone with 25 mm diameter was employed. A schematic view is reproduced in Figure 9.13.

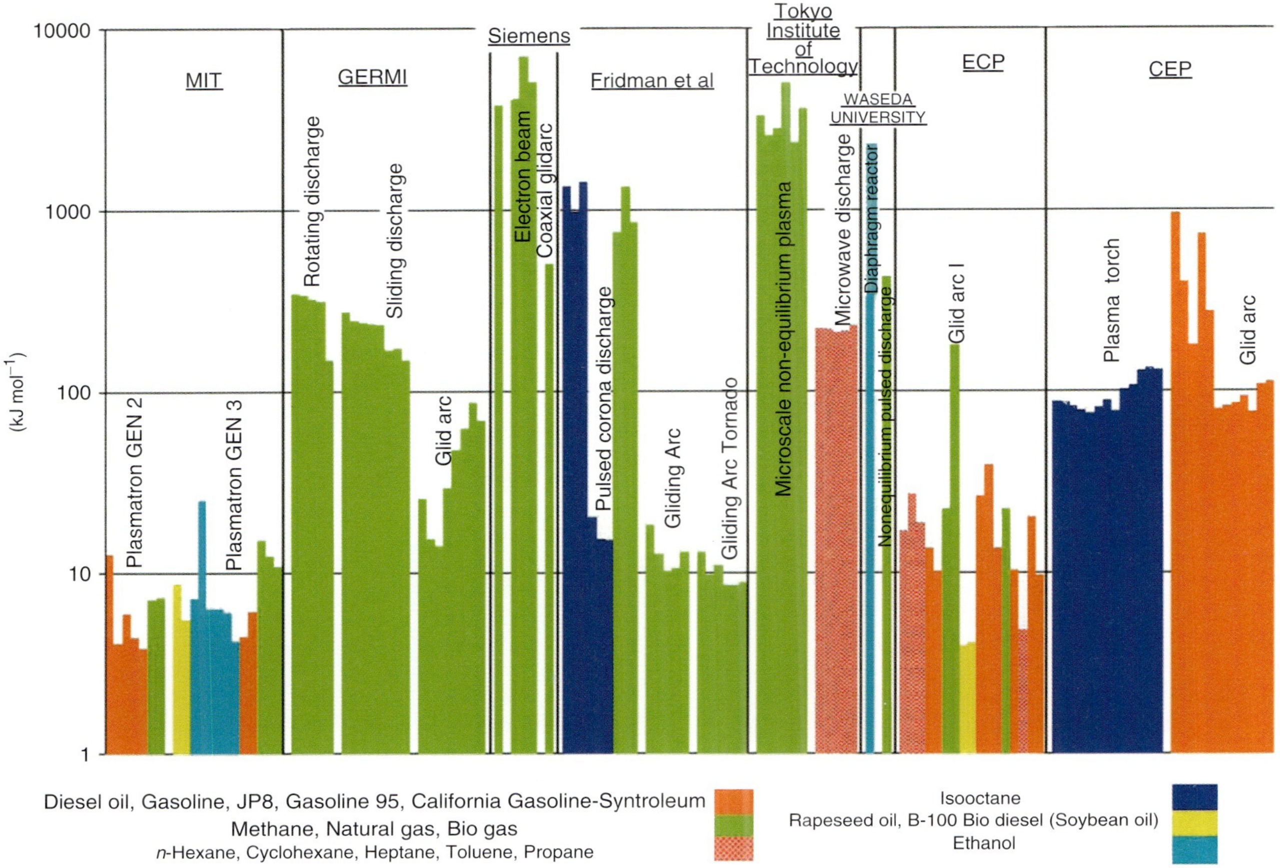

Figure 9.12 Graphic compilation of 121 SER values from the plasma reforming of a wide range of liquid fuels used for transportation in different plasma reactors [109]. Note the log scale.

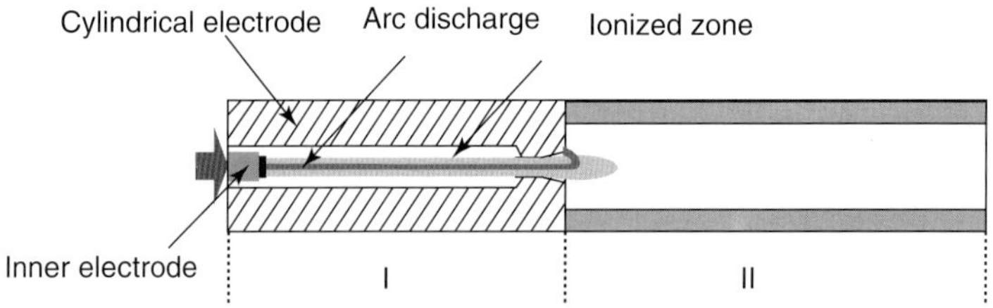

Figure 9.13 Nonthermal plasma torch reactor for the reforming of gasoline [110–112]. "I" refers to the plasma zone in the modeling and "II" to the postdischarge zone.

In this reactor, the arc was controlled by the gas flow, and the final electrode geometry allowing stable plasma operation resulted from a design optimization. The highest gasoline conversion rate was found at $O/C = 2.4$, corresponding to a reformer efficiency of 35%. The highest efficiency of 48% was achieved at $O/C = 1$ and a gasoline conversion of 81%. The addition of steam up to $H_2O/C = 0.2$ enhanced hydrogen formation, while at raised steam addition, the H_2 yield was reduced and increasing amounts of CO_2 and CH_4 were produced. This reactor combines the advantages of compactness, low electrode wear, and catalyst-free operation.

The reformer was addressed using three different mathematical models [111]: (i) thermodynamic modeling with minimization of Gibbs free energy allows determining the reactant conversion and product distribution finally available for long residence times, (ii) a perfectly stirred reactor is capable of showing how far conversion will proceed in a spatially confined reactor, and (iii) an adiabatic plug flow reactor demonstrates the spatial evolution of intermediates and products.

The temperature needed to achieve 98% conversion to syngas was shown to be constant at 1200 K for $H_2O/C + O/C > 1$. At higher oxidant concentration, the conversion temperature drops sharply when H_2O or CO_2 is formed. The thermodynamic model also provided predictions of the theoretical limit pertinent to reformer efficiency. From the modeling, it was concluded that the highest efficiency (up to 86%) was available in SR. However, SR is also the slowest reaction and remains incomplete in a compact reformer.

In a later paper, the authors combined the three models in order to model their plasma reformer as realistically as possible [112]. The arc reactor was modeled as two parallel reactors, to take into account the fact that only a part of the reactant gas stream passes through the discharge zone. This was modeled as a perfectly stirred reactor, into which the plasma power was introduced as heat. The postdischarge zone was modeled as a plug flow reactor and the two parallel gas streams were instantly mixed at the border between the plasma torch and the postdischarge zone. The authors point out that the flow ratio arc/bypass is not known a priori, but that it did not markedly affect the final result, because thermal equilibrium was rapidly established after mixing. The model did not take into account the ionic reactions specific to a gas discharge. This might be its main limitation. The parametric study predicts the best reformer performance at $O/C = 1.1$ without steam and electric power input of 20–25% fuel LHV. The model confirms some experimental findings, such as the relatively high reforming performance of rather small and dense plasma sources as compared to, for example, the DBD.

9.4.5.3 **Reforming with Carbon Dioxide and Oxygen**

CDR and POX were combined for assessing the reforming of natural gas containing high amounts of CO_2 and for the valorization of biogas. In the first study on CO_2 containing natural gas performed at the Chulalongkorn University, Bangkok [113], the authors compared the conversion of pure methane and its mixtures with He, higher hydrocarbons, and CO_2 in a GA reactor. Considerable formation of higher hydrocarbons was observed at the conditions chosen, and the addition of CO_2 significantly helped to increase H_2 yield and to reduce the specific energy requirement. Both chain growth to C_4H_{10} and dehydrogenation to C_2H_4 and C_2H_2 were observed. The conversions decreased in the sequence of propane > ethane > methane > CO_2. Unfortunately, the authors did not quantify soot formation.

In later studies, a simulated natural gas was used, having a $CH_4/C_2H_6/C_3H_8/CO_2$ molar ratio of 70/5/5/20% [114]. Both pure oxygen and air were investigated as oxidants. The reformer was operated at relatively low methane conversion values between 6 and 24%. The addition of oxidant markedly improved the conversion and selectivity to CO and H_2. Air was found to be more effective than pure oxygen, and an SER as low as 17 eV was reported. The reaction was investigated in a multistage GA reactor with the performance being studied in one to four stages, keeping either the residence time or the total flow constant. Methane conversion was raised up to about 40%, and optimum SER was 20 eV [115]. The selectivity to C_2 hydrocarbons was rather high ($S_{acetylene} \approx 30\%$), which would not be desirable in a practical fuel reformer.

9.4.5.4 **Reforming with Carbon Dioxide and Steam**

The reforming with carbon dioxide and steam could be of practical interest on a relatively small scale for the hydrogen production from biogas by SR. Some work was reported recently, but the results were not conclusive [116]. The quantitative description of this process is particularly challenging, since both CO_2 and steam may be the reactant or product, and hydrogen or CO selectivities might be based either on methane only or on combined $CH_4 + H_2O$ or $CH_4 + CO_2$, respectively.

9.4.5.5 **Other Feedstock**

The reforming of ethanol combining SR and POX was studied in a plasma–liquid reactor, similar to the work described in Section 9.4.1.3 on SR. A spark discharge was operated in a tank containing a mixture of ethanol and water, shown in Figure 9.14. The two electrodes were equipped with a quartz glass envelope, through which air was blown into the discharge zone [117, 118]. Minor by-products were methane, and C_2 hydrocarbons, and a fairly high reformer efficiency of 50% was reported.

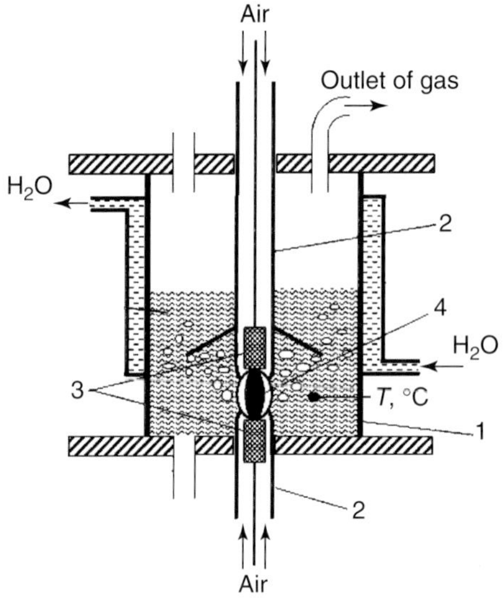

Figure 9.14 Reactor with spark discharge operated in direct contact with liquid reactant [117].

In recent years, several studies are reported on the plasma gasification of municipal waste or biomass [119–121]. Several studies are reported in a pilot scale to demonstrate the economic viability. Usually, commercially available thermal plasma torches are employed, so that the investigations, strictly speaking, do not fall into the scope of this review; however, nonthermal plasma sources could be an interesting alternative in the future. Thermochemical and electrochemical modeling indicates 33% electric efficiency for the integrated plasma gasification/fuel cell system shown in Figure 9.15 [122].

9.5 Summary of the Results and Outlook

Of the research published in recent years on hydrocarbon conversion using plasmas, about 74% started from methane, 17% from higher hydrocarbons, and 9% from oxygenates. About 18% investigated SR, 17% POX, 30% CDR, and 35% pyrolysis. This is somewhat surprising, since plasma POX or SR appears to be the most promising candidates for small-scale applications in fuel-cell-based auxiliary power units (APUs). For all four processes, demonstrator or prototype scale units have been presented recently, underlining the persistent interest in the field.

The best reforming performance was observed from the nonthermal arc discharges and microwave plasmas. The latter are, however, relatively expensive because of the magnetrons required for excitation. Also, maximum powers and efficiencies available from magnetrons are restricted (about 10 kW and 60% at 2.45 GHz and 75 kW and 80% efficiency at 915 MHz), and currently, there is no

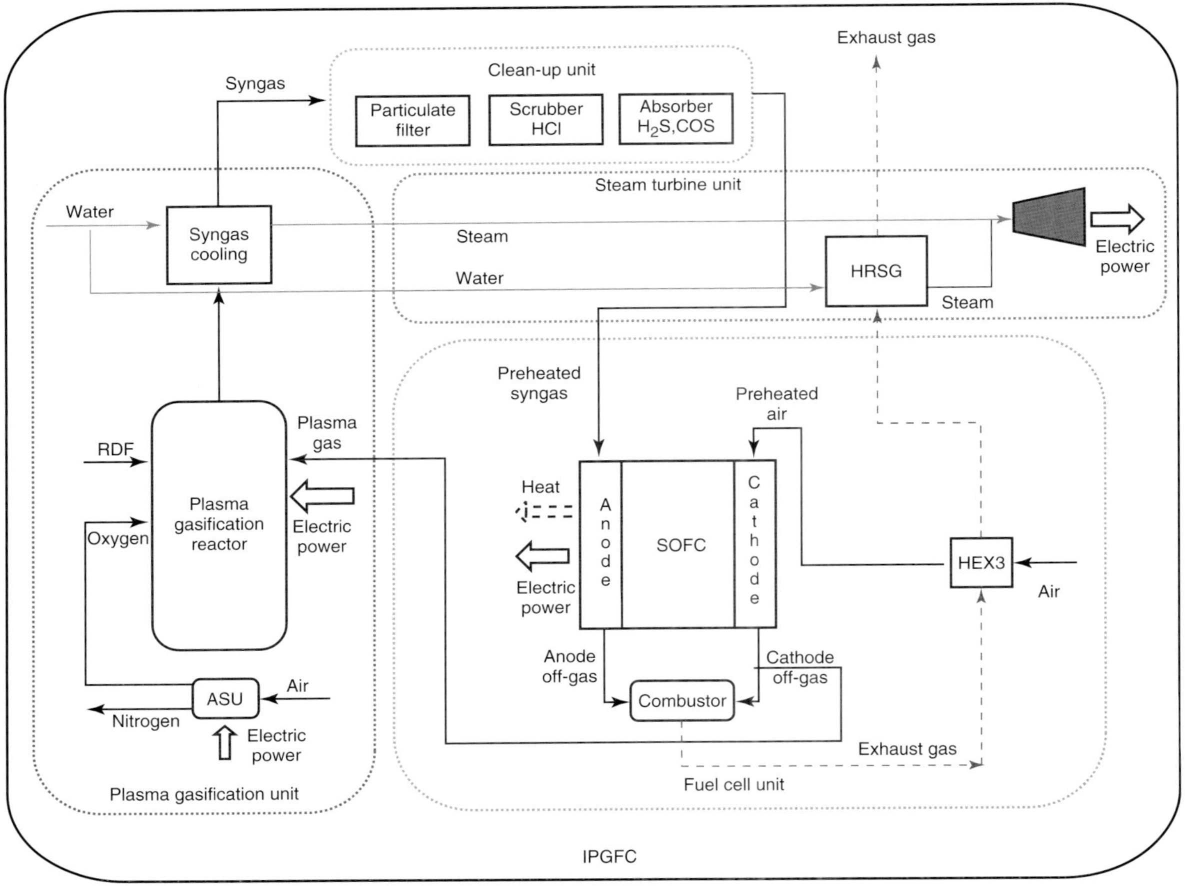

Figure 9.15 Integrated plasma gasification/fuel cell system proposed in thermochemical model [122]. Note that electric power is produced by the fuel cell and by a turbine running on steam generated from process heat. (ASU = air seperation unit, HRSG = high rate steam generator).

development to surpass these boundaries. As for arcs, whether operated at AC, high frequency or pulsed, in the future, their performance will profit from the ongoing development of power-conditioning electronics.

A recent study by authors from Drexel University and Chevron [123] uses for such dense, yet nonthermal plasmas, the term *warm* plasmas and gives an explanation for their high efficiency when compared to cold plasmas on one hand and hot (thermal) plasmas on the other: in warm plasmas, the temperature is high enough to support chemical reactions forming excited molecules in high concentrations, and hence stepwise ionization becomes important as compared to direct ionization. Such an understanding may in the future help to develop optimum plasma reformers.

Plasma hydrocarbon reforming is a field of research that has received considerable attention over a long time. Several studies in recent years only reiterate what has been found in previous work. In addition to this, the analysis of the products is often incomplete because of its complexity so that the relevant process characteristics reported are incomplete or inaccurate. Plasma hydrocarbon reforming appears somehow just as the search for the Holy Grail with great hopes but only few really promising approaches.

Several efforts have succeeded in making the transition from research to prototype development. Generally, in such cases, the developers keep quiet about their progress until the introduction of the final product to the market. However, some large prototype units have been constructed in the past 30 years, which have not led to commercialization. This may be partly due to the fact that power conditioning for nonthermal plasmas in the megawatt scale was expensive in the past, but here, much progress is to be expected in the future. The most probable breakthrough might be in the field of small APUs. Their success, however, also depends on another technology, that is, on fuel cells.

In the late 1990s, the prevailing opinion in the oil-refining industry was that at stable crude oil prices below US$ 20 per barrel, there was no economically reasonable scope for GTL or hydrogen as energy carrier. At present, the continuing research into new and unconventional reforming technologies is also motivated by the drastic increase in crude oil prices, and it remains the hope that one or another practical application of plasma technology in this field will emerge in the near future.

References

1. von Siemens, W. (1857) Über die elektrostatische Induction und die Verzögerung des Stroms in Flaschendrähten (On the electrostatic induction and time lag of current in glass jar wires). *Ann. Phys.*, **102**, 66–120.
2. Miller, S.L. (1953) A production of amino acids under possible primitive earth conditions. *Science, New Series*, **117** (3046), 528–529.
3. Drost, H. (1978) *Plasmachemie. Prozesse der chemischen Stoffwandlung unter Plasma-Bedingungen*, Akademie-Verlag, Berlin.
4. McTaggart, F.K. (1967) *Plasma Chemistry in Electrical Discharges. Topics in Inorganic and General*

Chemistry, Elsevier, New York, p. 9.
5. Eremin, E.N. (1968) *Elementy gazovoi elektrochimii (The Elements of Gas Electrochemistry)*. Izd. Moskov. Gos. Univ.
6. Fridman, A. (2008) *Plasma Chemistry*, Cambridge University Press, Cambridge.
7. Veprek, S. and Venugopalan, M. (1983) *Plasma chemistry, in Topics in Current Chemistry: Plasma Chemistry IV* (ed. F.L. Boschke), Springer-Verlag, New York.
8. Badyal, J.P.S. (1996) Catalysis and plasma chemistry at solid surfaces. *Top. Catal.*, **3** (1–2), 255–264.
9. Kitzling, M.B. and Järås, S.G. (1996) A review of the use of plasma techniques in catalyst preparation and catalyst reactions. *Appl. Catal., A Gen.*, **147**, 1–21.
10. Chen, H.L., Lee, H.M., Chen, S.H., Chao, Y., and Chang, M.B. (2008) Review of plasma catalysis on hydrocarbon reforming for hydrogen production-interaction, integration, and prospects. *Appl. Catal., B Environ.*, **85** (1–2), 1–9.
11. Wagner, U., Richter, S. 7 Hydrogen technologies. (ed. K. Heinloth), *SpringerMaterials-The Landolt-Börnstein Database http://www.springermaterials.com/*. DOI: 10.1007/10858992_16 (accessed 11.3.2011).
12. Bičáková, O. and Straka, P. (2010) The resources and methods of hydrogen production. *Acta Geodyn. Geomater.*, **7** (2), 175–188.
13. *http://www1.qatarairways.com/ae/en/csr-fuel.html* (accessed 4 February 2012).
14. Lu, Y. and Lee, T. (2007) Influence of the feed gas composition on the fischer-tropsch synthesis in commercial operations. *J. Nat. Gas Chem.*, **16** (4), 329–341.
15. Fan, M.-S., Abdullah, A.Z., and Bhatia, S. (2009) Catalytic technology for carbon dioxide reforming of methane to synthesis gas. *ChemCatChem*, **1** (2), 192–208.
16. Abbas, H.F. and Wan Daud, W.M.A. (2010) Hydrogen production by methane decomposition: a review. *Int. J. Hydrogen Energy*, **35** (3), 1160–1190.
17. Gladisch, H. (1962) How huels makes acetylene by DC Arc. *Hydrocarbon Process. Pet. Refiner*, **41**, 159–164.
18. *http://www.isp-marl.de/* (accessed 25 February 2011).
19. Ghorbanzadeh, A.M., Lotfalipour, R., and Rezaei, S. (2009) Carbon dioxide reforming of methane at near room temperature in low energy pulsed plasma. *Int. J. Hydrogen Energy*, **34** (1), 293–298.
20. Rusu, I. and Cormier, J.-M. (2003) On a possible mechanism of the methane steam reforming in a gliding arc reactor. *Chem. Eng. J.*, **91** (1), 23–31.
21. Kappes, T., Schiene, W., and Hammer, T. (2002) Energy balance of a dielectric barrier discharge reactor for hydrocarbon steam reforming. Proceeding 8th International Symposium on High Pressure Low Temperature Plasma Chemistry (HAKONE 8), at Pühajärve ESTONIA July 21-25, 2002.
22. Hammer, T., Kappes, T., and Baldauf, M. (2004) Plasma catalytic hybrid processes: gas discharge initiation and plasma activation of catalytic processes. *Catal. Today*, **89** (1–2), 5–14.
23. Zhang, X., Wang, B., Liu, Y., and Xu, G. (2009) Conversion of methane by steam reforming using dielectric-barrier discharge. *Chin. J. Chem. Eng.*, **17** (4), 625–629.
24. Wang, Y.-F., Tsai, C.-H., Chang, W.-Y., and Kuo, Y.-M. (2010) Methane steam reforming for producing hydrogen in an atmospheric-pressure microwave plasma reactor. *Int. J. Hydrogen Energy*, **35** (1), 135–140.
25. Nozaki, T., Muto, N., Kado, S., and Okazaki, K. (2004) Minimum energy requirement for methane steam reforming in plasma-catalyst reactor. *Prepr. Pap.-Am. Chem. Soc., Div. Fuel Chem.*, **49** (1), 179–180.
26. Nozaki, T., Hiroyuki, T., and Okazaki, K. (2006) Hydrogen enrichment of low-calorific fuels using barrier discharge enhanced Ni/γ-Al_2O_3 bed

reactor: thermal and nonthermal effect of nonequilibrium plasma. *Energy Fuels*, **20** (1), 339–345.
27. Nozaki, T., Tsukijihara, H., Fukui, W., and Okazaki, K. (2007) Kinetic analysis of the catalyst and nonthermal plasma hybrid reaction for methane steam reforming. *Energy Fuels*, **21** (5), 2525–2530.
28. Frost, L., Elangovan, S., and Hartvigsen, J. (2009) Non-thermal plasma reforming For TARDEC. 2009 Fuel Cell Seminar and Exposition, Palm Springs, CA, November 16–19, 2009, *http://www.fuelcellseminar.com/assets/2009/HRD44-3_0500PM_Frost.pdf* (accessed 3 March 2011).
29. Tang, L., Huang, H., Zhao, Z., Wu, C.Z., and Chen, Y. (2003) Pyrolysis of polypropylene in a nitrogen plasma reactor. *Ind. Eng. Chem. Res.*, **42** (6), 1145–1150.
30. Hueso Jose, L., Rico Victor, J., Cotrino, J., Jimenez-Mateos, J.M., and Gonzalez-Elipe Agustin, R. (2009) Water plasmas for the revalorisation of heavy oils and cokes from petroleum refining. *Environ. Sci. Technol.*, **43** (7), 2557–2562.
31. Ouni, F., Khacef, A., and Cormier, J.M. (2009) Syngas production from propane using atmospheric non-thermal plasma. *Plasma Chem. Plasma Process.*, **29** (2), 119–130.
32. Sekiguchi, H. and Mori, Y. (2003) Steam plasma reforming using microwave discharge. *Thin Solid Films*, **435** (1–2), 44–48.
33. Nakanishi, S. and Sekiguchi, H. (2005) Comparison of reforming behaviors of hexane and isooctane in microwave steam plasma. *J. Jpn. Pet. Inst.*, **48** (1), 22–28.
34. Tao, K., Ohta, N., Liu, G., Yoneyama, Y., Wang, T., and Tsubaki, N. (2010) Plasma enhanced catalytic reforming of biomass tar model compound to syngas. *Fuel*, (Article in press, Corrected Proof, Available online 12 June 2010).
35. Rico, V.J., Hueso, J.L., Cotrino, J., Gallardo, V., Sarmiento, B., Brey, J.J., and Gonzalez-Elipe, A.R. (2009) Hybrid catalytic-DBD plasma reactor for the production of hydrogen and preferential CO oxidation (CO-PROX) at reduced temperatures. *Chem. Commun.*, **41**, 6192–6194.
36. Sekine, Y., Asai, S., Urasaki, K., Matsukata, M., Kikuchi, E., Kado, S., and Haga, F. (2005) Hydrogen production from biomass-ethanol at ambient temperature with novel diaphragm reactor. *Chem. Lett.*, **34** (5), 658–659.
37. Aubry, O., Met, C., Khacef, A., and Cormier, J.M. (2005) On the use of a non-thermal plasma reactor for ethanol steam reforming. *Chem. Eng. J.*, **106** (3), 241–247.
38. Yan, Z., Chen, L., and Wang, H. (2009) Hydrogen generation by glow discharge plasma electrolysis of methanol solutions. *Int. J. Hydrogen Energy*, **34** (1), 48–55.
39. Yan, Z., Chen, L., and Wang, H. (2008) Hydrogen generation by glow discharge plasma electrolysis of ethanol solutions. *J. Phys. D: Appl. Phys.*, **41** (15), 155205/ 1–155205/7.
40. Sekine, Y., Haraguchi, M., Tomioka, M., Matsukata, M., and Kikuchia, E. (2010) Low-temperature hydrogen production by highly efficient catalytic system assisted by an electric field. *J. Phys. Chem. A*, **114** (11), 3824–3833.
41. Hoang, T.Q., Zhu, X., Lobban, L.L., and Mallinson, R.G. (2011) Effects of gap and elevated pressure on ethanol reforming in a non-thermal plasma reactor. *J. Phys. D: Appl. Phys*, **44** (27), 274003.
42. Chao, Y., Lee, H.-M., Chen, S.-H., and Chang, M.-B. (2009) Onboard motorcycle plasma-assisted catalysis system–role of plasma and operating strategy. *Int. J. Hydrogen Energy*, **34** (15), 6271–6279.
43. Tsai, C.-H., Huang, K.-L., Hsieh, L.-T., Chao, H.-R., and Fang, K.-C. (2005) New approach for methane conversion using an rf discharge reactor. 2. Characteristic of polycyclic aromatic hydrocarbon emissions. *Ind. Eng. Chem. Res.*, **44** (17), 6566–6571.
44. Nozaki, T., Agiral, A., Yuzawa, S., Han Gardeniers, J.G.E., and Okazaki, K. (2011) A single step

methane conversion into synthetic fuels using microplasma reactor. *Chem. Eng. J.*, **166** (1), 288–293.

45. Tsai, C.-H., Hsieh, T.-H., Shih, M., Huang, Y.-J., and Wei, T.-C. (2005) Partial oxidation of methane to synthesis gas by a microwave plasma torch. *AIChE J.*, **51** (10), 2853–2858.
46. Horng, R.-F., Chang, Y.-P., Huang, H.-H., and Lai, M.-P. (2006) A study of the hydrogen production from a small plasma converter. *Fuel*, **86** (1–2), 81–89 (Volume Date 2007).
47. Luche, J., Aubry, O., Khacef, A., and Cormier, J.-M. (2009) Syngas production from methane oxidation using a non-thermal plasma: experiments and kinetic modelling. *Chem. Eng. J.*, **149** (1–3), 35–41.
48. Lee, D.H., Kim, K.-T., Cha, M.S., and Song, Y.-H. (2010) Plasma-controlled chemistry in plasma reforming of methane. *Int. J. Hydrogen Energy*, **35** (20), 10967–10976.
49. Sreethawong, T., Thakonpatthanakun, P., and Chavadej, S. (2007) Partial oxidation of methane with air for synthesis gas production in a multistage gliding arc discharge system. *Int. J. Hydrogen Energy*, **32** (8), 1067–1079.
50. Horng, R.-F., Huang, H.-H., Lai, M.-P., Wen, C.-S., and Chiu, W.-C. (2008) Characteristics of hydrogen production by a plasma-catalyst hybrid converter with energy saving schemes under atmospheric pressure. *Int. J. Hydrogen Energy*, **33** (14), 3719–3727.
51. Heintze, M. and Pietruszka, B. (2004) Plasma catalytic conversion of methane into syngas: the combined effect of discharge activation and catalysis. *Catal. Today*, **89** (1–2), 21–25.
52. Khassin, A.A., Pietruszka, B.L., Heintze, M., and Parmon, V.N. (2004) The impact of a dielectric barrier discharge on the catalytic oxidation of methane over Ni-containing catalyst. *React. Kinet. Catal. Lett.*, **82** (1), 131–137.
53. Kim, S.C. and Chun, Y.N. (2008) Experimental study on partial oxidation of methane to produce hydrogen using low-temperature plasma in AC Glidarc discharge. *Int. J. Energy Res.*, **32** (13), 1185–1193.
54. Horng, R.-F., Lai, M.-P., Huang, H.-H., and Chang, Y.-P. (2009) Reforming performance of a plasma-catalyst hybrid converter using low carbon fuels. *Energy Convers. Manage.*, **50** (10), 2632–2637.
55. Horng, R.-F., Lai, M.-P., Chang, Y.-P., Yur, J.-P., and Hsieh, S.-F. (2009) Plasma-assisted catalytic reforming of propane and an assessment of its applicability on vehicles. *Int. J. Hydrogen Energy*, **34** (15), 6280–6289.
56. Gallagher, M.J., Geiger, R., Polevich, A., Rabinovich, A., Gutsol, A., and Fridman, A. (2010) On-board plasma-assisted conversion of heavy hydrocarbons into synthesis gas. *Fuel*, **89** (6), 1187–1192.
57. Bibikov, M.B., Demkin, S.A., Zhivotov, V.K., Konovalov, G.M., Moskovskii, A.S., Potapkin, B.V., Smirnov, R.V., and Strelkova, M.I. (2007) Partial oxidation of kerosene in plume microwave discharge. *High Energy Chem.*, **41** (5), 361–365.
58. Nikipelov, A.A., Popov, I.B., Correale, G., Rakitin, A.E., and Starikovskii, A.Y. (2010) Compact catalyst-free liquid fuel to syngas reformer with plasma-assisted flame stabilization. 48th AIAA Aerospace Sciences Meeting Including the New Horizons Forum and Aerospace Exposition, Art. No. 2010-1343.
59. Hartvigsen, J., Czernichowski, P., Hollist, M., Frost, L., Elangovan, S., and Nickens, A. (2010) Scale-up of plasma catalyzed logistics fuel reformer for navy shipboard fuel cell systems. 2010 Fuel Cell Seminar and Exposition – Henry B. Gonzalez Convention Center, San Antonio, TX, October 18–21.
60. The 2009 Spring National Meeting, Tampa, FL. session #17 - Fuel Processing for Hydrogen Production From Fossil Fuels: Plasma Reforming (TB001). *http://aiche.confex.com/aiche/s09/techprogram/MEETING.HTM.* (accessed 1 March 2011).

61. Song, L., Li, X., and Zheng, T. (2008) Onboard hydrogen production from partial oxidation of dimethyl ether by spark discharge plasma reforming. *Int. J. Hydrogen Energy*, **33** (19), 5060–5065.
62. Czernichowski, A., Czernichowski, M., and Sessa, John P. (2008) Waste glycerol conversion into syngas. *Prepr. Symp. -Am. Chem. Soc., Div. Fuel Chem.*, **53** (1), 427–428.
63. Istadi, I. and Amin, N.A.S. (2006) Co-generation of synthesis gas and C2+ hydrocarbons from methane and carbon dioxide in a hybrid catalytic-plasma reactor: A review. *Fuel*, **85** (5–6), 577–592.
64. Ghorbanzadeh, A.M., Norouzi, S., and Mohammadi, T. (2005) High energy efficiency in syngas and hydrocarbon production from dissociation of CH4-CO2 mixture in a non-equilibrium pulsed plasma. *J. Phys. D: Appl. Phys.*, **38** (20), 3804–3811.
65. Oberreuther, T., Wolff, C., and Behr, A. (2003) Volumetric plasma chemistry with carbon dioxide in an atmospheric pressure plasma using a technical scale reactor. *IEEE Trans. Plasma Sci.*, **31** (1), 74–78.
66. Behr, A., Wolff, C., and Oberreuther, T. (2004) Erzeugung von synthesegas durch trockenes plasma-reforming. *Chem. Ing. Tech.*, **76** (7), 951–955.
67. Oberreuther, T., Wolff, C., and Behr, A. (2004) Volumetric plasma chemistry in a technical scale: producing synthesis gas from carbon dioxide and hydrocarbons. *Galvanotechnik*, **95** (2), 438–443.
68. Behr, A., Oberreuther, T., and Wolff, C. (2004) Großtechnisches konzept für die erzeugung von synthesegas durch trockenes plasma-reforming. *Chem. Ing. Tech.*, **76** (7), 946–950.
69. Long, H., Shang, S., Tao, X., Yin, Y., and Dai, X. (2008) CO2 reforming of CH4 by combination of cold plasma jet and Ni/g-Al2O3 catalyst. *Int. J. Hydrogen Energy*, **33** (20), 5510–5515.
70. Tao, X., Qi, F., Yin, Y., and Dai, X. (2008) CO2 reforming of CH4 by combination of thermal plasma and catalyst. *Int. J. Hydrogen Energy*, **33** (4), 1262–1265.
71. Tsai, H.-L. and Wang, C.-S. (2008) Thermodynamic equilibrium prediction for natural gas dry reforming in thermal plasma reformer. *J. Chin. Inst. Eng.*, **31** (5), 891–896.
72. Wang, Q., Cheng, Y., and Jin, Y. (2009) Dry reforming of methane in an atmospheric pressure plasma fluidized bed with Ni/γ-Al2O3 catalyst. *Catal. Today*, **148** (3–4), 275–282.
73. Wang, Q., Yan, B.-H., Jin, Y., and Cheng, Y. (2009) Dry reforming of methane in a dielectric barrier discharge reactor with Ni/Al2O3 catalyst: interaction of catalyst and plasma. *Energy Fuels*, **23** (8), 4196–4201.
74. Sekine, Y., Yamadera, J., Kado, S., Matsukata, M., and Kikuchi, E. (2008) High-efficiency dry reforming of biomethane directly using pulsed electric discharge at ambient condition. *Energy Fuels*, **22** (1), 693–694.
75. Sekine, Y., Yamadera, J., Matsukata, M., and Kikuchi, E. (2010) Simultaneous dry reforming and desulfurization of biomethane with non-equilibrium electric discharge at ambient temperature. *Chem. Eng. Sci.*, **65** (1), 487–491.
76. Tao, X., Bai, M., Li, X., Long, H., Shang, S., Yin, Y., and Dai, X. (2011) CH4-CO2 reforming by plasma – challenges and opportunities. *Prog. Energy Combust. Sci.*, **37** (2), 113–124.
77. Liu, C.-J., Xue, B., Eliasson, B., He, F., Li, Y., and Xu, G.-H. (2001) Methane conversion to higher hydrocarbons in the presence of carbon dioxide using dielectric-barrier discharge plasmas. *Plasma Chem. Plasma Process.*, **21** (3), 301–310.
78. Kraus, M., Efli, W., Haffner, K., Eliasson, B., Kogelschatz, U., and Wokaun, A. (2002) Investigation of mechanistic aspects of the catalytic CO2 reforming of methane in a dielectric-barrier discharge using optical emission spectroscopy and kinetic modelling. *Phys. Chem. Chem. Phys.*, **4**, 668–675.
79. Sentek, J., Krawczyk, K., Młotek, M., Kalczewska, M., Kroker, T., Kolb, T., Schenk, A., Gericke, K.-H.,

and Schmidt-Szałowski, K. (2010) Plasma-catalytic methane conversion with carbon dioxide in dielectric barrier discharges. *Appl. Catal., B Environ.*, **94** (1–2), 19–26.

80. Goujard, V., Tatibouet, J.-M., and Batiot-Dupeyrat, C. (2009) Use of a non-thermal plasma for the production of synthesis gas from biogas. *Appl. Catal., A Gen.*, **353** (2), 228–235.

81. Bo, Z., Yan, J., Li, X., Chi, Y., and Cen, K. (2008) Plasma assisted dry methane reforming using gliding arc gas discharge: effect of feed gases proportion. *Int. J. Hydrogen Energy*, **33** (20), 5545–5553.

82. Futamura, S. and Annadurai, G. (2005) Plasma reforming of aliphatic hydrocarbons with CO_2. *IEEE Trans. Ind. Appl.*, **41** (6), 1515–1521.

83. Futamura, S. and Annadurai, G. (2008) Effects of temperature, voltage properties, and initial gas composition on the plasma reforming of aliphatic hydrocarbons with CO_2 *IEEE Trans. Ind. Appl.*, **44** (1), 53–60.

84. Zhang, X., Zhu, A., Li, X., and Gong, W. (2004) Oxidative dehydrogenation of ethane with CO_2 over catalyst under pulse corona plasma. *Catal. Today*, **89** (1–2), 97–102.

85. Bauer, M., Schwarz-Selinger, T., Kang, H., and von Keudell, A. (2005) Control of the plasma chemistry of a pulsed inductively coupled methane plasma. *Plasma Sources Sci. Technol.*, **14**, 543–548.

86. Bakken, J.A., Jensen, R., Monsen, B., Raaness, O., and Waernes, N. (1998) Thermal plasma process development in Norway. *Pure Appl. Chem.*, **70** (6), 1223–1228.

87. Cho, W., Kim, Y.C., and Kim, S.-S. (2010) Conversion of natural gas to C_2 product, hydrogen and carbon black using a catalytic plasma reaction. *J. Ind. Eng. Chem.*, **16** (1), 20–26.

88. Jasinski, M., Dors, M., Nowakowska, H., and Mizeraczyk, J. (2008) Hydrogen production via methane reforming using various microwave plasma sources. *Chem. Listy*, **102**, 1332–1337.

89. Jasinski, M., Dors, M., and Mizeraczyk, J. (2009) Application of atmospheric pressure microwave plasma source for production of hydrogen via methane reforming. *Eur. Phys. J. D*, **54**, 179–183.

90. Stoknes, P.E. and Dohmen, J.R. (2009) Gasplas low energy microwave plasma reactors. Poster on the Conference Hydrogen and Fuel Cells in the Nordic Countries, Oslo, November 24–26, 2009. *http://www.gasplas.com/w3/index.php* (accessed 27 January 2011).

91. Leins, M., Schaefer, T., Baumgärtner, K.-M., Walker, M., Schulz, A., Schumacher, U., and Stroth, U. (2008) Methane pyrolysis with a microwave plasma source for application in space. The Eleventh International Conference on Plasma Surface Engineering -PSE 2008 – Garmisch Partenkirchen, September 15–19, 2008, Poster 1028. *http://www.pse2008.net* (accessed 1 March 2011).

92. LÜ, J. and Li, Z. (2010) Conversion of natural gas to C_2 hydrocarbons via cold plasma technology. *J. Nat. Gas Chem.*, **19** (4), 375–379.

93. Fridman A., Babaritskyi A., Jivotov V., Dyomkin S., Nester S., and Rusanov V. (1991) Methane conversion in acetylene in the nonequilibrium MCW-discharge. Proceedings of ISPC-10 (ed. U. Ehlemann, H.G. Lergon, and K. Wiesemann Bochum), pp. 1–6. *http://134.147.148.178/ispcdocs/ispc10 /DB2.html.* (accessed 1 March 2011).

94. Yao, S.L., Suzuki, E., Meng, N., and Nakayama, A. (2002) A high-efficiency reactor for the pulsed plasma conversion of methane. *Plasma Chem. Plasma Process.*, **22** (2), 225–237.

95. Yao, S., Nakayama, A., and Suzuki, E. (2001) Acetylene and hydrogen from pulsed plasma conversion of methane. *Catal. Today*, **71** (1–2), 219–223.

96. Heintze, M. and Magureanu, M. (2002) Methane conversion into acetylene in a microwave plasma: optimization of the operating parameters. *J. Appl. Phys.*, **92** (5), 2276–2283.

97. Heintze, M., Magureanu, M., and Kettlitz, M. (2002) Mechanism of C-2 hydrocarbon formation from methane in a pulsed microwave plasma. *J. Appl. Phys.*, **92** (12), 7022–7031.

98. Heintze, M. and Magureanu, M. (2002) Efficient methane conversion to acetylene in a microwave plasma. Proceeding 8th International Symposium on High Pressure Low Temperature Plasma Chemistry (HAKONE 8) at Pühajärve ESTONIA, July 21–25, 2002, p. 3.6.
99. Schmidt-Szałowski, K., Krawczyk, K., and Mlotek, M. (2007) Catalytic effects of metals on the conversion of methane in gliding discharges. *Plasma Processes Polym.*, **4** (7–8), 728–736.
100. Lee, H. and Sekiguchi, H. (2011) Plasma-catalytic hybrid system using spouted bed with a gliding arc discharge: CH_4 reforming as a model reaction. *J. Phys. D: Appl. Phys*, **44** (27), 274008.
101. Heintze, M. and Magureanu, M. (2002) Methane conversion into aromatics in a direct plasma-catalytic process. *J. Catal.*, **206** (1), 91–97.
102. Li, X.-S., Shi, C., Xu, Y., Wang, K.-J., and Zhu, A.-M. (2007) A process for a high yield of aromatics from the oxygen-free conversion of methane: combining plasma with Ni/HZSM-5 catalysts. *Green Chem.*, **9** (6), 647–653.
103. Rico, V.J., Hueso, J.L., Cotrino, J., and González-Elipe, A.R. (2010) Evaluation of different dielectric barrier discharge plasma configurations as an alternative technology for green C_1 chemistry in the carbon dioxide reforming of methane and the direct decomposition of methanol. *J. Phys. Chem. A*, **114** (11), 4009–4016.
104. Zhu, X., Hoang, T., Lobban, L.L., and Mallinson, R.G. (2009) Plasma reforming of glycerol for synthesis gas production. *Chem. Commun.*, **20**, 2908–2910.
105. Ouni, F., Khacef, A., and Cormier, J.M. (2006) Effect of oxygen on methane steam reforming in a sliding discharge reactor *Chem. Eng. Technol.*, **29** (5), 604–609.
106. Kim, S.C. and Chun, Y.N. (2008) Production of hydrogen by partial oxidation with thermal plasma. *Renewable Energy*, **33** (7), 1564–1569.
107. Pietruszka, B. and Heintze, M. (2004) Methane conversion at low temperature: the combined application of catalysis and non-equilibrium plasma. *Catal. Today*, **90** (1–2), 151–158.
108. Nair, S.A., Nozaki, T., and Okazaki, K. (2007) Methane oxidative conversion pathways in a dielectric barrier discharge reactor-Investigation of gas phase mechanism. *Chem. Eng. J.*, **132** (1–3), 85–95.
109. Petitpas, G., Rollier, J.-D., Darmon, A., Gonzalez-Aguilar, J., Metkemeijer, R., and Fulcheri, L. (2007) A comparative study of non-thermal plasma assisted reforming technologies. *Int. J. Hydrogen Energy*, **32** (14), 2848–2867.
110. Rollier, J.-D., Gonzalez-Aguilar, J., Petitpas, G., Darmon, A., Fulcheri, L., and Metkemeijer, R. (2008) Experimental study on gasoline reforming assisted by nonthermal arc discharge. *Energy Fuels*, **22** (1), 556–560.
111. Rollier, J.-D., Petitpas, G., Gonzalez-Aguilar, J., Darmon, A., Fulcheri, L., and Metkemeijer, R. (2008) Thermodynamics and kinetics analysis of gasoline reforming assisted by arc discharge. *Energy Fuels*, **22** (3), 1888–1893.
112. Gonzalez-Aguilar, J., Petitpas, G., Lebouvier, A., Rollier, J.-D., Darmon, A., and Fulcheri, L. (2009) Three stages modeling of n-octane reforming assisted by a nonthermal arc discharge. *Energy Fuels*, **23** (10), 4931–4936.
113. Rueangjitt, N., Akarawitoo, C., Sreethawong, T., and Chavadej, S. (2007) Reforming of CO_2-containing natural gas using an ac gliding arc system: effect of gas components in natural gas. *Plasma Chem. Plasma Process.*, **27** (5), 559–576.
114. Rueangjitt, N., Sreethawong, T., and Chavadej, S. (2008) Reforming of CO_2-containing natural gas using an ac gliding arc system: effects of operational parameters and oxygen addition in feed. *Plasma Chem. Plasma Process.*, **28** (1), 49–67.
115. Rueangjitt, N., Jittiang, W., Pornmai, K., Chamnanmanoontham, J., Sreethawong, T., and Chavadej, S. (2009) Combined reforming and partial

oxidation of CO_2-containing natural gas using an ac multistage gliding arc discharge system: effect of stage number of plasma reactors *Plasma Chem. Plasma Process.*, **29** (6), 433–453.

116. Chun, Y.N., Yang, Y.C., and Yoshikawa, K. (2009) Hydrogen generation from biogas reforming using a gliding arc plasma-catalyst reformer. *Catal. Today*, **148** (3–4), 283–289.

117. Shchedrin, A.I., Levko, D.S., Chernyak, V.Y., Yukhimenko, V.V., and Naumov, V.V. (2008) Effect of air on the concentration of molecular hydrogen in the conversion of ethanol by a nonequilibrium gas-discharge plasma. *JETP Lett.*, **88** (2), 99–102.

118. Chernyak, V.Y., Olszewski, S.V., Yukhymenko, V.V., Solomenko, E.V., Prysiazhnevych, I.V., Naumov, V.V., Levko, D.S., Shchedrin, A.I., Ryabtsev, A.V., Demchina, V.P., Kudryavtsev, V.S., Martysh, E.V., and Verovchuck, M.A. (2008) Plasma-assisted reforming of ethanol in dynamic plasma-liquid system: experiments and modelling. *IEEE Trans. Plasma Sci.*, **36** (6), 2933–2939.

119. Van Oost, G., Hrabovsky, M., Kopecky, V., Konrad, M., Hlina, M., and Kavka, T. (2008) Pyrolysis/gasification of biomass for synthetic fuel production using a hybrid gas-water stabilized plasma torch. *Vacuum*, **83** (1), 209–212.

120. Dollard, J. (2010) Hot fix for renewable energy. *Pollut. Eng.*, **42** (9), 22–29.

121. Pourali, M. (2010) Application of plasma gasification technology in waste to energy – challenges and opportunities. *IEEE Trans. Sustainable Energy*, **1** (3), 125–130.

122. Galeno, G., Minutillo, M., and Perna, A. (2011) From waste to electricity through integrated plasma gasification/fuel cell (IPGFC) system. *Int. J. Hydrogen Energy*, **36** (2), 1692–1701.

123. Gutsol, A., Rabinovich, A., and Fridman, A. (2011) Combustion-assisted plasma in fuel conversion. *J. Phys. D: Appl. Phys*, **44** (27), 274001.

Index

Plasma Chemistry and Catalysis in Gases and Liquids, First Edition.
Edited by Vasile I. Parvulescu, Monica Magureanu, and Petr Lukes.

n